Biological Signal Transduction

NATO ASI Series

Advanced Science Institutes Series

A series presenting the results of activities sponsored by the NATO Science Committee, which aims at the dissemination of advanced scientific and technological knowledge, with a view to strengthening links between scientific communities.

The Series is published by an international board of publishers in conjunction with the NATO Scientific Affairs Division

A Life Sciences B Physics	Plenum Publishing Corporation London and New York
C Mathematical and Physical Sciences D Behavioural and Social Sciences E Applied Sciences	Kluwer Academic Publishers Dordrecht, Boston and London
F Computer and Systems Sciences G Ecological Sciences H Cell Biology I Global Environmental Change	Springer-Verlag Berlin Heidelberg New York London Paris Tokyo Hong Kong Barcelona Budapest

NATO-PCO DATABASE

The electronic index to the NATO ASI Series provides full bibliographical references (with keywords and/or abstracts) to more than 30000 contributions from international scientists published in all sections of the NATO ASI Series. Access to the NATO-PCO DATABASE is possible in two ways:

- via online FILE 128 (NATO-PCO DATABASE) hosted by ESRIN, Via Galileo Galilei, I-00044 Frascati, Italy.
- via CD-ROM "NATO-PCO DATABASE" with user-friendly retrieval software in English, French and German (© WTV GmbH and DATAWARE Technologies Inc. 1989).

The CD-ROM can be ordered through any member of the Board of Publishers or through NATO-PCO, Overijse, Belgium.

Series H: Cell Biology Vol. 52

Biological Signal Transduction

Edited by

Elliott M. Ross

Dept. of Pharmacology
University of Texas
Southwestern Medical Center
5323 Harry Hines Boulevard
Dallas TX 75235-9041
USA

and

Karel W. A. Wirtz

Centre for Biomembranes and Lipid Enzymology
State University of Utrecht
Padualaan 8
3508 TB Utrecht
The Netherlands

Springer-Verlag
Berlin Heidelberg New York London Paris Tokyo
Hong Kong Barcelona Budapest
Published in cooperation with NATO Scientific Affairs Division

Proceedings of the NATO Advanced Research Institute on Biological Signal Transduction held on the Island of Spetsai (Greece), August 6–17, 1990.

ISBN-13: 978-3-642-75138-7 e-ISBN-13: 978-3-642-75136-3

DOI: 10.1007/978-3-642-75136-3

Library of Congress Cataloging-in-Publication Data
NATO Advanced Research Institute on Biological Signal Transduction (1990 : Nísos Spétsai, Greece) Biological signal transduction / edited by Elliott M. Ross and Karel W. A. Wirtz. (NATO ASI series. Series H, Cell biology ; vol. 52)
"Proceedings of the NATO Advanced Research Institute on Biological Signal Transduction held on the Island of Spetsai (Greece), August 6–17, 1990"—T.p. verso. "Published in cooperation with NATO Scientific Affairs Division." Includes index.
1. Cellular signal transduction—Congresses. 2. G proteins—Congresses. 3. Cell receptors—Congresses. 4. Second messengers (Biochemistry)—Congresses. I. Ross, Elliott Morton II. Wirtz, Karel W. A. III. North Atlantic Treaty Organization. Scientific Affairs Division. IV. Title. V. Series. [DNLM: 1. Cell Communication—physiology—congresses. 2. Guanine Nucleotide Regulatory Protein—physiology—congresses. 3. Receptors, Endogenous Substances—physiology—congresses. 4. Signal Transduction—physiology—congresses. QP 517.C45 N279b 1990] QP517 C45N38 1990 574.87—dc20 DNLM/DLC for Library of Congress 91-4733 CIP

Softcover reprint of the hardcover 1st edition 1991

31/3140-543210 – Printed on acid-free-paper

PREFACE

A NATO Advanced Study Institute on "Biological Signal Transduction" was held on the Island of Spetsai from August 6-17, 1990. This Institute was organized to consider modern experimental approaches to the biochemistry, molecular biology and cell biology of eukaryotic signal transduction. Presentations and discussions focussed on fundamental mechanisms of receptor function, organization of second messenger pathways and the integration of cellular signals to regulate cell growth, cell differentiation and specialized functions.

A major focus was the mechanism of receptor - G-protein coupling; interaction of signaling at the receptor - G-protein interface; the complex system of second messengers derived from inositol lipids and the regulation of intracellular Ca^{2+}-levels. Signaling through the growth factors was discussed with an emphasis on the diversity of factors that influence protein tyrosine kinase activities, and the determination of relevant substrates for tyrosine phosphorylation. Overall integration of signaling pathways were generally directed toward questions of growth and exocytosis. Structure and function of the nicotinic acetylcholine receptor were discussed. Both the channel-forming proteins that confer ionic specificity and voltage dependence, and their long-term regulation by ligands, phosphorylation and subunit selection were covered in detail. These topics were extended to the related fields of genomic regulation of cell growth by the steroid receptor family; mechanism of endocytosis, sorting of membrane proteins among intracellular organelles and intracellular transport of inositol lipids; structure and function of cell adhesion molecules; membrane organization and structure of inositol lipids and glycosyl-phosphatidylinositol membrane anchors.

This book presents the content of the major lectures and of a selection of posters presented during the course of the Institute. This book is intended to make the proceedings of the Institute accessible to a larger audience and to offer a comprehensive account of the topics that were discussed in great depth for a most stimulating twelve-day period.

January 1991 The Editors

CONTENTS

I. RECEPTOR CHARACTERISTICS, FUNCTION AND REGULATION

II. G-PROTEINS AND PROTEIN KINASES

III. SURFACE MEMBRANE COMPONENTS

IV. INOSITOL PHOSPHATE AND CALCIUM

V. RELEASE REACTIONS AND PROTEIN SORTING

VI. ADDENDUM

THE NICOTINIC ACETYLCHOLINE RECEPTOR : A MEMBER OF THE SUPERFAMILY OF LIGAND-GATED ION CHANNELS

Jean-Luc Galzi and Jean-Pierre Changeux
UA CNRS D1284 "Neurobiologie Moléculaire"
Département des Biotechnologies,
Institut Pasteur,
25 rue du Dr Roux,
75724 Paris Cedex 15, France.

In the recent years, the methods of recombinant DNA technology have led to the identification of a large collection of amino acid sequences from several ligand gated ion channels, such as the nicotinic acetylcholine receptor from electric organ, muscle and brain, as well as glycine, GABA and glutamate receptors from brain. Structural and functional evidence support the view that these allosteric membrane proteins are pentameric hetero-oligomers, their distinctive pharmacological and physiological properties being associated with a defined subunit composition. Yet, still little is known about the actual three-dimension architecture of these oligomers and about the detailed structural mechanisms of the transitions which mediate fast and slow regulation of ion channel opening.

One of the best characterized member of this family is the peripheral nicotinic acetylcholine receptor from Torpedo electric organ and muscle (Changeux, 1981, 1990 ; Hucho, 1986 ; Karlin, 1980). It is composed of four distinct transmembrane subunits assembled into a heterologous $\alpha_2\beta\gamma\delta$ pentamer and contains the agonist/competitive antagonist binding sites, the cation selective ion channel and the mechanisms which mediate a fast coupling between them.

The purified and membrane-bound AChR from fish electric organ appears, on electron micrographs, as a ring-like particle 60-70 Å in diameter with a central 25 Å wide depression and five peaks of electron density, assigned to the five subunits, arranged around an axis of pseudo-symmetry perpendicular to the plane of the membrane (Brisson and Unwin, 1985). On side views, it looks like a 110 Å long transmembrane cylinder which protrudes by 55-65 Å into the synaptic cleft and by 15-20 Å into the cytoplasm (Toyoshima and Unwin, 1988 ; Mitra et al., 1989).

NATO ASI Series, Vol. H 52
Biological Signal Transduction
Edited by E. M. Ross and K. W. A. Wirtz

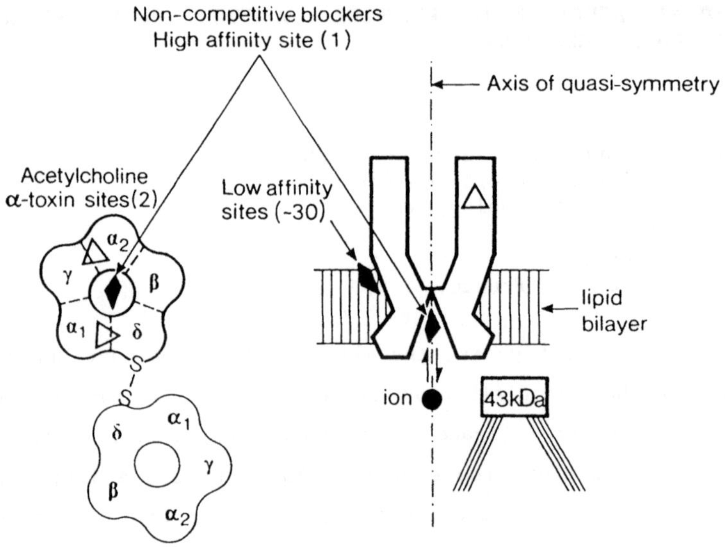

FIGURE 1 : Schematic organization of the muscle AChR. There are two ACh binding sites. The allosteric sites for the noncompetitive blockers are subdivided into two main categories. In this model, the unique high affinity site, sensitive to histrionicotoxin or phencyclidine, is located in the axis of quasi-symmetry of the molecule.

Complete cDNA coding sequences have been established in T. californica for the precursors of the α-, β-, γ- and δ-(Noda et al., 1983a ; Claudio et al., 1983) subunits as well as the complete nucleotide sequences (Devillers-Thiéry et al., 1983) for the α-subunit from T. marmorata. Subsequently, genes and cDNAs coding for muscle and neuronal AChR subunits were isolated.

The aligned sequences of the various cloned receptor subunits show similar hydrophobicity profiles which justify a common subdivision of the homologous chains into : 1) a large hydrophilic amino-terminal domain of 210-220 amino acids ; 2) a compact hydrophilic region of 70 residues subdivided into three segments of 19-27 uncharged amino acids (MI, MII and MIII) ; 3) a second hydrophilic domain of variable length ; 4) a carboxy-terminal segment of 20 hydrophobic residues (MIV).

On the basis of these primary sequence data, several models of transmembrane organization, common to the four homologous subunits in Torpedo, have been proposed (see Changeux et al., 1984 ; Guy and Hucho, 1987). Their common features are : 1) the orientation of the large hydrophilic domain towards the synaptic cleft ; 2) the orientation of the small hydrophilic domain towards the cytoplasm ; 3) the assignment of the four hydrophobic segments MI-MIV to transmembrane α-helices. All of the proposed models assume that : 1) the acetylcholine binding site is located in the large hydrophilic domain on the α-subunit, and 2) that the walls of the ionic channel lie in the axis of quasi-symmetry of the AChR molecule and are delineated by homologous portions of each subunit.

Tests for these models have been developed in several laboratories using different experimental approaches which include : 1) site directed mutagenesis and expression of cloned cDNAs for the α subunit (Mishina et al., 1985) ; 2) immunochemical studies using antibodies directed against peptides of defined sequences (review Lindstrom, 1986) ; 3) covalent labeling of specific sites of the AChR with radioactive affinity (or photoaffinity) reagents and identification of the labeled peptides. The results obtained in our laboratory by this last method are summarized in the next paragraphs.

STRUCTURE OF THE CHOLINERGIC LIGAND BINDING SITES

There are two main acetylcholine (ACh) binding sites per muscle receptor molecule (Reynolds and Karlin, 1978) which interact in a positively cooperative manner (review in Changeux et al., 1984, 1990). The positive cooperative interactions between ACh molecules takes place, on Torpedo AChR, between non equivalent sites, as assessed by their distinct affinities for various competitive antagonists (Weber and Changeux, 1974 ; Neubig and Cohen, 1979 ; Damle and Karlin, 1978, and some antibodies (Dowding and Hall, 1987).

Since the two α-subunits are encoded by a single gene in Torpedo (Klarsfeld et al., 1984), they are, most likely, identical in primary structure. However, within the $\alpha_2\beta\gamma\delta$ oligomer, the two α-subunits cannot be equivalent in their mode of interaction with other subunits, and the possibility was considered that other subunits contribute domains for ligand binding (Oswald and Changeux, 1982 ; Langenbuch-Cachat et al., 1988 ; Blount and Merlie, 1989 ; Pedersen and Cohen, 1990).

Early affinity labeling experiments of the reduced receptor with a maleimide reagent (MBTA) lead to the identification of Cys 192 and Cys 193 in or near the acetylcholine binding site in the α-subunit (Kao and Karlin, 1986). The utilisation of another reagent p-(N,N)-dimethylamino benzenediazonium fluoroborate (DDF), a competitive antagonist in the dark which can be photoactivated by energy transfer from the protein (Langenbuch-Cachat et al., 1988), revealed a more complex organisation of the cholinergic binding area (Dennis et al., 1988 ; Galzi et al., 1990). DDF labels amino acids from three different regions of the large NH2 terminal domain : α-Tyr 93, α-Trp 149 and a larger set which includes α-Cys 192, 193 and α-Tyr 190, an amino acid which also reacts with a derivative of the coral toxin lophotoxin (Abramson et al., 1989). None of them are aspartic and glutamic acids often assumed to contribute by their carboxylate anion to a "negative subsite" of the ACh binding site. But two (and possibly four) are tyrosines, which exhibit the electronegative character required to complement the quaternary ammonium of ACh. Tyrosines have been commonly found exposed to solvents in protein, in particular in antibodies combining sites (Chothia et al., 1989), and their presence in the ACh binding area may reflect a rather general feature of ligand recognition sites in proteins.

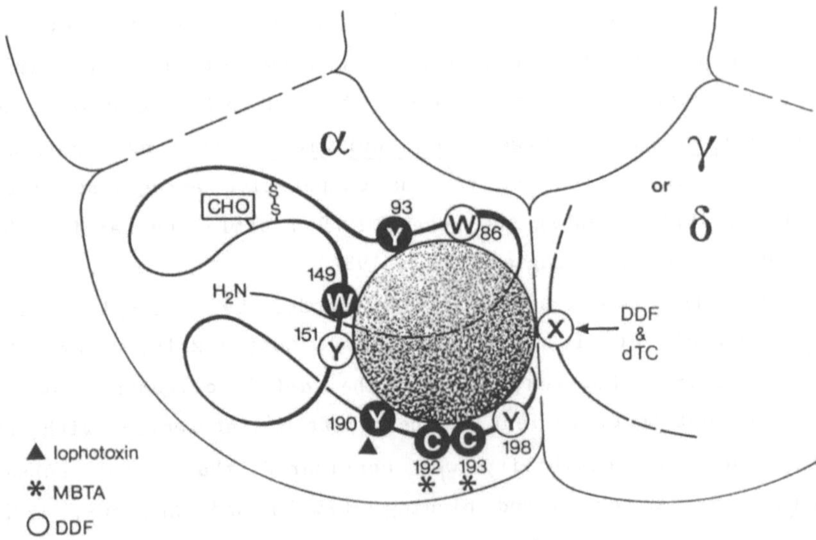

FIGURE 2 : Model of the agonist/competitive antagonists binding site(s) [^{3}H]DDF. The large sphere represents the space occupied by a molecule of DDF. Circles denote residues labeled. The amino acids labeled by [^{3}H]MBTA (*) and [^{3}H]- lophotoxin (▲) are also indicated.

None of the identified amino acids is present in muscle non α-subunits, but all are conserved in all α-subunits from <u>Torpedo</u> to man and from muscle to brain (except α_5) (review Galzi et al., 1990), and may thus be determinant for the ACh binding specificity. Yet, the neighboring amino acids significantly vary in the sequences of the various α-subunits and may contribute to the diversity of pharmacological properties noted among the several nicotinic AChR, in particular from brain. Moreover, comparison of the deduced amino acid sequences of all cloned nicotinic, GABA and glycine receptor subunits reveals comparable hydrophilic and hydrophobic distribution of amino acids suggesting similar tertiary architectural design (Greningloh, 1987 ; Shivers, 1989). This notion is further supported by the presence of several strictly conserved "canonical" amino acids. For instance, the loop which contains the DDF labeled α Tyr-93 from AChR α-subunit, also contains the canonical amino acids α Trp 86, Pro-88, Asp-89, Asn-94. Also, another DDF labeled amino acid α Trp 149 is located in the vicinity of the canonical α Cys 128 - Cys 142 loop and α Tyr-151. Furthermore, in the same region of the glycine receptor α-subunit, mutation at position 167 (homologous to AChR α-150) strikingly modifies the binding of the competitive antagonist strychnine (Kuhse et al., 1990). Thus, as observed for immunoglobulins (Chothia et al., 1989), all these ligand gated ion channels may adopt common main chain

	84 * 96	128 * 153
nAChRα	DV**W**L**PD**LVLY**N**NA	**C**EIIVTHF**P**F**D**QQN**C**TMKLGIWT**Y**DG
nAChRβ	DV**W**Q**PD**IVLM**N**NN	**C**TIKVMYF**P**F**D**WQN**C**TMVFKSYT**Y**DT
nAChRγ	LL**W**L**PD**VVLE**N**NV	**C**PIAVTYF**P**F**D**WQN**C**SLVFRSQT**Y**NA
nAChRδ	LV**W**I**PD**IVLQ**N**NN	**C**PINVLYF**P**F**D**WQN**C**SLKFTALN**Y**DA
GABA α	KI**W**T**PD**TFFH**N**GK	**C**PMHLEDF**P**M**D**AHA**C**PLKFGSYA**Y**TR
GABA β	QL**W**V**PD**TYFL**N**DK	**C**MMDLRRY**P**L**D**EQN**C**TLEIESYG**Y**TT
GABA γ2	KI**W**I**PD**TFFR**N**SK	**C**QLQLHNF**P**M**D**EHS**C**PLEFSSYG**Y**PR
GABA δ	KL**W**L**PD**TFIV**N**AK	**C**DMDLAKY**P**M**D**EQE**C**MLDLESYG**Y**SS
		*
GLY α2	SI**W**K**PD**LFFA**N**EK	**C**PMDLKNF**P**L**D**VQT**C**IMQLESFG**Y**TM

FIGURE 3 : Comparison of the amino acid sequences of the two peptide loops containing the [³H]DDF labeled and "canonical" amino acids. Sequence data from Noda et al., 1983 ; Shivers et al., 1989, Kuhse et al., 1990. Canonical amino acids are indicated in bold face type. (*) denotes amino acids involved in cholinergic ligand binding (nAChR α-subunit) or strychnine binding (Gly α_2-subunit).

structures in the regions homologous to those containing α Tyr 93 and α Trp 149. Accordingly, it is possible that these loops play a critical role in ligand binding which may be common to all the members of the superfamily of ligand gated ion channels. Discrete variations of key amino acid side chains within a conserved main chain skeleton, as well as the multiple subunits interactions could then govern the particular pharmacological specificity and the regulatory properties of each of these individual ligand gated ion channels.

IDENTIFICATION OF THE ION CHANNEL

The permeability response of muscle AChR to agonists can be blocked by an heterogeneous group of pharmacological agents referred to as non competitive blockers (NCBs). All of them decrease the amplitude of the permeability response to agonists without significantly affecting the agonist site (reviews in Heidmann et al., 1983 ; Adams, 1981). Equilibrium binding studies with a variety of NCBs showed that these compounds bind to populations of sites topographically distinct from the ACh binding sites but positively interacting with them in an allosteric manner (Weber and Changeux, 1974 ; Cohen and Boyd, 1979, see Changeux, 1990). Further detailed studies on T. marmorata AChR-rich membranes led to the distinction of 1) a high-affinity site, sensitive to histrionicotoxin, and present at a single copy per receptor pentamer and 2) low-affinity sites, insensitive to histrionicotoxin, present in large numbers (10-30 sites per receptor molecule, and lipid-dependent), possibly located at the interface of the receptor with membrane lipids (Heidmann et al., 1983).

An unambiguous demonstration that the high-affinity site for NCBs belongs to the $\alpha_2\beta\gamma\delta$ oligomer arose from labeling studies with a radioactive photoaffinity derivative of the potent local anesthetic trimethisoquin, $[^3H]$ 5-azido trimethisoquin. This compound labels almost exclusively the δ-subunit from T. marmorata receptor (Oswald et al., 1980) and this labeling is enhanced by carbamylcholine and inhibited by histrionicotoxin.*

*With quinacrine mustard (Kaldany and Karlin, 1983), the labeling primarily occurs at the level of the α-and β-subunits of T. Californica AChR. Using triphenylmethylphosphonium (Muhn and Hucho, 1983) or azidophencyclidine (Haring et al., 1983), differences in the labeling pattern between the receptor proteins from T. californica, T. marmorata and T. ocellata were noted.

Interestingly, $[^3H]$-chlorpromazine labeled all four chains of the T. marmorata receptor (Heidmann et al., 1983). On this basis, it was proposed that the unique high-affinity site for NCBs, to which all subunits may contribute, lie in the axis of pseudo-symmetry of the AChR oligomer where the distance between subunits is minimal.

The rate of covalent attachment of $[^3H]$chlorpromazine to the high affinity site resolved in the 10 msec-sec time-scale increases 100-1000-fold ($K_{on}=10^7$ $M^{-1}s^{-1}$) upon addition of acetylcholine. This effect is blocked by competitive antagonists such as d-tubocurarine, flaxedil and α toxin. The rate of $[^3H]$-chlorpromazine incorporation declines upon prolonged exposure of the receptor to acetylcholine, with time-course and concentration-dependence characteristics (Heidmann and Changeux, 1986) closely related to those reported for the rapid desensitization of the ion flux response of T. californica native membranes (Sakman et al., 1980 ; Hess et al., 1982) or reconstituted microsacs (Walker et al., 1982). The characteristics of this rapid labeling process support the notion that, under the conditions tested, $[^3H]$chlorpromazine binds to its high-affinity site without restriction to diffusion, and covalently reacts with this site while the channel opens (Heidmann and Changeux, 1986).

The Torpedo AChR amino acids photolabeled by $[^3H]$-chlorpromazine under equilibrium conditions were identified by peptide mapping and sequencing experiments. In the δ-subunit, the amino acid δ Ser 262 was unambiguously shown to be labeled (Giraudat et al., 1986) and this labeling was prevented by phencyclidine. Analysis of the other subunits resulted in the identification of Ser 254 and Leu 257 for the β subunit, Ser 248 for the α subunit (Giraudat et al., 1987, 1989) and Thr 253, Ser 257 and Leu 260 for the γ subunit (Revah et al., 1990).

The labeled serines on all AChR subunits have homologous positions and belong to the hydrophobic stretch MII. Interestingly, this MII sequence is highly conserved between subunits and throughout evolution from Torpedo to brain. Using triphenymethyl-phosphonium (TPMP), Oberthür et al. (1986) have also identified δ Ser 262 as the labeled amino acid, and Hucho et al. (1986) have obtained evidence also pointing at Ser 248 and Ser 254 as the amino acids labeled respectively on the α and β subunits. Also, using another noncompetitive blocker,

Pedersen and Cohen (1990) have reported that meproadifen mustard affinity labels the α subunit amino acid Glu-262, an amino acid located at the border of MII segment.

Homologous regions of different receptor subunits thus contribute to the unique high affinity site for NCBs within the ion channel, a finding consistent with the proposal that this site is located in the axis of quasi-symmetry of the receptor molecule. The distance between the agonist binding sites and the high-affinity site for NCB was evaluated by energy transfer using a fluorescent agonist and ethidium bromide within 21-40 Å (Herz et al., 1989). The positive interaction between the acetylcholine binding sites and the high-affinity site for NCB thus takes place between topographically distinct (and distant) sites. The contribution of segment MII to the ion channel has received further support from electrophysiological experiments carried out with functional AChRs channels expressed in *Xenopus* oocytes. The physiological importance of the ring of serines homologous to *Torpedo* α Serine 248 was demonstrated by site-directed mutagenesis of mouse receptor subunit mRNAs. Receptors with serine-to-alanine mutations displayed a selective decrease in outward single channel potassium currents. In addition, these mutations decrease the residence time (and thus the affinity) of the channel blocker QX 222 (Leonard et al., 1988), while serine-(or threonine)-to-alanine mutation in the ring homologous to *Torpedo* α serine 252 leads to a "stronger and longer binding" (Charnet et al., 1990). These results are consistent with the notion that the polar end of QX 222 binds to the polar ring formed by the [^{3}H]chlorpromazine labeled serines which provide a water like region.

In another study, point mutations were introduced into the *Torpedo* receptor subunit cDNAs to alter the net charge of the three rings of negatively charged or glutamine residues α-Asp 238, α-Glu 241 and α-Glu 262 and homologs which frame the MII segment (Imoto et al., 1988). On the basis of the sidedness of Mg^{2+} effects, the anionic ring between MII and MIII [α-Glu 262 and homologs] was located on the extracellular side, the two other ones between MI and MII [α-Asp 238 and homologs] and [α-Glu 241 and homologs], on the cytoplasmic side. Thus, the segment MII spans the membrane, and the site for [^{3}H]chlorpromazine and TPMP binding in MII is framed by the anionic rings within the membrane.

MODEL FOR THE ACETYLCHOLINE-GATED ION CHANNEL

As discussed above, models inferred from the primary structure of the transmembrane organization of the AChR subunits assume that the ion channel lies in the axis of quasi-symmetry of the receptor oligomer. Convergent results from affinity labeling studies and site-directed mutagenesis experiments point to the hydrophobic MII

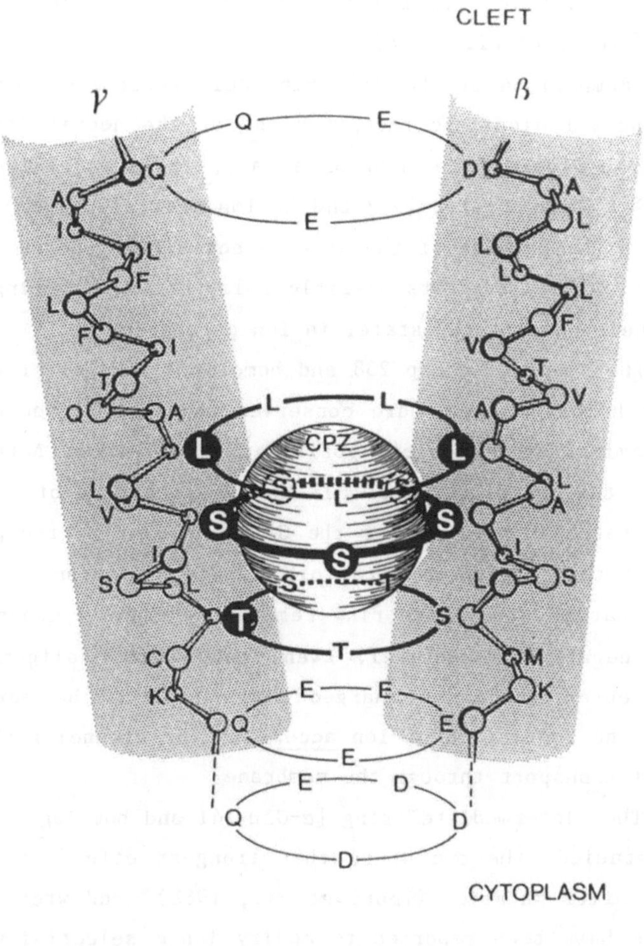

FIGURE 4 : Model of the high-affinity site for chlorpromazine within the AChR ion channel [taken from Revah et al. (1990)]. The MII segments, arranged as transmembrane α-helices, are quasi-symmetrically organized around the central axis of the molecule. The amino acids labeled by chlorpromazine are represented by black circles. The negatively charged rings which mutation causes conductance changes frame the chlorpromazine binding site.

segment as a critical component of the channel. The pattern of [^{3}H]-chlorpromazine labeled amino acids in the β and γ subunits is consistent with the organization of MII into an α helix. In such a configuration, the α carbons of the labeled amino acids of the β and α subunits are aligned on the same meridian on adjacent turns of the helix (see Giraudat et al., 1987 ; Revah et al., 1990), and also aligned with α Glu 241 (and homologs) but not with α Lys 242, a disposition consistent with the results of site directed mutagenesis experiments (Imoto et al., 1988).

The comparison of the MII amino acid sequences from the known muscle and neural nicotinic receptors and of the neural GABA (Shivers et al., 1989), Glycine (Grenningloh et al., 1987) and kainate (Gregor et al., 1989 ; Wada et al., 1989 and Hollmann et al., 1989) receptors, together with the results of the site directed mutagenesis experiments (Imoto et al., 1988) suggests possible roles of several rings of amino acids, in their non-mutated state, in ion permeation :

a) The "inner" [α asp 238 and homologs] and "outer charged" [α Glu 262 and homologs] rings are conserved (as aspartic acid, glutamic and/or glutamine) at homologous positions in all known AChR subunits. Despite that the contribution of MII to the transport of ions has not been demonstrated in the case of the GABA and glycine receptor, it is worth noting that in these receptors, where the transported ion is negatively charged, the inner ring remains negatively charged and the outer one, negative or neutral. Even though other alignment may be found, the outer and inner charged rings seem to be more directly involved in the regulation of ion access to the channel rather than in their actual transport through the membrane

b) The "intermediate" ring [α-Glu 241 and homologs] are, of all the rings studied, the one where the strongest effect on conductance takes place after mutation (Imoto et al., 1988) and where amino acid permutations have been reported to modify ionic selectivity (Imoto et al., unpublished). This negatively charged ring is conserved in all nicotinic receptors but absent from GABA or glycine receptor where, depending on the alignment, it is replaced, either by an alanine or by an arginine ring. Thus, the "intermediate ring" may be viewed as the site probing for the charge of the transported ion.

c) The serine and threonine "hydrophilic rings" are located within the MII segment and not at its borders and could act as "cation-binding macrocycles" (Perutz, 1989). Interestingly, the "ring of serines" [homologous to α ser 248] is conserved through all AChR subunits. The amino acid at homologous position is a serine in the frog and chick kainate receptors and a threonine in the Glycine and GABA receptors. Similarly, the second annulus of polar amino acids, the "ring of threonines" homologous to α Thr 244 is conserved in AChR subunit sequences. Mutations in this ring dramatically affect the conductance of mouse AChR (Charnet et al., 1990). Therefore, the presence of internal hydrophilic rings appear of more general importance than the charged rings, and may play a more fundamental role in the general process of ion permeation. They may simply constitute the narrowest region of the transmembrane channel and their hydrophilic character may be related to the state of hydration of the transported ion (Ascher and Nowak, 1988). A possibility is that the "hydrophilic rings" contribute to the exchange of water molecules surrounding the transported ion and both probe the size of the dehydrated (or partially hydrated) ion and facilitate (or even "catalyze") its translocation (Changeux, 1990).

d) The ring of leucines homologous to α Leu 251 is conserved in all known subunits of AChR, Glycine and GABA. The role of such bulky and hydrophobic residues in ion transport remains unclear. It has however been proposed that these residues could play a role in the gating mechanisms of the channel by sterically inhibiting ion flux in the closed conformation of the channel (Unwin, 1989).

Further work combining, for instance, site-directed mutagenesis and measurements of permeability ratios and/or streaming potentials should lead to a closer definition of the respective physiological roles of the "charged and hydrophilic" rings.

CONCLUSION

The structural and functional studies of the Torpedo and muscle AChR have provided extensive data on their quaternary organization, on the molecular mechanisms of the interaction between the various ligand binding sites they carry and on the regulation of ion permeation through the associated channel. The proposed models derived from these data show predictive properties which may apply to the neuronal AChR and to the other ligand-gated ion channels. For instance, the combination of site directed mutagenesis and labeling experiments have allowed the analysis of the amino acid level of the interactions between a permeant ion and the walls of the channel. It is now tempting to predict from primary amino acid sequences some of the properties of a channel, and to extend to the whole superfamily of ligand gated ion channels (AChR, GABA, glycine and kainate receptors) the interpretations and models derived from electric organ and muscle AChRs. For instance all these receptors may display a similar organization of the ion channel and similar allosteric coupling mechanisms between agonist binding sites and ion channel. They may also possess additional allosteric sites with pharmacological specificities distinct from that of the agonist binding sites. Such sites may be responsible for the pharmacological action of barbiturates, benzodiazepines, and β-carbolines for the GABA receptor, or of glycine for the glutamate-NMDA receptor. Yet, minor variations in the structure of the subunits and their multiple combinations (see e.g. Pritchett et al., 1988) may confer to the different members of the super-family important differences in ligand binding properties or ion channel function and distinctive regulatory properties, such as the parameters of the fast and slow allosteric transitions and, in particular, their sign : sensitization vs desensitization.

REFERENCES

Abramson SN, Li Y, Culver P, Taylor P (1989) An analog of lophotoxin reacts covalently with Tyr 190 in the α-subunit of the nicotinic acetylcholine receptor. J. Biol. Chem. 264:12666-12672.

Adams PR (1981) Acetylcholine receptor kinetics. J. Membrane Biol. 58:161-174.

Ascher P, Nowak L (1988) The role of divalent cations in the responses of central neurons to N-methyl-D-aspartate. J. Physiol. London, 399:247-266.

Blount P, Merlie JP (1989) Molecular basis of the two nonequivalent ligand binding sites of the muscle nicotinic acetylcholine receptor. Neuron 3:349-357.

Brisson A, Unwin PNT (1985) Quaternary structure of the acetylcholine receptor. Nature 315:474-477.

Changeux JP (1981) The acetylcholine receptor : An "allosteric" membrane protein. "Harvey Lectures". Acad. Press Inc. 75:85-254.

Changeux JP, Devillers-Thiéry A, Chemouilli P (1984) Acetylcholine receptor : an allosteric protein. Science 225:1335-1345.

Changeux JP (1990) Functional architecture and dynamics of the nicotinic acetylcholine receptor : an allosteric ligand-gated ion channel. Fidia Research Foundation Neuroscience Award Lectures, volume 4. JP Changeux, RR Llinas, D Purves, FE Bloom eds, Raven Press Ltd, pp. 21-168.

Charnet P, Labarca C, Leonard RJ, Vogelaar NJ, Czyzyk L, Gouin A, Davidson N, Lester HA (1990) An open-channel blocker interacts with adjacent turns of α-helices in the nicotinic acetylcholine receptor. Neuron 2:87-95.

Chothia C, Lesk AM, Tramontano A, Levitt M, Smith-Gill SJ, Air G, Sheriff S, Padlan EA, Davies D, Tulip WR, Comman PM, Spinelli S, Alzari PM, Poljak RJ (1989) Conformations of immunoglobulin hypervariable regions. Nature 342:877-883.

Claudio T, Ballivet M, Patrick J, Heinemann S (1983) Nucleotide and deduced amino acid sequences of Torpedo californica acetylcholine receptor gamma-subunit. Proc. Natl. Acad. Sci. USA 80:1111-1115.

Cohen JB, Boyd ND (1979) "Catalysis in Chemistry and Biochemistry", B. Pulmann and O. Ginsburg, Eds (D. Eidel, Publ.).

Damle VN, Karlin A (1978) Affinity labeling of one of two α-neurotoxin binding sites in acetylcholine receptor from Torpedo californica. Biochemistry 17:2039-2045.

Dennis M, Giraudat J, Kotzyba-Hibert F, Goeldner M, Hirth C, Chang JY, Lazure C, Chrêtien M, Changeux JP. (1988) Amino acids of the Torpedo marmorata acetylcholine receptor α subunit labeled by a photoaffinity ligand for the acetylcholine binding site. Biochemistry 27:2346-2357.

Devillers-Thiery A, Giraudat J, Bentaboulet M, Changeux JP (1983) Complete mRNA coding sequence of the acetylcholine binding alpha subunit of Torpedo marmorata acetylcholine receptor : A model for the transmembrane organization of the polypeptide chain. Proc. Natl. Acad. Sci. USA 80:2067-2071.

Dowding AJ, Hall ZW (1987) Monoclonal antibodies specific for each of the two toxin-binding sites of Torpedo acetylcholine receptor. Biochemistry 26:6372-6381.

Galzi JL, Revah F, Black D, Goeldner M, Hirth C, Changeux JP. (1990) Identification of a novel amino acid α Tyr 93 within the active site of the acetylcholine receptor by photoaffinity labeling : additional evidence for a three-loop model of the acetylcholine binding site. J. Biol. Chem. 265:10430-10437.

Giraudat J, Dennis M, Heidmann T, Chang JY, Changeux JP (1986) Structure of the high affinity binding site for noncompetitive blockers of the acetylcholine receptor : Serine-262 of the delta subunit is labeled by ^{3}H chlorpromazine. Proc. Natl. Acad. Sci. USA 83:2719-2723.

Giraudat J, Dennis M, Heidmann T, Haumont PT, Lederer F, Changeux JP (1987) Structure of the high-affinity binding site for noncompetitive blockers of the acetylcholine receptor : ^{3}H chlorpromazine labels homologous residues in the beta and delta chains. Biochemistry 26: 2410-2418.

Giraudat J, Galzi JL, Revah F, Changeux JP, Haumont PY, Lederer F (1989) The noncompetitive blocker [^{3}H]chlorpromazine labels segment M2 but not segment M1 of the nicotinic acetylcholine receptor alpha-subunit. FEBS Lett. 253:190-198.

Gregor P, Mano I, Maoz I, McKeown M, Teichberg VI (1989) Molecular structure of the chick cerebellar kainate-binding subunit of a putative glutamate receptor. Nature 342:689-692.

Grenningloh G, Rienitz A, Schmitt B, Methfessel C, Zensen M, Beyreuther K, Gundelfinger ED, Betz H (1987) The strychnine binding subunit of the glycine receptors shows homology with nicotinic acetylcholine receptors. Nature 328:215-220.

Guy RH, Hucho F (1987) The ion channel of the nicotinic acetylcholine receptor. Trends Neurosci. 10:318-321.

Haring R, Kloog Y, Kalir A, Sokolovsky M (1983) Species differences determine azido phencyclidine labeling pattern in desensitized nicotinic acetylcholine receptors. Biochem. Biophys. Res. Comm. 113:723-729.

Heidmann T, Changeux JP (1986) Characterization of the transient agonist-triggered state of the acetylcholine receptor rapidly labeled by the noncompetitive blocker [^{3}H]chlorpromazine : additional evidence for the open channel conformation. Biochemistry 25:6109-6113.

Heidmann T, Oswald RE, Changeux JP (1983) Multiple sites of action for noncompetitive blockers on acetylcholine receptor rich membrane fragments from Torpedo marmorata. Biochemistry 22:3112-3127.

Herz JM, Johnson DA,. Taylor P (1989. Distance between the agonist and noncompetitive inhibitor sites on the nicotinic acetylcholine receptor. J. Biol. Chem. 264:12439-12448.

Hess GP, Pasquale EB, Walker JW, McNamee MG (1982) Comparison of acetylcholine receptor-controlled cation flux in membrane vesicles form Torpedo californica and Electrophorus electricus : chemical kinetic measurements in the millisecond region. Proc. Natl. Acad. Sci. USA 79: 963-967.

Hollmann M, O'Shea-Greenfield, Rogers SW, Heinemann S (1989) Cloning by functional expression of a member of the glutamate receptor family. Nature 342:643-648.

Hucho FL (1986) The nicotinic acetylcholine receptor and its ion channel. Eur. J. Biochem. 158:211-226.

Hucho FL, Oberthür W, Lottspeich F (1986) The ion channel of the nicotinic acetylcholine receptor is formed by the homologous helices MII of the receptor subunits. FEBS Lett. 205:137-142.

Imoto K, Busch C, Sakmann B, Mishina M, Konno T, Nakai J, Bujo H, Mori Y, Fukuda K, Numa S (1988) Rings of negatively charged amino acids determine the acetylcholine receptor channel conductance. Nature 335: 645-648.

Kaldany RRJ, Karlin A (1983) Reaction of quinacrine mustard with the acetylcholine receptor from Torpedo californica : Functional consequences and sites of labeling. J. Biol. Chem. 258:6232-6242.

Kao PN, Dwork AJ, Kaldany RRJ, Silver ML, Wideman J, Stein S, Karlin A (1984) Identification of the alpha-subunit half-cystine specifically labeled by an affinity reagent for the acetylcholine receptor binding site. J. Biol. Chem. 259:11662-11665.

Karlin A (1980) Molecular properties of nicotinic acetylcholine receptors. "Cell Surface reviews" (Poste G, Nicolson, GL, Cotman, CW, Eds) 6, New York, 191-260.

Klarsfeld A, Devillers-Thiery A, Giraudat J, Changeux JP (1984) A single gene codes for the nicotinic acetylcholine receptor alpha-subunit in Torpedo marmorata : Structural and developmental implications. EMBO J. 3:35-41.

Kuhse J, Schmieden V, Betz H (1991) A single amino acid exchange alters the pharmacology of neonatal rat glycine receptor subunit. Neuron (in press).

Langenbuch-Cachat J, Bon C, Goeldner M, Hirth C, Changeux JP (1988) Photoaffinity labeling by aryldiazonium derivatives of Torpedo marmorata acetylcholine receptor. Biochemistry 27:2337-2345.

Leonard RJ, Labarca CG, Charnet P, Davidson N, Lester HA (1988) Evidence that the M2 membrane-spanning region lines the ion channel pore of the nicotinic receptor. Science 242:1578-1581.

Mishina M, Tobimatsu T, Imoto K, Tanaka K, Fujita Y, Fukuda K, Kurasaki M, Takahashi H, Morimoto Y, Hirose T, Inayama S, Takahashi T, Kuno M, Numa S (1985) Location of functional regions of acetylcholine receptor alpha-subunit by site-directed mutagenesis. Nature 313:364-369.

Mitra AK, McCarthy MP, Stroud RM (1989) Three-dimensional structure of the nicotinic acetylcholine receptor and location of the major associated 43-kD cytoskeletal protein, determined at 22 Å by low dose electron microscopy and X-ray diffraction to 12.5 Å. J. Cell Biol. 109:755-774.

Muhn P, Hucho F (1983) Covalent labeling of the acetylcholine receptor from Torpedo electric tissue with the channel blocker [^{3}H] triphenylmethylphosphonium by ultraviolet irradiation. Biochemistry 22:421-425.

Neubig RR, Cohen JB (1979) Equilibrium binding of (^{3}H) tubocurarine and [^{3}H] acetylcholine by Torpedo postsynaptic membranes : stoichiometry and ligand interactions. Biochemistry 18:5464-5475.

Noda M, Takahashi H, Tanabe T, Toyosato M, Kikyotani S, Furutani Y. Hirose T, Takashima H, Inayama S, Miyata T, Numa S (1983) Structural homology of Torpedo californica acetylcholine receptor subunits. Nature 302:528-532.

Oberthür W, Muhn P, Baumann H, Lottspeich F, Wittmann-Liebold B, Hucho F (1986) The reaction site of a noncompetitive antagonist in the delta-subunit of the nicotinic acetylcholine receptor. EMBO J. 5: 1815-1819.

Oswald RE, Changeux JP (1982) Crosslinking of alpha-bungarotoxin to the acetylcholine receptor from Torpedo marmorata by ultraviolet light irradiation. FEBS Lett. 139:225-229.

Oswald R, Sobel A, Waksman G, Roques B, Changeux JP (1980) Selective labeling by [^{3}H-]trimethisoquin azide of polypeptide chains present in acetylcholine receptor rich membranes from Torpedo marmorata. FEBS Lett. 111:29-34.

Pedersen SE, Cohen JB (1990a) d-tubocurarine binding sites are located at α-γ and α-δ subunit interfaces of the nicotinic acetylcholine receptor. Proc. Natl. Acad. Sci. USA 87:2785-2789.

Pedersen SE, Cohen JB. (1990b) [^{3}H]-mepraodifen mustard reacts with glu-262 of the nicotinic acetylcholine receptor (AChR) α-subunit. Biophy. J. 57:126a.

Perutz MF (1989) Mechanisms of cooperativity and allosteric regulation in proteins. Quarterly Rev. Biophys. 22:139-236.

Pritchett DB, Sontheimer H, Gorman CM, Kettenmann H, Seeburg PH, Schofield PR (1988) Transient expression shows ligand gating and allosteric potentiation of $GABA_A$ receptor subunits. Science 242:1306-1308.

Revah F, Galzi JL, Giraudat J, Haumont PY, Lederer F, Changeux, JP (1990) The noncompetitive blocker [^{3}H]chlorpromazine labels three amino acids of the acetylcholine receptor γ subunit : implications for the α helical organization of the MII segment and the structure of the ion channel. Proc. Natl. Acad. Sci. USA 87:4675-4679.

Reynolds JA, Karlin A (1978) Molecular weight in detergent solution of acetylcholine receptor from Torpedo californica. Biochemistry 17: 2035-2038.

Sakman B, Patlak J, Neher E (1980) Single acetylcholine activated channels show burst-kinetics in presence of desensitizing concentrations of agonist. Nature 286:71-73.

Shivers BD, Killisch I, Sprengel R, Sontheimer H, Köhler M, Schofield PR, Seeburg PH (1989) Two novel $GABA_A$ receptor subunits exist in distinct neuronal subpopulations. Neuron 33:327-337.

Toyoshima C, Unwin N (1988) Ion channel of acetylcholine receptor reconstructed from images of postsynaptic membranes. Nature 336: 247-250.

Unwin N (1989) The structure of ion channels in membranes of excitable cells. Neuron 3:665-676.

Wada K, Dechesne CJ, Shimasaki S, King RG, Kusano K, Buonanno A, Hampson DR, Banner C, Wenthold RJ, Nakatani Y (1989) Sequence and expression of a frog brain complementary DNA encoding a kainate-binding protein. Nature 342:684-689.

Walker JW, Takeyasu K, McNamee MG (1982) Activation and inactivation kinetics of Torpedo californica acetylcholine receptor in reconstitued membranes. Biochemistry 21:5384-5389.

Weber M, Changeux JP (1974) Binding of Naja nigricollis 3H-alpha-toxin to membrane fragments from Electrophorus and Torpedo electric organs. 2. Effect of the cholinergic agonists and antagonists on the binding of the tritiated α-neurotoxin. Mol. Pharmacol. 10:13-34.

EPIDERMAL GROWTH FACTOR RECEPTOR AUTOPHOSPHORYLATION IS NOT REQUIRED FOR NORMAL MITOGENIC OR TRANSFORMING SIGNAL GENERATION

Axel Ullrich, Alane Gray[1], Susan Munemitsu[1], Joseph Schlessinger[2], and Sharon Massoglia[1]

Max-Planck-Institut für Biochemie
Am Klopferspitz 18A
8033 Martinsried, F.R.G.

Abstract

The oncogene of the avian erythroblastosis virus, v-erbB, encodes a mutated epidermal growth factor (EGF) receptor. In addition to a deletion of sequences coding for the EGF binding domain, v-erbB exhibits a variety of other mutations in its cytoplasmic portion. We have investigated the functional properties of erbB mutations within a ligand-responsive, chimeric EGF receptor construct. The autophosphorylation activities of the cytoplasmic domains of HER-erbB-H and HER-erbB-ES4 receptors were found to be fully ligand-dependent in intact cells. However, the signalling properties of an EGF receptor carrying v-erbB-ES4 (but not v-erbB-H) cytoplasmic sequences were ligand independent. Structural alterations in the cytoplasmic domain of growth factor receptor tyrosine kinases may therefore induce constitutive signalling activity without autophosphorylation. These findings provide new insight into the mechanism of receptor-mediated signal transduction and suggest a novel alternative for subversion of cellular control mechanisms and proto-oncogene activation.

Introduction

Avian erythroblastosis virus (AEV) exerts its oncogenic effects in chickens by expressing the v-erbB gene in infected cells. This leads to the development of erythroleukemia and fibrosarcomas (Graf et al., 1976; Yamamoto et al., 1983a). V-erbB encodes a membrane-associated tyrosine kinase that originated by recombination of chicken EGF receptor (CER)

---

[1] Department of Developmental Biology, Genentech, Inc., 460 Point San Bruno Boulevard, South San Francisco, California 94080

[2] Department of Pharmacology, New York University Medical Center, 550 First Avenue, New York, New York 10016

NATO ASI Series, Vol. H 52
Biological Signal Transduction
Edited by E. M. Ross and K. W. A. Wirtz

sequences with the viral genome (Downward et al., 1984a; Ullrich et al., 1984; Nilsen et al., 1985). When compared with EGF receptor, the erbB proteins of all currently characterized AEV strains have lost most of their amino-terminal, extracellular ligand binding domain, which is thought to result in constitutive activation of its cytoplasmic tyrosine kinase activity (Yamamoto et al., 1983b; Choi et al., 1986; Damm et al., 1987; Raines et al., 1988). Furthermore, during and/or subsequent to the recombination event, other structural alterations that include cytoplasmic C-terminal truncations, internal deletions and point mutations were introduced into the virus-acquired CER sequences which differ between AEV strains (Yamamoto et al., 1983b; Choi et al., 1986; Gammett et al., 1986; Damm et al., 1987; Raines et al., 1988; Yarden and Ullrich, 1988). These alterations are likely to be responsible for the characteristic properties of distinct virus isolates in terms of their pathogenic potential in vivo and transforming activity in cultured cells (Gammett et al., 1986; Beug et al., 1986; Raines et al., 1988). Therefore, naturally selected v-erbB variants represent an excellent system with which to evaluate the structural basis for altered biochemical properties and biological signalling activities of growth factor receptor derivatives involved in a variety of distinct neoplastic phenotypes.

We have previously shown that erbB ligand responsiveness can be restored by precise reattachment of human EGF receptor (HER) extracellular sequences to repair the amino-terminal deletion. This type of HER-erbB chimeric protein exhibited EGF-responsive functions at the surface of transfected RatI fibroblasts (Riedel et al., 1987). We have extended this approach to compare the biological signalling activities of the cytoplasmic domains of two distinct, well-characterized v-erbB oncogene products: v-erbB-H and v-erbB-ES4. Parallel analyses of the signalling characteristics of the chimeric receptors have revealed pronounced differences between erbB-H and -ES4 cytoplasmic domains on the growth of transfected cells and the potential to transform mouse NIH-3T3 fibroblasts. While both erbB-H and erbB-ES4 autophosphorylation activities become ligand-dependent after reattachment of HER extracellular sequences, the normal and transforming signals mediated by the HER-erbB-ES4 chimera remain largely EGF-independent, in contrast to HER-erbB-H. This demonstrates that structural alterations within the cytoplasmic domain can generate a ligand-independent receptor and furthermore, that changes other than autophosphorylation can render the EGF receptor competent for the generation of normal mitogenic and transforming signals.

Results

V-erbB oncogene products share the same 554 amino acid amino-terminal deletion, but differ in their cytoplasmic sequence alterations (Figure 1; Yamamoto et al., 1983b; Choi et al., 1986). In addition to C-terminal deletions of 34 and 72 amino acids in v-erbB-H and v-erbB-

ES4, respectively, these oncogenes exhibit characteristic point mutations and internal sequence deletions. The only cytoplasmic sequence alteration shared between v-erbB-H and v-erbB-ES4, Phe 669, is located within the ATP binding site consensus sequence (GXGXXG) between the first two glycine residues. Further mutations are found in both oncogenes near this location (Ile 705/erbB-H; Thr 718/erbB-ES4) and downstream in the erbB-ES4 sequence (Figure 1).

To focus on the effects of cytoplasmic mutations, we eliminated the influence of the amino-terminal deletion by repairing it with intact human EGF receptor ligand binding sequences. CER and v-erbB cytoplasmic sequences were fused to human EGF receptor sequences using conserved restriction sites. Figure 1 shows schematically the structural differences between wild-type CER cytoplasmic sequences (Nilsen et al., 1985; Lax et al., 1988) and those of v-erbB-H and v-erbB-ES4 oncogenes, and the design of the EGF receptor

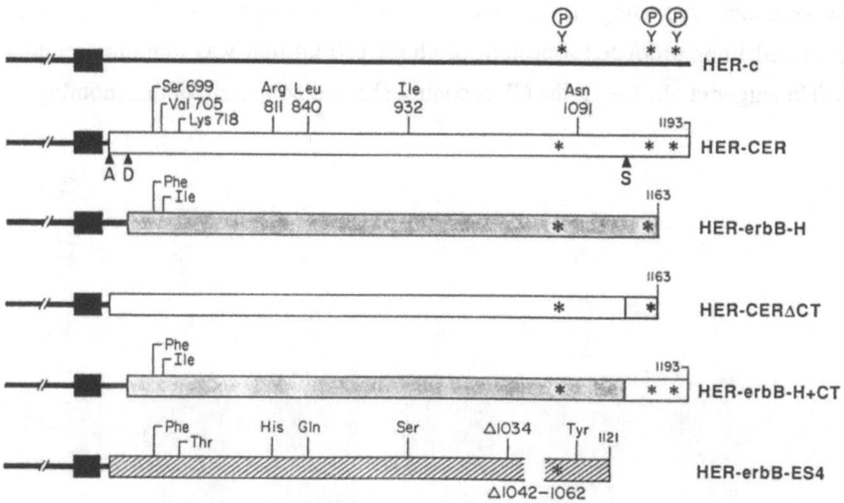

Figure 1. Structure of EGF receptor chimerae. Plasmids were constructed as described in the text. The transmembrane domains and specific point mutations are indicated, as well as the chimeric makeup of each construct. Δ indicates internal deletions of one (Δ1034) and 21 (Δ1042-1062) amino acids in the HER-erbB-ES4 sequence. Positions of human EGF receptor (HERc) tyrosine phosphorylation sites at positions 1068, 1148, and 1173 are indicated by asterisks and corresponding tyrosine residues in chicken EGF receptor are marked. HER-erbB-H and HER-CERΔCT constructs encode 4 retroviral *env* protein amino acids at their C-termini in addition to the 1155 HER-CER amino acids. Amino acid numbering is based on CER sequence (Lax et al., 1988). The actual coding sequences of chimeric constructs are 3 amino acids shorter due to differences in HER vs. CER extracellular sequences and 22 residues shorter in HER-erbB-ES4 due to internal deletions. [Reprinted from Massoglia *et al.*, 1990]

constructs employed here. The chimeric receptors were designated HER-erbB-H, HER-erbB-ES4, and HER-CER. Two additional expression vectors were constructed in which the C-termini of v-erbB-H and CER were exchanged. The portions exchanged include the CER tyrosine residue (Tyr 1179) that corresponds to Tyr 1173, the major HER autophosphorylation site (Downward et al., 1984b) (HER-erbB-H+CT), and allowed us to examine the significance of the two erbB-H point mutations within the ATP binding domain [at positions 699 (Phe vs. Ser) and 705 (Ile vs. Val)] and the C-terminal deletion itself (HER-CERΔCT). All chimeric receptor constructs were inserted into an SV40-based expression vector that included SV40 early promoter controlled genes for neomycin (G418) resistance and dihydrofolate reductase (DHFR), which allowed amplification of plasmid sequences in transfected cells under methotrexate selection (Crowley et al., 1983; Simonsen and Levinson, 1983; Hudziak et al., 1987). Mouse NIH 3T3 fibroblasts were transfected and amplified using standard procedures.

To examine whether transfected NIH 3T3 cells correctly synthesized and processed the various chimeric receptors and transported the properly folded molecules to the cell surface, cell lines were analyzed using an EGF receptor specific antibody. As shown in Figure 2, all transfected cell lines produced a protein of about 170 kd that was immunoprecipitated by Mab108 (Honegger et al., 1987); the CVN control showed no protein of this mobility.

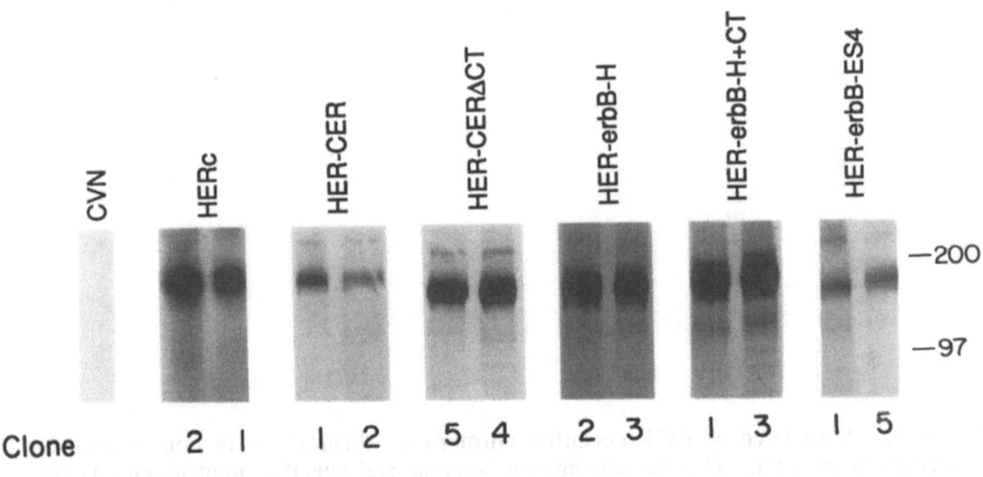

Figure 2. Immunoprecipitation of metabolically labeled EGF receptors. Subconfluent cells on 6 cm dishes were labeled for 18 hrs with 300 µCi ^{35}S-methionine in three mls of methionine-free MEM with 5% fetal calf serum. Labeled cells were solubilized, immunoprecipitated with Mab 108.1, and analyzed by 8% SDS-PAGE and fluorography. Positions of molecular weight markers are indicated in kilodaltons. ^{125}I-EGF surface receptor numbers per cell were determined by saturation binding and Scatchard plot analysis: CVN/3T3: $2x10^4$; HER-c/CL1: $4.3x10^5$; HER-c/CL2: $1.2x10^6$; HER-CER/CL1: $2.8x10^5$; HER-CER/CL2: $1.2x10^5$; HER-erbB-H/CL2: $4.7x10^5$; HER-erbB-H/CL3: $8x10^5$; HER-CERΔCT/CL4: $5.4x10^5$; HER-CER/CL5: $4.7x10^5$; HER-erbB-H+CT/CL1: $8.5x10^5$; HER-erbB-H+CT/CL3: $8.7x10^5$; HER-erbB-ES4/CL1: $1.2x10^5$; HER-erbB-ES4/CL5: $1.6x10^5$.

Ligand binding to the extracellular domain of the EGF receptor triggers rapid phosphorylation of several tyrosine residues, three of which are located within the EGF receptor C-terminal tail region (Downward et al., 1984b). Kinase activity is essential for most receptor functions (Honegger et al., 1987; Chen et al., 1987; Moolenaar et al., 1988) and can be detected in detergent lysates as well as in intact cells. We tested whether HER-erbB chimerae demonstrated EGF-dependent autophosphorylation to determine if fusion of the HER ectodomain to the cytoplasmic domain of erbB oncogene products was sufficient to reinstate this important signalling control. Furthermore, it was possible that the response of distinct chimerae to ligand stimulation would provide clues with respect to their distinct biological signalling properties.

EGF strongly stimulated the phosphorylation of all chimeric proteins both *in vitro* and *in vivo*. Under *in vitro* conditions, all chimeric receptors and the HERc control exhibited an EGF-inducible autophosphorylation activity (Figure 3). The basal activity detected *in vitro* for all receptor constructs was virtually undetectable in intact cells (see below), and is likely to be a consequence of the experimental protocol employed. To examine directly the tyrosine phosphorylation state of the receptors in intact cells, we carried out immunoblot analyses of cell

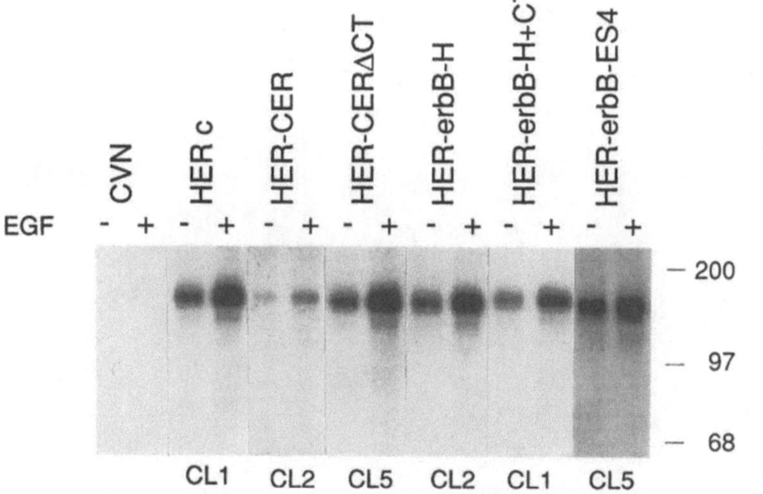

Figure 3. EGF-induced receptor autophosphorylation in vitro. Detergent cell lysates were incubated with (+) or without (-) 5 µg/ml EGF, immunoprecipitated and autophosphorylated. All samples were immunoprecipitated with MabRI, except those indicated for HER-erbB-ES4, which were immunoprecipitated with Mab 108.1. Samples were analyzed on 8% SDS-PAGE and the dried gel exposed for 2 hrs., except for HER-erbB-ES4, which was exposed overnight. Sizes of molecular weight markers are indicated in kilodaltons.

lysates, both before and after ligand exposure, using an anti-phosphotyrosine monoclonal antibody. As shown in Figure 4, both HER-erbB chimera and control receptors displayed a very strong induction of autophosphorylation upon EGF addition. Similar results were obtained for the HER-erbB-H+CT and HER-CERΔCT constructs (data not shown). Together, these experiments demonstrate that the human EGF binding domain controls the autophosphorylation activity of normal and mutated CER cytoplasmic domains, including the erbB-H and erbB-ES4 oncogene kinases, in the chimeric receptor constructs.

The mitogenic response of mammalian cells is thought to be strictly controlled by signals that are generated upon interaction of growth factors with specific cell surface receptors (Yarden and Ullrich, 1988). Subversion of this control at any stage of signal transmission may lead to aberrant behavior of cells in culture and possibly tumorigenesis *in vivo*. To further ascertain the signalling differences between various v-erbB isolates, we investigated the growth requirements and transformed state of the chimera-transfected cells lines.

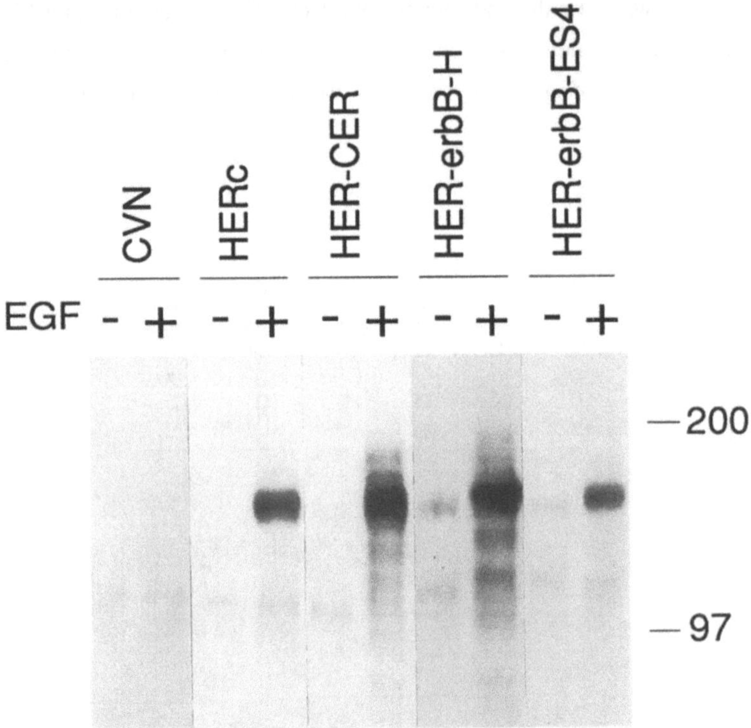

Figure 4. Receptor autophosphorylation in intact cells. Serum-starved confluent monolayers were incubated 30 min. with (+) or without (-) EGF. After electrophoresis on 8% SDS-PAGE and transfer to nitrocellulose, bands were detected with antiphosphotyrosine antibody 5E2 and ^{125}I-protein A. *In vivo* autophosphorylation of HER-CERΔCT and HER-erbB-H+CT was fully ligand-dependent (not shown). CVN is a vector-transfected control cell line. Sizes of molecular weight markers are indicated in kilodaltons. [Reprinted from Massoglia *et al.*, 1990]

Figure 5 shows the effect of EGF on [^{3}H]-thymidine incorporation into DNA of serum-starved, confluent cell monolayers. Control plasmid (CVN) transfected, NIH 3T3 control cells showed no mitogenic response to EGF, yet this growth factor stimulated ^{3}H-thymidine incorporation into all of the other transfected cell lines. However, while HERc, HER-CER, HER-CERΔCT, HER-erbB-H and HER-erbB-H+CT lines were each dependent on EGF stimulation and displayed a similar dose-response profile (Figure 5A and B), HER-erbB-ES4-expressing lines appeared to be EGF-independent. These cells could not be grown to quiescence and responded only marginally to increasing EGF concentrations (Figure 5B).

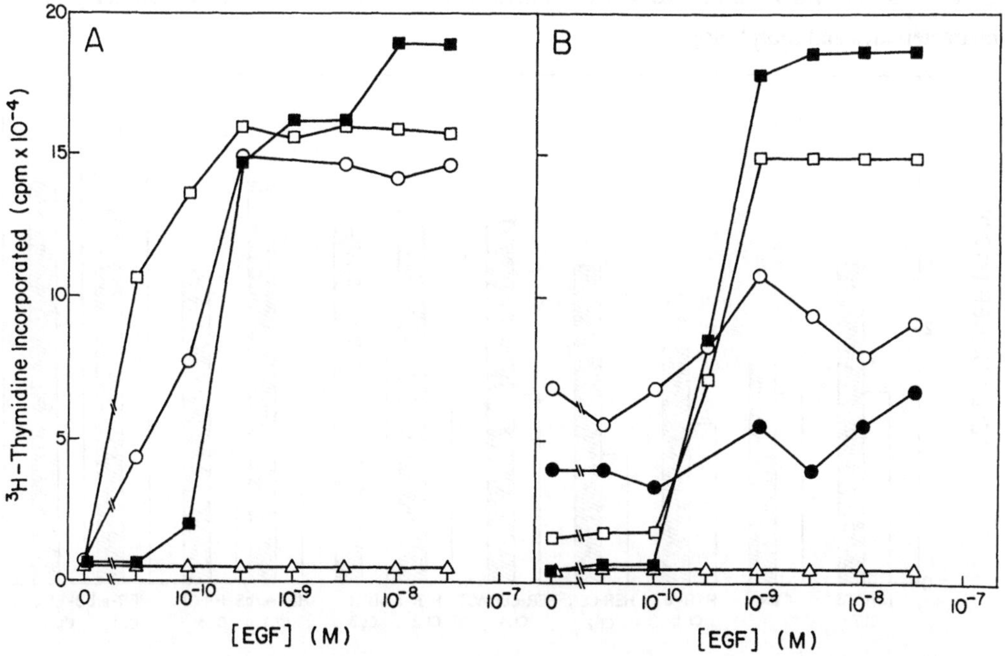

Figure 5. Mitogenic response of transfected NIH 3T3 fibroblasts. A) HERc CL2 ○, HER-CER CL1 □, HER-CERΔCT CL5 ■, CVN control Δ. **B)** HER-erbB-H CL3 □, HER-erbB-H+CT CL1 ■, HER-erbB-ES4 CL1 ●, HER-erbB-ES4 CL5 ○, CVN control Δ. Confluent cell monolayers in 24-well dishes were brought to quiescence for 24 hours in medium containing 0.5% fetal calf serum. Stimulation of DNA synthesis by various concentrations of receptor-grade EGF was monitored 18 hours after growth factor addition by 4-hour pulse labeling with 0.5 μCi methyl-^{3}H-thymidine. Cells were then washed three times with PBS and soluble radioactivity was extracted with 1 ml of 5% TCA for 30 min. at 4°C. Cells were washed twice with PBS, solubilized in 500 μl of 0.2 N NaOH and incorporated radioactivity was determined. The absence of an EGF response for CVN-transfected NIH 3T3 cells (Δ) in spite of the presence of EGF receptors has been observed in several independently isolated cell lines and cannot be explained at this point. [Reprinted from Massoglia *et al.*, 1990]

Mutant receptors expressing NIH 3T3 cell growth characteristics were also compared under serum free conditions with only insulin and transferrin added (Figure 6). Under these stringent conditions essentially the same picture emerged: while HER-erbB-ES4 cells grew vigorously even without EGF addition, all other cell lines, including the controls, required ligand activation of their receptors to achieve maximal growth response. In some cases, the relative growth inductions and maximal responses were different from those found at 0.5% FCS (not shown), suggesting that unknown serum components may cooperate or inhibit, via their own receptors, the signals generated by the mutant chimerae used here. Thus, despite the fact that the HER-erbB-ES4 hybrid receptor displays the same phosphorylation properties as HER-erbB-H in vitro and in vivo, these chimerae had very different effects on cellular growth characteristics and morphology.

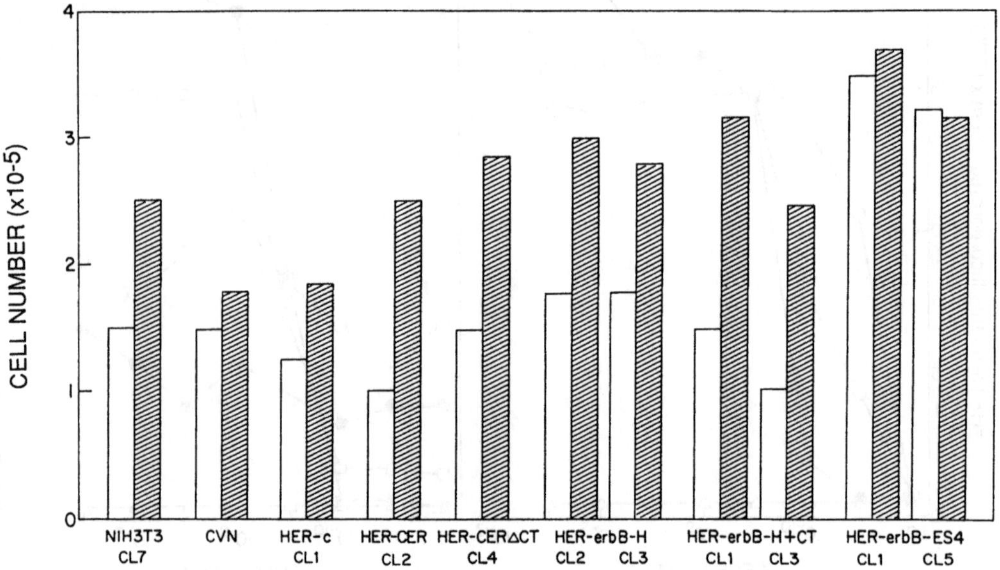

Figure 6. Effect of EGF on the proliferation of EGF receptor wild type and mutant lines in serum-free medium. 40,000 mock-transfected control NIH 3T3 cells and cells expressing each receptor construct were grown in 35 mm culture dishes in serum free DMEM in the presence of either transferrin (10 μg/ml) and insulin (1 μg/ml) (open box), or transferrin, insulin and EGF (10 ng/ml) (hatched box). Cell numbers were determined after six days using a Coulter counter. A representative result of three independent experiments is shown.

The loss of contact inhibition is a commonly used criterion to assess a cell's transformed state. To examine whether the transforming potential of v-erbB oncogene sequences was retained in the HER ectodomain fusions and to test whether this activity had become ligand-dependent, we subjected all chimeric EGF receptor mutant lines and the corresponding expression constructs to focus formation assays (Figure 7).

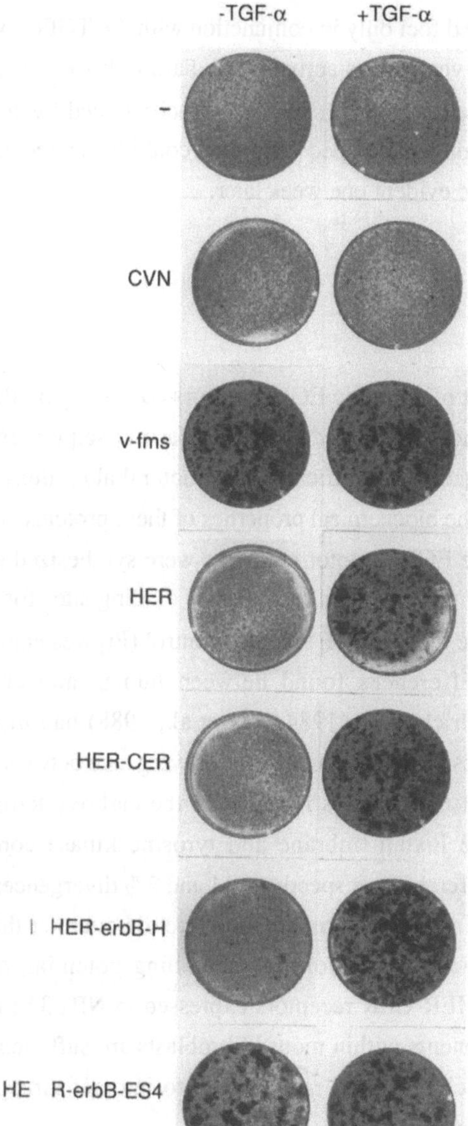

Figure 7. NIH 3T3 focus formation. Subconfluent NIH 3T3 monolayers were transfected with expression plasmids. In parallel, cells were cotransfected with a TGFα expression construct. After 21 days in culture, cell foci were stained with crystal violet. Only the TGFα cotransfection results are shown. No foci were detected in -TGFα controls except for HER-erbB-ES4 and v-*fms*. Several foci of each transfection experiment were picked and found to express both receptor and TGFα proteins (data not shown).

In cotransfection experiments with a TGFα expression construct (Rosenthal et al., 1986), transformed phenotype was induced in this stringent assay for all receptor expression plasmids (Figure 7). Yet, while HERc, HER-CER, HER-CERΔCT, HER-erbB-H, and HER-erbB-H+CT plasmids induced foci only in conjunction with the TGFα expression plasmid, the HER-erbB-ES4 expression vector was active in mediating this effect even without TGF-α, similar to the v-fms positive control. Interestingly, foci induced by cotransfection of HER-erbB-ES4 with TGFα plasmids developed faster and could be scored after two weeks, while TGFα-independent foci were evident one week later.

Discussion

We constructed a set of chimeric EGF receptors consisting of identical ligand binding and transmembrane sequences combined with cytoplasmic sequences of either cellular or retroviral origin, to investigate the significance of structural alterations on the mitogenic and transforming activities and the biochemical properties of these proteins. After introduction into mouse fibroblasts, all of the EGF receptor chimerae were synthesized and transported to the cell surface, where they formed high and low affinity binding sites for EGF and TGFα with properties comparable to the human EGF receptor control (Prywes et al., 1986). Apparently, the extensive sequence differences found between human and chicken EGF receptor cytoplasmic sequences (Ullrich et al., 1984; Lax et al., 1988) had little or no influence on ligand binding characteristics. Most of the sequence divergence between the EGF receptors of these two evolutionarily distant species is found in the carboxy-terminal-most tail region (33.5% divergence), while juxtamembrane and tyrosine kinase core regions are highly conserved (only 2 and 7 differences, respectively; 4 and 2% divergence; Lax et al., 1988). In addition, these differences in receptor primary structure did not alter the normal biochemical characteristics or mitogenic and transforming signalling potential, as demonstrated by a comparison of HERc and HER-CER receptors expressed in NIH 3T3 cells. These findings suggest that cellular components within mouse fibroblasts are sufficiently conserved in their structure between human, chicken and mouse cells to interact directly with these foreign receptors and mediate cellular responses.

The chimeric molecules, HER-CERΔCT, HER-erbB-H, HER-erbB-H+CT and HER-erbB-ES4, contain mutated chicken EGF receptor cytoplasmic domains that led to the generation of altered receptor signals and consequently, altered cellular responses. In addition to previously identified truncations and deletions within the C-terminal tail regions of erbB-H (34 amino acid truncation) and erbB-ES4 (72 amino acid truncation; 1 and 22 amino acid deletions), several point mutations have been identified in the tyrosine kinase core regions of these receptor-derived oncogenes (Yamamoto et al., 1983b; Nilsen et al., 1985; Choi et al., 1986; Lax et al., 1988). All of the point mutations in erbB-H and -ES4 sequences are in positions that are conserved between normal chicken and human EGF receptors, and even in the related but distinct HER2/neu receptor homolog (Coussens et al., 1985). This suggests an important role for these residues in the overall conformation of this receptor domain and perhaps also in the control and definition of receptor signals.

The most surprising aspect of our findings was the dramatic difference between the erbB-ES4 mutations compared with erbB-H sequences or variants thereof. At first, it appeared that faithful reconstitution of a ligand binding function, and thus elimination of the extracellular domain deletion, resulted in fully ligand-responsive chimeric receptors. Autophosphorylation experiments in detergent lysates of transfected cells indicated that all receptors showed varying degrees of basal activity in vitro. However, in intact cells, constitutive autophosphorylation was either barely detectable or not detected for HER-erbB-H and HER-erbB-ES4. In all cases, the basal activity was induced more than 10-fold in vivo.

Although all chimeric constructs showed EGF-inducible kinase activity, the structural alterations in the erbB-ES4 cytoplasmic domain generated proliferation signals that could not be controlled by a human EGF binding domain. In contrast to the HER-erbB-H receptor, expression of the HER-erbB-ES4 receptor consistently rendered cells ligand-independent in proliferation and transformation assays. When compared with the ligand-induced cell responses mediated by the other receptors, the EGF-independent proliferation signals generated by HER-erbB-ES4 were more effective under serum-free or low-serum conditions, while thymidine incorporation was constitutive and did not reach the maximal values of EGF-stimulated, ^{3}H-thymidine incorporation in other cell lines. Whether the lower HER-erbB-ES4 effects in this assay correlate with the lower expression levels of the two lines tested is not known with certainty.

These findings add a novel alternative to the currently known possibilities of growth factor receptor activation in mitogenesis and transformation. Most relevant for our studies on the biological effects of erbB mutations is a comparison with the tumorigenic *neu* gene (Bargmann et al., 1986; Muller et al., 1988). While a single point mutation in the transmembrane domain of the *neu* oncogene product (Bargmann et al., 1986; Bargmann and Weinberg, 1988) activates its cytoplasmic signalling functions independent of ligand, the same effect is achieved by erbB-ES4 cytoplasmic domain alterations. Most importantly, in contrast to *neu* (Bargmann and Weinberg, 1988), we have shown that these cytoplasmic domain mutations can render the EGF receptor active in the absence of ligand without autophosphorylation.

An intriguing interpretation of our findings is that structural alterations in the EGF receptor cytoplasmic domain direct this protein to assume the same conformation that the

EGF receptor acquires upon ligand-activated autophosphorylation; this conformation would represent an active state. Such a constitutively active kinase would be expected to catalyze the phosphorylation of the same repertoire of cellular substrates that is normally phosphorylated in response to ligand stimulation. According to this model, receptor autophosphorylation would play a secondary role, perhaps further decreasing the Km for phosphorylation of exogenous substrates. It remains likely that the phosphorylation of endogenous substrates plays a crucial role in mitogensis and transformation.

References

Bargmann CI, Hung M-C, Weinberg RA (1986) The *neu* oncogene encodes an epidermal growth factor receptor-related protein. Nature 319: 226-229

Bargmann CI, Weinberg RA (1988) Increased tyrosine kinase activity associated with the protein encoded by the activated *neu* oncogene. Proc Natl Acad Sci USA 85:5394-5398

Beug H, Hayman MJ, Raines MA, Kung H-J, Vennström B (1986) RAV-1-induced erythroleukemia cells exhibit a weakly transformed phenotype in vitro and release c-erbB-containing retroviruses unable to transform fibroblast. J Virol 57:1127-1133

Chen SW, Lazar SC, Poenie M, Tsien RY, Gill G, Rosenfeld GM (1987) Requirement for intrinsic protein tyrosine kinase in the immediate and late actions of the EGF receptor. Nature 328:820-823

Choi O-R, Trainor C, Graf T, Beug H, Engel JD (1986) A single amino acid substitution in v-erbB confers a thermolabile phenotype to ts167 avian erythroblastosis virus-transformed erythroid cells. Mol Cell Biol 6:1751-1759

Coussens L, Yang-Feng TL, Liao Y-C, Chen E, Gray A, McGrath J, Seeburg PH, Libermann TA, Schlessinger J, Francke U, Levinson A, Ullrich A (1985) Tyrosine kinase receptor with extensive homology to EGF receptor shares chromosomal location with *neu* oncogene. Science 230:1132-1139

Crowley CW, Liu C-C, Levinson AD (1983) Plasmid-directed synthesis of hepatitis B surface antigen in monkey cells. Mol Cell Biol 3:44-55

Damm K, Beug H, Graf T, Vennstrom B (1987) A single point mutation *erbA* restores the erythroid transforming potential of a mutant AEV defective in both *erbA* and *erbB* oncogenes. EMBO J 6:375-382

Downward J, Yarden Y, Mayes E, Scrace G, Totty N, Stockwell P, Ullrich A, Schlessinger J, Waterfield MD (1984a) Close similarity of epidermal growth factor receptor and v-*erbB* oncogene protein sequences. Nature 307:521-527

Downward J, Parker P, Waterfield MD (1984b) Autophosphorylation sites on the epidermal growth factor receptor. Nature 311:483-485

Gammett DC, Tracy SE, Robinson HL (1986) Differences in sequences encoding the carboxyl-terminal domain of the epidermal growth factor receptor correlate with differences in the disease potential of viral *erbB* genes. Proc Natl Acad Sci USA 83:6053-6057

Graf T, Royer-Pokora B, Schubert GE, Beug H (1976) Evidence for the multiple oncogenic potential of cloned leukemia virus: in vitro and in vivo studies with avian erythroblastosis virus. Virology 71:423-433

Honegger AM, Dull TJ, Felder S, Van Obberghen E, Bellot F, Szapary D, Schmidt A, Ullrich A, Schlessinger J (1987) Point mutation at the ATP binding site of EGF receptor abolishes protein-tyrosine kinase activity and alters cellular routing. Cell 51:199-209

Hudziak RM, Schlessinger J, Ullrich A (1987) Increased expression of the putative growth factor receptor p185HER2 causes transformation and tumorigenesis of NIH 3T3 cells. Proc Natl Acad Sci USA 84:7159-7163

Lax I, Johnson A, Howk R, Sap J, Bellot F, Winkler M, Ullrich A, Vennström B, Schlessinger J, Givol D (1988) Chicken epidermal growth factor (EGF) receptor: cDNA cloning, expression in mouse cells, differential binding of EGF and transforming growth factor alpha. Mol Cell Biol 8:1970-1978

Massoglia S, Gray A, Dull TJ, Munemitsu S, Kung HJ, Schlessinger J, Ullrich A (1990) Epidermal growth factor receptor cytoplasmic domain mutations trigger ligand-independent transformation. Mol Cell Biol 10:3048-3055

Moolenaar WH, Bierman AJ, Tilly BC, Verlaan I, Defize LHK, Honegger AM, Ullrich A, Schlessinger J (1988) A point mutation at the ATP-binding site of the EGF-receptor abolishes signal transduction. EMBO J 7:707-710

Muller WJ, Sinn E, Pattengale PK, Wallace R, Leder P (1988) Single-step induction of mammary adenocarcinoma in transgenic mice bearing the activated c-*neu* oncogene. Cell 54:105-115

Nilsen TW, Maroney PA, Goodwin RG, Rottman FM, Crittenden LB, Raines MA, Kung H-J (1985) c-erbB activation in ALV-induced erythroblastosis: novel RNA processing and promoter insertion result in expression of an amino-truncated EGF receptor. Cell 41:719-726

Prywes R, Livneh E, Ullrich A, Schlessinger J (1986) Mutations in the cytoplasmic domain of EGF receptor affect EGF binding and receptor internalization. EMBO J 5:2179-2190

Raines MA, Maihle NJ, Moscovici C, Moscovici MG, Kung HJ (1988) Molecular characterization of three *erbB* transducing viruses generated during ALV induced erythroleukemia: Extensive internal deletion near the kinase domain activates the fibrosarcoma and hemangioma-inducing potential of *erbB*. J Virol 62:2444-2452

Riedel H, Schlessinger J, Ullrich A (1987) A chimeric, ligand-binding v-*erb*B/EGF receptor retains transforming potential. Science 236:197-200

Rosenthal A, Lindquist PB, Bringman TS, Goeddel DV, Derynck R (1986) Expression in rat fibroblasts of a human transforming growth factor-α cDNA results in transformation. Cell 46:301-309

Simonsen CC, Levinson AD (1983) Isolation and expression of an altered mouse dihydrofolate reductase cDNA. Proc Natl Acad Sci USA 80:2495-2499

Ullrich A, Coussens L, Hayflick JS, Dull TJ, Gray A, Tam AW, Lee J, Yarden Y, Libermann TA, Schlessinger J, Downward J, Mayes ELV, Whittle N, Waterfield MD, Seeburg PH (1984) Human epidermal growth factor receptor cDNA sequence and aberrant expression of the amplified gene in A431 epidermoid carcinoma cells. Nature 309:418-425

Yamamoto T, Hihara H, Nishida T, Kawai S, Toyoshima K (1983a) A new avian erythroblastosis virus, AEV-H, carries *erbB* gene responsible for the induction of both erythroblastosis and sarcomas. Cell 34:225-234

Yamamoto T, Nishida T, Miyajima N, Kawai S, Ooi T, Toyoshima K (1983b) The *erbB* gene of avian erythroblastosis virus is a member of the *src* family. Cell 35:71-78

Yarden Y, Ullrich A (1988) Growth factor receptor tyrosine kinases. Ann Rev Biochem 57:443-478

THE EPIDERMAL GROWTH FACTOR RECEPTOR: RECEPTOR DIMERIZATION AND SIGNAL TRANSDUCTION

Gijs F. Verheijden, Ingrid Verlaan and Wouter H. Moolenaar
Division of Cellular Biochemistry
The Netherlands Cancer Institute
Plesmanlaan 121
1066 CX, Amsterdam
The Netherlands

INTRODUCTION

The epidermal growth factor (EGF) receptor is a single chain transmembrane glycoprotein with ligand-dependent protein tyrosine kinase activity (Carpenter,1987). Activation of the receptor tyrosine kinase is essential, although not necessarily sufficient, for the multiple biological effects of EGF (Chen et al.,1987; Moolenaar et al.,1988). Early events induced by EGF include tyrosine-specific protein phosphorylations, inositol lipid breakdown, changes in cytoplasmic pH and free Ca^{2+} and alterations in membrane potential (Carpenter,1987; Chen et al,1987; Moolenaar et al,1986 and 1988; Rothenberg et al,1982; Tilly et al.,1988; Pandiella et al.,1989). This chapter briefly summarizes our recent findings on two aspects of EGF receptor function: ligand-dependent receptor activation with emphasis on the proposed role of 'receptor dimerization' and post-receptor signal transduction, particularly the activation of phosphoinositide-specific phospholipase C.

RECEPTOR DIMERIZATION

Several models have been proposed to explain how EGF binding results in activation of the intracellular kinase domain. An intramolecular mechanism of receptor activation would require that ligand-induced conformational changes in the extracellular domain are transmitted to the intracellular kinase domain through the single transmembrane spanning region of the receptor. Evidence has been presented that such an intramolecular mechanism of EGF receptor kinase activation may indeed occur under certain experimental conditions (Koland et al.,1988; Northwood et al.,1988). On the other hand an intermolecular activation model has been proposed in which EGF promotes the formation of receptor homodimers (Yarden and Schlessinger,1987; Schlessinger,1988 for review). The dimeric state would provide a conformation that favors receptor activation, resulting in receptor

NATO ASI Series, Vol. H 52
Biological Signal Transduction
Edited by E. M. Ross and K. W. A. Wirtz

autophosphorylation, substrate phosphorylation and signal transduction. EGF-induced receptor dimerization has been claimed to occur both in purified receptor preparations and in intact cells (Yarden et al.,1987; Cochet et al., 1988). The dimerization model is supported by the observation that purified EGF receptor dimers exhibit increased tyrosine kinase activity when compared to receptor monomers (Boni-Schnezler and Pilch,1987). In addition, recent evidence suggests that autophosphorylation of EGF receptors is mediated, at least in part, by intermolecular cross-phosphorylation, a situation that is necessarily restricted to EGF receptor oligomers (Honegger et al.,1989, 1990). However, to date the monomer/dimer ratio of receptors at the cell surface has not yet been established, raising the possibility that EGF induces an allosteric activation of receptors which exist already as dimers at the cell surface.

Various agents, in particular phorbol esters and second messengers such as Ca^{2+} and cAMP, are known to affect EGF receptor activity and to modulate signal transduction. Perhaps the best documented way of regulation is via protein kinase C. The EGF receptor is phosphorylated on threonine and serine residues by protein kinase C resulting in an inhibition of tyrosine kinase activity of the receptor (Cochet et al.,1984; Hunter et al.,1984; Friedman et al.,1984; Davis,1988; Livneh et al.,1988). EGF can mimic phorbol esters in stimulating phosphorylation of residue Thr654, presumably as a result of increased diacylglycerol production and consequent activation of protein kinase C (Whiteley and Glaser,1986; King and Cooper,1986). Furthermore, down-regulation of protein kinase C leads to enhanced EGF receptor kinase activity in A431 cells (Sturani et al.,1988). Taken together, the available data suggest feedback regulation of EGF receptor functions via protein kinase C.

Among the earliest detectable events in EGF action is a rapid increase in intracellular [Ca^{2+}] (Chen et al.,1987; Moolenaar,1986 and 1988), which may lead to activation of a Ca^{2+}/calmodulin-dependent protein kinase and contribute to activation of protein kinase C, while artificially raising intracellular [Ca^{2+}] affects high-affinity EGF binding and endocytosis and reduces receptor kinase activity (Korc et al.,1984, Fearn and King, 1985; Friedman et al.,1989).

Another means of EGF receptor regulation may involve cAMP. EGF receptors can be phosphorylated by cAMP-dependent protein kinase <u>in vitro</u> (Gosh-Dastidar and Fox,1984), and elevation of intracellular cAMP levels has been reported to reduce affinity for EGF (Pessin et al.,1983) and affect EGF receptor signal transduction (Iwashita et al,1990).

It is not clear at what level regulation of EGF receptor kinase activity and signal transduction occurs. Using covalent crosslinkers and immunoblotting techniques we could monitor EGF-induced receptor kinase activity concomittant with the appearence of a -360 kDa band, which we and others believe to represent EGF receptor homodimers. In the following we will refer to this phenomenon as receptor dimer stabilization since it is not clear yet wether the appearence of this band represents EGF-induced receptor dimer formation or results from an EGF-induced conformational change, which allows preexisting receptor dimers to be crosslinked.

We have investigated wether parameters, involved in EGF receptor kinase activity and signal transduction can modulate EGF-induced receptor dimer stabilization.

Covalent crosslinking in vivo and in vitro

Stabilization of EGF receptor dimers in vivo and in vitro was studied using various covalent crosslinking agents. Following SDS-PAGE and immunoblotting we were able to monitor EGF-induced stabilization of receptor dimers in A431 cells, transfected NIH 3T3 cells and HeLa cells, expressing about $2x10^6$, $3x10^5$ and $3x10^4$ human EGF receptors/cell, respectively (Verheijden et al.,1990). In the immunoblots, the high molecular weight band of about 360 kDa, corresponding to the molecular weight of EGF receptor homodimers, was only detectable when cells were treated with both EGF and crosslinker. The intensity of this band varied between different experiments but was always maximal at 100-200 ng/ml EGF. The appearance of the -360 kDa band we and others found (Cochet et al.,1988; Kashles et al.,1988; Northwood and Davis,1989 Fanger et al.,1989) is most readily explained in terms of stabilization of receptor homodimers. However, it remains formally possible that the receptor is cross-linked to an other membrane protein. To investigate this point in further detail we followed dimer stabilization in purified EGF receptor preparations. We were able to show that the EGF-induced -360 kDa band of intact A431 cells comigrates with the dimer band of purified EGF receptors, suggesting that the -360 kDa band in intact cells indeed represents EGF receptor homodimers. In the purified receptor preparation, a low amount of dimers was detectable in the absence of EGF, which was not the case in intact cells. Importantly, dimer stabilization of isolated receptors occured in the absence of ATP, supporting the view that receptor tyrosine kinase activity is not essential this process. We further note that EGF also evokes the appearance of a -520 kDa band in purified receptors (but usually not in intact cells), presumably representing receptor trimers.

Effect of elevated intracellular Ca^{2+}

EGF evokes a rapid increase in cytoplasmic free Ca^{2+} in its target cells (Chen et al.,1987; Moolenaar et al.,1986 and 1988). Ca^{2+} functions as a second messenger that regulates various cellular functions. In particular, raising intracellular [Ca^{2+}] with an ionophore modulates EGF receptor binding, endocytosis and tyrosine kinase activity of the receptor (Friedman et al.,1984 and 1989; Korc et al.,1984; Fearn and King,1985). Hence, Ca^{2+} is a potential candidate for feedback regulation of EGF receptor functions.

We investigated whether an elevated Ca^{2+} level is capable of modulating EGF receptor dimer stabilization. The Ca^{2+}-ionophore ionomycin was added shortly before addition of EGF both in the presence and absence of EGTA and then stabilization of receptor dimers was followed as described. From these experiments it became clear that ionomycin did not interfere with stabilization of EGF receptor dimers, whereas EGF-induced receptor autophosphorylation was almost completely abolished. Furthermore, phosphorylation of several endogenous substrates was also inhibited by ionomycin. However, when added in the presence of EGTA to prevent Ca^{2+}-influx, ionomycin had no detectable effect on the phosphorylation state of the EGF receptor.

Studies in permeabilized cells

To further examine the effect of Ca^{2+} on EGF receptor function, A431 cells were permeabilized with streptolysin-O and tested for EGF receptor autophosphorylation in the presence or absence of Ca^{2+}. A dramatic reduction in EGF-induced tyrosine kinase activity was observed when Ca^{2+} was present (1mM), but not in nominally Ca^{2+}-free media. Ionomycin alone did not have a detectable effect in permeabilized cells. A Ca^{2+}-sensitive protease, has been described (Cassel and Glaser,1982) that degrades the EGF receptor into a doublet of Mr 150 and 170 kD. However, in our experiments neither the doublet receptor bands nor the 20 kDa breakdown product could be detected. Thus, reduction in EGF receptor kinase activity caused by Ca^{2+} cannot be due to activation of the Ca^{2+}-dependent protease.

Effect of phorbol ester

Brief treatment of cells with TPA is known to result in a significant inhibition of EGF receptor kinase activity (Cochet et al.,1984; Friedman et al.,1984; Davis,1988). To determine whether inhibition of EGF receptor activity by protein kinase C correlates with altered receptor dimer stabilization, cells were briefly treated with TPA and receptor dimer stabilization and autophosphorylation were monitored by immunoblotting. At each EGF concentration (0-100 ng/ml), the intensity of both monomeric and dimeric receptor bands in TPA-treated cells was not significantly different from that in untreated cells. However, when the blot was probed with anti-phosphotyrosine antibody, TPA treatment was found to result in a marked reduction in intensity of both monomer and dimer bands. These results indicate that, although activation of protein kinase C by TPA inhibits EGF-induced tyrosine kinase activity in vivo, stabilization of EGF receptor dimers is not affected. We conclude that the inhibitory action of protein kinase C on EGF receptor signal transduction does not occur at the level of dimer stabilization.

Studies in protein kinase C-down regulated cells

At high Ca^{2+} concentrations, protein kinase C may be activated without requirement for diacylglycerol or phorbol ester. To examine whether the observed effects of ionomycin/Ca^{2+} are mediated by protein kinase C, A431 cells were treated for 72 h with phorbol ester (0.5μg/ml) to down-regulate functional protein kinase C activity. Indeed, such chronic treatment of the cells with TPA lead to complete loss of [^{3}H]-phorboldibutyrate binding sites (G.F.V. unpublished results). TPA failed to reduce EGF receptor autophosphorylation in such protein kinase C-depleted cells. In marked contrast to TPA, ionomycin was still capable of abolishing receptor kinase activity when protein kinase C was down-regulated. Taken together, these results indicate that the inhibitory effect of Ca^{2+} on EGF receptor function is independent of functional protein kinase C. We are currently investigating the possibility that the Ca^{2+} induced reduction in EGF-induced tyrosine phosphorylation is mediated by (a) tyrosine specific phosphatase(s).

Effects of cAMP, pH, membrane potential and hyperosmolarity

To determine whether increases in cAMP and pH can modulate EGF receptor dimer stabilization, cells were treated with the cell permeant cAMP analog 8-bromo-cAMP and the Na^+/H^+ionophore monesin, respectively. Neither treatment was shown to affect stabilization of receptor dimers.

EGF has been reported to transiently alter the membrane potential in cells of epithelial origin (Rothenberg et al.,1982; Pandiella et al.,1989). It is conceivable that the electric field across the plasma membrane may act on charged residues of the receptor located near the inner side of the plasma membrane or that it may modulate putative non-covalent interactions that could be important for dimer stabilization. To test the possible effects of alterations in transmembrane potential, dimer stabilization was followed in cells incubated in medium containing 150 mM K^+, which abolishes the transmembrane potential. However in vivo stabilization of dimers in this high K^+ buffer was comparable to control, indicating that the electric field across the plasma membrane is not important for receptor dimer stabilization.

In a recent study, the EGF receptor was found to be phosphorylated by an unknown mechanism, not involving protein kinase C, when A431 cells were exposed to a hyperosmotic shock (Whitely and Glaser,1986). In addition, we have observed that hyperosmotic shock, induced by adding 100 mM NaCl to the incubation medium, inhibits EGF-induced inositol phosphate formation (G.F.V., unpublished results). However, we and others (Kashles et al. 1988) found that under these conditions, receptor dimer stabilization is comparable to control, indicating that hyperosmolarity does not affect this process.

Effect of treatment with glycosylation inhibitors and neuraminidase

The mature EGF receptor (170 kDa) contains a relatively large amount of N-linked oligosaccharides (Carpenter,1987;Soderquist and Carpenter,1984; Cummings et al.,1985; Slieker et al.,1986), but whether the glycosylation state of the receptor has a role in dimerization is not known. Addition of core oligosaccharides to the peptide backbone (130 kDa) is essential for EGF binding, but terminal processing of the carbohydrate moieties is not. We tested the effect of three different agents that interfere with carbohydrate processing on EGF receptor dimer stabilization. We used swainsonine, an inhibitor of cis Golgi a-mannosidase II, which blocks normal processing of high mannose chains to complex forms but leaves EGF binding almost unaffected (Soderquist et al.,1984), 1-deoxymannojirimycin (DMM), which inhibits mannosidase I in the rough endoplasmatic reticulum (Fuhrmann et al.,1984), and neuraminidase, which removes sialic acids. In the presence of EGF, receptor dimers and autophosphorylation were clearly detectable after pretreatment of A431 cells with DMM, swainsonine or neuraminidase, while the apparent molecular mass of both monomers and dimers slightly decreased, most likely as a result of changes in the carbohydrate moiety. These observations indicate that terminal processing of N-linked oligosaccharides is not essential for EGF receptor dimerization.

In conclusion, our results strongly suggest that modulation of EGF-receptor function and post-receptor signalling occurs at a step different from dimer stabilization.

In previous reports, we and others interpreted the appearance of the -360 kDa as to represent the proces of receptor dimerization. However our findings and those of others are also consistent with the possibility that the EGF-induced -360 kDa band represents pre-existing EGF receptor dimers which can only be trapped by crosslinking upon binding of EGF. In this respect, very recently, interesting experiments have been preformed with truncated EGF receptors which were only expressed at the cell surface if wild type receptors were present. This suggests that receptor dimerization precedes cell surface expression (Lax et al.,1990). Thus EGF receptor dimers may already be present and EGF induces only a conformational change which results in allosteric activation of the kinase domain and allows the dimers to be crosslinked. We are currently investigating this possibility in more detail. Furthermore, since the -360 kDa band seen in intact cells has not yet been formally proven to represent EGF receptor homodimers, this band could conceivably reflect dimer formation between EGF receptors and other molecules with similar electrophoretic mobility. It is of obvious importance to determine whether experimental protocols can be developed that inhibit EGF receptor dimer formation, and if so, whether post-receptor signalling is similarly affected. Furthermore, the exact domain(s) of the EGF receptor which interact with its counterpart to form a homodimer have yet to be identified.

EGF-INDUCED PHOSPHOINOSITIDE HYDROLYSIS

Many growth factors, hormones and neurotransmitters are known to stimulate the breakdown of phosphoinositides. This receptor-mediated response gives rise to at least two second messengers, inositol 1,4,5-trisphosphate, which mediates Ca^{2+} release from internal stores, and diacylglycerol, which activates protein kinase C (Berridge,1987; Nishizuka,1986). The hydrolysis of phosphoinositides can be catalyzed by any of the phospholipase C (PLC) isozymes that, in many cases, are thought to be stimulated via an as yet unidentified guanine-nucleotide binding protein (putative G_p; Cockcroft,1987). In some cells, receptor-mediated phosphoinositide hydrolysis is inhibited by pertussis toxin, which inactivates certain guanine-nucleotide binding proteins of the $\alpha\beta\gamma$ heterotrimeric form (G proteins), whereas in other cells this toxin has no effect, suggesting that at least two different types of guanine-nucleotide binding proteins are involved (Boyer et al.,1989). It appears that guanine-nucleotide binding proteins mediate PLC activation by a wide variety of agonists, including many neurohormones (Uhing et al.,1986; Smith et al.,1987; Hasegawa-Sasaki et al.,1988; Tilly et al.,1990) and the mitogenic phospholipid LPA (van Corven et al.,1989), as assessed by several criteria, in particular the ability of stable guanine nucleotide analogs to modulate PLC activity in cell-free preparations. A characteristic feature of receptors that utilize heterotrimeric G proteins for signal transduction is that they are integral membrane proteins with seven membrane-spanning segments (reviewed Hanley and Jackson,1987). Whether growth factor receptors of the protein tyrosine kinase family with a single transmembrane domain, such as the EGF receptor, may similarly couple to guanine-nucleotide binding proteins to stimulate PLC is uncertain.

In case of the PDGF receptor, contrasting results have been reported with respect to the mechanism of activation of PLC. Hasegawa-Sasaki et al. (1988) suggested that the PDGF receptor does not couple to a G protein to activate PLC, whereas two other groups reported that PDGF-induced phosphoinositide hydrolysis in membranes has an absolute requirement for GTPγS (Kamata and Kung,1988; Huang and Ives,1989). Phospholipid turnover induced by the insulin receptor tyrosine kinase has also been claimed to proceed via a specific pertussis toxin-sensitive G protein (Luttrell et al.,1988).

EGF stimulates phosphoinositide turnover in various cell types, including A431 cells (Pike and Eakes,1987; Tilly et al.,1988; Wahl and Carpenter,1988) and 3T3 transfectants (Moolenaar et al. 1989). Receptor tyrosine kinase activity is essential, although not necessarily sufficient, for this response (Moolenaar et al.,1989), while recent evidence indicates that the EGF receptor phosphorylates the γ-form of PLC at tyrosine residues in vivo (Margolis et al.,1989; Meisenhelder et al.,1989; Wahl et al.,1990). However, whether tyrosine phosphorylation of PLC affects its enzymatic activity is currently impossible to determine. On the other hand, from studies on hepatocytes there is evidence that EGF utilizes a pertussis toxin-sensitive G protein to stimulate phosphoinositide metabolism (Johnson and Garrison,1987). Furthermore, recent studies indicate that EGF stimulates adenylate cyclase in cardiac membranes and phospholipase A2 in kidney cells through stimulatory GTP-binding proteins (Nair et al.,1989; Teitelbaum,1989). Thus, the emerging picture is by no means complete, and the role of guanine-nucleotide binding proteins in EGF-stimulated second messenger formation needs to be clarified.

We have examined PLC activation in permeabilized NIH-3T3 cells expressing the human EGF receptor, using LPA as a control stimulus known to activate PLC in a GTP-dependent manner (van Corven et al.,1989). In NIH-3T3 cells overexpressing the wild-type human EGF receptor, both EGF (100 ng/ml) and the mitogenic phospholipid LPA (10 μg/ml) activated PLC, as revealed by a 2.5-5.0 fold increase in the level of total inositol phosphates. EGF-induced inositol phosphate formation was not affected by preincubating the NIH-3T3 cells with pertussis toxin for two hours. In contrast, LPA-induced inositol phosphate formation was significantly reduced by pertussis toxin in these cells. It is noteworthy that in other cells, such as human foreskin fibroblasts, pertussis toxin has no effect on LPA-induced inositol phosphate formation (van Corven et al.,1989), suggesting a degree of cell-specific heterogeneity in the guanine-nucleotide binding proteins involved.

Phosphoilipase C activation in permeabilized cells

To explore the possible involvement of a GTP-binding protein in the activation of PLC by EGF, we made use of the streptolysin-O permeabilization protocol. This method allows direct access to intracellular compartments while leaving guanine-nucleotide binding protein-mediated breakdown of inositol lipids unaffected (Howell and Gomperts,1987; Tilly et al.;1990). After treatment with streptolysin-O for 5 minutes, about 95% of the NIH-3T3 cells were leaky and remained permeable for at least 30 minutes, as determined by trypan blue staining. We confirmed that EGF-induced receptor tyrosine kinase activity is not affected by this permeabilization method as revealed by immunoblotting of cell lysates using anti-phosphotyrosine antibodies. Although permeabilization reduces the absolute amount of total radiolabeled inositol phosphates at each EGF concentration tested, the cells clearly remain responsive to EGF. The concentration dependency was similar to that in intact cells, with half-maximal stimulation of inositol phosphate formation observed at about 10 ng/ml EGF.

Using this procedure we could show that the non-hydrolyzable GTP-analogue GTPγS activates PLC within 10 min. in a dose-dependent manner. Half-maximal stimulation was observed at a concentration of about 1μM. It was confirmed that GTPγS does not affect inositol phosphate metabolism when added to intact cells.

We also investigated the kinetics of PLC activation by EGF and LPA in the presence or absence of GTPγS. The inositol phosphate response to a half-maximal effective concentration of GTPγS (1μM) developed rather slowly with time, significant stimulation being observed only after 5-10 min of addition of the nucleotide. Addition of EGF (100 ng/ml) resulted in increased inositol phosphate formation within a few minutes, without requirement of exogenously added GTP-analog. When EGF was added together with GTPγS, the initial rate of inositol phosphate formation was not significantly increased. The response to EGF and GTPγS together was simply additive rather than synergistic, strongly suggesting that EGF does not rely on a guanine-nucleotide binding protein to stimulate phospholipase C activity.

The situation is completely different with LPA, a mitogenic phospholipid that is a potent activator of phsophoinositide breakdown (van Corven et al.,1989; Jalink et al.,1990). It was clear that, when LPA and GTPγS are added together, the initial rate of inositol phosphate formation was dramatically increased. After 5 min, there was a several-fold increase above basal levels, while LPA alone caused a barely detectable stimulation over this time period. This kinetic behavior of the LPA response, unlike that of EGF, resembles that of a G protein-coupled receptor, where receptor activation accelerates the exchange of bound GDP for GTP on the G protein α subunit.

Further support for the notion that the EGF receptor does not couple to a stimulatory guanine-nucleotide binding protein comes from experiments using the GDP analog GDPβS, which keeps guanine-nucleotide binding proteins in their inactive GDP-bound state. GDPβS failed to affect inositol phosphate formation induced by EGF, whereas it completely blocked the response to LPA/GTPγS.

Tyrosine-containing peptides inhibit EGF-induced PLC activation

Synthetic peptides belonging to the family of angiotensins (angiotensin I, DRVYIHPFHL; angiotensin II, DRVYIHPF) are known to be readily phosphorylated by receptor tyrosine kinases with a K_m of 0.9 mM (Hunter and Cooper, 1985). Such tyrosine-containing peptide substrates have been shown to act as competitive inhibitors of EGF receptor autophosphorylation in vitro (Honegger et al.,1989). Using [γ-^{32}P] ATP, we confirmed that [Val^5]-angiotensin II (DRVYVAPF) is rapidly phosphorylated and inhibits autophosphorylation of the EGF receptor in membrane preparations.

We then studied whether the peptide substrates are capable of inhibiting EGF receptor-mediated activation of PLC in permeabilized cells. We found that the EGF-induced increase in inositol phosphates is inhibited up to 50% by the presence of angiotensin (at 500μM), while basal inositol accumulation production is not significantly affected. In contrast, inositol phosphate formation induced by LPA/GTPγS was not significantly altered by the substrate peptide. These results suggest that tyrosine-containing substrate peptides can act as competitive inhibitors not only of receptor autophosphorylation but also of receptor-mediated activation of effector enzymes such as PLC.

In our study, we have taken advantage of a recently developed permeabilized cell system for the study of phosphoinositide breakdown evoked by EGF and other stimuli. Our results and those of others indicate that activation of PLC can be mediated by at least two different mechanisms (Boyer et al.,1989). One mechanism involves an as yet unidentified guanine-nucleotide binding protein (G_p). In some cell types this way of stimulation is sensitive to pertussis toxin (Martin,1989), which ADP-ribosylates and thereby inactivates certain G proteins. Another means of PLC regulation does not depend on a stimulatory guanine-nucleotide protein, as shown by us and by others (Hepler et al.,1990) to be the case for the activated EGF receptor. The γ-form of PLC is phosphorylated on tyrosine residues following activation of EGF and PDGF receptors, and it seems plausible to assume that these phosphorylations are essential for enhancing the enzymatic activity of PLC-γ in a guanine-nucleotide binding protein-independent way.

Furthermore, we have employed commonly used peptide substrates to specifically inhibit EGF receptor-mediated activation of PLC. The results obtained are consistent with a mechanism in which exogenous peptide substrates and PLC-γ compete for the protein tyrosine kinase active site, but further kinetic studies are required to firmly establish this point. Nevertheless, the present data suggest that the use of small phosphorylatable peptides in permeabilized cell systems is a promising approach in the study of EGF receptor signal transduction.

REFERENCES

Berridge, M.J. (1987). Ann. Rev. Biochem. **56**, 159-193

Boni-Schnetzler, M. & Pilch, P.F. (1987). Proc.Natl.Acad.Sci.USA **84,** 7832-78362

Boyer, J.L., Hepler, J.R., and Harden T.K. (1989). Trends Pharmacol. Sci. **10**, 360-364

Carpenter, G. (1987). Ann. Rev. Biol. Chem. **56**, 881-914

Cassel, D. & Glaser, L. (1982). J. Biol. Chem. **257**, 9845-9848

Chen, W.S., Lazar, C.S., Poenie, M., Tsien, R.Y., Gill, G.N. & Rosenfeld, M.G. (1987). Nature **328**, 820-823

Cochet, C., Gill, G.N., Meisenhelder, J., Cooper, J.A. & Hunter, T. (1984). J.Biol.Chem. **259**, 2553-2558

Cochet, C., Kashles, O., Chambaz, E.M., Borrello, I., King, C.R., & Schlessinger, J. (1988). J.Biol.Chem. **26**, 33290-3295

Cockcroft, S. (1987). Trends Biochem. Sci. **12**, 75-78

Cummings, R.D., Soderquist, A.M. & Carpenter, G. (1985). J.Biol. Chem. **260**, 11944-11952

Davis, R.J. (1988). J.Biol.Chem **263**, 9462-94694

Fanger, B.D., Stephens, J.E. & Staros, J. (1989). FASEB J. **3**, 71-75

Fearn, J.C. & King, A.C. (1985). Cell **40**, 991-1000

Friedman, B., van Amsterdam, J., Fusiki, H. & Rosner, M.R. (1989). Proc. Natl. Acad. Sci. USA. **86**, 812-816

Friedman, B., Franckelton Jr., A.R., Ross, A.A., Conners, J.M., Fujiki, H., Sugimura, T. & Rosner, M.R. (1984). Proc.Natl.Acad.Sci.USA. **81**, 3034-3038

Fuhrmann, U., Bause, E., Legler, G. & Ploegh, H.L. (1984). Nature **307**, 755-758

Ghosh-Dastidar, P. & Fox, C.F. (1984). J.Biol.Chem. **259**, 3864-3869

Hanley, M.R., and Jackson, T. (1987). Nature **329**, 766-767

Hasegawa-Sasaki, H., Lutz, F., and Sasaki, T. (1988). J.Biol. Chem. **263**, 12970-12976

Hepler, J.R., Jeffs, R.A., Huckle, W.R., Outlaw, H.E., Rhee, S.G., Earp, H.S., and Harden, T.K. (1990). Biochem. J. **279**, 337-344

Honegger, A.M., Dull, T.J., Felder, S., Van Obberghen, E., Bellot, F., Szapary, D., Schmidt, A., Ullrich, A., and Schlessinger, J. (1987). Cell **51**, 199-209

Honegger, A.M., Kris, R.M., Ullrich, A. & Schlessinger, J. (1989). Proc. Natl.Acad.Sci. USA. **86**, 925-929

Honegger, A.M., Schmidt, A., Ullrich, A., and Schlessinger, J.(1990). Mol. Cell. Biol. **10**, 4035-4044

Howell, T.W., and Gomperts, B.D. (1987). Biochim. Biophys. Acta. **927**, 177-183

Huang, C. and Ives, H.E. (1989). J. Biol. Chem. **264,** 4391-4397

Hunter, T., and Cooper, J.A. (1985). Ann. Rev. Biochem. **54**, 897-930

Hunter, T., Ling, N., & Cooper, J.A. (1984). Nature **311**, 480-483

Iwashita S., Mitsui, K-i., Shoji-Kasai and Sensu-Miyaike M., (1990). J. Biol. Chem. **265**, 10702-10708

Johnson, R.M., and Garrison (1987). J. Biol. Chem. **262**, 17285-17293

Jalink K., van Corven, E., and Moolenaar W.H. (1990). J. Biol. Chem. **265**, 12232-12239

Kamata, T., and Kung, H.F. (1988). Proc. Natl. Acad. Sci. USA **85**, 5799-5803

Kashles, O., Szapary, D., Bellot, F., Ullrich, A., Schlessinger, J. & Schmidt,A. (1988). Proc. Natl.Acad.Sci.USA. **85**, 9567-9571

King, C.S. & Cooper, J. (1986). J.Biol.Chem. **261**, 10073-10078

Koland, J.G. & Cerione, R.A. (1988). J. Biol. Chem. **263**, 2230-2237

Korc, M., Matrisian, L.M. & Magun, B.E. (1984). Proc.Natl.Acad.Sci.USA. **81**, 461-465

Lax, I., Bellot, F., Honegger, A.M., Schmidt, A., Ullrich, A., Givol, D., and Schlessinger, J. (1990). Cell Regulation, **1**, 173-188

Livneh, E., Dull, T.J., Berent, E., Ullrich, A. & Schlessinger, J. (1988). Mol.Cell.Biol. **8**, 48-53

Luttrell, L.M., Hewlett, E.L., Romero, G., and Rogol, A.D. (1988). J. Biol. Chem. **263**, 6134-6141

Margolis, B., Rhee, S.G., Felder, S., Mervic, M., Lyall, R., Levitzki, A., Ullrich, A., Zilberstein, A., and Schlessinger, J. (1989). Cell **57**, 1101-1107

Martin, T.F.J. (1989). in "Inositol lipids in cell signalling" Michell, R.H., Drummond, A.H., and Downes, C.P. eds. pp 81-112, Acad. Press, New York.

Meisenhelder, J., Suh, P-G., Rhee, S.G., and Hunter, T. (1989). Cell **57**, 1109-1122

Moolenaar, W.H., Aerts, R.J., Tertoolen, L.G.J. & de Laat, S.W. (1986). J. Biol. Chem **261**, 279-284

Moolenaar, W.H., Bierman, A.J., Tilly, B.C., Verlaan, I., Honegger, A.M., Ullrich, A., and Schlessinger, J. (1988). EMBO **7**, 707-710

Nair, B.G., Rashed, H.M. and Patel, T.B. (1989). Biochem. J. **264**, 563-571

Nishizuka, Y. (1986). Science **233**, 305-312

Northwood, I.C. & Davis, R.J. (1988). J. Biol. Chem. **263**, 7450-7453

Northwood, I.C. & Davis, R.J. (1989). J.Biol. Chem. **264**, 5746-5750

Pandiella A., Magni,M., Lovisolo, D. and Meldolesi, J. (1989). J. Biol. Chem 264, 12914-12921

Pessin, J.E., Gitomer, W., Oka, Y., Oppenheimer, L. & Czech, M.P. (1983). J. Biol.Chem. **25**, 7386-7394

Pike, L.J., and Eakes, A.T. (1987). J. Biol. Chem. 262, 1644-1651

Rothenberg, P., Reuss, L. & Glaser, L. (1982). Proc.Natl.Acad.Sci.USA. **79**, 7783-7787

Schlessinger, J. (1988). TIBS **13**, 443-447

Slieker, L., Todd, J., Martensen, M. & Lane, M. (1986). J. Biol.Chem. **261**, 15233-15241

Smith, C.D., Uhing, R.J., and Snyderman, R. (1987). J. Biol. Chem **262**, 6121-6127

Soderquist, A.M. & Carpenter, G. (1984). J.Biol.Chem. **259**, 12586-12594

Sturani, E., Zippel, R., Toschi, L., Morello, L., Comoglio, P.M. & Alberghina, L. (1988). Mol.Cell.Biol. **8**, 1345-1351

Teitelbaum, I. (1990). J. Biol. Chem. **265**, 4218-4222

Tilly, B.C., Tertoolen, L.G.J., Lambrechts, A.C., Remorie, R., de Laat, S.W. & Moolenaar, W.H. (1990). Biochem. J. **266**, 235-243

Tilly, B.C., van Paridon, P.A., Verlaan, I., de Laat, S.W., and Moolenaar, W.H. (1988). Biochem. J. **252**, 857-863

Uhing, R.J., Prpic, V., Jiang, H., and Exton, J.H. (1986). J. Biol. Chem. **261**, 2140-2146

van Corven, E.J., Groenink, A., Jalink, K., Eichholtz, T. and Moolenaar, W.H. (1989). Cell **59**, 45-54

Verheijden, G. F., Verlaan, I., van Iersel, M., and Moolenaar, W.H. (1990). Biochem. J. **271**, 215-221

Verheijden, G.F., Verlaan, I., and Moolenaar, W.M. (1990). Cell Regulation **1**, 615-620

Wahl, M.I., and Carpenter, G. (1988). J. Biol. Chem. **263**, 7581-7590

Wahl, M.I., Nishibe, S., Kim, J.W., Kim, H., Rhee, S.G. and Carpenter, G. (1990). J. Biol. Chem. **265**, 3944-3948

Whiteley, B. & Glaser, L. (1986). J.Cell. Biol. **103**, 1355-1362

Yarden, Y. & Schlessinger J. (1987). Biochemistry **26**, 1434-1442

Yarden, Y. & Schlessinger J. (1987). Biochemistry **26**, 1443-1451

DIMERIZATION ACTIVATES THE EPIDERMAL GROWTH FACTOR RECEPTOR TYROSINE KINASE

Marcel Spaargaren[1,2], Libert H.K. Defize[1], Johannes Boonstra[2], and Siegfried W. de Laat[1]

[1]*Hubrecht Laboratory, Netherlands Institute for Developmental Biology, Uppsalalaan 8, 3584 CT Utrecht, The Netherlands*
[2]*Department of Molecular Cell Biology, University of Utrecht, Padualaan 8, 3584 CH Utrecht, The Netherlands*

INTRODUCTION

The epidermal growth factor receptor (EGF-R) is a transmembrane glycoprotein of 170 kDa with an extracellular ligand binding domain, a single hydrophobic transmembrane stretch, and an intracellular domain containing protein tyrosine kinase activity (Ullrich et al., 1984). Binding of EGF to its receptor enhances this tyrosine kinase activity, thus inducing tyrosine phosphorylation of several protein substrates including the EGF-R itself (Hunter and Cooper, 1985). Activation of the EGF-R tyrosine kinase results in a cascade of biochemical and physiological responses, finally leading to stimulation of DNA synthesis and cell division in most cells (Carpenter, 1987). Using EGF-R mutants, the protein tyrosine kinase activity was shown to be essential for mitogenic signaling (Schlessinger, 1988a).

The mechanism whereby extracellular EGF binding activates the intracellular EGF-R tyrosine kinase remains to be elucidated. Three models have been proposed (Fig. 1). According to the intramolecular activation model (Fig. 1; 1), EGF binding leads to a conformational change in the extracellular EGF-R domain which is transmitted through the hydrophobic transmembrane stretch to the cytoplasmic tyrosine kinase domain, resulting in activation of the EGF-R tyrosine kinase (Bertics and Gill, 1985; Gill et al., 1987; Koland and Cerione, 1988; Weber et al., 1984). This model implicates that EGF-R monomers act as transmembrane signal transducers. According to the intermolecular activation model, EGF causes a shift in a hypothetical equilibrium between EGF-R monomers and dimers, resulting in enhanced tyrosine kinase activity (Fig. 1; 2). Two variations of the intermolecular activation model have been proposed. Using Triton X-100 solubilized, purified receptor material, Das and

NATO ASI Series, Vol. H 52
Biological Signal Transduction
Edited by E. M. Ross and K. W. A. Wirtz

colleagues interpreted their data as EGF binding resulting in dissociation of inactive receptor oligomers, thus shifting the equilibrium towards catalytically active monomers (Fig. 1; 2a)(Basu et al., 1986; Basu et al., 1989; Biswas et al., 1985). Schlessinger, however, proposed a model in which EGF binding shifts the equilibrium towards the dimeric form, which possesses an elevated kinase activity (Fig. 1; 2b)(Schlessinger, 1986+1988b).

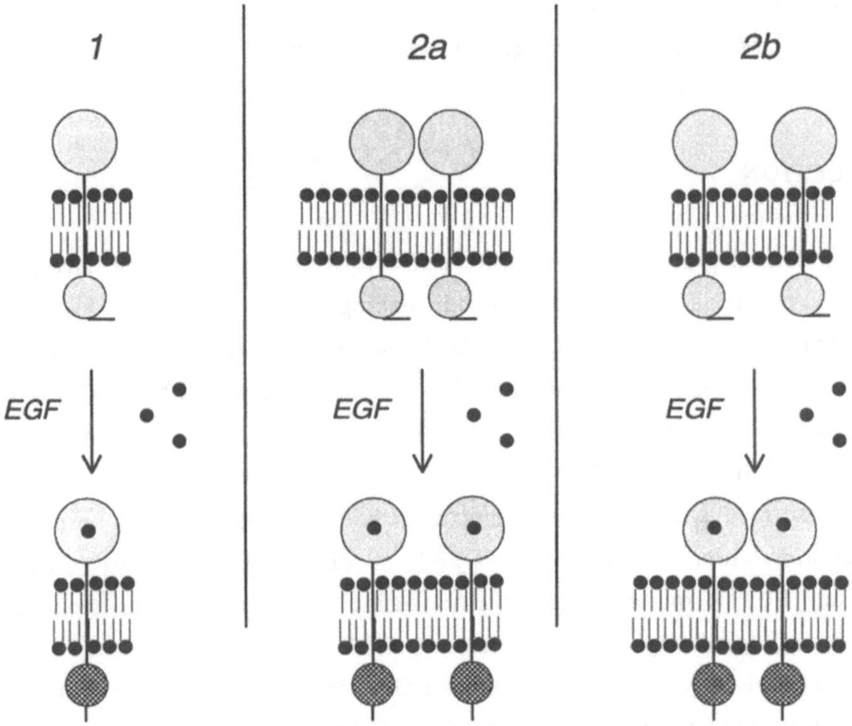

Fig. 1. **The three main EGF-R tyrosine kinase activation models.** 1) The intramolecular activation model. 2a) The intermolecular, active monomer, activation model. 2b) The intermolecular, active dimer, activation model. Dotted EGF-R represents a tyrosine kinase inactive EGF-R, crossed EGF-R represents a tyrosine kinase active EGF-R.

In accordance with the latter, a number of data demonstrate EGF-induced dimerization of receptors. A reversible receptor oligomerization after EGF treatment of solubilized immunoaffinity purified EGF receptors was shown (Yarden and Schlessinger, 1987a+b). Furthermore, EGF-induced dimerization of receptors could be detected by chemical cross-linking of receptors in Triton X-100 solubilized receptor preparations (Boni-Schnetzler and Pilch, 1987; Fanger et al., 1988), membrane preparations (Northwood and Davis, 1988), and intact cells (Cochet et al., 1988; Defize et al., 1989; Fanger et al., 1986). In several cases enhanced autophosphorylation in receptor dimer complexes has been reported (Boni-Schnetzler and Pilch, 1987; Fanger et al., 1988). Taken together, these data indicate that, in agreement with the intermolecular, active dimer, activation model, EGF binding results in EGF-R dimerization. However, a direct cause and effect relationship between EGF-R dimerization and activation of the tyrosine kinase has not been established yet. In this report we describe and discuss our main experimental evidence, as previously published (Spaargaren et al., 1990+1991), demonstrating that antibody-induced dimerization activates the EGF-R tyrosine kinase, which is in accordance with an intermolecular, active dimer, activation model.

RESULTS AND DISCUSSION

To study the relationship between EGF-R dimerization and tyrosine kinase activation, we made use of anti-EGF-R antibodies. It has been reported that under certain conditions a number of such antibodies mimic EGF in activating the EGF-R tyrosine kinase (Defize et al., 1986+1987; Gill et al., 1984; Schreiber et al., 1981+1983; Yarden and Schlessinger, 1987). Concerning the intermolecular activation model, antibody-induced receptor activation was suggested to be the result of bivalent antibody-induced cross-linking, resulting in activated receptor dimers (Yarden and Schlessinger, 1987a). However, the ability of these antibodies to activate the EGF-R could also be due to intramolecular activation, by mimicking an EGF-induced conformational change, or intermolecular activation, by physical blocking of receptor association sites, resulting in activated monomers (Basu et al., 1986). We investigated the mechanism of antibody-induced EGF-R tyrosine kinase activation, using A431 membrane preparations as well as permeabilized A431 cells.

Antibody-induced EGF-R autophosphorylation

Anti-EGF-R antibodies were reported to be able to induce EGF-R autophosphorylation and tyrosine kinase activation in *in vitro* phosphorylation assays (Defize et al., 1986; Gill et al., 1984; Yarden and Schlessinger, 1987a), but not in intact cells (Defize et al., 1989; Gill et al., 1984). This discrepancy turned out to be due to the presence of detergent in the *in vitro* phosphorylation assays. We have shown that anti-EGF-R antibody 2E9, directed against the extracellular domain, and antiserum 281-7, directed against the intracellular domain of the EGF-R, were able to induce EGF-R autophosphorylation and activate the tyrosine kinase in membrane preparations and whole cells, in the presence of detergent (Spaargaren et al., 1990). In addition, monoclonal antibodies 2E9, 2D11, 528, and 225, which have all been reported to be unable to activate the receptor in intact cells (Defize et al., 1989; Gill et al., 1984), were all able to induce EGF-R autophosphorylation in the presence of detergent. Fig. 2 shows that all antibodies, although directed against different receptor epitopes, stimulate EGF-R phosphorylation in saponin-permeabilized cells. 2-D phosphoamino acid analysis, shown in the lower panel for control, EGF and 2E9 treated cells, demonstrates that phosphorylation was almost exclusively on tyrosine residues. The data provide evidence in favor of an intermolecular activation model, as it is highly unlikely that all antibodies induce an EGF-like conformational change, necessary for activation of the kinase in an intramolecular fashion. Moreover, the detergent dependency emphasizes the fact that antibody binding as such is not sufficient for receptor activation, this in contrast to EGF-induced receptor activation.

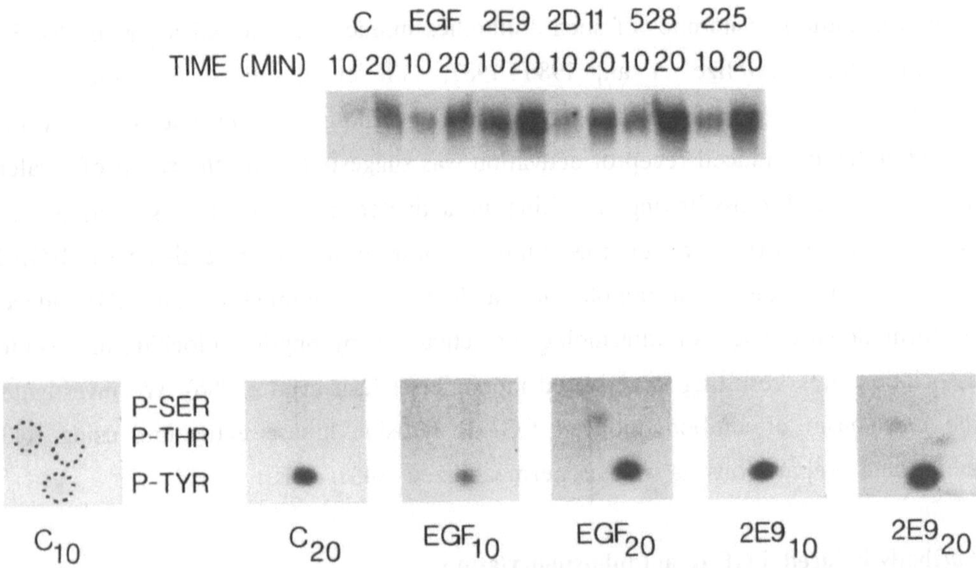

Fig. 2. **Antibody-induced EGF-R autophosphorylation in saponin-permeabilized cells.** Intact A431 cells were treated with the indicated antibodies (50 μg/ml) or EGF (50 ng/ml) in permeabilization buffer containing [γ-^{32}P]ATP and 0.005% saponin for 10 or 20 min on ice. Subsequently, the cells were lysed in RIPA buffer, the EGF-R was immunoprecipitated, subjected to 5-15% SDS-PAGE and autoradiographed (Upper panel). EGF-R-containing gel bands were excised and protein was eluted and 2-D phosphoamino acid analysis was performed by high voltage thin layer electrophoresis, as shown for control (C), EGF, and 2E9 (Lower panel).

The observation that all five before mentioned anti-EGF-R antibodies stimulated EGF-R autophosphorylation, in spite of the fact that they are directed against different receptor epitopes, indicates that a common mechanism is involved. Therefore, we have studied the activation mechanism for one of the antibodies, i.e. 2E9, in more detail.

It has been suggested that the bivalent property of the antibodies may play an important role in their receptor kinase activation (Gill et al., 1984; Yarden and Schlessinger, 1987a). The result from the next experiment was in line with such a suggestion. A431 membrane preparations were incubated with different concentrations 2E9 for 20 min at room temperature, and EGF-R phosphorylation was determined by addition of [γ-^{32}P]ATP-containing phosphorylation assay mix. As well as in the saponin-permeabilized cells, EGF-R phosphorylation induced by 2E9 specifically occurs on tyrosine residues as determined by 2-D phosphoamino acid analysis (Spaargaren et al., 1990). Interestingly, maximal EGF-R autophosphorylation occurred at a narrow antibody concentration range around 50 μg/ml, whereas at lower or higher antibody concentrations the level of EGF-R phosphorylation decreased (Spaargaren et al., 1991). This result was consistent with data obtained with other antibodies (Gill et al., 1984; Yarden and Schlessinger, 1987a). At low antibody concentrations, binding will occur in a bivalent fashion but is unsaturated, and therefore the EGF-R phosphorylation is sub-maximal. The decrease of EGF-R phosphorylation at high concentrations supports the assumption that antibody bivalence is indeed an important parameter in EGF-R

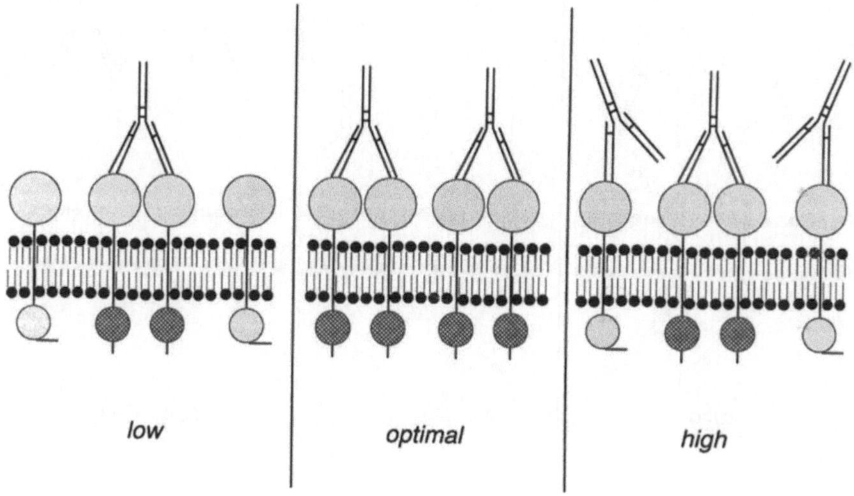

Fig. 3. **The effect of different antibody concentrations on EGF-R.** Schematic model for the effect of low, optimal, and high antibody concentrations as interpreted according to the intermolecular, active dimer, activation model.

phosphorylation, since at high antibody concentrations the binding will be oversaturated and occurs mainly in a monovalent fashion (Fig. 3). The same result was obtained using the saponin-permeabilized cell system (Spaargaren et al., 1991). These observations suggest that antibody-induced dimerization may be the cause of EGF-R activation.

Bivalent antibody binding results in EGF-R dimerization

The data thus far suggest that antibodies activate the EGF-R tyrosine kinase through induction of receptor dimerization. In order to detect possible EGF-R dimer complexes, the chemical cross-linker EDAC (1-ethyl-3-(3-dimethyl aminopropyl)-carbodiimide) was used. EDAC covalently links receptor dimers, thus enabling the use of SDS-PAGE for separation of receptor monomers and dimers. Furthermore, to investigate directly the importance of antibody bivalence in the formation of receptor dimers, monovalent 2E9-Fab fragments were prepared by papain digestion and affinity purification. If antibody bivalence is an essential feature of the antibodies in dimerization and activation of the EGF-R, these monovalent Fab fragments should be unable to do so unless they are cross-linked by a second antibody, thus restoring the bivalent character of the binding complex (Fig. 4).

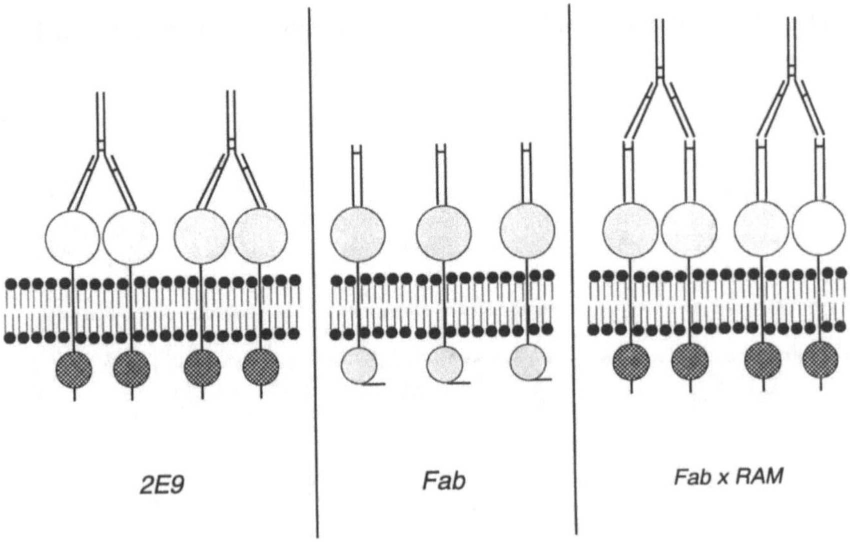

Fig. 4. **The effect of antibody bivalence on EGF-R.** Schematic model for the binding of a bivalent antibody (2E9), monovalent Fab fragments (Fab), or monovalent Fab fragments followed by binding of a second bivalent antibody (Fab x RAM), as interpreted according to the intermolecular, active dimer, activation model.

Membrane preparations were incubated with the indicated ligands (Fig. 5) for 20 min at room temperature, phosphorylation assay mix containing [γ-^{32}P]ATP was added and phosphorylation was allowed to proceed for 10 min on ice and stopped by the addition of 10 mM EDTA. Subsequently, 15 mM EDAC was added for 15 min at 30°C and samples were applied to SDS-PAGE. As receptors are detected by means of autophosphorylation, the increase of phosphorylation of the receptors after different ligand incubations was normalized to control levels in order to quantify dimerization. As can be seen in Fig. 5, SDS-PAGE revealed that antibody 2E9, like EGF, induced EGF-R dimerization.

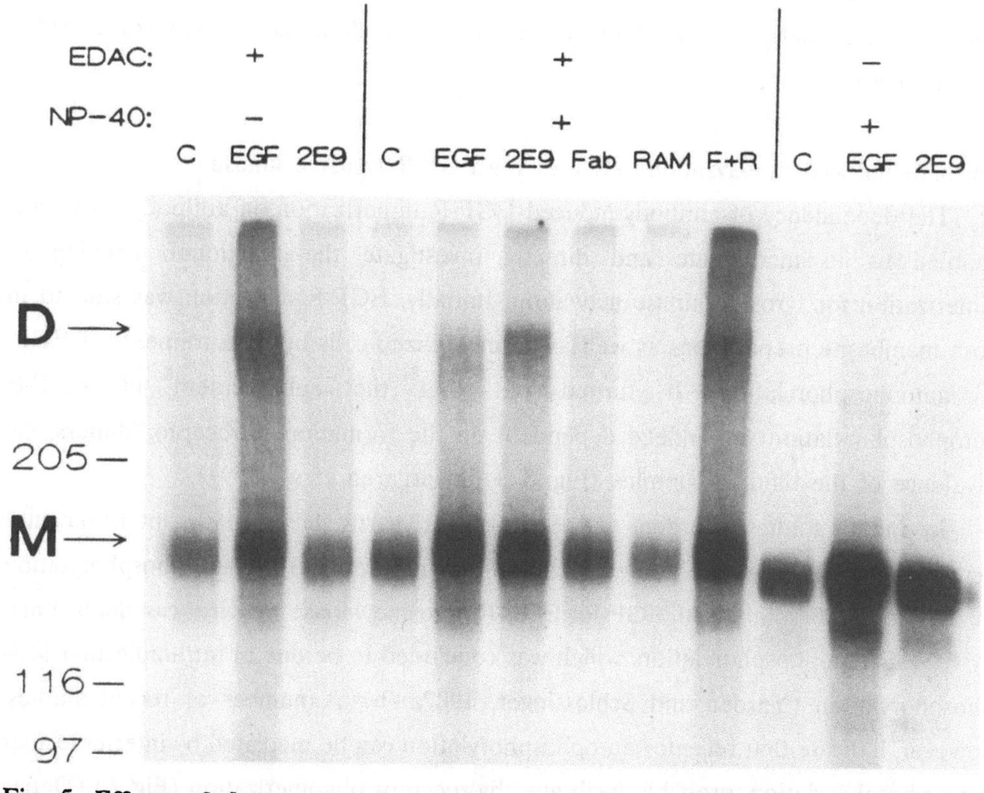

Fig. 5. **Effects of detergent and antibody bivalence on EGF-R dimerization.** A431 membrane preparations (20 μg) were preincubated as indicated with 100 μg/ml monovalent Fab fragments on ice for 2 hours (F, F+R), and washed by centrifugation. Subsequently, membranes were incubated for 20 min at room temperature with buffer alone (C), EGF (1 μg/ml), 2E9 (50 μg/ml), or RAM (500 μg/ml, R, F+R). After ligand incubation, membranes were phosphorylated by the addition of [γ-^{32}P]ATP containing phosphorylation assay mix with (+) or without NP-40 (-) for 10 min on ice. After quenching the phosphorylation reaction by the addition of 10 mM EDTA, receptor dimers were cross-linked by the addition of 15 mM EDAC (+) or buffer alone (-) for 15 min at 30°C. Samples were boiled in sample buffer, applied to 3-10% SDS-PAGE, and the gel was autoradiographed. Receptor dimerization was quantitated by excision of monomer (M)-, dimer (D)-, and background (in between)-containing gel bands and liquid scintillation of these bands.

No 2E9-induced receptor dimerization is observed when detergent is omitted from the phosphorylation mix. This observation is in line with the disability of the antibodies to stimulate the tyrosine kinase in the absence of detergent (Spaargaren et al., 1990) and corroborates a role for receptor dimerization in kinase activation. Moreover, as shown in Fig. 5, the Fab fragments (100 μg/ml) were unable to induce EGF-R dimerization in A431 membrane preparations. Similar results were obtained using a Fab concentration range from 10-200 μg/ml (not shown). However, preincubation of membranes with Fab fragments (100 μg/ml), followed by an incubation with RAM antiserum, indeed restored the ability of EGF-R dimer formation (Fig. 5). These data clearly indicate that antibody binding results in EGF-R dimerization, as a consequence of antibody bivalence.

Antibody-induced dimerization activates the EGF-R tyrosine kinase

The dependency of antibody-induced EGF-R dimerization on antibody bivalence, enabled us to manipulate and directly investigate the functional meaning of dimerization for tyrosine kinase activation. Initially, EGF-R activation was studied in both membrane preparations as well as permeabilized cells by measurements of EGF-R autophosphorylation. It turned out that the enhancement of EGF-R autophosphorylation was indeed dependent on the formation of receptor dimers, i.e. bivalence of the binding complex (Fig. 5 + Spaargaren et al., 1991).

In initial studies, Yarden and Schlessinger partly based their intermolecular activation model on data obtained with antibody-induced EGF-R autophosphorylation in Triton X-100 solubilized EGF-R. EGF-R tyrosine kinase activity was determined by EGF-R autophosphorylation, which was concluded to be due to intramolecular self-phosphorylation (Yarden and Schlessinger, 1987a+b). A number of recent studies, however, indicate that receptor autophosphorylation can be mediated by intermolecular cross-phosphorylation, probably facilitated by receptor oligomerization (Fig. 6) (Defize et al., 1989; Honegger et al., 1989+1990). Thus, detection of enhanced EGF-R autophosphorylation does not necessarily reflect tyrosine kinase activation. To eliminate the possibility that enhanced EGF-R autophosphorylation is merely due to facilitated intermolecular cross-phosphorylation, we performed exogenous and endogenous substrate phosphorylation experiments in membrane preparations and permeabilized cells, respectively.

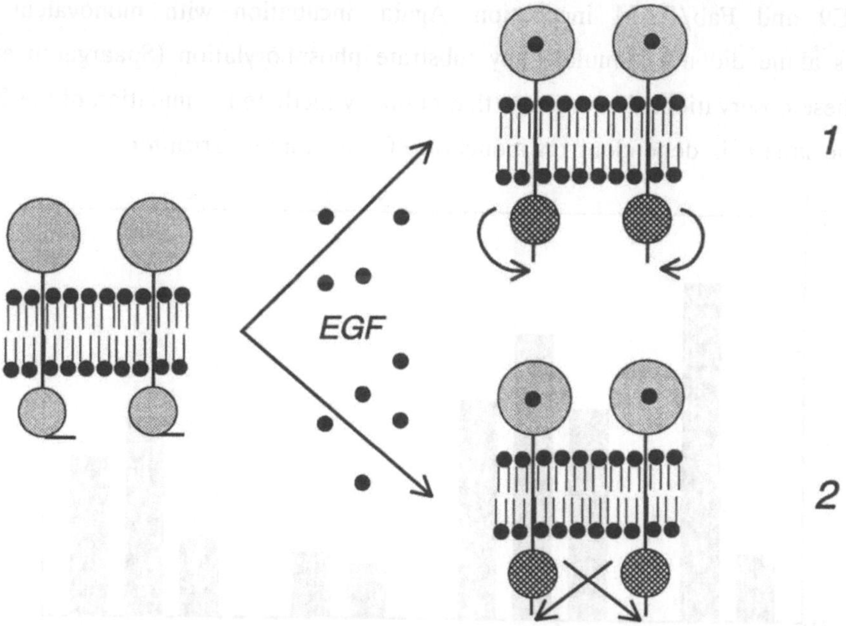

Fig. 6. **The mechanisms of EGF-R autophosphorylation.** 1) Intramolecular self-phosphorylation. 2) Intermolecular cross-phosphorylation.

A431 membranes were preincubated with or without Fab fragments and subsequently incubated with ligands for 20 min at room temperature as indicated (Fig. 7). After this incubation a phosphorylation mixture was added containing [γ-^{32}P]ATP, and 2 mM Angiotensin I as an exogenous substrate. Incubation of membrane preparations with EGF or 2E9 results in an increase of Angiotensin I phosphorylation, demonstrating activation of the tyrosine kinase under these conditions (Fig. 7). When the membranes are incubated with the monovalent Fab fragments, no induction of tyrosine kinase activity was observed, unless followed by treatment with RAM as a second antibody. Furthermore, also in this case, tyrosine kinase activation was strongly concentration dependent for 2E9, or in case of Fab preincubation, for RAM antiserum (Fig. 7). In accordance with the previously mentioned detergent dependency, 2E9 did not induce EGF-R tyrosine kinase activation in the absence of detergent (Spaargaren et al., 1990).

Using saponin-permeabilized cells we were able to measure endogenous substrate phosphorylation as well. It was found that several substrates could be detected upon

EGF, 2E9 and Fab/RAM incubation. Again, incubation with monovalent Fab fragments alone did not stimulate any substrate phosphorylation (Spaargaren et al., 1991). These observations demonstrate that antibody-mediated stimulation of the EGF-R tyrosine kinase is dependent on induction of receptor dimerization.

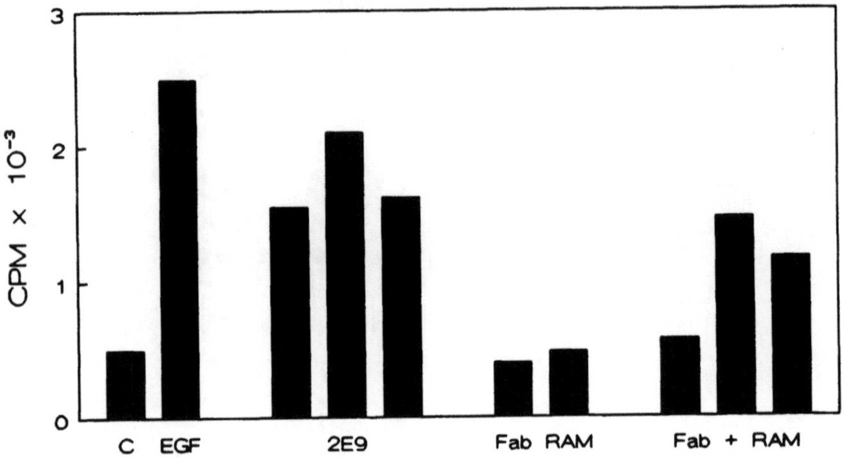

Fig. 7. **Effects of antibody bivalence and concentration on EGF-R tyrosine kinase activity towards exogenous peptide in membrane preparations.** A431 membrane preparations (20 μg) were preincubated as indicated with 100 μg/ml monovalent Fab fragments on ice for 2 hours, and washed by centrifugation. Subsequently, membranes were incubated for 20 min at room temperature with buffer (C), EGF (1 μg/ml), 25, 50 or 75 μg/ml 2E9, Fab fragments (100 μg/ml), RAM (500 μg/ml), or 250, 500 or 750 μg/ml RAM after Fab fragment preincubation (Fab+RAM). Phosphorylation assay mix containing [γ-^{32}P]ATP and Angiotensin I was added for 10 min on ice. Subsequently, the reaction was stopped by addition of TCA, and the exogenous peptide Angiotensin I phosphorylation was measured by collecting the peptide on phosphocellulose filters.

Additional evidence for dimerization-induced EGF-R tyrosine kinase activation

In order to establish further the mutual relationship between EGF-R dimerization and kinase activation, we have determined the effect of the specific EGF-R tyrosine kinase inhibitor tyrphostin AG 213 (a kind gift of Dr. A. Levitzki)(Lyall et al., 1989; Yaish et al., 1988) on EGF-induced receptor dimerization. After preincubation of surface-iodinated A431 cells with the kinase inhibitor, EGF still induces receptor dimerization to the same extent as in control cells (Spaargaren et al., 1991).

Furthermore, using the A431 membrane preparations, it was observed that receptor dimers are also formed by EGF as well as 2E9 prior to addition of the ATP-containing phosphorylation mixture (membranes were incubated with ligand, followed by cross-linking, and subsequently phosphorylated). This indicates that dimerization occurs in the absence of ATP and is thus not dependent on the phosphorylation reaction (Spaargaren et al., 1991).

These results show that receptor dimerization occurs efficiently in the absence of receptor phosphorylation, as well as at reduced tyrosine kinase activity. Taken together with the other data this demonstrates a causative role for dimerization in activation of the EGF-R protein-tyrosine kinase.

EGF-R tyrosine kinase activation: a model

In the present study we used anti-EGF-R antibodies to examine the relationship between antibody bivalence, EGF-R dimerization and EGF-R tyrosine kinase activity in both membrane preparations and permeabilized cells. Modification of antibody bivalence, the use of a specific EGF-R tyrosine kinase inhibitor, and the detergent dependency of antibody-induced EGF-R activation, enabled us to manipulate EGF-R dimerization as well as tyrosine kinase activity. We have demonstrated that (1) Anti-EGF-R antibodies activate the EGF-R tyrosine kinase irrespective of the recognized epitope; (2) antibody binding induces EGF-R dimerization; (3) antibody-induced EGF-R dimerization as well as tyrosine kinase activation are dependent on antibody bivalence and the presence of detergent; and (4) EGF-R dimerization is not dependent on receptor autophosphorylation and tyrosine kinase activity. These data provide strong evidence in favor of the intermolecular, active dimer, EGF-R tyrosine kinase activation model, as originally proposed by Schlessinger (Schlessinger 1986+1988b).

The efficiency of several ligands in activation of receptors containing a tyrosine kinase domain has been shown to be dependent on their bivalence. Monoclonal antibodies stimulated the insulin receptor tyrosine kinase by cross-linking receptor sub-unit molecules (O'Brien et al., 1987). The *neu* protooncogene product, closely related to the EGF-R, can be converted into an oncogene by a point mutation in the transmembrane region. This mutation results in induction of receptor aggregation and an increased tyrosine kinase activity (Weiner et al., 1989). Interestingly, the *neu* protooncogene encoded kinase, but not the constitutively active oncogenic mutant, can be stimulated by agonistic antibodies but not by monovalent antibody-derived Fab fragments (Yarden, 1990). Furthermore, activation of the PDGF receptor by PDGF involves receptor dimerization and is dependent on PDGF bivalence (Bishayee et al., 1989; Hammacher et al., 1989; Heldin et al., 1989).

Unlike PDGF or antibodies, EGF is a monovalent molecule that binds only one receptor (Weber et al., 1984). This raises the question as to what can be the driving force for EGF-induced receptor dimerization. We favor the hypothesis that EGF-R dimerization occurs through an EGF-induced conformational change in the extracellular domain of the EGF-R molecule, which causes enhanced affinity for other EGF receptors (Ullrich and Schlessinger, 1990). Evidence for such a conformational change has been reported recently in a study using purified EGF-R ectodomains (Greenfield et al., 1989). Subsequently, EGF-R tyrosine kinase activation may occur as a consequence of receptor-receptor interactions at the cytoplasmic domains. In this view the EGF-induced enhanced affinity for receptor-receptor interactions constitutes the driving force for EGF-R dimerization and subsequent tyrosine kinase activation (Fig. 8).

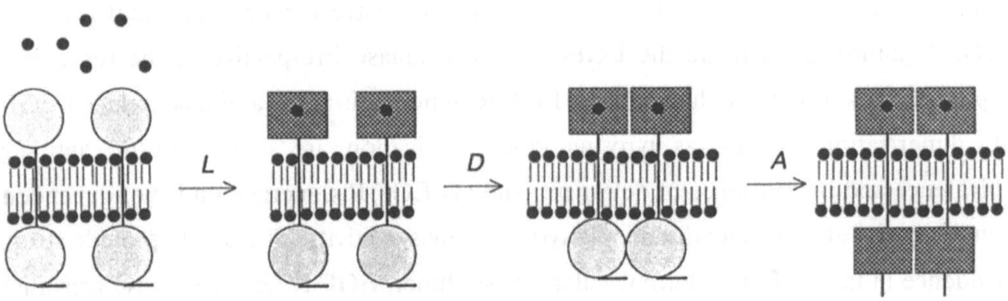

Fig. 8. **The proposed EGF-R tyrosine kinase activation model.** L = Ligand binding, inducing a conformational change of the extracellular EGF-R domain, which results in enhanced affinity of receptors for each other, thus resulting in EGF-R dimerization. D = Dimerization, which results in an interaction between cytoplasmic EGF-R domains, thus inducing a conformational change of the intracellular EGF-R domain, resulting in EGF-R tyrosine kinase activation. A = Activation of the tyrosine kinase domain, resulting in EGF-R autophosphorylation and subsequent substrate phosphorylation.

Acknowledgments: This work was supported by the Center for Developmental Biology, Utrecht, the Netherlands.

REFERENCES

Basu, A., Raghunath, M., Bishayee, S., and Das, M. (1989) Mol Cell. Biol. 9, 671-677.

Basu, M., Sen-Majumdar, A., Basu, A., Murthy, U., and Das, M. (1986) J. Biol. Chem. 261, 12879-12882.

Bertics, P.J., and Gill, G.N. (1985) J. Biol. Chem. 260, 14642-14647.

Bishayee, S., Majumdar, S., Khire, J., and Das, M. (1989) J. Biol. Chem. 264, 11699-11705.

Biswas, R., Basu, M., Sen-Majumdar, A., and Das, M. (1985) Biochemistry 24, 3795-3802.

Boni-Schnetzler, M., and Pilch, P.F. (1987) Proc. Natl. Acad. Sci. USA 84, 7832-7836.

Carpenter, G. (1987) Annu. Rev. Biochem. 56, 881-914.

Cochet, C., Kashles, O., Chambaz, E.M., Borello, I., King, C.R., and Schlessinger, J. (1989) J. Biol. Chem. 263, 3290-3295.

Defize, L.H.K., Moolenaar, W.H., van der Saag, P.T., and de Laat, S.W. (1986) EMBO J. 5, 1187-1192.

Defize, L.H.K., Mummery, C.L., Moolenaar, W.H., and de Laat, S.W. (1987) Cell Diff. 20, 87-102.

Defize, L.H.K., Boonstra, J., Meisenhelder, J., Kruijer, W., Tertoolen, L.G.J., Tilly, B.C., Hunter, T., van Bergen en Henegouwen, P.M.P., Moolenaar, W.H., and de Laat, S.W. (1989) J. Cell Biol. 109, 2495-2507.

Fanger, B.O., Austin, K.S., Earp, H.S., and Cidlowski, J.A. (1986) Biochemistry 25, 6414-6420.

Fanger, B.O., Stephens, J.E., and Staros, J.V. (1989) FASEB J. 3, 71-75.

Gill, G.N., Kawamoto, T., Cochet, C., Le, A., Sato, J.D., Masui, H., McLeod, C., and Mendelsohn, J. (1984) J. Biol. Chem. 259, 7755-7760.

Gill, G.N., Bertics, P.J., and Santon, J.B. (1987) Mol. Cell. Endocrinol. 51, 169-186.

Greenfield, C., Hiles, I., Waterfield, M.D., Federwisch, M., Wollmer, A., Blundell, T.L., and McDonald, N. (1989) EMBO J. 8, 4115-4123.

Hammacher, A., Mellstrom, K., Heldin, C-H., and Westermark, B. (1989) EMBO J. 8, 2489-2495.

Heldin, C-H., Ernlund, A., Rorsman, C., and Ronnstrand, L. (1989) J. Biol. Chem. 264, 8905-8912.

Honegger, A.M., Kris, R.M., Ullrich, A., and Schlessinger, J. (1989) Proc. Natl. Acad. Sci. USA 86, 925-929.

Honegger, A.M., Schmidt, A., Ullrich, A., and Schlessinger, J. (1990) Mol. Cell. Biol. 10, 4035-4044.

Hunter, T., and Cooper, J.A. (1985) Annu. Rev. Biochem. 54, 897-930.

Koland, J.G., and Cerione, R.A. (1988) J. Biol. Chem. 263, 2230-2237.

Lyall, M., Zilberstein, A., Gazit, A., Gilon, C., Levitzki, A., and Schlessinger, J. (1989) J. Biol. Chem. 264, 14503-14509.

Northwood, I.C., and Davis, R.J. (1988) J. Biol. Chem. 263, 7450-7453.

O'Brien, R.M., Soos, M.A., and Siddle, K. (1987) EMBO J. 6, 4003-4010.

Schlessinger, J. (1986) J. Cell Biol. 103, 2067-2072.

Schlessinger, J. (1988a) Biochemistry 27, 3119-3123.

Schlessinger, J. (1988b) Trends Biochem. Sci. 13, 443-447.

Schreiber, A.B., Lax, I., Yarden, Y., Eshhar, Z., and Schlessinger, J. (1981) Proc. Natl. Acad. Sci. USA 78, 7535-7539.

Schreiber, A.B., Libermann, T.A., Lax, I., Yarden, Y., and Schlessinger, J. (1983) J. Biol. Chem. 258, 846-853.

Spaargaren, M., Defize, L.H.K., de Laat, S.W., and Boonstra, J. (1990) Biochem. Biophys. Res. Commun. 171, 882-889.

Spaargaren, M., Defize, L.H.K., Boonstra, J., and de Laat, S.W. (1991) J. Biol. Chem. 266, in press.

Ullrich, A., Coussens, L., Hayflick, J.S., Dull, T.J., Gray, A., Tam, A.W., Lee, J., Yarden, Y., Libermann, T.A., Schlessinger, J., Downward, J., Mayes, E.L.V., Whittle, N, Waterfield, M.D., and Seeburg, P.H. (1984) Nature 309, 418-425.

Ullrich, A., and Schlessinger, J. (1990) Cell 61, 203-212.

Weber, W., Bertics, P.J., and Gill, G.N. (1984) J. Biol. Chem. 259, 14631-14636.

Weiner, D.B., Liu, J., Cohen, J.A., Williams, W.V., and Greene, M.I. (1989) Nature 339, 230-231.

Yaish, P., Gazit, A., Gilon, C., and Levitzki, A. (1988) Science 242, 933-935.

Yarden, Y., and Schlessinger, J. (1987a) Biochemistry 26, 1434-1442.

Yarden, Y., and Schlessinger, J. (1987b) Biochemistry 26, 1443-1451.

Yarden, Y. (1990) Proc. Natl. Acad. Sci. USA 87, 2569-2573.

INSULIN-RELATED PEPTIDES AND THEIR RECEPTOR(S) IN THE MOLLUSC, *LYMNAEA STAGNALIS*

Edwin Roovers[1, 2], August B. Smit[2], Wijnand P.M. Geraerts[2], Joos Joosse[2], Rudi J. Planta[1], Erno Vreugdenhil[1] and Harm van Heerikhuizen[1]
Departments of Biochemistry[1] and Biology[2]
Free University
De Boelelaan 1083
1081 HV Amsterdam
The Netherlands

Insulin is a protein hormone with a long record in scientific research. In 1922 it was discovered that insulin is a hypoglycaemic hormone in dogs (Banting and Best, 1922). The succesful treatment of diabetes mellitus by injections of insulin is regarded as one of the major medical triumphs of this century. In 1955, insulin was the first protein whose primary structure was fully identified (Ryle *et al.*, 1955). In 1972, another landmark was set by Blundell and coworkers (Blundell *et al.*, 1972), who succeeded in crystallizing 2Zn-insulin, and revealing its three dimensional structure (3D) by röntgen-diffraction. Insulin, can therefore be considered as an outstanding model both in the field of endocrinology and molecular biology. The study of insulin has further developed during the last two decades. Various functions of insulin were discovered, in addition to its control of glucose metabolism. These "novel" functions concern the regulation of energy metabolism in general, of growth-associated processes, and of the development of the CNS (Czech *et al.*, 1985; Froesch *et al.*, 1985; Girbau *et al.*, 1987).

In contrast to the situation in vertebrates, little is known about the presence and physiological significance of insulin-related molecules in invertebrates. Most of our understanding of invertebrate insulin-related molecules, is derived from immunocytochemical experiments in which antibodies, raised against vertebrate insulin, were used to identify insulin-producing cells in different invertebrate species. However, we and other groups recently succeeded in isolating cDNA clones that encode insulin-related preprohormones from a mollusc, *Lymnaea stagnalis* (Smit *et al.*, 1988) and two insect species, *Bombyx mori* (Adachi *et al.*, 1989; Kawakami *et al.*, 1989) and *Locusta migratoria* (Lageux *et al.*, 1990). In this chapter we wish to discuss the presence and physiological significance of these molecules in the freshwater snail, *L. stagnalis*.

NATO ASI Series, Vol. H 52
Biological Signal Transduction
Edited by E. M. Ross and K. W. A. Wirtz

Insulin-related peptides (MIPs) in *L. stagnalis*

Growth and associated processes in *Lymnaea* are thought to be controlled by neuroendocrinal giant neurons, the light green cells (LGCs). The LGCs are present as two paired groups of together 200 cells in the cerebral ganglia (Joosse, 1964; Wendelaar Bonga, 1970). The axons of the LGCs extend to the peripheries of the median lip nerves, which are used as neurohaemal areas (Fig. 1) (Benjamin *et al.*, 1976; Van Minnen *et al.*, 1979; Roubos *et al.*, 1980). In addition, it has been demonstrated that the LGCs are responsive to dopamine (De Vlieger *et al.*, 1987; Werkman *et al.*, 1988), neuropeptides (*e.g.*, FMRF amide), and metabolites in the haemolymph (*e.g.*, glucose) (Kits, unpublished results). The LGCs may be involved in a broad spectrum of processes associated to growth such as the control of body growth (Geraerts, 1976), formation of the shell (Dogterom *et al.*, 1979), and protein and glycogen metabolism (Dogterom, 1980; Veldhuijzen and Dogterom, 1976; Veldhuijzen and Van Beek, 1976). In addition, the LGCs may also produce neuropeptides that are involved in the outgrowth of neurites (Kits *et al.*, 1990).

Immunocytochemical experiments showed that the LGCs can be stained with vertebrate

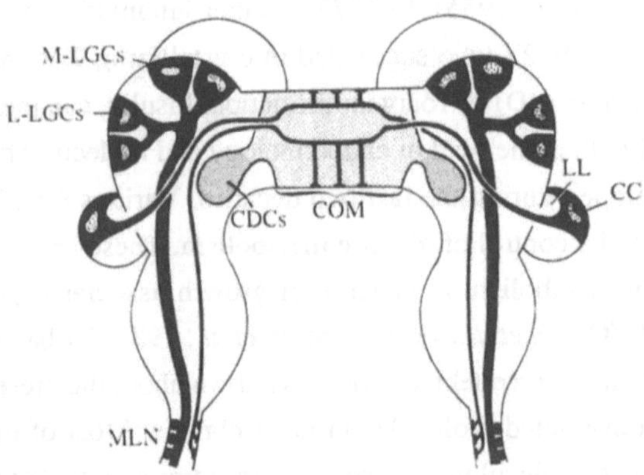

Fig. 1. Schematic representation of the anatomical organization of the LGC system. In each cerebral ganglion two bundles of axons running from the M-LGC and L-LGC clusters, respectively, meet and run together to the ipsilateral median lip nerves (MLN). In the lateral lobes (LL), small ganglia connected with the cerebral ganglia, the canopy cells (CCs), are located (Van Minnen *et al.*, 1979). The axon of each CC runs straight to and through the cerebral commissure (COM) to the contralateral MLN, where it ramifies and forms neurohaemal axon terminals in the periphery of the proximal part of the MLN. In the COM, the crossing axon gives rise to collaterals that run to the periphery all around the COM, forming neurohaemal axon terminals. These axon terminals are located in between those of the caudodorsal cells (CDCs, controlling egglaying in *Lymnaea* (Vreugdenhil *et al.*, 1985).

insulin antibodies indicating that insulin-related molecules are involved in the control of growth and associated processes in *Lymnaea* (Van Minnen, personal communication). To get a further insight into the presence and structural features of such molecules, we used the differential screening method to isolate and sequence several cDNA and genomic clones that encode insulin-related peptides, which are called molluscan insulin-related peptides (MIPs) (Smit *et al.*, 1988). The structural features of these clones and the predicted MIPs are as follows.

In contrast to the number of vertebrate insulin genes (one or two per species), genomic Southern blot analysis, extensive screening of genomic libraries, and analysis of genomic clones indicated the presence of six MIP genes in *Lymnaea*. The MIP genes are present in the genome as three couples of closely linked genes, which are transcribed in the same direction. All these genes show the architecture typical of the vertebrate insulin genes, with three exons and two introns at similar positions (*i.e.* one within the leader sequence and one within the C peptide domain; see Fig. 2). This is surprising, taking into consideration that invertebrates and vertebrates diverged 6×10^8 years ago, and it indicates an early genesis of the basic insulin gene organization. This genomic organization, however, is in contrast to that of the bombyxin insulin-related genes of the insect, *Bombyx mori* (Kawakami *et al.*, 1989). Bombyxin genes do not contain introns and are arranged as units of paired genes transcribed in the opposite direction. The MIP genes IV and VI are likely pseudogenes, as they are not detectably transcribed and lack a TATA box at the appropriate position. Moreover, the MIP VI gene contains an insertion of 7 nucleotides in exon 2, which causes a frame shift that results in a stop codon at the end of exon 2.

The identified MIP cDNA clones encode preproMIPs, I, II, III and V, which are very

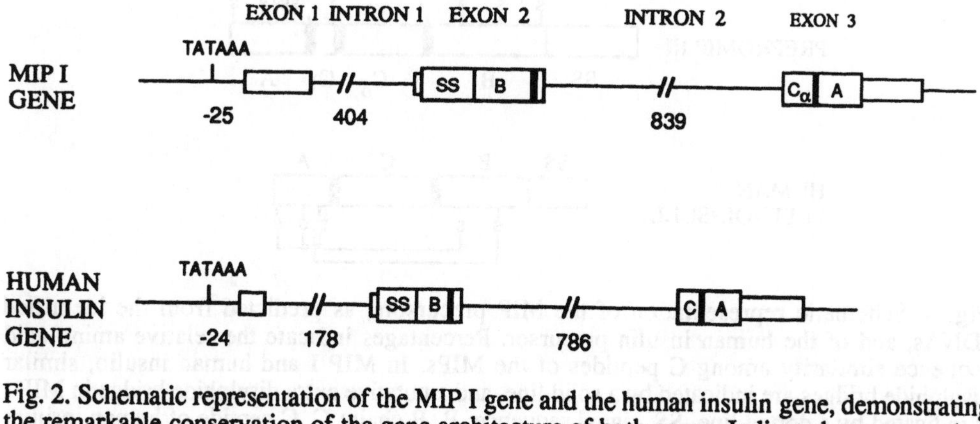

Fig. 2. Schematic representation of the MIP I gene and the human insulin gene, demonstrating the remarkable conservation of the gene architecture of both genes. Indicated are: exons 1, 2 and 3 and the introns 1 and 2. SS, signal sequence; B, B chain; Cα, Cα peptide; C, C peptide. Indicated is further: the TATAAA box and the total length in nucleotides of the introns.

reminiscent to the structure of vertebrate preproinsulin. A signal peptide is followed by a B chain, a C peptide and an A chain, with dibasic residues flanking the C peptide domain (see Fig. 3). Recent peptide sequence data indeed show that the preproMIPs are processed to form characteristic 2-chain MIP molecules (Li, unpublished results). Comparison of the predicted amino acid residues of the A and B chains of MIPs I, II, III and V revealed that they have a low sequence identity of about 45-75%. However, all MIPs share residues that are thought to be important in adopting the basic insulin configuration.When compared with members of the insulin superfamily, the A and B chains of the MIPs contain cysteine residues at positions typical for the members of this superfamily. This suggests that the three characteristic disulphide bridges in the MIP molecules have been conserved. In addition, the important

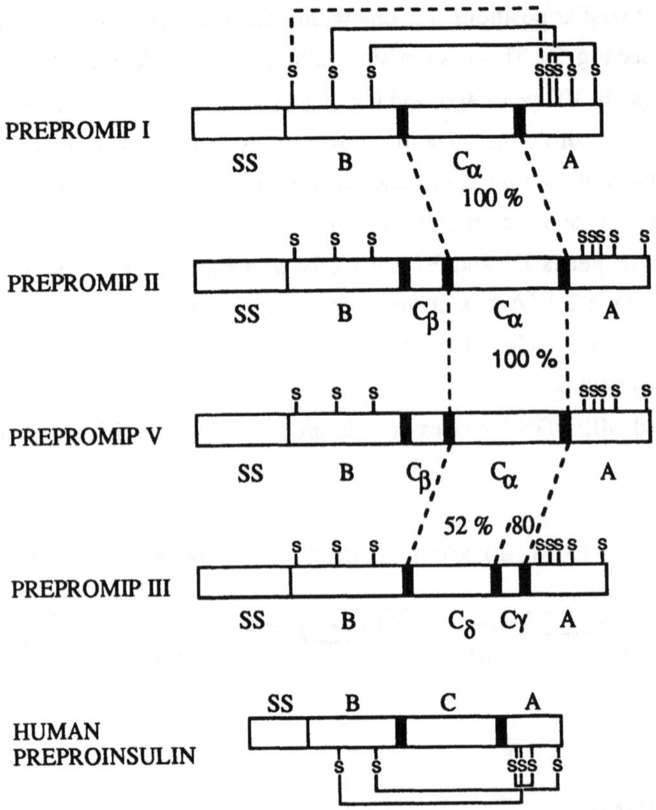

Fig. 3. Schematic representation of the MIP precursors, as predicted from the identified cDNAs, and of the human insulin precursor. Percentages indicate the relative amino acid sequence similarity among C peptides of the MIPs. In MIP I and human insulin, similar disulphide bridges are indicated by a solid line, and a putative extra disulphide bridge in MIP I is indicated by a dotted line. SS, signal sequence; B, B chain; C, C peptide of human insulin; Cα, Cβ, Cγ, Cδ, the various C peptides of the MIPs; A, A chain; cysteine residues are indicated by an S; thick vertical bars indicate the position of dibasic proteolytic processing sites.

hydrophobic core residues of the globular insulin structure are either conserved, or are replaced by residues with an equal hydrophobic character. Also α-helices, which are present in the A and B chains of vertebrate insulins, can be predicted for the MIPs (Chou *et al.*, 1974). Because MIPs possess the typical features of the insulin core structure, they must be considered members of the insulin superfamily (Steiner *et al.*, 1985).

Although the MIPs are 2-chain peptides that fit the insulin core structure, they cannot be considered as genuine insulin molecules, because they exhibit additional characteristics, that classify them as a distinct group of the superfamily. Both the A and B chains are N-terminally extended, and the B-chain contains an extra cysteine at B -6 that may form a third interchain disulphide bridge with an extra cysteine at A4. Due to these properties, the MIPs are the most complexely folded molecules of the insulin superfamily. Furthermore, MIPs exhibit strong divergence of surface residues, which are important in many functions of the vertebrate insulin molecules (Blundell *et al.*, 1975), such as receptor binding, solubility of the molecule, processing of the precursor, and monomer, dimer and hexamer interactions. A striking example is the almost complete divergence of the amino acid residues involved in receptor binding, including residues at A1-A5, A8 and B23-26, which may play a key role in insulin-specific metabolic effects (Blundell *et al.*, 1975). Together, these features suggest that MIP molecules adopt a specific 3D-structure, which may deviate from that of insulin. Indeed, 3D-structure predictions of the MIPs support this suggestion. Based on the primary structure and the predicted 3D-structure, it is unlikely that the extracellular binding domain of the MIP receptor is of the vertebrate type. Interestingly, the putative receptor binding residues are divergent among MIPs, which suggests that each MIP binds a different receptor, or alternatively, that various MIPs bind the same receptor with different affinities.

The predicted C peptides of the various preproMIPs do not show sequence similarity with their vertebrate and insect counterparts. However, in contrast to the A and B chains, the predicted C_{α} peptides of MIPs I, II and V are completely identical. In addition, some of the preproMIPs, i.e. II, III and V, contain additional dibasic residues in the C peptide region, which offers the possibility to generate additional C peptides (called C_{β}, C_{γ} and C_{δ}; see Fig. 3). The predicted C_{β} peptides exhibit 100% sequence identity amongst each other, while the C_{γ} and C_{δ} peptides show 80% and 52% identity, respectively, to the corresponding domains in the C_{α} peptide. This strong conservation of the C_{α} and C_{β} peptides is different from the situation in vertebrates, where C peptides appear to be quite diverse. This might indicate that the *Lymnaea* C peptides may serve important biological functions.

MIP gene expression in the adult *Lymnaea* CNS was demonstrated, using both Northern blotting and *in situ* hybridization techniques, and is entirely restricted to the LGCs and the canopy cells of the lateral lobes. The lateral lobes are small ganglia attached to the cerebral ganglia (see Fig. 1). Detection of MIPs using immunocytochemistry agreed completely with these observations.

The prepro MIPs very likely are loaded into secretory granules, where the C peptides are removed from the MIP precursors. The granules are transported to the neurohaemal area, where their contents are released by exocytosis (Van Heumen and Roubos, manuscript submitted).

The peptide sequence data of MIPs I, II, III and V are in agreement with the sequence data derived from the cDNAs (Li and Geraerts, unpublished results). The peptide chemistry further indicated that the A and B chains of MIPs I, II and V are N-terminally blocked, and the B chain N-terminus of MIP III is 2 amino acids longer than predicted by the cDNA sequence. The two C-terminal amino acids of the B-chains of all four MIPs are posttranslationally removed. The trimming of MIPs is not unusual as truncated forms of other members of the insulin superfamily, such as insulin, relaxin and IGF, have been reported (Huth and Rapoport, 1982; Bullesbach and Schwabe, 1985; Conlon *et al.*, 1988; O'Byrne *et al.*, 1989).

<u>Gut insulin-like substance (ILS) in *L. stagnalis*</u>

Apart from structural differences between *Lymnaea* MIPs and the vertebrate insulins, there is also a marked difference in the tissues, in which they are expressed. MIP expression seems to be restricted to the CNS of *Lymnaea*, whereas vertebrate insulins are produced in the pancreas. Furthermore, MIPs are thought to act as growth factors, while vertebrate insulin is primarily known as a hypoglycaemic hormone. It might be possible, therefore, that other insulin-related peptides are produced in non-neuronal tissues of *Lymnaea*. Indeed, there is biochemical and pharmacological evidence for the presence of such a peptide in the gut of *Lymnaea* (Hemminga, 1984). This peptide was named, insulin-like substance (ILS). It appears to have a molecular weight similar to that of mammalian insulin. ILS is also able to compete with ^{125}I-labelled insulin for the insulin receptors (IRs) of isolated rat fat cells (Ebberink and Joosse, 1985). Interestingly, the competition by ILS is almost as strong as that by porcine insulin, indicating that ILS can bind to rat fat cell IRs with similar affinity as porcine insulin. Another indication for the structural similarity of *Lymnaea* ILS to mammalian insulins can be deduced from the observation that *Lymnaea* ILS stimulates the uptake of 2-deoxyglucose by isolated rat fat cells, although less well than porcine insulin does (Hemminga, 1984). Comparison of the amino acid composition of ILS with that of MIPs shows that the ILS molecule is different from the MIPs (Ebberink *et al.*, 1989). Moreover, Northern blot analysis of gut mRNA failed to detect MIP transcripts.

In conclusion, it seems likely that an insulin-related peptide with hypoglycaemic activity is produced in the gut of *Lymnaea* and that this peptide differs from the MIPs.

<u>Characterization of insulin receptor-related protein(s) in molluscs</u>

Cloning of insulin receptor (IR) genes and their corresponding cDNAs from molluscan species will be of significant interest for a number of reasons. Firstly, the vertebrate IR belongs to a superfamily of structurally related transmembrane proteins, that all contain a tyrosine kinase moiety located in the cytoplasmic part of the receptor. Members of this superfamily include the receptors for insulin, insulin-like growth factor I, nerve growth factor, epidermal growth factor, fibroblast growth factor and platelet derived growth factor (Ullrich and Schlessinger, 1990). The structural and sequence similarities suggest that these proteins may have evolved from a common ancestor. It is likely that the MIP receptor(s) will also belong to this superfamily. Therefore, comparisons of molluscan MIP receptor sequences with their vertebrate counterparts will be of evolutionary interest. Secondly, despite the elucidation of the primary structure of the human insulin receptor, and the expression of its cDNA, the precise mechanism by which this receptor operates is still unknown. Comparisons of molluscan insulin receptor sequences with that of vertebrates may help to correlate structural domains with specific functions. Thirdly, in contrast to the situation in most vertebrates, where the insulin gene is present in a single copy, *Lymnaea* expresses at least four different MIP genes. The primary structure of the four MIPs suggests that they contain different receptor binding sites, which may either mean that each MIP binds a different receptor, or alternatively, that various MIPs bind the same receptor with different affinities.

The method we have used for cloning the Lymnaean MIP receptor was based on the assumption that certain parts of a putative ancestral IR were conserved during evolution and therefore are present in mammalian IRs and MIP receptors. This can be deduced from the observation that IRs in human (hIR) and *Drosophila* (two organisms that share the same common ancestor with *Lymnaea*) share a region of very strong amino acid sequence identity (*i.e.* 70% identity within a stretch of 180 amino acids; see Petruzzelli *et al.*, 1986). The mammalian IR and its *Drosophila* homologue (dIRh) are integral membrane glycoprotein oligomers consisting of two α subunits capable of insulin binding, and two β subunits spanning the membrane and containing a cytoplasmic tyrosine kinase domain capable of insulin-dependent autophosphorylation (Czech *et al.*, 1985; Ullrich *et al.*, 1985; Ebina *et al.*, 1985; Fernandez-Almonacid and Rosen, 1987; Nishida *et al.*, 1986; Petruzelli *et al.*, 1985;

1986). A functional tyrosine kinase is absolutely essential to IR function in mammals. It is the tyrosine kinase domain that comprises the region that was found to be highly identical in hIR and dIRh. From this region we have selected a number of amino acid stretches that were identical in hIR and dIRh and deduced from them 4 sets of completely degenerated oligonucleotides that could be used in polymerase chain reactions (PCR) of Lymnaean cDNA (see Fig. 4).

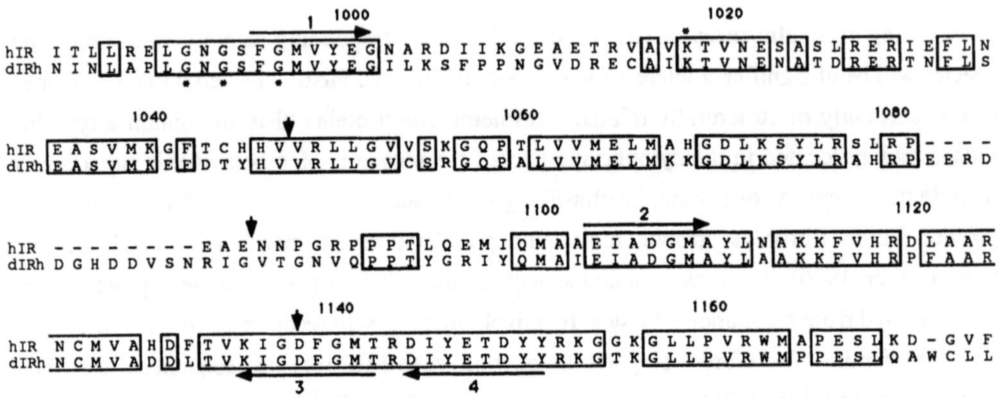

Fig. 4. Comparison of the amino acid sequence of the hIR and dIRh in a region comprising the major part of the tyrosine kinase domain. Amino acids identical in both sequences are boxed. The numbering of the hIR sequence is as in Ullrich *et al.* (1985).The positions of the four sets of degenerated oligonucleotides used for PCR are indicated by numbered arrows; the direction of the arrows is in the 5' to 3' direction of the oligonucleotide. Asterisks below the sequence mark the position of the (GXGXXG) sequence implicated in ATP binding, the Lys residue that serves as the actual ATP binding site is marked by an asterisk above the sequence. Positions where introns are present in the hIR genomic sequence are indicated by arrow heads.

We chose mRNA isolated from egg masses of *Lymnaea* as a template to generate cDNA, since we argued that the level of MIP receptor expression would possibly be high during embryogenesis, the same way as was observed for the expression of the dIRh (Garofalo and Rosen, 1988). All possible combinations of the four primers yielded PCR products having approximately the same length as the PCR products generated by using the same primers in a reaction on cloned hIR cDNA (Ullrich *et al.*, 1985; Ebina *et al.*, 1985). The ethidium bromide stained gel shown in Fig. 5A illustrates, the length distribution of the PCR fragments generated by using a combination of the primers 1 and 3. The major band of ~470 bp (Fig. 5A) clearly represents a *Lymnaea* fragment homologous to part of the hIR tyrosine kinase domain since it appears to hybridize to oligonucleotide 2 (a primer located in between the primers 1 and 3; see Fig. 4) (Fig. 5B).

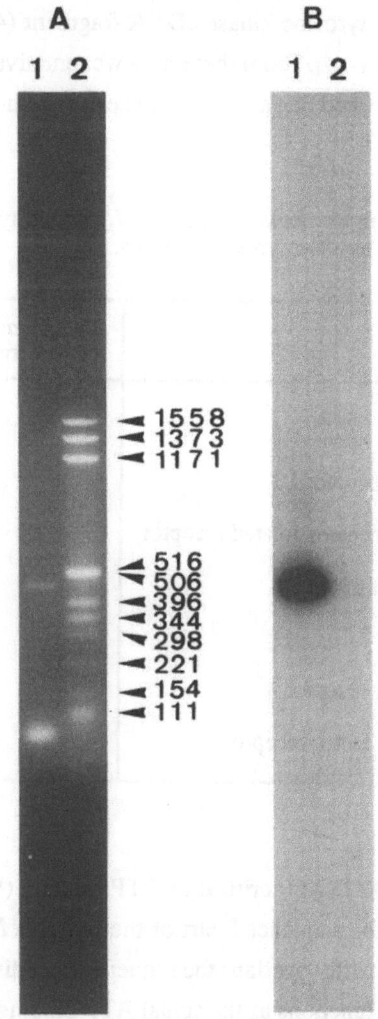

Fig.5. PCR amplification of *Lymnaea* egg mass cDNA with primers 1 and 3. (A) Products were separated on a 2% agarose gel (lane 1) and stained with ethidium bromide. The length of marker fragments (lane 2) is given in bp. The ~470bp fragment in lane 1 is clearly visible. (B) The gel shown in panel A was blotted and the blot was hybrized to ^{32}P-labeled primer 2 as a probe. Hybridization was performed at 42^{0}C, and washing with 1xSSC, 0.1% SDS at 42^{0}C. The ~470bp fragment strongly hybridizes with oligo 2.

This 470 bp PCR generated fragment was directly sequenced. The amino acid sequence deduced from the nucleotide sequence exhibits a high degree of identity to the homologous region of several IRs, including the hIR, the IGF-I receptor and the insulin receptor related receptor (IRR) (73-75%; see Table 1). The identity to the pertinent dIRh fragment is slightly lower (59%), but the degree of identity to other tyrosine protein kinases is clearly lower (28 and 38%). The avian sarcoma virus UR2 oncogene v-*ros* shows a significantly higher degree of homology with the Lymnaean tyrosine kinase cDNA fragment (48%), but this is also the case for comparisons of IRs with v-*ros*. From these data we tentatively conclude that the 470 bp PCR generated cDNA fragment will be part of a Lymnaean insulin-like receptor.

Table I. Amino acid sequence identity between part of a *Lymnaea* tyrosine kinase, putatively encoding an insulin-like receptor, to other tyrosine kinases.

	Lymnaea ~470bp tyrosine kinase cDNA fragment
human insulin receptor	75%
human insulin growth factor I receptor	75%
human and guinea pig insulin receptor-related receptor	73%
Drosophila insulin receptor homologue	59%
v-*ros* oncogene	48%
human epidermal growth factor receptor	38%
human platelet derived growth factor receptor	28%

The consensus sequence (GXGXXG) critical to ATP binding (Wierenga and Hol, 1983; Russo, 1985) is not present on the sequenced part of the putative *Lymnaea* tyrosine kinase domain fragment, since primer 1 partly overlaps the sequence encoding the (GXGXXG) motif. However, the lysine residue that functions as the actual ATP binding itself is conserved in the *Lymnaea* sequence. Also, all the tyrosine residues within the hIR tyrosine kinase domain that have been implicated in autophosporylation (*i.e.*, residues 1075, 1110, 1146, 1150, 1151 and 1198) appear to be conserved in the *Lymnaea* sequence.

In order to clone a full length cDNA encoding the *Lymnaea* MIP receptor, we decided to use a modification of the PCR technique, the RACE strategy (Rapid Amplification of cDNA Ends; Frohman *et al.*, 1988; and 1989; Ohara *et al.*, 1989; Kaiser, 1990). This RACE strategy (*i.e.*, the version as we have used it to generate 3'-terminal cDNA fragments) is outlined in Fig. 6. Based on the nucleotide sequence encoding part of the *Lymnaea* tyrosine kinase domain we synthesized two additional oligonucleotides 5 and 6. First strand synthesis was carried out using the oligo dT-adapter primer. Subsequent PCR amplification was per-formed by using both primer 5 and the adapter part of the oligo dT-adapter primer (Fig. 6).

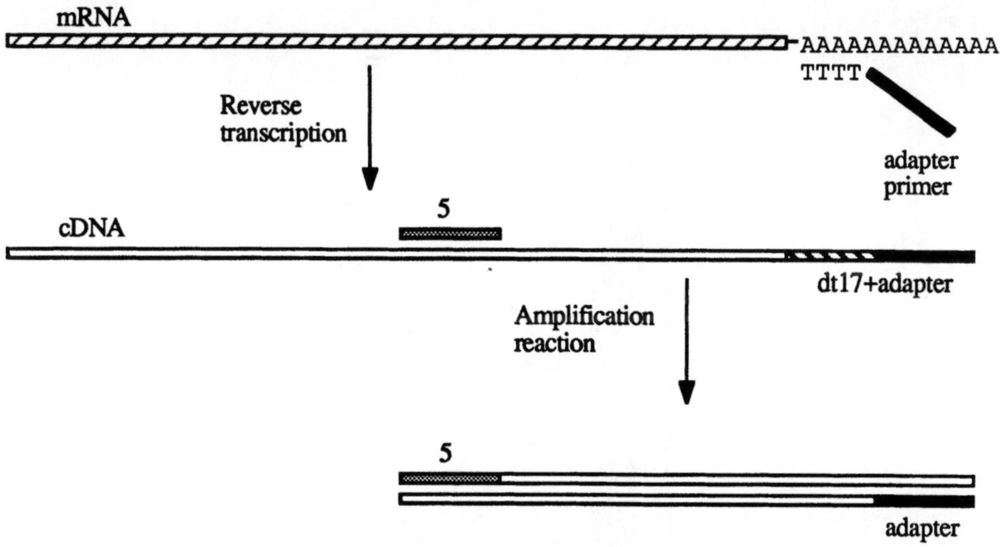

Fig. 6. Schematic representation of the RACE 3' ends strategy. First mRNA is reversed transcribed using a $(dT)_{17}$-adapter primer. From the resulting product only the cDNA strand is drawn. PCR amplification of the resulting cDNA is carried out using primer 5 (stippled bar) and the adapter primer (black bar).

Gel electrophoresis of the PCR generated material shows the presence of several DNA fragments of different length (Fig. 7A). Blotting of the gel and hybridization to oligonucleotide 6 (synthesized according to the sequence of the *Lymnaea* tyrosine kinase domain immediately downstream of the location of primer 5) indicates that the band of 2.1 kb is the amplified fragment looked for (Fig. 7B).

We are currently in the process of sequencing this 2.1 kb fragment, and, in addition, are performing a similar RACE strategy to amplify, clone and sequence the 5'-part of the cDNA under study. Whether this cDNA indeed codes for a MIP receptor (or alternatively for an ILS receptor, or another type of tyrosine kinase containing protein) cannot be answered as yet. A definitive conclusion has to await cloning of a full length cDNA and expression of the cDNA in a suitable expression system in order to allow for binding studies using purified MIPs.

A (preliminary) genomic Southern analysis (not shown), using the 470 bp PCR generated fragment as a probe, performed under stringent hybridization and washing conditions, indicates the possible presence of more than one gene encoding tyrosine kinase containing proteins. This may indicate, that there is indeed more than one MIP receptor in *Lymnaea,* and possibly that there will be a different MIP receptor for each of the four MIPs. Since the genomic region covered by the (cDNA) probe used may contain several introns, the conclusion regarding the number of genes, must be viewed with caution.

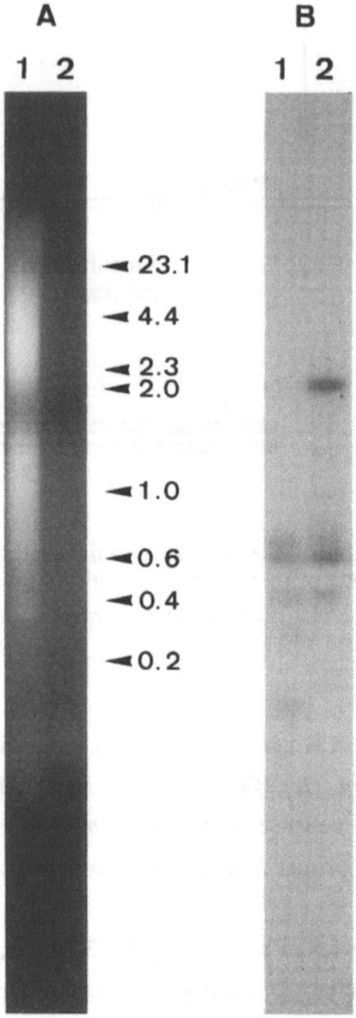

Fig. 7. Amplification of *Lymnaea* egg mass cDNA using a RACE 3' ends protocol. The amplification was carried out as described in Fig. 6. (A) Products of the amplification reaction were separated on a 1.6% agarose gel and stained with ethidium bromide. The PCR was performed in the presence of 3 mM $MgCl_2$ (lane 1) and 1.5 mM $MgCl_2$ (lane 2). The length of marker DNA fragments is indicated (kb). In lane 2 a weak ~2.1kb fragment is visible. (B) Southern blot of the gel shown in panel A using ^{32}P-labeled oligonuceotide 6 (*i.e.* an oligonucleotide, which is located immediately downstream of primer 5) as a probe.The blot was hybridized at 55^0C and washed with 1xSSC, 0.1% SDS at 55^0C. The ~2.1kb fragment hybridizes with oligo 6 as well as with some smaller fragments. It appeared that the smaller fragments were due to aspecific annaeling of the (dT_{17}-adapter primer) at A-rich regions in the first strand synthesis at 42^0C.

References

Adachi T et al. (1989). J Biol Chem 264 : 7681-7685
Banting FG and Best CH (1922). J Lab Clin Med 7 : 251-266
Benjamin PR, Swindale NV and Slade CT (1976) In: Neurobiology of invertebrates Gastropoda brain (ed J Salanki). pp 85-100. Budapest, Akademio Kiado
Blundell TL, Dodson GG, Mercola D and Hodgkin DC (1972). Adv Protein Chem 26 : 279 402
Blundell TL and Wood SP (1975). Nature 257 : 197-203
Bullesbach EE and Schwabe C (1985) Biochemistry 24 : 7717-7722
Chou PY and Fasman GD (1974). Biochemistry 13 : 211-245
Conlon M, Andrews PC, Falkmer S and Thim L (1988). Gen Comp Endocrinol 72 : 154-160
Czech MP et al (ed) (1985) In: Molecular basis of insulin action. pp 473 Plenum Press New York
De Vlieger TA, Lodder JC, Werkman TR and Stoof JC (1987). In: Neurobiology. Molluscan Models (eds Boer HH, Geraerts WPM and Joosse J). pp 172-175 North Holland Publ Cy Amsterdam
Dogterom AA Van Loenhout H and Van Der Schors RC (1979). Gen Comp Endocr 39 : 63-68
Dogterom AA (1980). Comp Biochem Physiol B65 : 163-167
Ebberink RHM and Joosse J (1985). Peptides 6 : 451-457
Ebberink RHM, Smit AB and Van Minnen J (1989). Biol Bull 177 : 176-182
Ebina Y, Ellis L, Jarnagin K, Edery M, Graf L, Clauser E, Ou J, Masiarz F, Kan YW, Fernandez-Almonacid R and Rosen OM (1987). Mol Cell Biol 7 : 2718-2727
Froesch ER, Schmidt C, Schwander J and Zapf JA (1985). Rev Physiol 47 : 443-467
Frohman MA, Dush MK and Martin GR (1988). Proc Natl Acad Sci USA 85 : 8998-9002
Frohman MA and Martin GR (1989). J Meth Cell Mol Biol 1 : 165-170
Garafalo RS and Rosen OM (1988). Mol Cell Biol 8 : 1638-1647
Geraerts WPM (1976). Gen Comp Endocr 29 : 61-71
Girbau M, Gomez JA, Lesniak MA and de Pablo F (1987). Endocrinology 121 : 1477-1482
Hemminga MA (1984). Thesis, Vrije Universiteit Amsterdam
Huth A, Rapoport TA (1982). Gen Comp Endocrinol 46 : 158-167
Joosse J (1964). Arch Neerl Zool 15 : 1-103
Kaiser K (1990). J Meth Cell Mol Biol 2 : 51-64
Kawakami A, Iwami M, Nagasawa H, Suzuki A and Ishizaki H (1989). Proc natl Acad Sci USA 84 : 6843-6847
Kits KS, De Vries NJ and Ebberink RHM (1990). Neurosci Lett 109 : 253-258
Lageux M, Lwoff L, Meister M, Goltzené F and Hoffmann J (1990). J Eur J Biochem 187 : 249-254
Nishida Y, Hata M, Nishizuka Y, Rutter WJ and Ebina Y (1986). Biochem Biophys Res Commun 141 : 474-481
O'Byrne EM Tabachnik M, Anderson LL and Steinetz BG (1989). Endocrinology 124 : 2920 2927
Ohara O, Dorit RL and Gilbert W (1989). Proc Natl Acad Sci USA 86 : 5673-5677
Petruzelli L, Herrera R, Garcia-Arenas R and Rosen OM (1985). J Biol Chem 260 : 16072-16075
Petruzelli L, Herrera R, Garcia-Arenas R, Fernandez R, Birnbaum MJ and Rosen OM (1986). Proc Natl Acad Sci USA 83 : 4710-4714
Roubos EW, Geraerts WPM, Boerrigter GH and Van Kampen GPJ (1980). Gen Comp Endocr 40 : 446-454
Russo MW, Lukas TJ, Cohen S and Staros JV (1985). J Biol Chem 260 : 5205-5208
Ryle AP, Sanger F, Smith LF and Kitai R (1955). Biochem J 60 : 541-556
Smit AB, Vreugdenhil E, Ebberink RHM, Geraerts WPM, Klootwijk J and Joosse J (1988). Nature 331 : 535-538
Steiner DF, Chan SJ, Welsh JM and Kwok SCM (1985). A Rev Genet 19 : 463-484

Ullrich A, Bell JR, Chen EY, Herrera R, Petruzelli LM, Dull TJ, Gray A, Coussens L, Liao Y-C, Tsubokawa M, Mason A, Seeburg PH, Grunfeld C, Rosen OM and Ramachandran J (1985). Nature 313 : 756- 761
Ullrich A and Schlessinger J (1990). Cell 61 : 203-212
Van Heumen WRA and Roubos EW (1991). Submitted to Neuroscience
Van Minnen J, Reichelt D and Lodder JC (1979). Cell Tissue Res 214 : 453-462
Veldhuizen JP and Dogterom GE (1976). Neth J zool 25 : 125-137
Veldhuizen JP and Van Beek (1976). Neth J Zool 26 : 106-118
Vreugdenhil E, Geraerts WPM, Jackson JE and Joosse J (1985). J Peptides 6 : 465-470
Wierenga R and Hol W (1983). Nature 302 : 842-844
Wendelaar Bonga SE (1970). Z Zellforsch 108 : 190-224
Werkman TR, De Vlieger TA, Van Minnen J, Voorn P and Stoof JC (1988). Progress in catecholamine research, Part A: Basic aspects and peripheral Mechanisms, Neurology and Neurobiology (eds Dahlström A, Belmaker RH and Zandler M) pp 239-325 Allan R Liss Inc New York

HEPATIC VASOPRESSIN RECEPTOR: A KEY RECEPTOR OF PHOSPHOINOSITIDE METABOLISM

Z. Georgoussi*, S. J. Taylor, S. B. Bockkino and J. H. Exton
Howard Hughes Medical Institute and Department of Molecular Physiology and Biophysics, Vanderbilt University School of Medicine
Nashville, Tennessee 37232
United States of America

Institute of Biology*
National Research Center of Physical Sciences "Democritos"
Athens, Greece

It is generally accepted that many drugs, hormones and neurotransmitters elicit responses in cells by interacting with specific receptors located in the plasma membrane. A number of receptors for hormones and neurotransmitters have been purified to homogeneity and functionally reconstituted. Most of them mediate agonist actions by interacting with effector systems to modulate the intracellular levels of second messengers via members of a family of guanine nucleotide binding proteins (G proteins).

Vasopressin is a neurohypophysial hormone that promotes glycogenolysis in liver, induces antidiuresis and contracts vascular smooth muscle. It interacts with high affinity with two classes of receptors, V1 and V2 (Michell et al. 1979). The V2 receptor is present in the kidney and its effects are mediated by cAMP (Jard, 1983), whereas the V1 receptor has been found in vascular smooth muscle, platelets, liver and other tissues, and its effects are mediated by a mechanism involving the elevation of intracellular Ca^{2+} secondary to the breakdown of phosphoinositides (Michell et al. 1979, 1981). Figure 1 shows the reaction sequence involved in the action of vasopressin via the V1-receptor. Upon binding to the this receptor, vasopressin stimulates the hydrolysis of phosphatidylinositol 4,5-bisphosphate in the plasma membrane to yield two signalling molecules: inositol 1,4,5 trisphosphate, which releases intracellular Ca^{2+}, and 1,2 diacylglycerol, which activates protein kinase C (Berridge 1987). There is much evidence from a variety of experiments that the signal transduction involves a G protein (Blackmore et al. 1985; Fitzgerald et al. 1986; Bojanic and Fain, 1986; Uhing et al. 1986; Lynch et al. 1986).

Although considerable advances have recently been made in the understanding of the mechanisms involved in receptor - G protein coupling in the regulation of adenylate cyclase and retinal cGMP phosphodiesterase, much work needs to be done (Casey and Gilman, 1988; Weiss et al. 1988). In

NATO ASI Series, Vol. H 52
Biological Signal Transduction
Edited by E. M. Ross and K. W. A. Wirtz

particular, the mechanisms involving the receptors that stimulate phosphoinositide hydrolysis are poorly understood. Attempts to purify the hepatic vasopressin receptor have been limited by its low abundance and by the loss of hormone binding activity upon solubilization (Guillon et al. 1980; Boer et al. 1983).

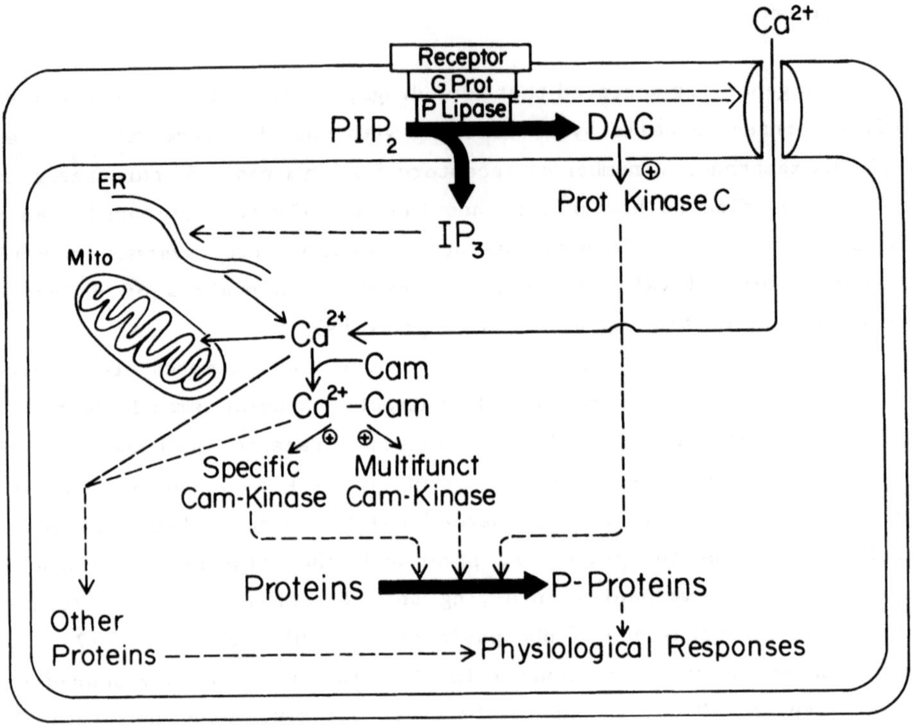

Figure 1: Mechanism by which Ca^{2+}-mobilizing agonists exert their effects. In order to define the properties of the hepatic receptor, we have developed a new purification procedure using a novel affinity column. Purification was facilitated by the use of a new and reproducible assay in which vasopressin binding was restored after solubilization, by incorporation of cholate extracts into phospholipid vesicles.

Rat liver plasma membranes (Prpic et al. 1984) were solubilized with 0.8% cholate. Prior to solubilization, the plasma membranes were incubated with 10 μM guanosine 5-(3-O-thio)triphosphate (GTPγS) in the presence of 10 mM $MgCl_2$. This treatment resulted in a dissociation of the receptor-G protein complex. Although this resulted in a low affinity state of the receptor, it facilitated purification. Since the vasopressin receptor loses its binding capacity upon solubilization, reconstitution of extracts with a mixture of bovine phosphatidylcholine and phosphatidylinositol (4:1) was performed on each fraction during all the steps of purification. As seen in Table I, approximately 90% of $[^3H]$vasopressin binding was recovered in the soluble extract.

Table I

PURIFICATION OF LIVER PLASMA MEMBRANE V_1-VASOPRESSIN RECEPTOR

STEP	PROTEIN (mg)	TOTAL ACTIVITY (pmoles)	SPECIFIC ACTIVITY (pmol/mg protein)	YIELD %
Membranes	260	661	2.5	--
Cholate extract	156	624	4	100
Q-Sepharose	5.85	340	58	54
Hydroxylapatite	0.45+	117	260	19
Affinity resin	0.0012+	20	16700	3

The results are from a preparation utilizing 40 rats. Where indicated (+), proteins were determined by the Protein Gold method.

To purify further the vasopressin receptor, the soluble extract was first applied to a heparin-Sepharose column. This step was included since the extract was also used as a source of PIP_2-phospholipase C and G proteins. The phospholipase bound to the column, whereas the vasopressin- and GTPγS-binding activities eluted in the flow-through. The flow-through fraction was applied to a Q-Sepharose column.

As shown in Figure 2, vasopressin binding eluted as a single peak at the beginning of the gradient. Most of the GTPγS-binding activity eluted later in the gradient. The separation of the G-proteins from the receptor was due to the pretreatment of the membranes with GTPγS, which did not affect the reconstitutable vasopressin binding activity.

Q-Sepharose chromatography yielded a 15-fold purification (Table I). When the membranes were not treated with GTPγS the majority of the vasopressin-binding activity eluted together with GTPYS-binding activity in a broader peak, yielding a lesser degree of purification.

The calculated specific binding activity of the eluate at this step was 58 pmol of [^{3}H]vasopressin/mg of protein (Table I). The peak of the vasopressin binding activity from Q-sepharose was further purified by hydroxylapatite chromatography (Figure 3). About 40% of the protein loaded onto the hydroxylapatite column remained after extensive washing. The column was then developed with a phosphate gradient. Most of the reconstitutable vasopressin binding elutes as a single peak at approximately 90 mM KH_2PO_4. At this stage the dissociation constant for vasopressin determined by equilibrium binding was 117 nM, as shown in Figure 4.

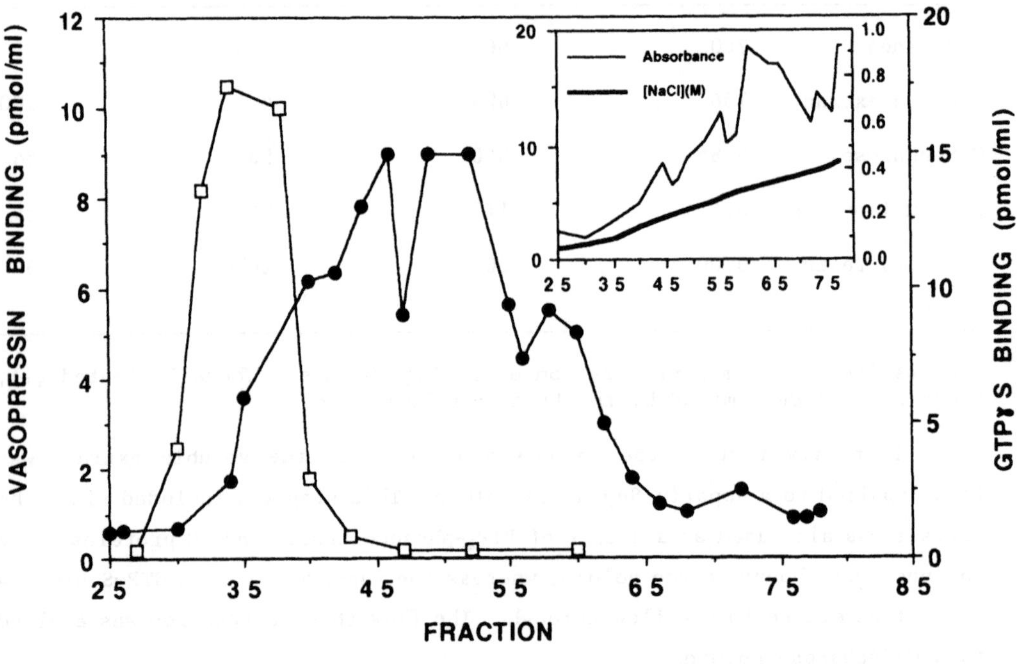

Figure 2: Q Sepharose chromatography of cholate solubilized plasma membranes. The heparin-Sepharose flow-through (120 ml) was applied to a 25 ml Q-Sepharose column. The column was washed with 25 mM Hepes buffer containing 0.5 mM DTT, 1 mM EGTA, 10 μg/ml aprotinin and leupeptin, 0.8% cholate, 10 mM $MgCl_2$ and 50 mM NaCl (Buffer A). Proteins were eluted with a gradient of 50-400 mM NaCl. Reconstitutable vasopressin binding and GTPγS binding were measured as described in Georgoussi et al., 1990. The inset shows the NaCl gradient (heavier line, units in M) and absorbance at 280 nM (lighter line).

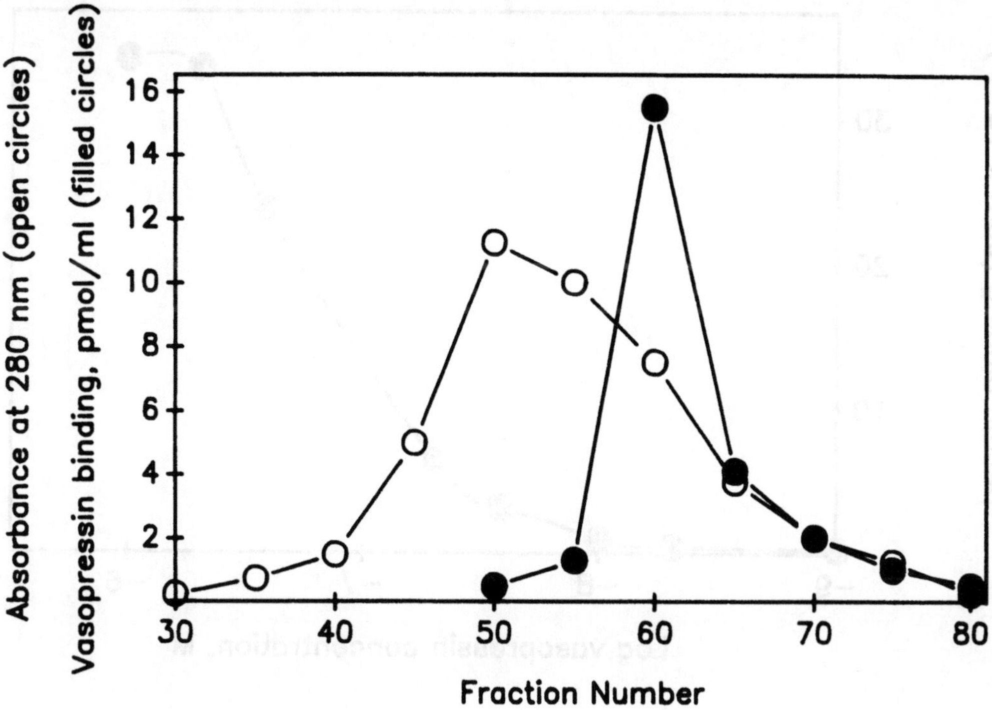

Figure 3: <u>Hydroxylapatite chromatography of Q Sepharose eluate.</u> An hydroxylapatite column (16 ml) was equilibrated with buffer A. The material eluted from Q sepharose was concentrated in a Centricell to 5 ml and diluted 1:1 with the above buffer without Mg^{2+} and NaCl. The protein was eluted with a linear gradient (35 ml) from 10 to 150 mM $KHPO_4$ at 0.5 ml/min. Fractions (1 ml) were collected and assayed for vasopressin binding.

The concentrated hydroxylapatite pool was further purified on an affinity column. The V1-vasopressin antagonist peptide, D-Tyr-Phe-Val-Asn-αAminobutyryl-Pro-Arg-GlyCOO⁻, was synthesized and coupled to Affigel, using ethyl-3-(3-dimethyl-aminopropyl) carbodiimide hydrochloride. Prior to chromatography, the resin was extensively washed with buffer A in the absence of detergent. The concentrated hydroxylapatite pool was incubated with a mixture of phosphatidylcholine and phosphatidylinositol, subjected to gel filtration, and mixed with the affinity matrix. Vasopressin binding was eluted with an increasing cholate gradient.

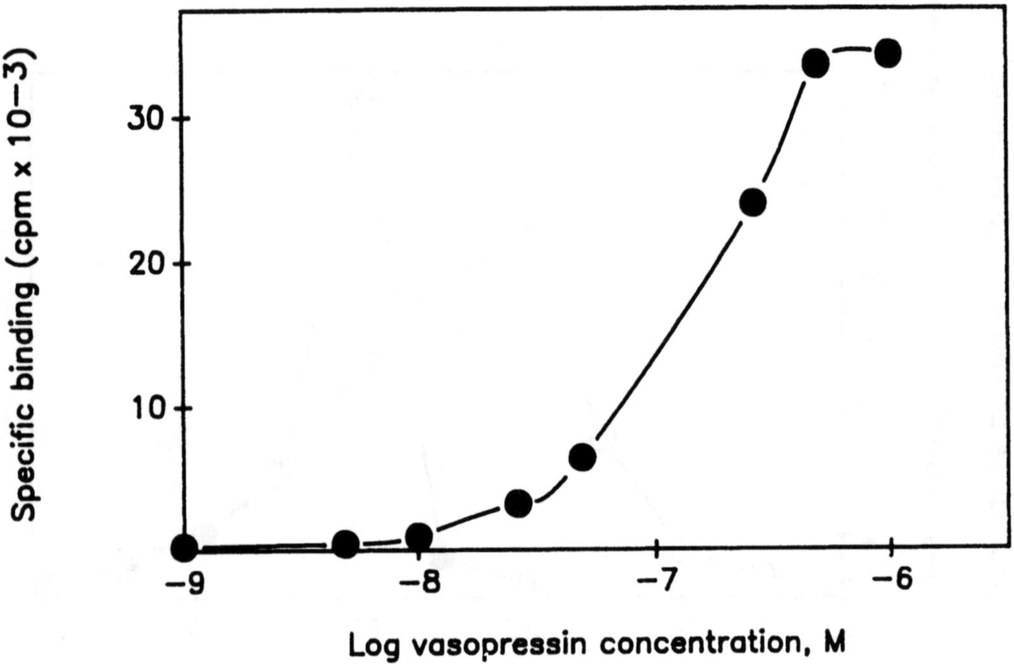

Figure 4: Equilibrium [^{3}H]vasopressin binding to partially purified reconstituted vasopressin receptor. The material eluted from the hydroxylapatite column was reconstituted with phosphatidylcholine and phosphatidylinositol (4:1) and, after gel filtration on sephadex G50 columns, the proteoliposomes were incubated with varying concentrations of radiolabeled vasopressin. Nonspecific binding was determined in the presence of 10^{-5} M unlabeled vasopressin. Bound ligand was separated from free by Millipore filtration on Whatman GF/F filters, which were washed with 12 ml cold buffer containing 50 mM Hepes, pH 7.5, 1 mM EGTA, 10 mM $MgCl_2$ and 100 mg/ml bovine serum albumin.

A sharp peak of binding activity was observed at 0.4% cholate (Figure 5). Although the receptor bound to the column in the presence of phosphatidylcholine alone, the omission of phosphatidylinositol from the applied mixture did not improve purification. Attempts to increase the lipid/protein ratio, or alter the detergent during the elution did not improve the degree of purification. This and other observations indicate that the reconstituted receptor binds to the affinity column via a specific interaction, which is impaired during the cholate gradient. It was not possible to measure any vasopressin binding in the absence of both lipids and detergents due to instability and possible denaturation of the receptor.

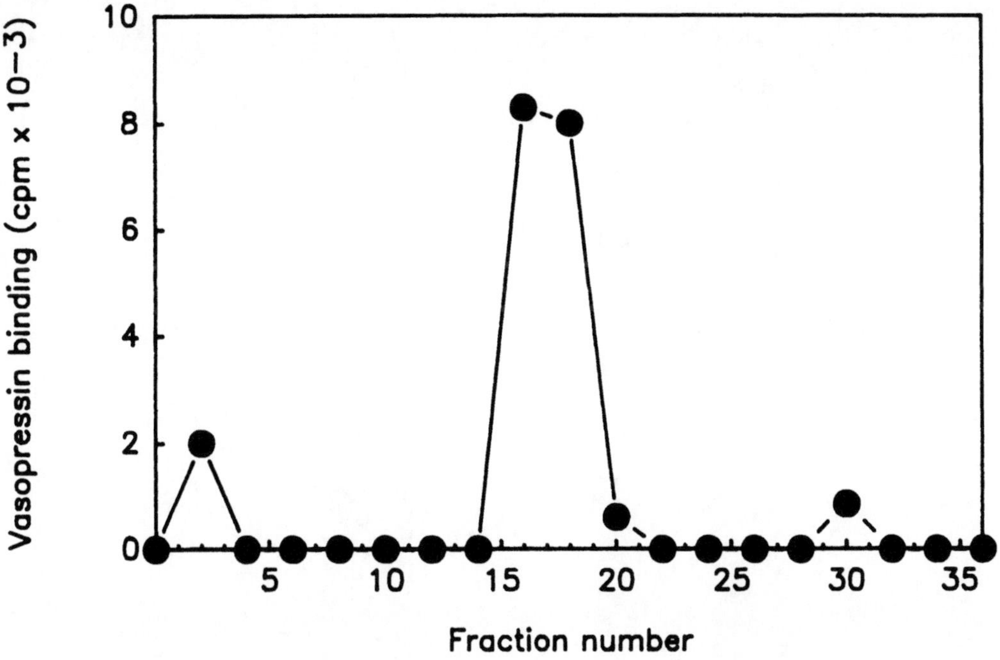

Figure 5: Affinity chromatography on the V1-vasopressin antagonist-agarose. The concentrated hydroxylapatite pool was reconstituted with phosphatidylcholine and phosphatidylinositol and mixed with affinity matrix after gel filtration on Sephadex G50. The column (10 ml) was equilibrated with 50 mM Hepes, 10 mM $MgCl_2$, 50 mM NaCl, 1 mM EGTA, 0.5 mM DTT, pH 7.4, and eluted with a 30 ml gradient of 0-1% cholate.

Figure 6 shows a silver-stained SDS gel of the peak of reconstituted vasopressin binding. A major band of Mr approx. 58,000 was present, demonstrating that the preparation was near-homogeneous.

The calculated specific binding activity of the peak fractions from the affinity chromatography was 16.7 nmol of [^{3}H]vasopressin bound/mg of protein. In order to demonstrate that the affinity-purified protein was the vasopressin receptor, cross linking of the purified material to [^{125}I] vasopressin in presence of disulfosuccinimidyl tartrate (1 mM) was performed. SDS-polyacrylamide gel electrophoresis and autoradiography showed that the Mr=58,000 protein bound [^{125}I] vasopressin. Addition of an excess of unlabelled vasopressin during crosslinking completely abolished the labelling (Figure 7).

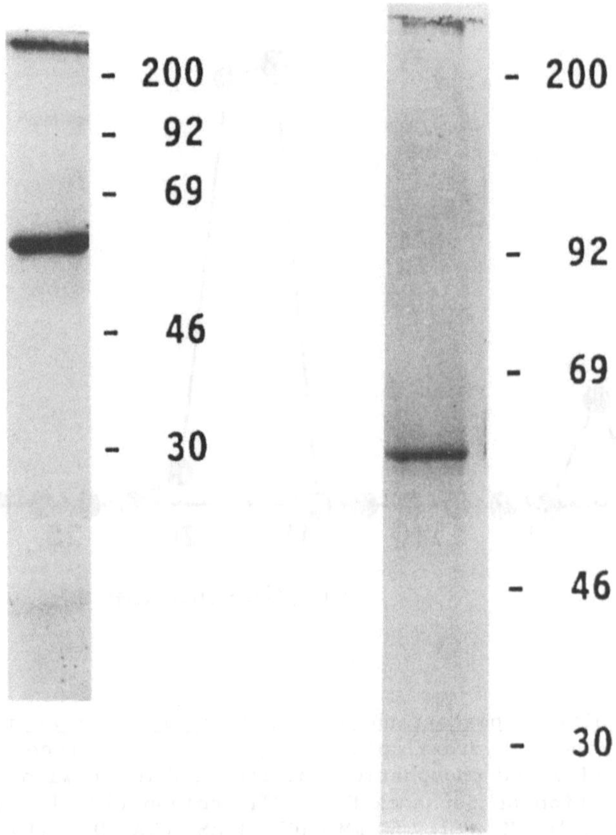

Figure 6: <u>SDS-polyacrylamide gel electrophoresis of the purified protein.</u> The peak binding fractions from affinity chromatography were precipitated with 10% trichloroacetic acid and gel electrophoresed. The gel on the left was run in 10% acrylamide and that on the right in 7.5%. Both were stained with silver (Merrill et al. 1982). Molecular weight markers (Amersham) were myosin, 200,000, phosphorylase b, 92,500; Bovine serum albumin, 69,000; ovalbumin, 46,000 and carbonic anhydrase 30,000.

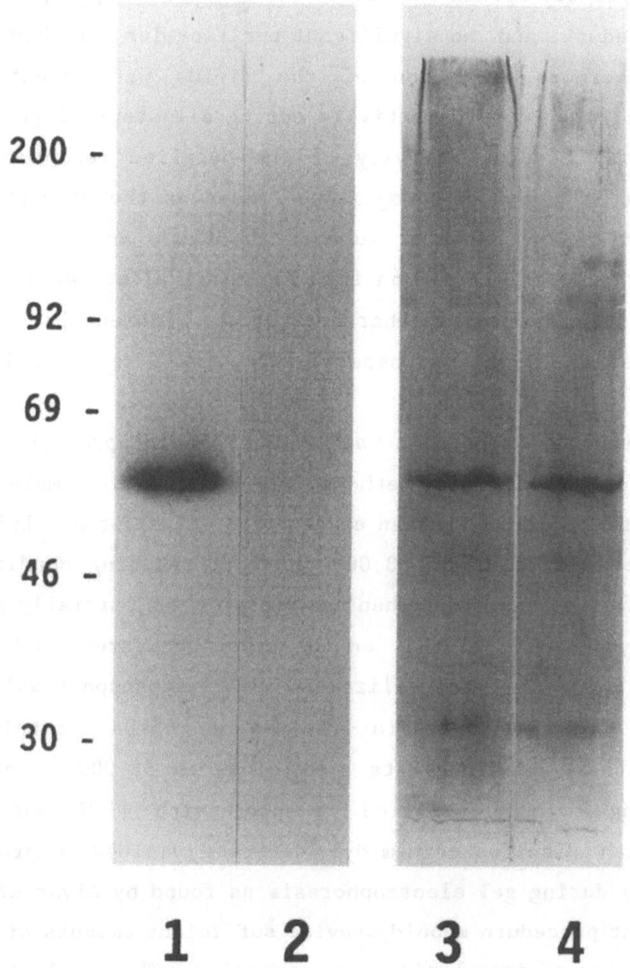

Figure 7: Cross-linking of [^{125}I]Vasopressin to purified preparation. An aliquot of eluate from the affinity column was reconstituted and cross linked. Unreacted [^{125}I]vasopressin was removed on PD10 columns prior to SDS-PAGE on 7.5% acrylamide gels. Lane 1, receptor crosslinked to 20 nM [^{125}I]vasopressin in absence of unlabelled ligand. Lane 2, vasopressin receptor crosslinked in the presence of 10 mM vasopressin. Lanes 3,4 are the corresponding silver stained gels. Molecular weight markers are as in Figure 6 legend.

The vasopressin-antagonist affinity chromatography played the major role in the present purification of the hepatic vasopressin receptor to apparent homogeneity. Since the procedure requires only two other chromatographic steps, it should facilitate further characterization of this protein. The success of the affinity column was due to the addition of phospholipids before the elution procedure, and no significant purification was observed in their absence. Furthermore, addition of the lipids was essential for the reconstitution of the receptor activity during all steps of purification.

The specific binding activity of the purified receptor preparation confirms that it is close to homogeneous, based on the assumption that the molecular weight of the receptor subunit is 58,000 and that there is one binding site per subunit. Based on its physiological properties (Michell et al. 1979, 1981) and binding characteristics (Fahrenholz et al. 1986; Gopalakrishnan et al. 1988) the hepatic vasopressin receptor is of the V1-subtype.

Previous reports of hepatic vasopressin binding proteins, as measured by crosslinking or photoaffinity methods, have given various molecular weights (Boer and Fahrenholz, 1985; Fishman et al. 1987; Aiyar et al. 1989). Fishman et al. (1987) claimed a Mr of 58,000 under nonreducing conditions and of 68,000 under reducing conditions when the receptor was partially purified from microsomes. Aiyar et al. (1989), on the other hand, reported that the rat liver plasma membranes solubilized with lysophosphatidylcholine and crosslinked to [^{125}I]-vasopressin yielded a 65kDa protein, although examination of their data suggests a Mr close to 58,000. In the present study, labelling of the purified receptor with [^{125}I]-vasopressin and crosslinking with disulfosuccinimidyl tartrate yielded a protein with a similar mobility during gel electrophoresis as found by Aiyar et al. (1989).

The present procedure should provide sufficient amounts of the receptor to begin the process of its cloning and sequencing. The availability of large amounts of the receptor should facilitate reconstitution studies with G proteins to examine the mechanisms of receptor effector coupling.

Acknowledgements

We would like to thank Prof. M. Manning, Medical College of Ohio, for suggesting the structure of the vasopressin antagonist affinity matrix. The help of Judy Childs in typing this manuscript is gratefully acknowledged.

REFERENCES

Aiyar, N., Bennet, F.C., Nambi, P., Valinski, W., Angioli, M., Minnich, M., and Crooke, S. T., 1989, Solubilization of rat liver vasopressin receptors as a complex with a guanine-nucleotide-binding protein and phosphoinositide-specific phospholipase C, Biochem.J. 261:63.

Berridge, M.J., 1987, Inositol trisphosphate and diacylglycerol: two interacting second messengers, Ann.Rev.Biochem. 56:759.

Berridge, M.J., and Irvine, R.F., 1984, Inositol trisphosphate, a novel second messenger in cellular signal transduction, Nature, 312:315.

Blackmore, P.F., Bocckino, S.B., Waynick, L. E., and Exton, J. H., 1985, Role of guanine nucleotide binding regulatory protein in the hydrolysis of hepatocyte phosphatidylinositol 4,5-bisphosphate by calcium mobilizing hormones and the control of cell Ca^{2+}: studies utilizing aluminum fluoride, J.Biol.Chem. 260:14477.

Boer, R., and Fahrenholz, F., 1985, Photoaffinity labeling of the V_1 vasopressin receptor in plasma membranes from rat liver, J.Biol.Chem. 260:15051.

Boer, R., Crause, P., and Fahrenholz, F., 1083, Solubilization of ligand-stabilized vasopressin receptors from plasma membranes of bovine kidney and rat liver, Biochem. Biophys.Res.Commun. 776:91.

Bojanic, D., and Fain, J. N., 1986, Guanine nucleotide regulation of [^{3}H] vasopressin binding to liver plasma membranes and solubilized receptors, Biochem.J. 240:361.

Casey, P. J., and Gilman, G. A., 1988, G protein involvement in receptor-effector coupling, J.Biol.Chem. 263:2577.

Fahrenholz, F., Kojro, E., Müller, M., Boer, R., Löhr, R., and Grzonka, Z., 1986, Iodinated photoreactive vasopressin antagonists: labelling of hepatic vasopressin receptor subunits, Eur.J.Biochem. 161:321

Fishman, J.B., Dickey, B. F., and Fine, R. E., 1987, Purification and characterization of the rat liver vasopressin (V_1) receptor, J.Biol.Chem. 262:14049.

Fitzgerald, T. J., Uhing, R. J., and Exton, J. H., 1986, Solubilization of the vasopressin receptor from rat liver plasma membranes, J.Biol.Chem. 261:16871.

Georgoussi, Z., Taylor, S. J., Bocckino, S. B., and Exton, J. H., 1990, Purification of the hepatic vasopressin receptor using a novel affinity column, Biochim.Biophys.Acta. in press.

Gopalakrishnan, V., McNeill, R. J., Sulakhe, V., and Triggle, C. R., 1988, Hepatic vasopressin receptor: Differential effects of divalent cations, guanine nucleotides, and N-ethylmaleimide on agonist and antagonist interactions with the V_1-subtype receptor, Endocrinology 123:922.

Guillon, G., Gouraud, P. O., Butlen, D., Cantau, B., and Jard, S., 1980, Size of vasopressin receptors from rat liver and kidney, Eur.J.Biochem. 111:287.

Jard, S., 1983, Vasopressin isoreceptors in mammals: relation to cyclic AMP-dependent and cyclic AMP-independent transduction mechanisms, Curr.Top.Membr.Transp. 18:255-285.

Lynch, C. J., Prpic, V., Blackmore, P. F., and Exton, J. H., 1986, Effect of islet-activating pertussis toxin on the binding characteristics of Ca^{2+}-mobilizing hormones and on agonist activation of phosphorylase in hepatocytes, Mol.Pharmacol. 29:196.

Michell, R. H., Kirk, C. J., and Billah, M.M., 1979, Hormonal stimulation of phosphatidylinositol breakdown, with particular reference to the hepatic effects of vasopressin, Biochem.Soc.Trans. 7:861-865.

Michell, R. H., Kirk, C. J., Jones, L. M., Downes, C. P., and Creba, J. A., 1981, Stimulation of inositol lipid metabolism that accompanies calcium mobilization in stimulated cells: Defined character and unanswered questions, Philos.Trans.R.Soc.Land.(Biol.) 296:123

Prpic, V., Green, K. C., Blackmore, P. F., and Exton, J. H., 1984, Vasopressin; angiotensin II and α_1-adrenergic-induced inhibition of Ca^{2+}-transport by rat liver plasma membrane vesicles, J.Biol.Chem. 259:1382.

Uhing, R. J., Prpic, V., Jiang, H., and Exton, J. H., 1986, Hormone stimulated polyphosphoinositide breakdown in rat liver plasma membranes: Roles of guanine nucleotides and calcium, J.Biol.Chem. 261:2140.

Weiss, E. R., Kelleher, D. J., Moon, C. W., Sopackar, S., Osawa, S., Heasley, L.E., and Johnson, G. L., 1988, Receptor activation of G proteins, FASEB J. 2:2841.

GENE TARGETING BY HOMOLOGOUS RECOMBINATION AS A TOOL TO STUDY THE BIOPHYSICAL ROLE OF THE INTERFERON-γ SIGNALLING PATHWAY

Wiljan Hendriks, Silvio Hemmi, Michel Aguet
Institute of Molecular Biology I
University of Zürich
Hönggerberg
CH-8093 Zürich
Switzerland

SUMMARY

Interferon-γ (IFN-γ) is a cytokine as much as an interferon and displays a wide variety of biological effects. IFN-γ exerts these effects through binding to a specific cell surface receptor, which is encoded by a single copy gene. To design novel models for investigating the physiological role of IFN-γ we plan to generate IFN-γ receptor deficient mice using recently established strategies for gene targeting by homologous recombination. In this chapter, the current strategies and achievements in gene targeting experiments are discussed and an outlook will be given on possible experiments with IFN-γ receptor deficient mice that could elucidate IFN-γ physiology.

INTRODUCTION

Interferons (IFNs) were discovered on the basis of their antiviral activity, but exert many additional biological effects as well, including inhibition of cell growth and modulation of various immune responses. The protein products of the vertebrate IFN genes form two structurally distinct families: type I IFNs comprise the related IFN-α and IFN-β molecules (which have around 25% identity at the amino acid level), while IFN-γ on its own makes up type II IFN (Weissmann and Weber, 1986; Pestka et al., 1987). IFN-γ, or immune interferon, is produced by activated T-cells and although the physiological significance is still not clear, it probably plays an important role as a macrophage activating factor and also in regulating antigen presentation (Trinchieri and Perussia, 1985). It is, for example, one of the few physiological inducers of major histocompatibility (MHC) class II antigens. Thus, IFN-γ is a cytokine as much as an interferon.

NATO ASI Series, Vol. H 52
Biological Signal Transduction
Edited by E. M. Ross and K. W. A. Wirtz

IFNs must first bind to specific cellular receptors in order to exert regulatory effects. Both IFN-α and β bind to a common type I receptor. In contrast, IFN-γ initiates its biological effects through binding to its own specific cell surface receptor (Langer and Pestka, 1989). It enters the cell via typical endocytotic pathways (Filgueira et al., 1989), yet the second messengers involved in signal transduction are not known. Consequently, the molecular basis for the sometimes synergistic and sometimes antagonistic effects of the two types of interferon (Pestka et al., 1987) is widely obscure. A similar situation is faced for the cytokine-like effects of IFN-γ. From studies of cytokine biology it has emerged that pleiotropy and redundancy is the overriding principle (Paul, 1989). For example, some antiproliferative and immunoregulatory effects of IFNs and tumor necrosis factor are overlapping (Beutler and Cerami, 1987). Here again the molecular mechanisms are not known.

From all this it follows that in vitro studies on the effects of IFN-γ have only limited value for a description of its biophysical role in vivo. Although recombinant IFN-γ as well as anti-IFN-γ antibodies have proved very valuable for modulating the IFN system in a variety of experimental mouse models (DeMaeyer and DeMaeyer-Guignard, 1988), these tools are inadequate for exploring certain aspects of the physiology of the IFN system. Administration of anti-IFN antibodies over prolonged periods for example is impracticable, as is neutralization of a putative IFN activity during embryonic development. Beside the case of Mx locus which is inactivated in certain mouse strains resulting in a specific susceptibility of these animals to influenza virus infection (Staeheli et al., 1986), naturally occurring deficiencies of the IFN system have not been described. To address the numerous questions on the physiological involvement of the IFN system in embryonic development, cell growth and differentiation control, and maturation of the immune system, we plan to generate mice that are lacking a functional IFN-γ receptor gene and are thus insensitive to the cognate IFN-γ. This will be achieved by embryonic stem (ES) cell mediated germ-line transmission of a targeted gene disruption (Capecchi, 1989[a]; Capecchi, 1989[b]; Porter, 1989).

GENE TARGETING BY HOMOLOGOUS RECOMBINATION

Much information on gene function and regulation in mice could be obtained by studying naturally occurring mutants. To generate additional and potentially interesting mutant mice, methods have been developed to introduce cloned genes into the mouse genome by micro-injection or infection of oocytes. The resulting transgenic mice have indeed provided a wealth of information about mammalian gene function and regulation (Hanahan, 1989). A limitation of the transgenic mice method is that it predominantly results in "gain

Table 1. Genes targeted succesfully in ES cells

Gene	Reference
HPRT	Thomas and Capecchi, 1987
	Doetschman et al., 1987
	Doetschman et al., 1988
int-2	Mansour et al., 1988
Hox 1.1	Zimmer and Gruss, 1989
c-fos	Johnson et al., 1989
Adipsin	Johnson et al., 1989
aP2	Johnson et al., 1989
β2M	Zijlstra et al., 1989
	Koller and Smithies, 1989
n-myc	Charron et al., 1990
c-abl	Schwartzberg et al., 1990
IGF-II	De Chiara et al., 1990
En-2	Joyner et al., 1990
int-1	Thomas and Capecchi, 1990
	McMahon and Bradley, 1990

of function" mutants. Occasionally a "loss of function" mutant is obtained, but this is the result of a random integration into and subsequent inactivation of a gene locus which is not related to the transgene. Nevertheless, such accidental "loss of function" mutants have been gratefully studied and have provided interesting novelties (Jaenisch, 1988). The chance creating a "loss of function" mutant for a specific gene by inactivation through random integration is much too low (on the order of 10^{-6}). Also the chance to create such mutant mice by injecting a construct in fertilized eggs and then to screen transgenic mice for homologous recombination (initially estimated to be around 10^{-3}) was too low to encourage such attempts. Nevertheless, this strategy has been used succesfully to correct a deletion in an MHC class II gene (Brinster et al., 1989).

Fortunately, another approach has become available. Along with the work on germ-line transformation of mice by DNA injection into zygotes, cell lines from embryonic stem (ES) cells were established (Evans and Kaufman, 1981; Martin, 1981). These ES cell lines can be cultured in vitro and will resume normal embryonic development following microinjection into a host embryo and subsequent incubation in a foster mother. The ES cells can contribute to somatic tissues as well as the germ cell lineage in the chimeric animals thus formed (Bradley et al., 1984). It was shown that also after in vitro cell transfection and selection procedures ES cells could still contribute to the germ-line, thereby demonstrating that they can be used as vehicles for transgenesis (Gossler et al., 1986). It follows that a plausible experimental strategy to create site-specific (targeted)

Table 2. Genes, succesfully targeted in ES cells, for which germ-line chimeric mice have been obtained

Gene	Reference
HPRT	Thompson et al., 1989
	Koller et al., 1989
c-abl	Schwartzberg et al., 1989
β2M	Zijlstra et al., 1989
	Koller and Smithies, 1989
IGF-II	De Chiara et al., 1990
int-1	Thomas and Capecchi, 1990
	McMahon and Bradley, 1990

alterations in the mouse genome would be to introduce homologous recombination constructs into ES cells, select in vitro the rare cells in which homologous recombination has occurred and use them to generate mouse strains heterozygous for the altered gene. Such mice can be interbred to produce offspring homozygous for the targeted genetic change and the functional consequence of the genetic alteration studied.

Initially, the hypoxanthine phosphoribosyl transferase gene (HPRT) was used for model studies because one can select directly for the targeting effect (Thomas and Capecchi, 1987; Doetschman et al., 1987; Doetschman et al., 1988). Indeed, the HPRT targeted ES cells could still give rise to germ-line chimeras (Thompson et al., 1989; Koller et al., 1989), thereby clearing the path for gene targeting by homologous recombination to introduce desired changes in the murine genome. Various genes have now been targeted (Table 1). In only a few cases was germline transmission of the targeted gene reported (Table 2), illustrating that still the most difficult step in targeting experiments is the reconstruction of mice. Even if germ-line chimeras are obtained, it might be that the desired disruption cannot be studied because no altered phenotype is obtained even in the homozygous state or because it demonstrates to be lethal already at the earliest stages of embryogenesis. Up to now only studies on mice homozygous for inactivated alleles of int-1 (Thomas and Capecchi, 1990; McMahon and Bradley, 1990) and β2-microglobulin (Zijlstra et al., 1990; Koller et al.,1990) have been reported. In the case of IGF-II, mice heterozygous for the targeted allele already showed a phenotypic deviation from wildtype mice (De Chiara et al., 1990). Nevertheless, this limited number of reports justify the high expectations and it is exciting to anticipate the wealth of information that will emerge from gene targeting experiments in the coming decade.

SELECTION STRATEGIES

The search for ES cell clones representing a successful targeted event can be much facilitated by appropriate selection. In selection strategies the transfecting DNA rescues specifically the homologous recombinants. The power of direct selection is most convincingly shown for HPRT (Table 1) but of course in the vast majority of cases a suitable selective medium cannot be compiled. Instead of selection based on the targeted gene function, one can use selectable genes strategically positioned in the transfecting construct. The principle in such a conditional positive selection is to make the activity of the selectable marker dependent on homologous recombination. For instance, if the gene of interest is expressed in ES cells, a construct with a strategically placed promoter-less selectable gene will after transfection and selection predominantly yield homologous recombinants. This because the chance is low that random integration of the construct will happen just downstream of an active promoter. This approach has been applied succesfully in ES cells for n-myc and c-abl (Charron et al., 1990; Schwartzberg et al., 1990). For genes that are not expressed in ES cells an alternative conditional positive selection can be applied. To limit the detection of nonhomologous integration events, a selectable gene can be used that does not contain a polyadenylation signal. Homologous recombination will combine the selectable gene with the target gene polyadenylation signal, while random integration will not always occur in the neighborhood of a suitable downstream polyadenylation signal. A two-fold enrichment was obtained in this way for targeting the En-2 gene (Joyner et al., 1989).

In positive and negative selection (PNS) two dominant selection-response genes are used: a gene to disrupt the target gene and to select for stably transfected cells, and a negatively selectable gene which is placed outside the region of homology with the target gene. Homologous recombination will result in the loss of the negative selectable gene and resulting clones are thus resistant to both selection procedures. Mansour et al. (1988), who developed the positive and negative selection, used the neomycin (neo) resistance gene for positive and the herpes simplex virus thymidine kinase (TK) gene for negative selection. The enrichment for homologous recombination events by PNS has been convincingly demonstrated in the cases of HPRT, int-2, int-1, IGF-II, and c-fos, Adipsin and aP2 (Mansour et al., 1988; Thomas and Capecchi, 1990; McMahon and Bradley, 1990; De Chiara et al., 1990; Johnson et al., 1989). Nevertheless, it should be mentioned that the PNS approach not always results in a considerable enrichment. First, it can be that there is random breakage of the DNA on transfection, resulting in the loss of flanking TK sequences in the illegitimate recombinants (Zijlstra et al., 1989). In addition, one should bear in mind that additional integrations of the targeting vector in a homologous recombinant will make it sensitive for the negative selection. Thus, PNS has the advantage that only cell lines are obtained that are pure homologous recombinants without additional integrations, but it can result in a low probability of finding the targeting event if multiple integrations are common.

SCREENING STRATEGIES

Following selection for stably transfected cells, eventually in combination with selection procedures described above, homologous recombinants must be identified in the population of stable transfectants of which the majority are still illegitimate recombinants. Southern blotting (Southern, 1975) is the usual method to confirm the homologous recombination event in obtained clones. It is also suitable to be used as a screening strategy for the desired genotype on individual clones or small pools of clones. So far it has only been used for screening in addition to a conditional selection procedure (e.g. Schwartzberg et al., 1989; Johnson et al., 1989; Thomas and Capecchi, 1990). Only in the case of the targeting of MHC class II Eα gene by DNA microinjected into mouse eggs (Brinster et al., 1989) Southern blot analysis was the sole selection/screening approach.

More common is the use of the polymerase chain reaction (Saiki et al., 1988; White et al., 1989) as a very sensitive assay, especially in situations where no prior conditional selection is or can be applied (Kim and Smithies, 1988; Udy et al., 1990). A pair of oligonucleotides is chosen that will amplify a portion of the modified genotype that is absent in both the original genotype and the targeting construct. One sequence is chosen from a region in the genome close to the target but not present in the targeting construct. The other sequence must be absent in the unmodified genome. Not only the extreme sensitivity makes the PCR assay attractive, but also its general applicability. In theory, targeting to any part of the genome that is available for recombination can be screened for in this way. PCR has been used succesfully to isolate ES cell clones targeted at the Hox1.1 (Zimmer and Gruss, 1989), En-2 (Joyner et al., 1989) and β2M (Zijlstra et al., 1989; Koller and Smithies, 1989) genes, and, in combination with positive and negative selection, for the int-1 (McMahon and Bradley, 1990) and IGF-II (De Chiara et al., 1990) genes.

As mentioned previously, for individual clones the correct targeting event is confirmed by Southern blot analysis of genomic DNA digests. Also the number of integrated targeting vectors can be determined in this way. Additional copies of the targeting vector can represent random integration events somewhere else in the genome that by chance resulted in the inactivation of other genes. Therefore, to be able to ascribe unambiguously eventual changes in the phenotype to the targeted gene, additional copies of the targeting vector should not be present. Nevertheless, targeted ES cells with multiple integrations of the vector can still be used to generate mice because there is a fair chance that the additional integrated copies will not cosegregate with the targeted allele in subsequent mouse generations and thus can be removed by breeding.

TARGETING THE IFN-γ RECEPTOR GENE

To generate mice that are deficient in IFN-γ signal transduction, we plan to inactivate the IFN-γ receptor gene by homologous recombination in ES cells. An advantage for ablating the receptor instead of the ligand gene (IFN-γ is also encoded by a single copy gene) is that in the obtained receptor-less mice one can reconstitute IFN-γ sensitivity compartmentwise through tissue specific transgenic expression of the IFN-γ receptor gene or grafting specific tissues or cell lineages. Transgenic expression of the IFN-γ gene would probably result in a systemic endocrine action. cDNAs encoding the human and murine IFN-γ receptor have been cloned recently (Aguet et al., 1988; Hemmi et al., 1989; Gray et al., 1989; Kumar et al., 1989). Using murine IFN-γ receptor cDNA probes, from a mouse genomic library clones were isolated that represent a region of 28 kb containing the complete single copy murine IFN-γ receptor gene. Like its human counterpart the murine IFN-γ receptor gene consists of seven exons and the membrane spanning region is encoded by exon VI (S.H. and M.A., manuscript in preparation).

IFN-γ receptor mRNA is not present in ES cells, thus a conditional positive selection based on the receptor gene promoter is not feasible. Also, the application of conditional selection by using a neo construct without a polyadenylation signal turned out not to be useful. First, no difference in number of neo resistant colonies was observed if targeting vectors with and without a poly(A) signal in the neo cassette were compared. Second, only in the experiments with the poly(A) signal-containing vector homologous recombinants could be detected. This can be explained by assuming that transcripts extending from the TK promoter to the IFN-γ receptor gene poly(A) signal for some reason can not be processed or transported properly or are instable.

For designing the targeting vector it was decided to disrupt the gene in the part encoding the extracellular domain of the receptor. In this way IFN-γ receptor transcripts, if stable and translatable at all, would at most result in a truncated receptor that is secreted and will not be able to participate in signal transduction. Most conveniently, a 2.1 kb Hind III fragment containing exons V and VI had a unique Aat II site present within exon V. At this Aat II site the neomycin phosphotransferase gene driven by the TK promoter was introduced as a selectable marker, thereby disrupting the exon (Fig. 1).

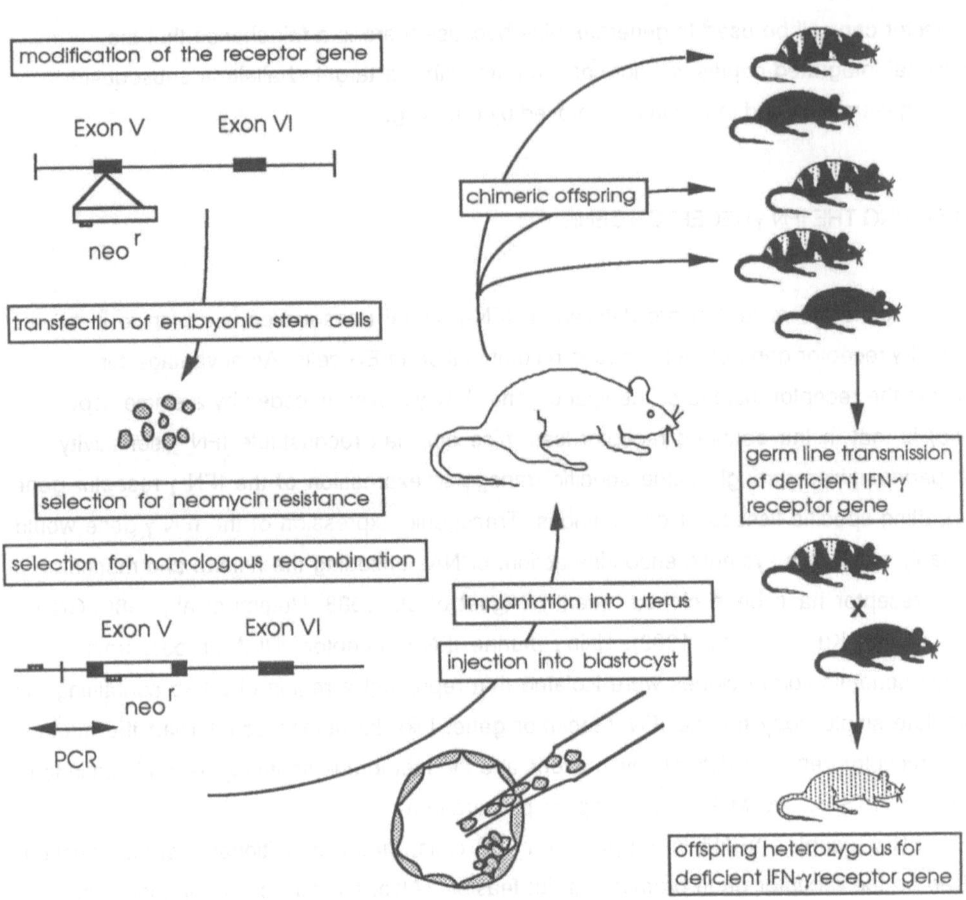

Figure 1. Strategy to generate IFN-γ receptor deficient mice. A targeting vector was constructed in which the IFN-γ receptor gene is disrupted by a TK promoter-driven neo gene which is inserted into exon V. Mouse ES cells are transfected by electroporation with the linear targeting vector. Selection with the drug G418 results in neo-resistant colonies. Identification of the targeted cells is done by PCR. One oligonucleotide primer binding site is chosen within the neo gene and a second one in the IFN-γ receptor gene just upstream of the fragment used for creating the targeting vector. Cell lines that appear positive for homologous recombination are in addition subjected to Southern blot analysis of genomic DNA digests to verify correct targeted gene disruption. The ES cell lines containing one inactive IFN-γ receptor allele are then microinjected into the blastocoel cavity of a 3,5 day mouse embryo. The blastocyst is then surgically transferred into the uterus of a pseudopregnant mouse. Since the ES cell lines are derived from male mice homozygous for the agouti locus (129 strain) and the blastocyst donor mice (C57Bl/6J) are nonagouti black, contribution by the injected line to the skin of the newborn mouse can be scored on the basis of agouti hair in the black background. Mating of the obtained chimeric male mice with nonagouti females should then reveal if germ-line transmission of the targeted ES cell genome occurs. If so, subsequent breeding should eventually result in mice homozygous for a disrupted IFN-γ receptor gene.

To identify homologous recombinants in a population of neo resistant colonies PCR-based detection was used. One oligonucleotide primer was chosen in the TK promoter part of the neo cassette, and the other primer in the receptor gene sequence just upstream of the 5' Hind III site of the fragment used for creating the targeting vector. To establish optimal conditions for the PCR and to determine sensitivity, this primer pair was tested on mixes of positive and negative control cell lines. As a negative control line, cells containing randomly integrated targeting vector (thus having both primer binding sites in the genome) were used. To create a positive control cell line, cells were transfected with a construct identical to the targeting vector but containing additional 5' flanking receptor gene sequences. Tests indicated that the correct fragment was detected on Southern blots following amplification of only 30 positive cell genomes (containing a single copy of the positive template) in the presence of 10^5 negative cell genomes.

Transfection of 20 million ES cells by electroporation (280 V, 500 µF in a Biorad Genepulser) with 20 µg of linearized targeting vector DNA usually yielded about 1000 neo resistant colonies. The frequency of gene targeting was found to be 1/250 neo resistant colonies. Both TG2a (Hooper et al., 1987) and D3 (Doetschman et al., 1985) ES cell lines were used and from each several independent cell lines were established that carry the homologous recombination in one allele as judged by PCR and by genomic Southern blot analysis. Currently, we are in the process of injecting these lines into blastocysts and subsequently reimplanting into pseudopregnant foster mice to generate chimeric offspring. Breeding should then reveal if the pluripotency of the ES cell lines was maintained during the various manipulations.

FUTURE PROSPECTS

It can merely be speculated on the phenotype to be expected from inbreeding offspring heterozygous for the disrupted IFN-γ receptor gene. Antiviral defense and the immune response will probably be affected, probably resulting in a pronounced immunodeficiency. Macrophage function and antigen presentation notably are expected to be impaired, but the defect might also result in abnormal maturation of the immune system or reflect unforeseen roles of IFN-γ in ontogeny. For type I IFNs evidence for a developmental role has been provided in studies on in vitro differentiation of teratocarcinoma cells (Burke et al., 1978) and by the discovery in sheep of an embryonic type I IFN believed to be the causal agent for maternal recognition of pregnancy (Imakawa et al., 1987). Both type I and type II IFNs have recently been identified in cultures of early porcine embryos (Lefèvre et al., 1990). There is, of course, a chance that the IFN-γ receptor gene deletion is lethal. In that case no offspring homozygous for the disrupted IFN-γ receptor gene will be obtained and investigations then have to focus on early embryogenesis to determine at which stage development is affected. If viable and fertile offspring homozygous for IFN-γ receptor deficiency are born, studies can focus on the effects of IFN-γ on cell growth and differentiation, and on immunomodulatory actions of IFN-γ.

Both type I and type II IFNs exert a direct inhibitory effect on in vitro cell growth of a variety of normal and tumor cells (DeMayer and DeMayer-Guignard, 1988). The action of several growth factors can be inhibited by IFNs (Taylor-Papadimitriou et al., 1981), suggesting that induction of IFNs by growth factors could play a role as a feedback signal. Colony stimulating factor 1 (CSF-1) can induce IFNs while addition of anti-IFN antibodies to CSF-1 stimulated cultures enhances the proliferative response (Moore et al., 1984). Likewise, myelosuppression in vivo by both murine type I and type II IFNs can be prevented by CSF-1 (Klimpel et al., 1982). These observations suggest that IFNs may act as natural regulators of cell growth and differentiation. To verify this hypothesis for IFN-γ, the receptor deficient mice can be carefully examined for abnormalities at various stages of development. Not only morphometric studies but also immunohistochemical and flow-cytometric analyses of cell lineage-specific differentiation markers of hemopoetic and lymphatic tissues can be used to detect alterations and deviations from normal development.

Among the multitude of reported immunoregulatory effects of IFN-γ, stimulation of cytotoxic lymphocytes and activation of macrophages has been most extensively documented (DeMayer and DeMayer-Guignard, 1988) but the physiological relevance of these effects remains uncertain. The most provoking immunoregulatory effect is the widespread enhancement and induction of MHC antigens. IFN-γ not only enhances MHC antigen expression but can induce de novo expression of MHC class II antigens on a variety of cells and tissues (Skoskiewicz et al., 1985). It is known that quantitative differences in expression of MHC antigens influences immune recognition (Unanue and Allen, 1987). Most interestingly, transgenic ectopic expression of IFN-γ in pancreatic islets can lead to autoimmune sensitization and destruction of islet cells (Sarvetnick et al., 1990) and it is believed that this recognition process is initiated by the aberrant induction of MHC class II antigens. For the study of immunoregulatory processes in the IFN-γ receptor deficient mice, the previously mentioned descriptive experiments using various lymphoid cell markers and MHC antigens will provide a first basis. If maturation defects were observed, these might be indicative of functional abnormalities. In addition, virus neutralization assays and cytotoxicity tests using virus-infected target cells should help to assess a possible impairment of both humoral and cellular immune response to virus infections in IFN-γ receptor deficient mice. Also, cells or cell lines can be isolated from the mice and in vitro cell functions such as cytotoxicity of macrophages and cytotoxic lymphocytes can be assayed.

From the abrogation of immunotolerance by constitutive ectopic expression of IFN-γ (Sarvetnick et al., 1990) one might argue that IFN-γ unresponsiveness results in immunosuppression. Accordingly, the fate of allografts in the receptor deficient mice will be interesting, as is the susceptibility to autoimmune disorders and parasitic infections. Also, the frequency of spontaneous tumor formation and the role of the IFN system in host defense against transplantable tumors (Gresser et al, 1988) will be interesting to study.

This short overview of possible experiments to establish the biophysical role of IFN-γ using IFN-γ receptor deficient mice obtained by gene targeting via homologous recombination can of course not be complete. Nevertheless, it shows the powerful potentials of gene ablation as a method to study complex signal transduction pathways. Eventually, as more and more mice carrying an inactivated cytokine (receptor) gene become available, intercrossing of mice carrying different defects will perhaps make it possible to unravel the redundancy and pleiotropy in the cytokine network.

ACKNOWLEDGEMENTS

We would like to thank Drs. D. Melton and R. Kemler for providing ES cell lines and W. Vree Egberts for typing the manuscript. This work was supported in part by the Kanton of Zürich and The Swiss National Science Foundation. W.H. is a recipient of a long-term fellowship of the European Molecular Biology Organization (EMBO).

REFERENCES

Aguet M, Dembic Z, Merlin G (1988) Molecular cloning and expression of the human interferon-γ receptor. Cell 55: 273-280

Beutler B, Cerami A (1987) Cachectin: More than a tumor necrosis factor. N. Engl. J. Med. 316: 379-385

Bradley A, Evans MJ, Kaufman MH, Robertson EJ (1984) Formation of germ-line chimaeras from embryo-derived teratocarcinoma cell lines. Nature 309: 255-256

Brinster RL, Braun RE, Lo D, Avarbock MR, Oram F, Palmiter RD (1989) Targeted correction of a major histocompatibility class II Eα gene by DNA microinjected into mouse eggs. Proc. Natl. Acad. Sci. USA 86: 7087-7091

Burke DC, Graham CF, Lehman JM (1978) Appearance of interferon inducibility and sensitivity during differentiation of murine teratocarcinoma cells in vitro. Cell 13: 243-248

Capecchi MR (1989[a]) The new mouse genetics: altering the genome by gene targeting. Trends Genet. 5: 70-76

Capecchi MR (1989b) Altering the genome by homologous recombination. Science 244: 1288-1292

Charron J, Malynn BA, Robertson EJ, Goff SP, Alt FW (1990) High-frequency disruption of the n-myc gene in embryonic stem and Pre-B cell lines by homologous recombination. Mol. Cell. Biol. 10: 1799-1804

De Chiara TM, Efstratiadis A, Robertson EJ (1990) A growth-deficiency phenotype in heterozygous mice carrying an insulin-like growth factor II gene disrupted by targeting. Nature 345: 78-80

DeMayer E, DeMayer-Guignard J (1988) Interferons and other regulatory cytokines. Wiley New York

Doetschman TC, Eistetter H, Katz M, Schmidt W, Kemler R (1985) The in vitro development of blastocyst-derived embryonic stem cell lines: formation of visceral yolk sac, blood islands and myocardium. J. Embryol. Exp. Morphol. 87: 27-45

Doetschman T, Gregg RG, Maeda N, Hooper M, Thompson DW, Smithies O (1987) Targeted correction of a mutant HPRT gene in mouse embryonic stem cells. Nature 330: 576-578

Doetschman T, Maeda N, Smithies O (1988) Targeted mutation of the HPRT gene in mouse embryonic stem cells. Proc. Natl. Acad. Si. USA 85: 8583-8587

Evans MJ, Kaufman MH (1981) Establishment in culture of pluripotential cells from mouse embryos. Nature 292: 154-156

Filgueira L, Groscurth P, Aguet M (1989) Binding and internalization of gold-labeled IFN-γ by human raji cells. J. Immunol. 142: 3436-3439

Gossler A, Doetschman T, Korn R, Serfling E, Kemler R (1986) Transgenesis by means of blastocyst-derived embryonic stem cell lines. Proc. Natl. Acad. Sci. USA 83: 9065-9069

Gray PW, Leong S, Fennie EH, Farrar MA, Pingel JT, Fernandez-Luna J, Schreiber RD (1989) Cloning and expression of the cDNA for the murine interferon γ receptor. Proc. Natl. Acad. Sci. USA 86: 8497-8501

Gresser I, Maury C, Vignaux F, Haller O, Belardelli F, Tovey MG (1988) Antibody to mouse interferon α/β abrogates resistance to the multiplication of Friend erythroleukemia cells in the livers of allogeneic mice. J. Exp. Med. 168: 1271-1291

Hanahan D (1989) Transgenic mice as probes into complex systems. Science 246: 1265-1275

Hemmi S, Peghini P, Metzler M, Merlin G, Dembic Z, Aguet M (1989) Cloning of murine interferon-γ receptor cDNA: Expression in human cells mediates high-affinity binding but is not sufficient to confer sensitivity to murine interferon-γ. Proc. Natl. Acad. Sci. USA 86: 9901-9905

Hooper M, Hardy K, Handyside A, Hunter S, Monk M (1987) HPRT-deficient (Lesh-Nyhan) mouse embryos derived from germline colonization by cultured cells. Nature 326: 292-295

Imakawa K, Anthony RV, Kazemi M, Marotti KR, Polites HG, Roberts RM (1987) Interferon-like sequence of ovine trophoblast protein secreted by embryonic trophectoderm. Nature 330: 377-379

Jaenisch R (1988) Transgenic animals. Science 240: 1468-1474

Johnson RS, Sheng M, Greenberg ME, Kolodner RD, Papaioannou VE, Spiegelman BM (1989) Targeting of nonexpressed genes in embryonic stem cells via homologous recombination. Science 245: 1234-1236

Joyner AL, Skarnes WC, Rossant J (1989) Production of a mutation in mouse En-2 gene by homologous recombination in embryonic stem cells. Nature 338: 153-156

Kim H-S, Smithies O (1988) Recombinant fragment assay for gene targeting based on the polymerase chain reaction. Nucl. Acids Res. 16: 8887-8903

Klimpel GR, FleischmannWR Jr, Klimpel KD (1982) γ-Interferon (IFN-γ) and IFN α/β suppress murine myeloid colony formation (CFU-C): Magnitude of suppression is dependent upon level of colony-stimulating factor (CSF). J. Immunol. 129: 76-80

Koller BH, Hagemann LJ, Doetschman T, Hagaman JR, Huang S, Williams PJ, First NL, Maeda N, Smithies O (1989) Germ-line transmission of a planned alteration made in a hypoxanthine phosphoribosyl transferase gene by homologous recombination in embryonic stem cells. Proc. Natl. Acad. Sci. USA 86: 8927-8931

Koller BH, Smithies O (1989) Inactivating the β2-microglobulin locus in mouse embryonic stem cells by homologous recombination. Proc. Natl. Acad. Sci. USA 86: 8932-8935

Koller BH, Marrack P, Kappler JW, Smithies O (1990) Normal development of mice deficient in β2M, MHC class I proteins and $CD8^+$ T cells. Science 248: 1227-1230

Kumar CS, Muthukumaran G, Frost LJ, Noe M, Ahn YH, Mariano TM, Pestka S (1989) Molecular characterization of the murine interferon γ receptor cDNA. J. Biol.Chem. 264: 17939-17946

Langer JA, Pestka S (1988) Interferon receptors. Immunol. Today 9: 393-400

Lefèvre F, Mège D, Martinat-Botté F, Guillomot M, Zouari K, La Bonnardière C (1990) Natural expression of gamma and alpha interferons by the porcine trophoblast in early gestation. J. Interferon Res. 10 Suppl. 1: 79

Mansour SL, Thomas KR, Capecchi MR (1988) Disruption of the proto-oncogene int-2 in mouse embryo-derived stem cells: A general strategy for targeting mutations to non-selectable genes. Nature 336: 348-352

Martin G (1981) Isolation of a pluripotent cell line from early mouse embryos cultured in medium conditioned by teratocarcinoma cells. Proc. Natl. Acad. Sci. USA 78: 7634-7638

McMahon AP, Bradley A (1990) The wnt-1 (int-1) proto-oncogene is required for development of a large region of the mouse brain. Cell 62: 1073-1085

Moore RN, Larsen HS, Horohov DW, Rouse BT (1984) Endogenous regulation of macrophage proliferative expansion by colony-stimulating factor-induced interferon. Science 223: 178-181

Paul WE (1989) Pleiotropy and redundancy: T cell-derived lymphokines in the immune response. Cell 57: 521-524

Pestka S, Langer JA, Zoon KC, Samuel CE (1987) Interferons and their actions. Ann. Rev. Biochem. 56: 727-777

Porter ACG (1989) Designer Genomes. Technique 1: 53-65

Saiki RK, Gelfland DH, Stoffel S, Scharf SJ, Higuchi R, Horn GT, Mullis KB, Erlich HA (1988) Primer-directed enzymatic amplification of DNA with a thermostable DNA polymerase. Science 239: 487-491

Sarvetnick N, Shizuru J, Liggit D, Martin L, McIntyre B, Gregory A, Parslow T, Stewart T (1990) Loss of pancreatic islet tolerance induced by β-cell expression of interferon-γ. Nature 346: 844-847

Schwartzberg PL, Goff SP, Robertson EJ (1989) Germ-line transmission of a c-abl mutation produced by targeted gene disruption in ES cells. Science 246: 799-803

Schwartzberg PL, Robertson EJ, Goff SP (1990) Targeted gene disruption of the endogenous c-abl locus by homologous recombination with DNA encoding a selectable fusion protein. Proc. Natl. Acad. Sci. USA 87: 3210-3214

Skoskiewicz MJ, Colvin RB, Schneeberger EE, Russell PS (1985) Widespread and selective induction of major histocompatibility complex-determined antigens in vivo by γ interferon. J. Exp. Med. 162: 1645-1664

Southern EM (1975) Detection of specific sequences among DNA fagments separated by gel electrophoresis. J. Mol. Biol. 98: 503-517

Staeheli P, Haller O, Boll W, Lindenmann J, Weissmann C (1986) Mx protein: Constitutive expression in 3T3 cells transformed with cloned Mx cDNA confers selective resistance to influenza virus. Cell 44: 147-158

Taylor-Papadimitriou J, Shearer M, Rozengurt E (1981) Inhibitory effect of interferon on cellular DNA synthesis: Modulation by pure mitogenic factors. J. Interferon Res. 1: 401-410

Thomas KR, Capecchi MR (1987) Site-directed mutagenesis by gene targeting in mouse embryo-derived stem cells. Cell 51: 503-512

Thomas KR, Capecchi MR (1990) Targeted disruption of the murine int-1 proto-oncogene resulting in severe abnormalities in midbrain and cerebellar development. Nature 346: 847-850

Thompson S, Clarke AR, Pow AM, Hooper M, Melton DW (1989) Germ-line transmission and expression of a corrected HPRT gene produced by gene targeting in embryonic stem cells. Cell 56: 313-321

Trinchieri G, Perussia B (1985) Immune interferon: a pleiotropic lymphokine with multiple effects. Immunol. Today 6: 131-136

Udy GB, Wilson VA, Evans MJ (1990) PCR amplification of DNA from replica plated ES cell colonies. Technique 2: 88-92

Unanue ER, Allen PM (1987) The basis for the immunoregulatory role of macrophages and other accessory cells. Science 236: 551-557

Weissmann C, Weber H (1986) The interferon genes. Prog. Nucl. Acids Res. Mol. Biol. 33: 251-300

White TJ, Arnheim N, Erlich HA (1989) The polymerase chain reaction. Trends Genet. 5: 185-189

Zijlstra M, Li E, Sajjadi F, Subramani S, Jaenisch R (1989) Germ-line transmission of a disrupted β2-microglobulin gene produced by homologous recombination in embryonic stem cells. Nature 342: 435-438

Zijlstra M, Bix M, Simister NE, Loring JM, Raulet DH, Jaenisch R (1990) β2-microglobulin deficient mice lack $CD4^{-}8^{+}$ cytolytic T cells. Nature 344: 742-746

Zimmer A, Gruss P (1989) Production of chimaeric mice containing embryonic stem (ES) cells carrying a homeobox Hox 1.1 allele mutated by homologous recombination. Nature 338: 150-153

EXPRESSION OF NOVEL CLONED RAT PITUITARY RECEPTORS IN MAMMALIAN CELLS

J. Zabavnik, P. L. Taylor, T. Peters and K. A. Eidne
Medical Research Council
Centre for Reproductive Biology
37 Chalmers Street
Edinburgh EH3 9EW
United Kingdom

Introduction

Considerable interest has been focussed on the superfamily of G-protein coupled receptors by pharmacologists, neurologists, endocrinologists and everyone concerned with mechanisms of cell signalling. Information about the structure of these receptors has become far more accessible using molecular cloning techniques, especially homology cloning strategies.

Our work has concentrated on the pituitary receptors within this family which recognize releasing hormones and neurotransmitters. These receptors constitute an important interface between the brain and the endocrine system; a knowledge of their detailed structure will enable us to produce strategies for pharmacological intervention at this point. All known G-protein coupled receptors to date characteristically have seven hydrophobic transmembrane regions in their primary sequence, and the second, third, sixth and seventh such regions show sequence homologies which extend across many otherwise unrelated receptor types. As we have already reported (Eidne, 1989) we used an approach described by Libert et al. (Libert, 1989) to clone G protein-coupled receptors. Using sequence information from stretches of amino acid sequence homology across some of the seven transmembrane regions of the previously described known receptors, degenerate oligonucleotides were synthesized and used as PCR primers to amplify a rat pituitary cDNA library. The major PCR product was isolated by agarose gel electrophoresis and used to probe a size selected rat pituitary cDNA library. Twelve positive clones were isolated and

NATO ASI Series, Vol. H 52
Biological Signal Transduction
Edited by E. M. Ross and K. W. A. Wirtz

subjected to restriction analysis and sequencing. Restriction analysis revealed that all the isolated clones represented the same sequence. The most complete cDNA sequence (clone pZ19.12 cDNA) was analysed further. It showed very close homology to the previously published rat brain dopamine D2 receptor sequence (Bunzow, 1988). However, clone pZ19.12 cDNA contained an 87 base pair insert in the coding region, potentially encoding an additional 29 amino acid residues. A hydropathicity plot of the pZ19.12 cDNA using the method of Kyte and Doolittle (Fig.1), showed that this putative 29 amino acid insert was in the functionally important third cytoplasmic loop domain, where it formed a hydrophobic spike in the predominantly hydrophilic region.

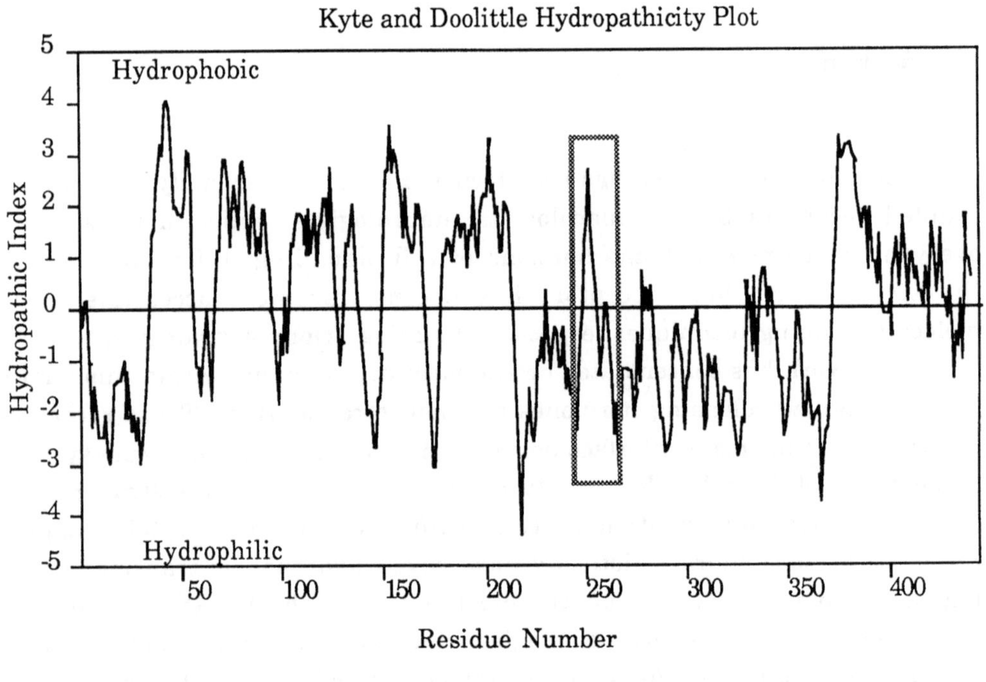

Figure 1: Hydropathic analysis of the protein encoded by the clone pZ19.12. The method of Kyte and Doolittle was used to generate the plot. Transmembrane domains I to VII are indicated and appear as hydrophobic regions on the plot. The inserted sequence is highlighted, and appears as a pronounced hydrophobic spike in the largest cytoplasmic loop.

The hydrophobic regions of the hydropathic analysis probably represent regions of the protein that are anchored in the membrane. Using this and a comparison with other receptor structures we propose this schematic representation of the receptor structure encoded by the clone pZ19.12 cDNA.

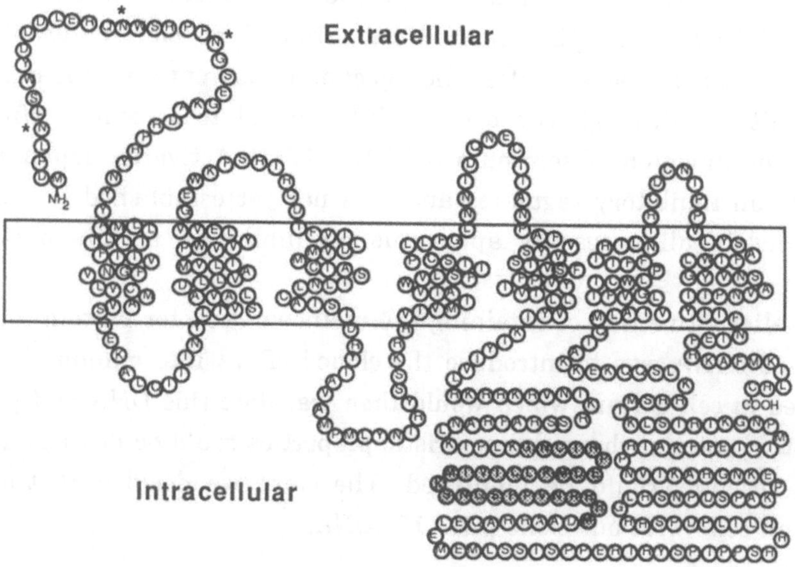

Figure 2. Schematic representation of the predicted receptor structure encoded by the clone pZ19.12 cDNA. The amino acid sequence is shown in one-letter code and the protein is displayed in the seven-transmembrane domain configuration. The inserted sequence in the third cytoplasmic loop is shaded. Potential N-glycosylation sites in the amino-terminal extracellular domain are represented by asterisks.

Having cloned a novel receptor, we wanted to ascertain whether our clone pZ19.12 cDNA encoded a functional dopamine D2 receptor. For identification it was necessary to obtain a translated fully functional protein receptor. The production of authentic, biologically active eukaryotic protein receptors from cloned DNA requires posttranslational modifications such as accurate disulphide bond formation, glycosylation, phosphorylation, correct folding and incorporation into the plasma membrane. In order to determine the functional expression of the receptor a system is needed in which the receptor is incorporated into the membrane and coupled to the G proteins and second messenger systems within the cell, which can produce measurable responses after the ligand activation of the receptor. Only eukaryotic cells in which the endoplasmic reticulum, Golgi apparatus and machinery associated with cell signalling are functioning, are able to fulfil all of these requirements. Many groups working on the molecular cloning of G-protein coupled receptors have been using a toad *Xenopus laevis* oocyte expression system to study the receptors. In this system stimulation of second messengers is measured using electrophysiology. Using this method we obtained no electrophysiological responses in *Xenopus* oocytes after the injection of our putative dopamine D2 receptor mRNA, although the oocytes did respond to serotonin after the simultaneous injection of serotonin 5HT-1c mRNA. Activated dopamine D2 receptor is an inhibitory regulator and it is not yet established if *Xenopus* oocytes contain all necessary apparatus for inhibition of this particular pathway.

An alternative way of obtaining the unknown receptor protein coded by the pZ19.12 cDNA was to introduce the cloned cDNA into mammalian cells maintained in cell culture which would then translate this DNA and produce the receptor protein, whose characteristic properties could be determined and thus the receptor could be identified. Therefore we decided to transfect mammalian cells with our clone pZ19.12 cDNA.

Materials and Methods

Transfection of mammalian cells in tissue culture by lipofection

Clone pZ19.12 cDNA was released from the plasmid pBluescript using Eco R1 restriction endonuclease and subcloned into the Eco R1 cloning site in the eukaryotic expression vector pcDNA 1.

RAG cells and 293 cells were plated in 175 cm tissue culture flasks in 25 ml Dulbecco's modified Eagle's medium supplemented with 5% fetal bovine serum, penicillin (100 iu/ml) and streptomycin (100µg/ml). The cells were incubated in a humidified, carbon dioxide environment (5%) at 37°C until they were 60% confluent.

pZ19.12 cDNA inserted in pcDNA 1 vector was prepared by miniplasmid preparations, column purified and linearized with Not 1 restriction endonuclease. pSP-neo plasmid was prepared in the same way and linearized with Bam HI enzyme. For each 175 mm tissue culture flask 6 µg of pZ19.12 cDNA inserted into the vector and 6 µg of plasmid pSP-neo was diluted in 150 µl of sterile H_2O. In a separate tube 80 µg of Lipofection Reagent (Bethesda Research Laboratories) was diluted in 150 µl of sterile water. The diluted DNA and Lipofection Reagent were then combined and left to stand at room temperature for 15 minutes. The cells were washed twice with serum-free DMEM. 12 ml of serum-free DMEM was added to the cells followed by the uniform dropwise addition of 300 µl of Lipofection Reagent-DNA complex. The cells were incubated at 37°C in a humidified, 5% carbon dioxide environment. After 15 hours incubation the medium was changed with fresh DMEM supplemented with 20% fetal bovine serum and incubated again in normal tissue culture conditions.

Selection of transfected cells with G418

Two days after transfection, transfected RAG cells were selected by the addition of 400 µg/ml geneticin sulphate G418 (Gibco) to the culture medium. After 2 weeks the concentration was increased to 800 µg/ml of the medium. Selection of transfected 293 cells was done in the first two weeks with 400 µg/ml of G418, this concentration was then increased to 600 µg/ml.

Subcloning of the colonies obtained after G418 selection

A month after the G418 selection separate colonies obtained on the surface of the tissue culture were subcloned. The top of the tissue culture flask was cut off. A single colony was then isolated with a cloning ring which was sealed to the surface of the flask with silicone grease. Cells in the sealed cloning ring were trypsinized. Then the cells from each colony were transferred to a 26 well tissue culture dish.

Preparation of DNA from transfected cells for the PCR reaction (whole cell lysates)

The transfected RAG cell lines 5 and 14 obtained from subcloning, and the untransfected cells used as a control, were trypsinized. A volume of 1 ml of the cells in DMEM medium was centrifuged for 5 minutes at 3000 rpm, the supernatant was decanted and the pellet was resuspended in 1 ml of PBS. The cells were counted using a haemocytometer and resuspended in PBS, to give a final concentration of 1000 cells/25µl of PBS. The cells were heated at 90°C for 10 minutes before adding 1µl of proteinase K (10mg/ml) (Stratagene) and incubating at 55°C for 30 minutes. After incubation the proteinase K was inactivated with heating at 90°C for 10 minutes. The cells were spun down at 600 rpm to loose debris, and the supernatant was taken for the PCR reaction.

Polymerase chain reaction

The PCR reaction was carried out using DNA prepared from whole cell lysates as a template and oligonucleotides 8 and 9 in reaction A and oligos 14 and 9 in reaction B as primers. Oligonucleotide 8 was constructed from base 300 to 320 of the dopamine D2 receptor cDNA, oligo 9 from base 733 to 755, and oligo 14 from base 502 to 520. The reaction mixture consisted of:

16 µl of dNTPs (40 mM), 10 µl of 10 x reaction buffer, 0.5 µl (50 ng) of each oligo, 0.5 µl (2.5 units) of Taq polymerase and 47 µl of H_2O per tube. The reaction consisted of a series of 30 cycles of denaturing temperature at 93°C for 3 minutes, annealing temperature at 37°C and extension temperature at 72°C for 5 minutes.

The obtained reaction products were run on a 0.8% agarose gel stained with ethidium bromide and visualized on UV transilluminator.

In situ hybridization

The plasmid pBluescript with the inserted pZ19.12 clone was prepared by miniplasmid preparation and column purified. The obtained plasmid was linearized with the enzymes Xba 1 and Kpn 1 for the transcription of antisense and sense riboprobes. The riboprobes were radiolabelled using 50 µCi of [^{35}S] uridine triphosphate with T7 and T3 RNA polymerases. After incubation at 37°C for 1 hour the probes were precipitated, rediluted and hydrolysed for 70 minutes.

RAG cells were grown in 8 well chamber slides (Lab-tek) under normal tissue culture conditions. The cells were washed with serum-free medium, the chambers removed and the slides with attached cells were frozen at -70°C. The cells were fixed in 4% paraformaldehyde for 5 minutes, followed by two 5 minute washes in phosphate buffered saline, treatment with 0.25% acetic anhydride in triethanolamine and rinsing in 2 x SSC. These washes were carried out at room temperature. The slides were incubated with [^{35}S] uridine triphosphate labelled riboprobes diluted in hybridization buffer under sterile Parafilm coverslips at 55°C for 14-16 hours. The coverslips were removed and the slides were rinsed twice in 4 x SSC at room temperature, followed by two 30 minutes washes in RNA buffer with RNase treatment at 37°C, 2 x SSC at room temperature for 45 minutes and 0.1 x SSC at 60°C for 30 minutes, before being dehydrated and air dried. The slides were dipped in Kodak NTB-2 emulsion and stored at 4°C for 14 days prior to development and counterstaining with haematoxylin and eosin.

Radioligand binding assay

Confluent transfected and untransfected RAG cells were harvested following incubation in 0.5 mM EDTA in phosphate buffered saline at 37°C and centrifuged at 800 g for 10 minutes at 4°C. The cell pellet was resuspended in 0.25 M sucrose, 25 mM Tris pH 7.4, 6 mM MgCl2, 1 mM EDTA, homogenized in a Dounce homogenizer and centrifuged at 10,000 g for 20 minutes. The membrane pellet was resuspended in assay buffer (50 mM Tris pH 7.4, 6 mM MgCl2, 0.001% bovine serum albumin (BSA) and 0.1 ascorbic acid). Radioligand binding assays were carried out in duplicate in a final volume of 1 ml, containing 70,000 cpm [3H] spiperone, 0.1 ml membrane preparation, and appropriate drugs. Incubations were carried out at 37°C for 50 minutes and stopped by the addition of 10 ml of ice-cold wash buffer (10 mM Tris pH 7.4 and 0.9% NaCl) to each assay. The samples were rapidly filtered through GF/B glass-fibre filters pre-soaked in 0.3% polyethylenamine at pH 10 (ref. In Neurochemistry a practical approach p.137-160). Radioactivity retained on this filter was counted using an LKB Rack Beta liquid scintillation counter. Non-specific binding was determined in the presence of 1 μM haloperidol. Data were analysed as previously described (Wormald, 1985)

Results and Discussion

For efficient expression of clone pZ19.12 cDNA in mammalian cells maintained in cell culture we subcloned our clone pZ19.12 cDNA into the eukaryotic expression vector pcDNA 1. This plasmid contains transcription elements from the human cytomegalovirus (HCMV) which enable transcription and thus expression of foreign DNA in eukaryotic cells. We decided to use HCMV promoter/enhancer elements because they are known as not very species and tissue specific transcription elements (Boshart, 1985).

However, they are most active in human and rodent cells, thus we transfected human embryonic kidney cell line (293 cells) and murine renal adenocarcinoma cell line (RAG cells). For introducing clone pZ19.12 cDNA into the cells we used a lipofection method with Lipofection Reagent (Felgner, 1987). This method is very simple, convenient and appears to be efficient, but is comparatively new and untested. In order to obtain stable transfections the cells were co-transfected with the plasmid pSPneo (a pSVneo derivative) carrying Geneticin G418 resistance, which enabled the selection of transfected cells with G418. A month after the treatment with selection media containing G418, all the control untransfected cells died. We assumed that all the surviving cells contained plasmid pSP-neo. Single transfected colonies were isolated and subcloned and transfected cell lines were obtained.

When co-transfecting mammalian cells with two unlinked plasmids, cells which take up one plasmid will have also taken up the other in 90% of cases (Boshart, 1985). To test if our cells contained clone pZ19.12 cDNA inserted into the genomic material of the geneticin resistant cell lines, PCR analysis was performed. Oligonucleotide primers based on the pZ19.12 cDNA nucleotide sequence were designed and used in a polymerase chain reaction to amplify genomic DNA prepared from whole cell lysates. Only in successfully transfected cells, whose DNA contained the inserted pZ19.12 cDNA clone, would these specific primers align to the DNA and generate a PCR product. Figure 3 shows the PCR products obtained after the amplification. Genomic DNA prepared from the transfected cell lines generated DNA bands which matched those DNA bands obtained after the amplification of clone pZ19.12 cDNA alone. When DNA prepared from untransfected cells was used as a template for DNA synthesis, the PCR reaction did not produce any DNA. These results confirmed that the selected cell lines had been successfully transfected with the clone pZ19.12 cDNA.

To ascertain whether the transcription apparatus in the transfected cells was able to transcribe the inserted clone pZ19.12 cDNA, mRNA transcribed from the inserted clone was determined by *in situ* hybridization. Sense and antisense riboprobes were transcribed from the pZ19.12 cDNA and used in this reaction. Positive hybridization with the antisense riboprobe was obtained in the transfected cells (Fig.4 and 5). This confirmed that the transfected cells were able to transcribe the inserted clone.

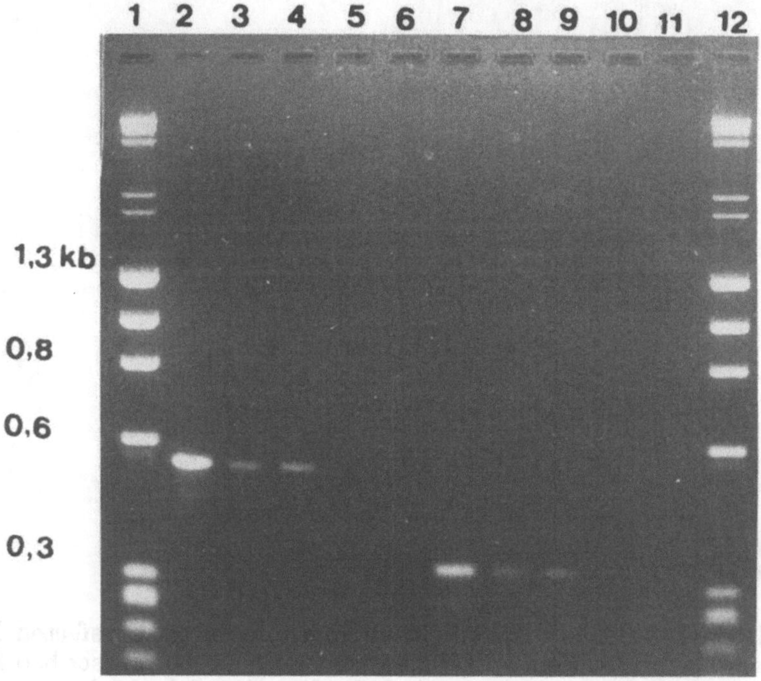

Figure 3: PCR analysis of the inserted clone pZ19.12 in transfected RAG cells. PCR was performed on genomic DNA prepared from whole cell lysates. Reaction products were run on a 0,8% agarose gel, stained with ethidium bromide and visualized on UV transilluminator.

Lane 1: DNA markers (λ and ϕ); Lane 2: Primers 8 and 9 used on pZ19.12 cDNA clone (positive control); Lane 3: Primers 8 and 9 used on genomic DNA prepared from transfected RAG cell-line 14; Lane 4: Primers 8 and 9 used on genomic DNA prepared from transfected RAG cell-line 5; Lane 5: Primers 8 and 9 used on genomic DNA prepared from untransfected control RAG cells; Lane 6: No template with primers 8 and 9 (negative control); Lane 7: Primers 9 and 14 used on pZ19.12 cDNA clone (positive control); Lane 8: Primers 9 and 14 used on genomic DNA prepared from transfected RAG cell-line 14; Lane 9: Primers 9 and 14 used on genomic DNA prepared from transfected RAG cell-line 5; Lane 10: Primers 9 and 14 used on genomic DNA prepared from untransfected control RAG cells; Lane 11: No template with primers 8 and 9 (negative control).

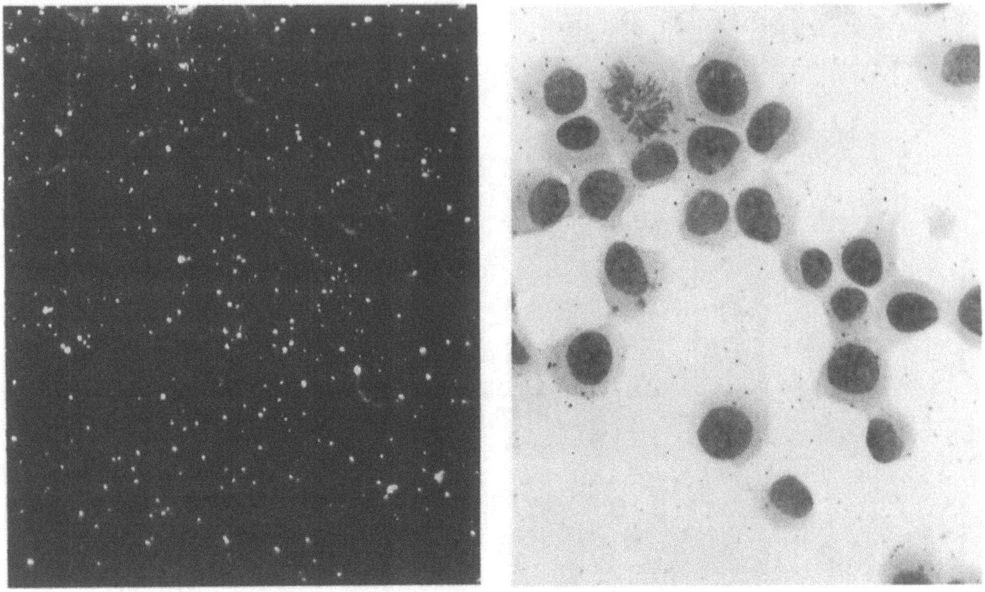

Figure 4: Dark field and light field micrographs of untransfected RAG cells after *in situ* hybridization with the antisense riboprobe transcribed from clone pZ19.12 cDNA show only non specific hybridization of the background.

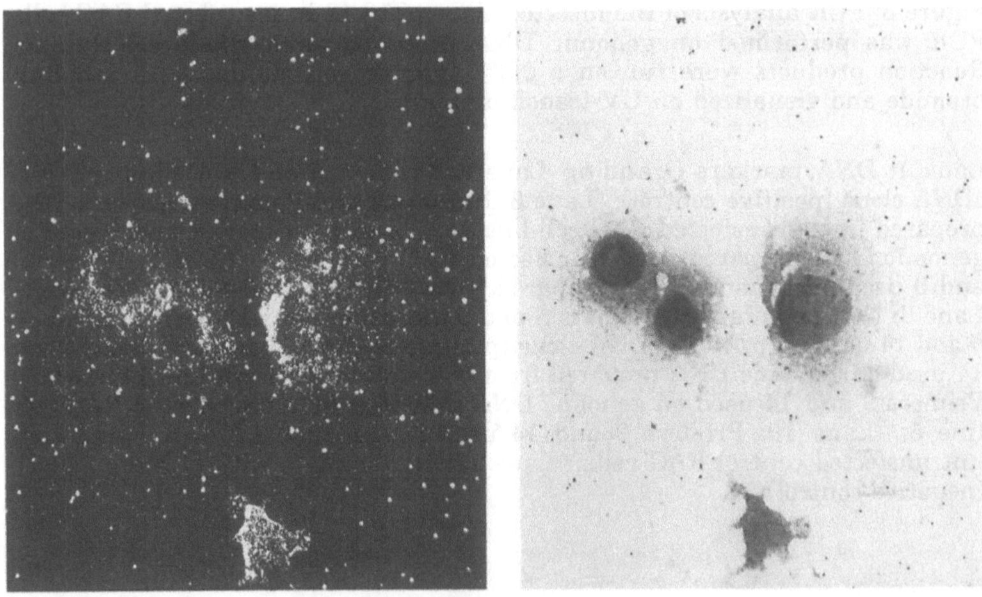

Figure 5: Dark field and light field micrographs showing expression of mRNA in transfected RAG cells after *in situ* hybridization with the antisense riboprobe transcribed from clone pZ19.12 cDNA.

Having confirmed the presence of the clone pZ19.12 cDNA in the transfected cells and its mRNA transcription, we performed studies on the proteins produced by this inserted DNA. Membranes were prepared from the transfected cells and analyzed by binding to a variety of dopamine D2 receptor ligands. Ligand competition of [^{3}H] spiperone binding was used to determine relative affinities of the dopamine antagonists haloperidol, domperidone and quinpirole and dopamine itself. The 50% inhibitory values (IC_{50}'s) were calculated (Wormald, 1985) from Fig.6 and these values converted to K_i's (nM) of 8.85×10^9, 1.47×10^9, 1.47×10^6 and 1.18×10^5 respectively. These are in close agreement with the rank orders and values expected for interaction with the dopamine D2 receptor by the above ligands (Senogles, 1988; Bunzow, 1988).

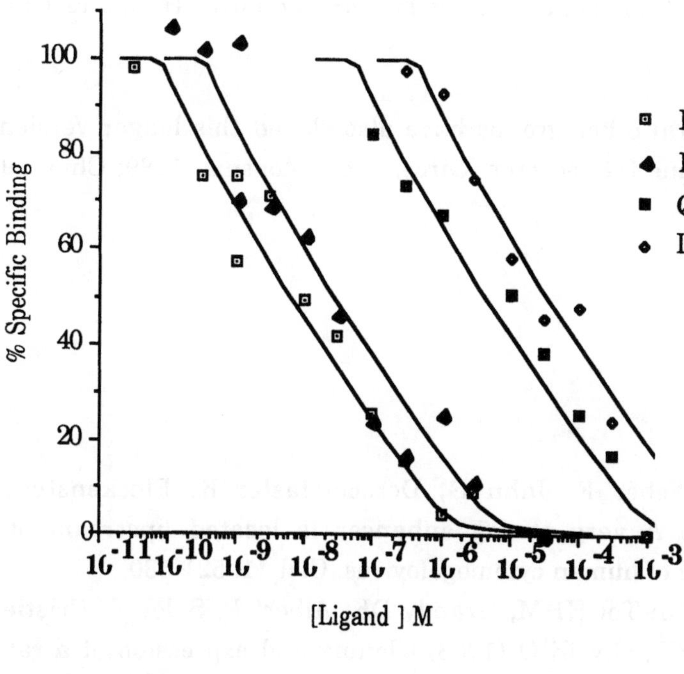

Figure 6: Pharmacological profile of the protein encoded by the clone pZ19.12 cD transfected RAG cell-line.
Competition binding curves showing the binding of [3H] spiperone (1 nM) assayed in the presence of different concentrations of dopamine, haloperidol, quinpirole (LY 171555) and dopamine.

Membranes prepared from untransfected control RAG cells showed no displacement of [^{3}H] spiperone by any of the above ligands. According to this difference in binding properties of proteins in transfected and in untransfected cells we could assume that clone pZ19.12 cDNA which was introduced into transfected cells codes for dopamine D2 receptor.

Conclusions

Receptor binding assays performed on the mammalian cell lines transfected with the clone pZ19.12 cDNA confirmed that this clone encodes a receptor with the binding properties known for the dopamine D2 receptor. Our stable transfected cell lines, which express high levels of the dopamine D2 receptor, have also provided us with a useful model for further studies. Using a similar approach other clones coding for further novel receptors can be identified.

Recently, several other groups have also cloned this longer version of sequences for dopamine D2 receptor (Giros, 1989; Monsma, 1989; Chio, 1990; Dal Toso, 1989).

References

Boshart M, Weber F, Jahn G, Dorsch-Hasler K, Fleckenstein B, Schaffner W (1985) A very strong enhancer is located upstream of an immediate early gene of human cytomegalovirus. Cell 41: 521-530

Bunzow JR, Van Tol HHM, Grandy DK, Albert P, Salon J, Cristie M, Machida CA, Neve KA, Civelli O (1988) Cloning and expression of a rat D2 dopamine receptor cDNA. Nature 336: 783-787

Chio CL, Hess GF, Graham RS, Huff RM (1990) A second molecular form of D2 dopamine receptor in rat and bovine caudate nucleus. Nature 343: 266-269

Dal Toso R, Sommer B, Ewert M, Herb A, Pritchett DB, Bach A, Shivers BD, Seeburg PH (1989) The dopamine D2 receptor: two molecular forms generated by alternative splicing. EMBO J 8: 4025-4034

Eidne KA, Taylor PL, Zabavnik J, Saunders PTK, Inglis JD (1989) Dopamine D2 receptor, a missing exon. Nature 342: 865

Felgner PL, Ringold GM (1989) Cationic liposome-mediated transfection. Nature 337: 387-388

Felgner PL, Gadek TR, Holm M, Roman R, Chan HW, Wenz M, Northrop JP, Ringold GM, Danielsen M (1987) Lipofection: a highly efficient, lipid-mediated DNA-transfection procedure. Proc. Natl. Acad. Sci. USA 84: 7413-7417

Giros B, Sokoloff P, Matres MP, Riou JF, Emorine LJ, Schwartz JC (1989) Alternative splicing directs the expression of two D2 dopamine receptor isoforms. Nature 342: 923-926

Kyte J, Doolittle RF (1982) A simple method for displaying the hydropathic character of a protein. J.Mol.Biol 157: 105-132

Libert F, Parmentuer M, Lefort A, Dinsart C, van Sande J, Maenhaut C, Simons M, Dumont JE, Vassart G (1989) Selective anplification and cloning of four new members of the G protein-coupled receptor family. Science 244: 569-572

Monsma FJ, McVittie LD, Gerfen CR, Mahan LC, Sibley DR (1989) Multiple D2 dopamine receptors produced by alternative RNA splicing. Nature 342: 926-929

Senogles SE, Amlaiky N, Falardeau P, Caron MG (1988) Purification and characterization of the D2-dopamine receptor from bovine anterior pituitary. J. Biol. Chem. 263: 18996-19002

Wigler M, Sweet R, Sim GK, Wold B, Pellicer A, Lacy E, Maniatis T, Silversten S, Axel R (1979) Transfection of mammalian cells with genes from procaryotes and eucaryotes. Cell 16: 777

Wormald PJ, Eidne KA, Millar RP (1985) Gonadotropin-releasing hormone receptors in human pituitary: Ligand structural requirements, molecular size and cationic effects. J. Clin. Endocr. Metab 61: 1190-1194 [illegible] Eidne KA, Taylor PL, Zabavnik J, Saunders PTK, Inglis JD (1989) Dopamine D2 receptor, a missing exon. Nature 342:865

Felgner PL, Ringold GM (1989) Cationic liposome-mediated transfection. Nature 337:387-388

Felgner PL, Gadek TR, Holm M, Roman R, Chan HW, Wenz M, Northrop JP, Ringold GM, Danielsen M (1987) Lipofection: a highly efficient, lipid-mediated DNA-transfection procedure. Proc Natl Acad Sci USA 84: 7413-7417

Giros B, Sokoloff P, Martres MP, Riou JF, Emorine LJ, Schwartz JC (1989) Alternative splicing directs the expression of two D2 dopamine receptor isoforms. Nature 342: 923-926

Kyte J, Doolittle RF (1982) A simple method for displaying the hydropathic character of a protein. J Mol Biol 157: 105-132

Libert F, Parmentier M, Lefort A, Dinsart C, van Sande J, Maenhaut C, Simons MJ, Dumont JE, Vassart G (1989) Selective amplification and cloning of four new members of the G protein-coupled receptor family. Science [illegible]

Monsma FJ, McVittie LD, Gerfen CR, Mahan LC, Sibley DR (1989) Multiple D2 dopamine receptors produced by alternative RNA splicing. [illegible]

Senogles SE, Amlaiky N, Falardeau P, Caron MG (1988) Purification and characterization of the D2-dopamine receptor from bovine anterior pituitary. J Biol Chem 263: 18996-19002

Wigler M, Sweet R, Sim GK, Wold B, Pellicer A, Lacy E, Maniatis T, Silverstein S, Axel R (1979) Transformation of mammalian cells with genes from procaryotes and eucaryotes. Cell 16: 777

Wormald PJ, Eidne KA, Millar RP (1985) Gonadotropin-releasing hormone receptors in human pituitary: ligand structural requirements, molecular size and cationic effects. J Clin Endocr Metab 61: 1190-1194

CELLULAR LOCALIZATION OF SEVEN TRANSMEMBRANE DOMAIN RECEPTOR mRNA'S BY IN-SITU HYBRIDIZATION

T. Peters, P. L. Taylor and K. A. Eidne
MRC Reproductive Biology Unit
Centre for Reproductive Biology
37 Chalmers Street
Edinburgh
EH3 9EW
United Kingdom

Introduction

Our research programme involves the study of G-protein coupled, neurotransmitter and neuropeptide receptors, which have been isolated from species specific cDNA libraries, using molecular cloning techniques. The primary structure of a number of G-protein coupled receptors has previously been analysed and it is generally accepted that the members of this receptor superfamily possess a characteristic structure consisting of seven hydrophobic transmembrane domains, generating three cytoplasmic loops, the third of which is the most variable. The extracellular domain probably embodies the ligand recognition site; the cytoplasmic domains are involved in G-protein recognition and coupling, and hence activation of second messenger systems.

An increased knowledge of receptor structure, functional characteristics and cellular distribution is vital if we are to further our understanding and aid in the treatment of clinical disorders which may be related to receptor dysfunction. A fundamental aspect of our project has been to establish in-situ hybridization as a technique for studying specificity of gene expression. This method can be used to identify and analyse specific messenger RNA transcripts within individual cells or populations of cells and as such is invaluable for investigating the distribution of genes and mRNA's encoding receptor molecules.

NATO ASI Series, Vol. H 52
Biological Signal Transduction
Edited by E. M. Ross and K. W. A. Wirtz

In-situ hybridization is an exact and sensitive technique and many variables must be controlled before one can obtain reproducible data. There are a vast array of apparently conflicting protocols available, the subtle differences between which are of critical importance to the success of each particular experiment. All the procedures, however, share the common goals of rendering the tissue permeable to a probe without the loss of RNA, tissue morphology and hybrid specificity.

There are three types of probe commonly in use, RNA probes (riboprobes), oligonucleotide probes and complimentary DNA (cDNA) probes. cDNA probes and riboprobes are products of cloning procedures, whereas oligonucleotide probes are constructed using a DNA synthesizer. The choice of label is usually a radioisotope (in particular ^{35}S, ^{32}P or ^{3}H), but non-isotopic labels are becoming more readily available. Each combination of probe and label offers advantages and disadvantages with respect to requirements such as sensitivity, anatomical resolution and speed of development. In these experiments ^{35}S labelled riboprobes were used because they form more stable hybrids than the other probes with a higher specific activity than with ^{3}H, better cellular localization than with ^{32}P and greater overall signal; but they are prone to increased levels of non-specific binding.

We have isolated several clones from a rat pituitary size selected cDNA library encoding the predominant form of the dopamine D_2 receptor (Eidne, 1989) and, more recently, a novel putative receptor R334. Our cloning strategy involved the application of the polymerase chain reaction (PCR). Degenerate oligonucleotides were synthesized from certain regions of transmembrane domains three and six, (These domains show the highest homology between members of the G-protein coupled receptor superfamily), and used as primers to amplify rat pituitary cDNA. The products of this amplification were then used as probes to screen the rat pituitary cDNA library.

The D_2 receptor is a G-protein coupled receptor and as such has seven hydrophobic transmembrane domains (For further structural information see the chapter by J. Zabavnik). Sequence analysis of our dopamine D_2 receptor clone revealed that it differed from the originally published sequence (Bunzow, 1988) in that it contained an additional 87 base pair insertion in the coding region. This 29 amino acid stretch is located in the functionally important third cytoplasmic loop and appears to be encoded by a separate exon. The function of this region is as yet unclear, but it is thought to be closely associated with G-protein coupling.

R334 has been identified as a seven transmembrane receptor on the basis of its hydropathicity profile and observed sequence homologies with previously published receptors. R334 is most closely related to, but clearly distinct from, the beta-adrenergic and muscarinic receptor families. As yet we have been unable to determine the ligand for this receptor. The tissue distribution of R334 has been studied using in-situ hybridization, in the hope that this will help with its formal identification.

Materials and methods

Riboprobe preparation

cDNA's coding for the D_2 receptor and R334 were individually cloned into pBluescript plasmid polylinkers, at the EcoR1 site. For transcription the circular plasmids were linearized by restriction endonuclease digestion. These enzymes cleave immediately downstream from the inserted sequences, with respect to the T7 and T3 promoters, forming the DNA templates. Complementary free nucleotides including ^{35}S radiolabelled UTP and T7 and T3 polymerases were added to the DNA template mixtures. Under appropriate reaction conditions T7 and T3 polymerase enzymes recognize their promotors and transcribe in the 5' to 3' direction until reaching the sites of linearization. This direction is defined by the orientation of the sugar-phosphate backbone of the RNA chain.

The DNA-RNA hybrids formed are treated with DNAse which destroys the DNA template, thus producing antisense and sense riboprobes of similar length, base composition and specific activity, but reading in opposite directions. The antisense riboprobe will hybridize to the cellular mRNA and the sense riboprobe will act as a negative control. The riboprobes are then hydrolysed at 60C for 1 hour. This shortens the probes and allows greater tissue penetration and thus higher hybridization signals.

Two of our D_2 clones were used for in-situ hybridization, clone 12 and clone 8 which were respectively 2.4 and 1.5 kilobases in length. Both clones included the extra 29 amino acid sequence, and so would hybridize to both the short and long forms of the dopamine D_2 receptor reported.

Tissue preparation

Adult human neural tissue and pituitaries were obtained at post-mortem performed 6 to 18h after death. Neural tissue and pituitaries were collected from adult rats (Wistar), sheep (Scottish Blackface ewes), and marmosets (Callithrix Jacchus). All tissue was frozen on dry ice. Cryostat sections were cut at 5µm (pituitary) and 8µm (neural tissue), mounted onto precleaned poly-l-lysine coated slides and stored at -70C.

Prehybridization washes

The sections were dried quickly and fixed in 4% paraformaldehyde for five minutes. This fixative promotes the cross linking of proteins and thus generally provides better RNA retention and cellular morphology. Fixation was followed by two 5min washes in 0.1M phosphate buffer, washing with DEPC treated water, treatment with 0.25% acetic anhydride in triethanolamine and rinsing in 2xSSC; all at room temperature.

Hybridization

The sections were incubated with [^{35}S] uridine triphosphate labelled riboprobes, (1 million counts per slide diluted in 60µl hybridization buffer), under sterile Parafilm coverslips, at 55C for 14-16h.

Posthybridization washes

The coverslips were removed and the sections rinsed twice in 4xSSC at room temperature, followed by two thirty minute washes in RNAse buffer with RNAse treatment at 37C, 2xSSC at room temperature for 45 minutes and 0.1xSSC at 60C for 30 minutes, before being dehydrated and air dried.

The pre and post hybridization washes are designed to lower the degree of non-specific background labelling, which may be due to a number of causes such as formation of imperfect hybrids, electrostatic attraction between the probe and basic proteins in the tissue, physical entrapment in the 3D lattice of the tissue section and artefacts of the detection system. The washes include measures to counteract these causes, namely, treatment with acetic anhydride and dextran sulphate (hybridization buffer), optimal concentration of hydrolysed probe, a gradual decrease in the salt concentration and, most importantly, post-hybridization treatment with RNAse which destroys any unhybridized riboprobe.

Autoradiography

The sections were dipped in Kodak NTB-2 emulsion and stored at 4C for 10-13 days prior to development and counterstaining with haematoxylin and eosin.

Results and Discussion

Dopamine D_2 receptor

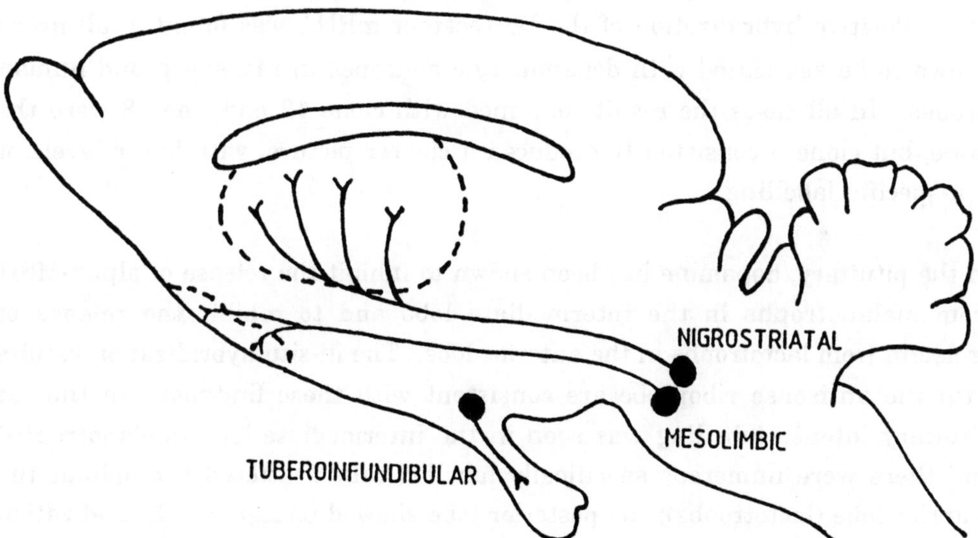

Figure 1: Schematic representation of the major dopaminergic pathways in the brain.

Dopamine cell bodies in the brain are located in three main sites, giving rise to three major dopaminergic pathways (Figure 1). Firstly, cell bodies in the substantia nigra project to the striatum giving rise to the nigrostriatal pathway. Degeneration of this pathway is associated with Parkinson's disease. Secondly, cell bodies in the ventral tegmental area project to the nucleus accumbens forming the mesolimbic pathway, which is known to be overactive in schizophrenia. The third system is known as the tuberoinfundibular pathway and these cell bodies are located in the arcuate nucleus with the axons projecting to the stalk (infundibulum), anterior and intermediate lobes of the pituitary. Moreover, neocortical dopaminergic projections from the substantia nigra and ventral tegmental area are found in regions of the deep cerebral cortex, where the dopamine fibres form a series of clusters principally located in layers III and V.

Emulsion autoradiographs probed with the labelled antisense riboprobe consistently showed a specific signal in the pituitary and neural tissue, while the sense (control) riboprobe produced non-specific background hybridization only. Positive hybridization of the D_2 receptor mRNA was found in all areas known to be associated with dopaminergic neurones in rat, sheep and human tissues. In all cases the results obtained with clone 12 and clone 8 were the same, but clone 8 consistently produced a clearer picture, with lower levels of non-specific labelling.

In the pituitary, dopamine has been shown to inhibit the release of alpha-MSH from melanotrophs in the intermediate lobe and to inhibit the release of prolactin from lactotrophs in the anterior lobe. The in-situ hybridization results with the antisense riboprobe are consistent with these findings. In the rat pituitary intense labelling was seen in the intermediate lobe (melanotrophs) and there were numerous specifically labelled cells scattered throughout the anterior lobe (lactotrophs); the posterior lobe showed background hybridization only (Fig 4a). In the sheep and human pituitaries the intermediate lobe is not obvious as a separate entity, as the cells of this lobe and the anterior lobe intermingle. In-situ hybridization showed expression of the dopamine D_2 receptor mRNA in the anterior lobes of these species. The posterior lobes depicted background hybridization only, as did all the pituitary tissue that was probed with the sense riboprobe (Fig 4b).

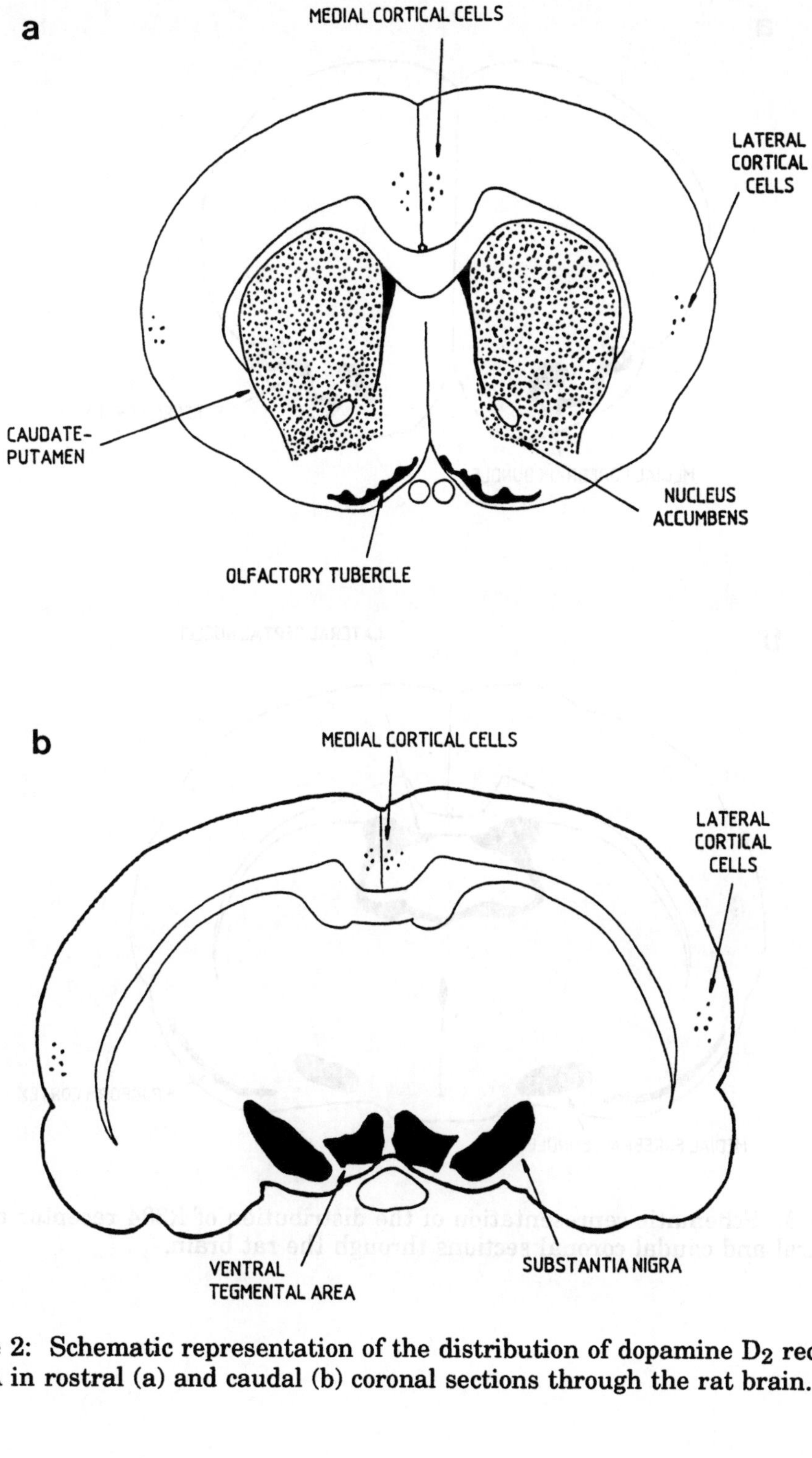

Figure 2: Schematic representation of the distribution of dopamine D_2 receptor mRNA in rostral (a) and caudal (b) coronal sections through the rat brain.

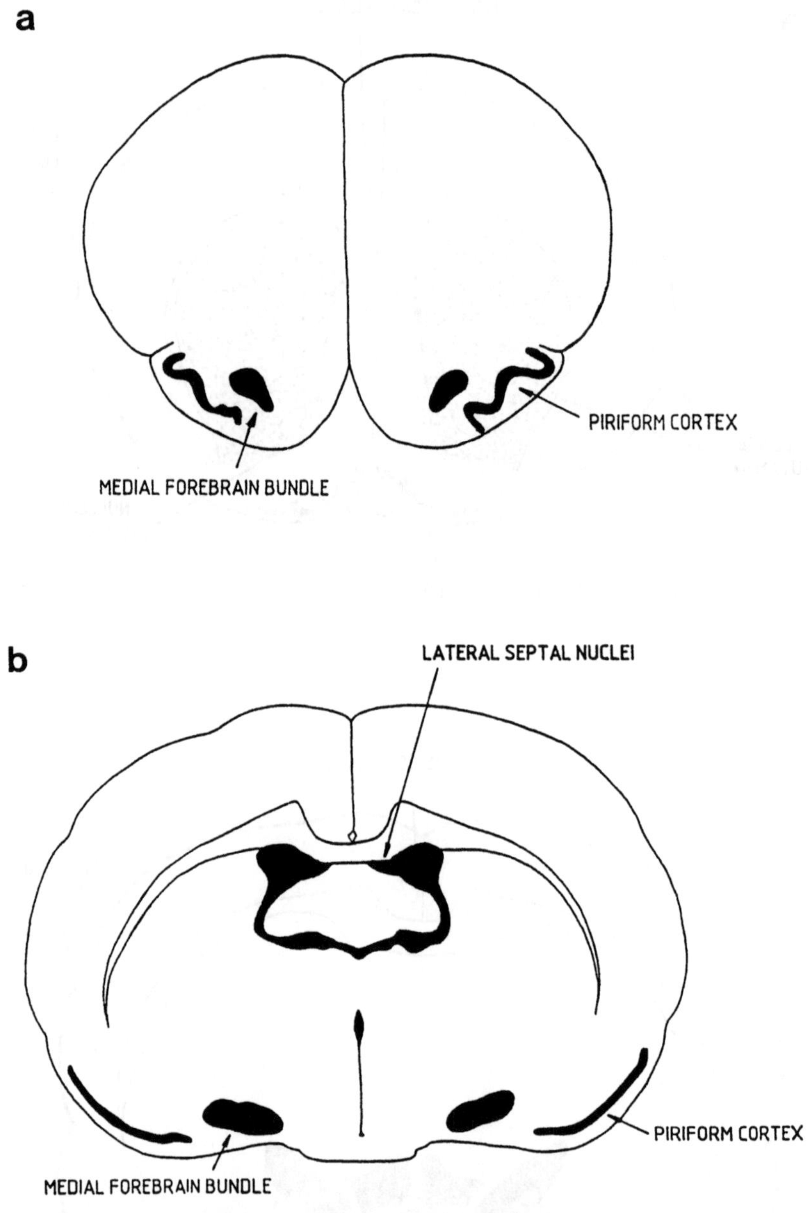

Figure 3: Schematic representation of the distribution of R334 receptor mRNA in rostral and caudal coronal sections through the rat brain.

In the neural tissues the antisense riboprobe detected expression of D_2 receptor mRNA both in regions of dopamine containing cell bodies such as the pars compacta of the substantia nigra (Fig 4c) and the ventral tegmental area, and in the projection sites of the dopaminergic neurones, which include the striatum (Fig 4e), the olfactory tubercle, the nucleus accumbens (Fig 4d), and lateral and medial cortical cells in the neocortex. The striatum contains both striatal neurones and neuroglial cells, which function as packing tissue. Under high power magnification with both light and dark field microscopy the hybridization in the caudate-putamen was seen to be specific to certain cells (Fig 4g and 4h). The glial cells did not express D_2 mRNA, and only 75% of the striatal neurones were observed to contain silver grains. A similar percentage of neurones in the substantia nigra showed hybridization to the antisense riboprobe. These nigral cells are thought to express pre-synaptic dopamine D_2 autoreceptors.

There are a number of possible reasons to explain why not all of the striatal neurones are expressing the D_2 receptor. It could be that some striatal neurones produce different peptides which would affect the type of receptors that are present. Or it could perhaps be a result of the anatomical structure of the caudate-putamen, which arises from two different embryological derivatives. These neurones then intermingle to create one nucleus made up of so called matrices and patches. It could be that only one of these compartments expresses D_2 receptor mRNA.

We have successfully localised dopamine D_2 receptor mRNA in the pituitary and neural tissues of a variety of species, using a rat cDNA clone isolated from a rat pituitary library. Out of the species studied only the human cDNA has been isolated and sequenced. The rat and human D_2 receptors are highly similar, (excepting the 29 amino acid chain in the third cytoplasmic loop of the rat), showing 96% homology (Dal Toso, 1989). The positive in-situ hybridization results would indicate that the D_2 receptor gene is probably also well conserved in the other species studied.

R334 receptor

In-situ hybridization with the antisense riboprobe revealed positive localization of R334 mRNA in rat pituitary and neural tissue. The pituitary appeared to be expressing R334 mRNA in the anterior, posterior and intermediate lobes, although the signal was obviously stronger in the anterior lobe. In the brain,

Figure 4: Autoradiographic detection of cells containing dopamine D_2 receptor mRNA in rat pituitary and neural tissue by in-situ hybridization.

a

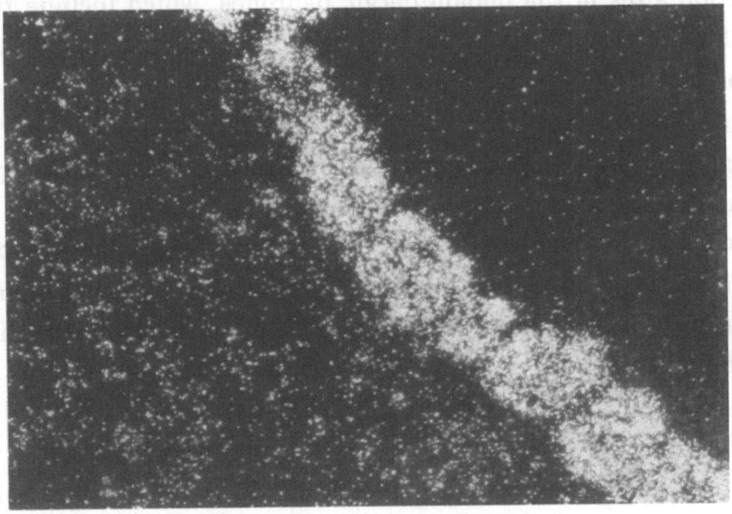

b

a and b: Dark field photomicrographs of a horizontal section through the pituitary (mag x10) hybridized to the antisense riboprobe (a) and the sense riboprobe (b). With the antisense probe there is intense hybridization of melanotrophs in the intermediate lobe and specific labelling of lactotrophs in the anterior lobe. The posterior lobe shows non-specific background hybridization only, as does the whole pituitary when probed with the sense riboprobe.

c

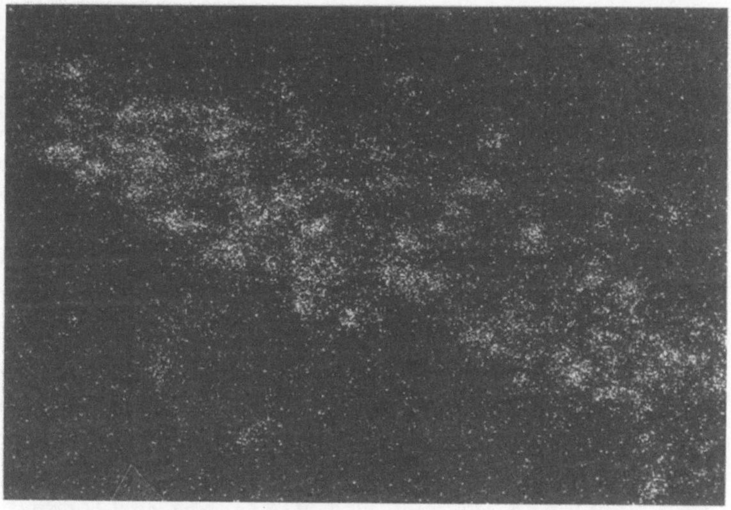

c: Dark field photomicrograph of a coronal section through the brain (mag x10) showing expression of D_2 receptor mRNA in the cells of the pars compacta of the substantia nigra. These are thought to be autoreceptors.

d

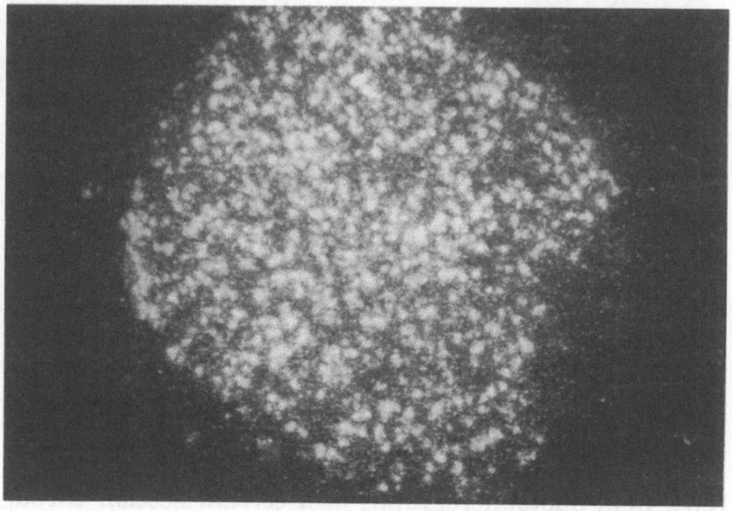

d: Dark field photomicrograph of a coronal section through the brain showing intense hybridization of the cells of the rat nucleus accumbens (mag x4).

e

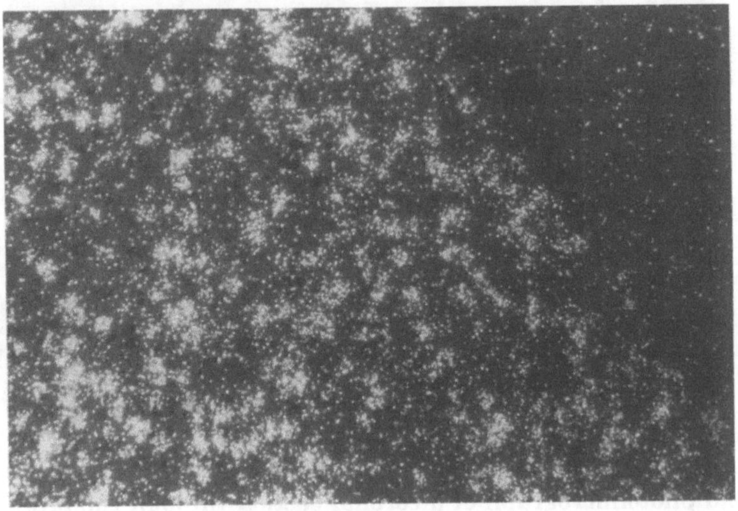

f

e and f: Dark field photomicrographs of a coronal section through the brain (mag x10) hybridized to the antisense (e) and sense (f) riboprobes. With the antisense riboprobe there is specific labelling of certain cells in the caudate putamen and non specific background labelling of cells in the adjacent cortex. The sense riboprobe does not hybridize to these cells.

g

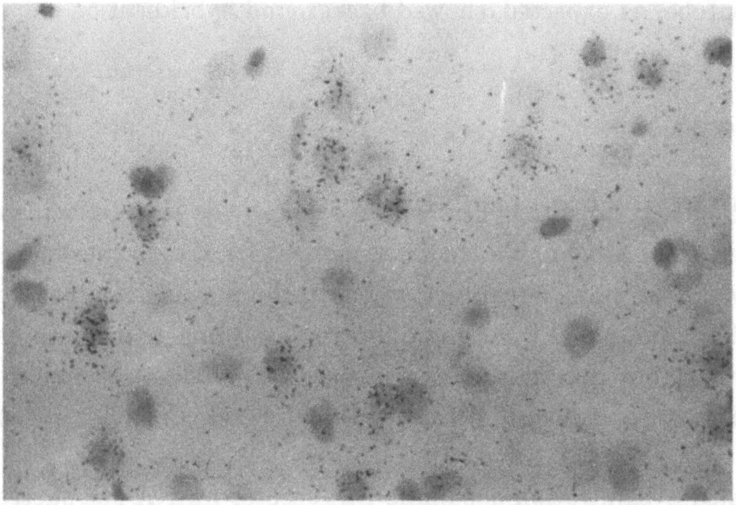

h

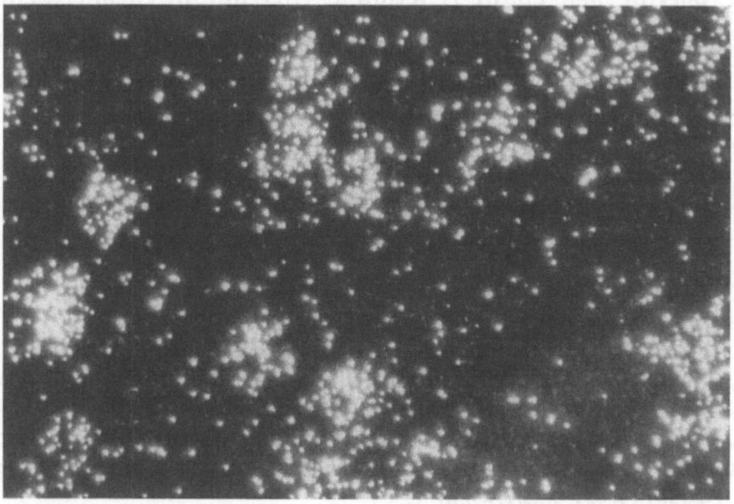

g and h: High power photomicrographs (mag x40) of the same section shown in 4e and 4f viewed through light field and dark field respectively. The light field view allows determination of the cell types expressing the dopamine D_2 receptor mRNA. The haematoxylin and eosin stained section depicts both striatal neurones (large, pale nuclei), and glial cells (small, dark nuclei). The hybridization signal is aggregated around the majority of the striatal neurones. The same region is shown in the dark field view.

however, the signal was restricted to several discrete areas, namely, the piriform cortex, the medial forebrain bundle and the lateral septal nuclei. The sense riboprobe, as expected, showed background hybridization only.

Conclusions

As a methodology in-situ hybridization can provide more exact and sensitive information than any simple chemical method for the measurement of genetic expression. The main advantage of the technique is its great specificity in terms of both the genetic information expressed and the identification of the cells in which that information is expressed. The potential importance of this method in receptor research, both scientifically and clinically, is enormous.

Using in-situ hybridization dopamine D2 receptor mRNA has successfully been located in the brain and pituitary of a number of different species, in regions that have previously been reported to be associated with dopaminergic pathways. As yet the identity of R334 and its ligand are unknown. However, the results of polymerase chain reaction (PCR) amplification techniques, the hydropathicity profiles and the sequence homology all suggest that R334 is indeed a seven transmembrane receptor. The cellular distribution of R334 mRNA would suggest that it is perhaps a neurotransmitter.

References

Bunzow JR, Van Tol HHM, Grandy DK, Albert P, Salon J, Cristie M, Machida CA, Neve KA, Civelli O; (1988); Cloning and expression of a rat D_2 dopamine receptor cDNA. Nature 336: 783-787.

Camps M, Cortes R, Gueye B, Probst A, Palacious JM; (1989); Dopamine receptors in human brain: Autoradiographic distribution of D_2 sites. Neuroscience 28, 2: 275-290.

Del Taso R, Sommer B, Ewert M, Herb A, Pritchett DB, Bach A, Shivers BD, Seeburg PH; (1989); The dopamine D_2 receptor: two molecular forms generated by alternative splicing. EMBO J 8: 4025-4034.

Eidne KA, Taylor PL, Zabavnik J, Saunders PTK, Inglis JD; (1989); Dopamine D_2 receptor, a missing exon. Nature 342a: 865.

Machida CA, Neve KA, Civelli O; (1989); Cloning and expression of a rat D_2 dopamine receptor cDNA. Nature 336: 783-787.

Meador-Woodruff JH, Mansour A, Bunzow JR, Van Tol HHM, Watson SJ, Civelli O; (1989); Distribution of D2 dopamine receptor mRNA in rat brain. Proc. Natl. Acad. Sci. USA, 86: 7625-7628.
Sibley DR, De Lean A, Creese I: (1982); Anterior pituitary dopamine receptors. J Biol Chem 257: 6351-6361.

THE T CELL ACTIVATION MOLECULE CD27 IS A MEMBER OF THE NERVE GROWTH FACTOR RECEPTOR GENE FAMILY

Wil A.M. Loenen, Rolien de Jong*, Loes A. Gravestein, René A.W. van Lier* and Jannie Borst
Division of Immunology
The Netherlands Cancer Institute
Plesmanlaan 121
1066 CX Amsterdam
The Netherlands

Summary

CD27 is a membrane glycoprotein expressed on human T lymphocytes. Antigen recognition by T cells transiently induces high CD27 RNA and protein levels. This is dependent on intracellular Ca^{2+} and cAMP concentrations, but not on protein kinase C activation. CD27 belongs to a newly defined gene family, that includes receptors for nerve growth factor and tumor necrosis factor. Homology is primarily located in the N-terminal cysteine-rich domain, thought to be involved in ligand binding. Certain members of this family share a role in induction of neuronal and/or lymphoid cell differentiation and survival. We speculate that CD27 is a receptor for a T cell growth and/or differentiation factor that acts as a progression signal during a late stage of T cell activation.

T cell activation

Resting T lymphocytes can respond to the presentation of foreign peptides in the context of molecules of the major histocompatibility complex (MHC) by activation, followed by proliferation and differentiation. Triggering of the T cell receptor (TCR)/CD3 complex results in activation of phospholipase C (PLC), that hydrolyses phosphatidyl inositol bisphosphate to form diacylglycerol (DAG) and

---

*Division of Clinical Viral Immunology, Central Laboratory of the Netherlands Red Cross Blood Transfusion Service, Amsterdam, The Netherlands.
DNA and protein analyses mentioned in this paper have been facilitated by the GCG software (Devereux et al (1984) Nucl Acid Res 12: 387-395

NATO ASI Series, Vol. H 52
Biological Signal Transduction
Edited by E. M. Ross and K. W. A. Wirtz

inositol trisphosphate (IP_3). DAG activates protein kinase C (PKC), thus initiating a signalling cascade, while IP_3 plays a role in releasing Ca^{2+} ions from intracellular stores that may subsequently function as second messenger (Weiss et al, 1986). Apart from the Ser/Thr kinase PKC, TCR/CD3 triggering also results in the activation of (a) tyrosine kinase(s), as evidenced by the tyrosine phosphorylation of the TCR-associated ζ chains (Patel et al, 1987). It is unclear whether the tyrosine kinase activity results in the activation of an independent signalling pathway. It may be instrumental in coupling the TCR to PLC, since tyrosine phosphorylation appears to be required for PLC activation and TCR-mediated signal transduction (June et al, 1990). The nature of the Tyr kinase responsible for ζ phosphorylation is unknown, but it has been demonstrated recently that *fyn*, a Tyr kinase of the *src* family, is physically associated with the TCR/CD3 complex (Samelson et al, 1990). A third type of kinase, activated as a result of TCR/CD3 triggering is protein kinase A (PKA) (Patel et al, 1987).

The various signal transduction routes thus activated ultimately affect the transcription of many genes (reviewed in Crabtree, 1989), including genes encoding transcription factors (c-fos, c-myc, NFAT-1), molecules involved in cell division (e.g. cyclin), factors that drive cell division and/or differentiation (e.g. interleukin (IL)-2,-3,-4,-5,-6, interferon (IFN)-γ, tumor necrosis factor (TNF), IL-2 receptor (R), Insulin R, Transferrin R) and late genes encoding proteins that play a role on the differentiated effector T cells (e.g. adhesion molecules, such as the β1 integrins and cytolytic effector molecules, such as serine proteases). Induction of IL-2 and IL-2 R proteins and subsequent receptor/ligand interaction is the major factor driving T cells to proliferate in an autocrine fashion (Smith, 1988).

T cell surface molecules known to increase in expression after T cell activation are CD2 (Meuer et al, 1984), CD5 (Ledbetter et al, 1986), CD7 (Jung & Fu, 1988), CD26 (Hafler et al, 1989), CD27 (Van Lier et al, 1987), CD28 (Lesslauer et al, 1986), CD45 (Ledbetter et al, 1985) and CD69 (Testi et al, 1989). Following activation via the TCR/CD3 complex, CD5, CD27, CD28 and CD45 play a supporting role in T cell proliferation as shown by the fact that antibodies directed at these molecules can enhance proliferative responses.

CD27 is a T cell differentiation and activation molecule

CD27 is a T cell differentiation marker. The molecule is not expressed on immature cortical thymocytes, but only at the plasma membrane of mature, medullary thymocytes and on about 75% of peripheral blood T lymphocytes, equally distributed over the CD4+ and CD8+ subsets (Bigler et al, 1986; Van Lier et al, 1987). Activation of peripheral blood T cells via the TCR/CD3 complex gives rise to a fivefold increase in CD27 membrane expression (Bigler et al, 1986, 1988; Van Lier et al, 1987, 1988). The enhancement of CD27 membrane expression is transient: it reaches a maximum at three to four days after T cell triggering and subsides during the subsequent four days. Since the activation related increase in CD27 protein expression occurs relatively late, it is unlikely that CD27 contributes to a more efficient interaction of resting T cells with antigen presenting cells. Rather, CD27 may be important in the events following antigen recognition, i.e. T cell proliferation and/or differentiation. An indication that CD27 is instrumental in cell division comes from the observation that the proliferative response of T cells activated via the TCR/CD3 complex is enhanced by the addition of anti-CD27 mAbs (Van Lier et al, 1987). Also, continuously growing malignant T cell lines as well as non-transformed T cell clones, usually express CD27 at very low levels, which would be in line with an effect of CD27 during the *transition* of resting peripheral T cells to differentiated effector cells.

It has recently been found that all CD27- T cells in peripheral blood express the CD45R0 isoform, a hallmark of memory cells, while CD27+ cells express either CD45R0 or an isoform characteristically expressed on naive T cells, CD45RA (De Jong et al, in preparation). CD27 expression defines functionally distinct subpopulations of T cells. Within the CD4+ subset the CD27+ cells provide help to B lymphocytes in IgM production. This is probably related to IL-2 production (Van Lier et al, 1988). Finally, only CD4+CD27- peripheral T cells are capable of IL-4 secretion (De Jong et al, submitted). CD27 expression apparently distinguishes between peripheral blood T lymphocytes at different stages of activation or differentiation.

CD27 protein

On resting T cells, CD27 is expressed as a disulphide-linked homodimer of 55 kDa (p55) chains (Van Lier et al, 1987, Bigler et al, 1988; Borst et al, 1989). The 55 kDa chains are constitutively phosphorylated at serine residues (De Jong et al, submitted). Upon T cell activation via the TCR/CD3 complex, concurrent with the increase in CD27 expression, a second component with a mol mass of 28-32 kDa appears at the cell surface (p32), that has extensive structural homology with the p55 CD27 form. Both proteins carry N-linked carbohydrate and most likely a high proportion of O-linked sugar groups containing sialic acid residues (Borst et al, 1989). Both p55 and p32 can be detected intracellularly at a very early timepoint in biosynthesis, when the proteins are expected to reside within the endoplasmic reticulum. The two protein forms are not derived from a common precursor protein by alternative glycosylation. Possibly p32 is generated from p55 by proteolytic processing. Alternatively, p32 may be translated from a different mRNA than p55. Since CD27 is encoded by a single copy gene (Camerini et al, submitted), this mRNA would have to arise as a result of an alternative splicing event.

A soluble form of CD27 has recently been detected in the supernatant of activated T cells as well as in serum and urine of heathy individuals and patients (Hintzen et al, submitted). This soluble CD27 form may be identical to the activation specific p32 protein.

Regulation of CD27 expression

We have analysed the effects of different signalling pathways, activated as a result of antigen receptor triggering, on CD27 protein and RNA expression (De Jong et al, submitted). Considering the time lapse between TCR/CD3 triggering and the upregulation of CD27 membrane expression, it is likely that induction involves transcriptional regulation. Indeed, parallel to the changes in cell surface expression, we find that TCR/CD3 triggering induces a strong increase in CD27 mRNA levels (Camerini et al, submitted; De Jong et al, submitted). It remains to be established whether the increase in mRNA levels is the result of transcriptional activation and/or increase in mRNA stability. Enhanced CD27 mRNA levels are detectable within 16 hours, peak between 2 and 4 days and decrease afterwards. This is late compared to the increase in IL2-R mRNA, that is detectable within 4-6 hours and peaks between 16-24 hours after TCR triggering and might imply that elevation of cytoplasmic Ca^{2+} and cAMP levels does not directly affect CD27 gene transcription. Rather, it may induce synthesis of other (cytokine) receptors and/or their ligands, which in turn provide the signal for enhancement of CD27 transcription.

Effect of Ca^{2+} mobilization

Both peripheral blood T cells and cells of the leukemic T cell line Jurkat can be activated with calcium ionophores. This enhances CD27 membrane expression on peripheral blood T cells 2 to 4-fold. The Jurkat cell line does not express CD27, but addition of Ca^{2+} ionophore induces high CD27 expression on these cells. This effect on Jurkat cells can be completely abolished by Cyclosporin A (CyA), an inhibitor of Ca^{2+} dependent intracellular events (Bloemena et al, 1989). Although less pronounced, CyA also reduces Ca^{2+} induced CD27 expression on primary T cells. Therefore, CD27 induction is dependent on Ca^{2+} mobilization.

Effect of PKC activation

The second pathway induced after TCR/CD3 triggering, that could influence CD27 expression, involves PKC. Phorbol esters efficiently activate T cells by direct stimulation of PKC, but do not cause a rise in $[Ca^{2+}]_i$ (Nishizuka, 1984). Addition of phorbol ester to peripheral blood T cells results in *down*-regulation of CD27 expression (Van Lier et al, 1987; De Jong et al, submitted). Moreover, this decrease in CD27 membrane expression also occurs when the cells are already activated via the TCR/CD3 complex. In line with these observations, a PKC inhibitor does not affect the increase in CD27 expression after TCR/CD3 triggering, although it causes a 50% reduction in T cell proliferation.

Treatment of T lymphocytes with phorbol ester is known to cause down-modulation of several cell surface antigens. Different mechanisms could account for this effect: enhanced internalization or shedding of membrane molecules, inhibition of transcription or destabilization of mRNA. There is no evidence for enhanced shedding of CD27 from the plasma membrane after T cell stimulation with phorbol ester (Hintzen et al, submitted). Downregulation of CD3 and CD4 molecules has been suggested to be a direct consequence of PKC-mediated phosphorylation of these proteins (Cantrell et al, 1985; Shin et al, 1990). However, in the case of CD27, the signalling pathways following direct PKC activation and those following TCR/CD3 triggering *both* lead to enhanced phosphorylation of CD27 on serine residues, while they have opposite effects on CD27 membrane expression (De Jong et al., submitted). This suggests that CD27 downregulation is not a direct consequence of phosphorylation.

Since PKC activation does not seem to affect CD27 membrane expression via direct protein phosphorylation, it may affect CD27 transcription. We have used preactivated lymphocytes, expressing CD27 at high level, to evaluate the effect of phorbol ester on CD27 mRNA expression. PKC activation causes a strong decrease in steady state CD27 mRNA levels of such cells. Therefore, the observed changes in CD27 protein expression are regulated at the RNA level. This decrease occurs rapidly, i.e. within 2 hours after treatment of T cells with phorbol ester, suggesting a direct effect of the PKC dependent signalling pathway on CD27 transcription or mRNA stability. Recently, Paillard et al (1990) found that T cell stimulation with phorbol ester and anti-CD3 mAb caused a decrease in TCR, CD4 and CD8 mRNA. This was due both to inhibition of transcription and to reduction of the halflife of the mRNA. Whether CD27 mRNA levels are regulated in a similar manner is currently under investigation.

Effect of intracellular cAMP

T cell activation via triggering of the TCR/CD3 complex also leads to activation of cAMP-dependent PKA. Both forskolin, which directly activates adenylate cyclase, as well as a cAMP derivative cause a further enhancement of CD27 expression on T cells stimulated with anti-CD3 mAb, resulting in an 8-9 fold increase in density of CD27 membrane molecules. Neither component can induce CD27 expression on non-activated T cells (De Jong et al, submitted). The combined findings strongly suggest that both elevation of intracellular cAMP and $[Ca^{2+}]_i$ levels play a role in the increase of CD27 expression, while on the other hand activation of PKC leads to a decrease. Since PKA can negatively regulate PKC activity (discussed in Patel et al, 1987), cAMP possibly exerts its positive effect on CD27 expression by inhibiting the downregulatory effect of PKC.

Martorell et al (1990) recently showed induction of CD27 on $CD27^-$ thymocytes by lectin stimulation, an effect abolished by addition of CyA. Treatment with phorbol ester, on the other hand, did not result in CD27 expression. These findings, together with our current data, point to similar control mechanisms for CD27 expression on thymocytes and peripheral blood T cells.

<u>CD27 belongs to a new family of growth factor receptors</u>

A cDNA clone encoding the 55 kDa CD27 protein form was isolated from transfected COS7 cells by means of selection with specific monoclonal antibody. Sequence analysis identifies p55 as a typical type I transmembrane protein, with a hydrophobic leader sequence, an extracellular region of 171 amino acids, a hydrophobic transmembrane domain and a cytoplasmic domain of 48 residues (Camerini et al, submitted). The extracellular region of the p55 protein can be divided into two domains: the "Cys" domain and the "stalk" domain. The N-terminal Cys domain contains a high proportion of cysteines (16 out of 80 residues) and histidines (10 out of 80 residues). Cys domains are found in the extracellular portions of many receptors, e.g. the epidermal growth factor R (EGF-R), the *neu* oncogene product, the low density lipoprotein R (LDL-R) and the insulin R (I-R), and have been shown to be involved in ligand binding (reviewed in Carpenter, 1987; Yarden & Ullrich, 1988). The cysteine residues are spaced differently in each receptor family. Formation of multiple intrachain disulphide bonds could give rise to tightly folded globular domains with a conformation that is conserved within each family. Cys domains are thought to confer stability to receptor proteins, allowing passage through acidic intracellular compartments without destruction of the ligand binding site during receptor recycling.

The spacing of the cysteine residues in the N-terminal domain places CD27 in a recently recognized receptor gene family. This family includes both rat and human nerve growth factor (NGF)-R (Johnson et al, 1986; Radeke et al, 1987), two types of human tumor necrosis factor (TNF)-R (Loetscher et al, 1990; Schall et al, 1990; Smith et al, 1990), the Shope fibroma virus protein T2 (Upton et al, 1987), the B lymphocyte and carcinoma membrane protein CD40 (Braesch-Andersen et al, 1989; Stamenkovic et al, 1990), the rat T cell-specific membrane molecule OX40 (Mallett et al, 1990) and the murine T cell-specific molecule 4-1BB (only characterized at the cDNA level; Kwon and Weissman, 1989). The overall homology between these proteins in the Cys domain is about 25%. The consensus sequence of this domain is shown in figure 1.

Figure 1.: The consensus sequence of the Cys domain of the members of the nerve growth factor receptor family (human and rat NGF-R, human CD27, human CD40, human TNF-R, Shope Fibroma virus T2, murine 4-1BB, rat OX40):

C 4 Y 5-8 **C** <u>**C** 2 **C**</u> 1 P G 2-6 **C** 5-7 <u>**C** 2 **C**</u> 4-5 F 8-9 <u>**C** 2 **C**</u> 2-3 **C** 7-9
C 7 <u>**C** 1 **C**</u> 2 G 2 **C** 3-7 <u>**C** 1-2 **C**</u> 5 **C** 2 G 12-14 <u>**C** 2 **C**</u> 2 G 2 S 4-8
C 5 **C**

The consensus sequence shows the characteristic spacing of the conserved cysteine residues (in bold letters) within this family in the region of the molecule, that is thought to be the ligand binding domain. In addition to the cysteine residues, amino acids, that are conserved in at least six members of the family are indicated. The cysteine C 2 C ( ) C 2 C motif, as found in "Zn fingers", is underlined.

The arrangement of the cysteine residues in the extracellular domain, C 2 C ( ) C 2 C, is reminiscent of "zinc finger" structures. Such fingers have first been found in the transcription factor TFIIIA (Miller et al, 1985) and appear to be present in many DNA binding proteins. The zinc finger is a unique protein domain because of its high stability and protease resistance (reviewed by Evans and Hollenberg, 1988). An exotic possibility is that CD27 and other members of this family of transmembrane receptors may bind nucleic acids after internalization of the receptor/ligand complex.

The last conserved cysteine of the N-terminal domain is followed in CD27, NGF-R, TNF-R II, 4-1BB and OX40 by a region of about 60 to 80 amino acids, that is rich in serine, threonine and proline residues (together about 30% of all residues). This membrane proximal domain is probably a flexible, extended, surface exposed "stalk". Such regions are found in many receptor molecules where they are usually heavily glycosylated with O-linked sugars and might confer stability to the receptors. CD27 (Borst et al, 1989), NGF-R (Grob et al, 1985) and TNF-R II (Hohmann et al, 1989) indeed contain O-linked carbohydrates.

The intracellular portions of the NGF-R related receptors do not show clear sequence homology to each other. According to the size of the cytoplasmic domains, the members of the NGF-R family fall into two groups: CD27, CD40, OX40, and 4-1BB have relatively short cytoplasmic domains of about 50 amino acids, while the NGF-R and the two TNF-R have more than 150 intracellular residues. None of these cytoplasmic regions show homology to well defined functional domains. Therefore, the structure provides no clue for the mechanism of signal transduction employed by these receptors. Although ligand binding has clearly demonstrable effects on a variety of cellular activities, there are also no indications from functional experiments how signal transduction would be mediated by these receptors. All receptors contain at least one site that may be subject to phosphorylation by PKC, according to the consensus sequence S/T-X-R/K. CD27 (De Jong et al, submitted), CD40 (Paulie et al, 1989) and the NGF-R (Grob et al, 1985) are indeed phosphorylated at serine residues. The functional significance of this phosphorylation is unclear. One other common feature of the cytoplasmic domains of all receptors in the NGF-R family is the high content of proline, serine, threonine and glutamic acid residues. The presence of these so-called PEST sequences has been shown to correlate with rapid intracellular protein degradation (Rogers et al, 1986). The Shope fibroma virus T2 molecule has a hydrophobic leader sequence, but no transmembrane region and is therefore probably a secretory molecule in contrast to the other members of the NGF-R family.

In an interesting analogy with CD27, both NGF-R (DiStefano & Johnson, 1989) and TNF-R (Heller et al, 1990) can occur in a truncated, soluble form. Truncated NGF-R and TNF-R can be purified from serum and urine. These soluble forms have maintained their ligand binding capacity and contain the cysteine-rich domain and part or all of the stalk domain. So far, a number of membrane bound receptor molecules that also occur in soluble form have been described, such as IL6-R, IFNγ-R, IL-4R, and IL-2Rα. In the case of the IL-4R, there is evidence for a separate mRNA encoding the soluble form (Mosley et al, 1989). Possibly, these truncated receptors are shed from the membrane as a mechanism for receptor downregulation. Alternatively, soluble receptors may function as carrier proteins for secreted ligands, or interact with cell-bound ligands and in that way play a role in intercellular communication.

Function of the members of the NGF-R family

The NGF-R is present on many cell types of both neuronal and non-neuronal origin. Interaction with its ligand, NGF β, is essential for nerve cell survival and differentiation (reviewed in Levi-Montalcini, 1987). NGF directs the growing of regenerating axons (neurotropism). Its action is mediated by receptors at the axon termini. The NGF-R is also present on peripheral blood cells, though at low levels compared to nerve cells. B cells express a high affinity NGF-R, T cells a low affinity NGF-R. Interestingly, NGF enhances IgM secretion by B cells in vitro (Otten et al, 1989). Therefore, in addition to its effects on the nervous system, NGF appears to be an immunoregulatory cytokine.

A wide variety of cell types have binding sites for TNF. Protein chemical evidence was provided for the existence of two TNF receptors, one of which would be preferentially expressed on myeloid cells, the other on epithelial cells (Hohmann et al, 1989). Indeed, two structurally distinct TNF receptors, binding both TNF α and TNF β, were cloned recently (Loetscher et al, 1990, Schall et al, 1990; Smith et al, 1990). Surprisingly, these two TNF-R do not have more homology to each other than to the other members of the NGF-R family. TNF α (cachectin) and TNF β (lymphotoxin) are produced by many cell types, e.g. granulocytes, activated monocytes and lymphocytes and have a wide variety of biological activities. TNFs can induce cell killing via the active process of apoptosis (involving DNA fragmentation), mediate differentiation of leukocytes and endothelial cells and modulate expression of proteins such as MHC class I and the cytokines IFN-β, IL-1 and IL-6 (reviewed in Beutler & Cerami, 1989). It has recently been shown that TNF increases IL-2Rα expression on activated T cells and activates NF-κB like transcription factors (Lowenthal et al, 1989).

CD40 is expressed on pre-B and mature B lymphocytes, but not on plasma cells (Paulie et al, 1985; Clark and Ledbetter, 1986; Ledbetter et al, 1987b). Analogous to CD27 on T cells, CD40 is upregulated at the mRNA level after antigenic stimulation of B cells (Stamenkovic et al, 1989). Anti-CD40 mAbs induce activated B cells to progress through the cell cycle and significantly enhance B cell proliferation (Clark and Ledbetter, 1986; Paulie et al, 1989). The latter property is reminiscent of that of anti-CD27 mAbs, but the effects of anti-CD40 mAbs are more pronounced. CD40 expression can be induced on epithelial cells by IFN-γ and has also been detected in the malignant counterparts of epithelial cells, i.e. carcinomas of various tissues (Paulie et al, 1985; Ledbetter et al, 1987a). It has been speculated that CD40 might be instrumental in the transformation process, possibly as a growth factor receptor enabling epithelial cells to proliferate according to an autocrine mechanism (Stamenkovic et al, 1989). B lymphocytes, explanted from lymph node germinal centers, rapidly kill themselves by apoptosis upon culture in vitro. Apoptosis can be prevented with a combination of anti-Ig antibodies and anti-CD40 mAb (Liu et al, 1989), suggesting that the CD40 ligand, like NGF, promotes the survival of mitotically competent cells.

Little is known about the function of the last three members of the NGF-R family, OX40, 4-1BB and T2. OX40 is expressed on activated $CD4^+$ T cells in the rat. Ligation of OX40 with specific antibody enhances the proliferative response of T cells to antigenic stimulation (Paterson et al, 1987). Possibly, OX40 is the rat homologue of the human CD27. A T cell specific cDNA clone, 4-1BB, has been isolated from an activated T cell library (Kwon & Weissman, 1989). However, our preliminary data indicate, that 4-1BB is not the murine homologue of CD27, but another, apparently T cell specific, relative of this family in the mouse. T2 has only been identified as an open reading frame in Shope fibroma virus, a pox virus that causes tumors in rabbits (Upton et al, 1987). It has particularly strong homology with the extracellular domain of the TNF-R from lung fibroblasts (TNF-R II, see fig 1).

In conclusion, members of the NGF-R family appear to be involved in proliferation, differentiation and survival or death of cells of neuronal and/or lymphoid origin. This fits in with the functional data accumulated on CD27 thusfar.

Ligands of the NGF-R family

Obviously, study of CD27 function would be greatly facilitated by the identification of its ligand. All members of the NGF-R family have strong sequence conservation in the Cys domain, which is thought to contain the ligand binding capacity. This suggests that their ligands may be related. Known ligands of the NGF-R family are NGF, TNF α and TNF β. TNF α and TNF β are highly homologous to each other. In addition, a lower but significant sequence conservation is detectable between all three cytokines. Recently, two molecules that are very closely related to NGF have been defined: brain-derived neurotrophic factor (BDNF; Leibrock et al, 1989) and neurotrophin-3 (NT-3; Maisonpierre et al, 1990). No receptors for these factors have been identified as yet. Like NGF, BDNF and NT-3 promote the survival of neuronal cells. The sequence conservation between the two groups (NGF, BDNF and NT-3 in one group, and TNF α and TNF β in the other) may imply, that the as yet unknown ligands for CD27, CD40, OX40 and 4-1BB also belong to this growth factor family.

A striking observation is that TNF α and β are only 35% identical at the amino acid level and yet can bind to the same receptor. In fact, they can each bind to either of these two different receptors, that in turn are only 25% homologous. Therefore, there may be significant promiscuity in receptor/ligand interactions within this family. For instance, it cannot be excluded that one of the five NGF related cytokines can bind to CD27. Perhaps the greatly divergent functional activities of TNF can in part be attributed to the effect of distinct receptor/ligand interactions. An additional complication in the seach for ligands of the NGF-R family is, that TNF, like other cytokines, has been shown to occur not only in secreted form, but in addition may function as a transmembrane molecule, that mediates cell killing in direct cell-to-cell contact (Perez et al, 1990).

In summary, the NGF receptor/ligand family presents a challenge for both immunologists and neurobiologists. Particularly intriguing is the potential for interaction between the immune system and the nervous system that is provided by members of this family.

Acknowledgements

We thank C.J.M. Melief for critical reading of the manuscript, R. Hintzen for communicating unpublished information and M.A. van Halem for help in preparing the manuscript.

References

Beutler B, Cerami A (1989) The biology of the cachectin/TNF-α primary mediator of the host response. Ann Rev Immunol 7: 625-655

Bigler RD, Chiorazzi N (1986) A novel disulfide-linked cell surface molecule present on resting and activated human T Lymphocytes. In: Leukocyte Typing II, Reinherz RL,Haynes BF, Nadler LM, Bernstein ID eds. Springer-Verlag, NY p. 503-512

Bigler RD, Bushkin Y, Chiorazzi N (1988) S152 (CD27) a modulating disulfide-linked T cell activation antigen. J Immunol 141: 21-28

Bloemena E, Van Oers RHJ, Weinreich S, Stilma-Meinesz AP, Schellekens PTA Van Lier RAW (1989) The influence of cyclosporin A on the alternative pathways of human T cell activation in vitro. Eur J Immunol 19: 943-946

Borst J, Sluyser C, De Vries E, Klein H, Melief CJM, Van Lier RAW (1989) Alternative molecular form of human T cell-specific antigen CD27 expressed upon T cell activation. Eur J Immunol 19: 357-364

Braesch-Andersen S, Paulie S, Koho H, Nika H, Aspenstrom P, Perlmann P (1989) Biochemical characteristics and partial amino acid sequence of the receptor-like human B cell and carcinoma antigen CDw40. J Immunol 142: 562-567

Cantrell DA, Davies AA, Crumpton MJ 1985. Activators of protein kinase C down-regulate and phosphorylate the T3/T-cell antigen receptor complex of human T lymphocytes. Proc Natl Acad Sci USA 82: 8158-8162

Carpenter G (1987) Receptors for epidermal growth factor and other polypeptide mitogens. Ann Rev Biochem 56: 881-914

Clark EA & Ledbetter JA (1986). Activation of human B cells mediated through two distinct cell surface differentiation antigens, Bp35 and Bp50. Proc. Natl. Acad. Sci. U.S.A. 83: 4494-98

Crabtree GR (1989). Contingent genetic regulatory events in T lymphocyte activation. Science 243: 355-361

DiStefano PS & Johnson EM (1988) Identification of a truncated form of the nerve growth factor receptor. Proc Natl Acad Sci USA 85: 270-274

Dubravec DB, Spriggs DR, Mannick JA & Rodrick ML (1990) Circulating human peripheral blood granulocytes synthesize and secrete tumor necrosis factor α. PNAS 87: 6758-6761

Evans RM & Hollenberg SM (1988) Zinc fingers: gilt by association. Cell 52: 1-3

Grob PM, Ross AH, Koprowski H, and Bothwell MA (1985) Characterization of the human melanoma nerve growth factor receptor. J Biol Chem 260: 8044-8049

Hafler DA, Chofflon M, Benjamin D, Dang NH, and Breitmeyer J (1989) Mechanisms of immune memory. T cell activation and CD3 phosphorylation correlate with Ta_1 (CDw26) expression. J Immunol 142: 2590-2596

Heller RA, Song K, Onasch M, Fischer WH Chang D & Ringold GM (1990) Complementary DNA cloning of a receptor for tumor necrosis factor and demonstration of a shed form of the receptor. Proc Natl Acad Sci 87: 6151-6155

Hohnmann H-P, Remy R, Brockhaus M & Van Loon APGM (1989) Two different cell types have different major receptors for human TNF α. J Biol Chem 264: 14927-14934

Johnson D, Lanahan A, Buck CR, Sehgal A, Morgan C, Mercer E, Bothwell M and Chao M (1986). Expression and structure of the human NGF receptor. Cell 47: 545-554

June CH, Fletcher MC, Ledbetter JA, Schieven GL, Siegel JN, Phillips AF & Samelson LE (1990) Inhibition of tyrosine phosphorylation prevents T cell receptor mediated signal transduction PNAS 87: 7722-7726

Jung LKL and Fu SM (1988). Effect of tumor promoter 12-O-tetradecanoyl-phorbol 13-acetate on CD7 expression by T lineage cells. Eur J Immunol 18: 711-715

Kwon BS and JWeissman SM (1989) cDNA sequences of two inducible genes. Proc Natl Acad Sci USA 86: 1963-1967

Ledbetter JA, Rose LM, Spooner CE, Beatty PG, Martin PJ, & Clark EA (1985). Antibodies to common leukocyte antigen p220 influence human T cell proliferation by modifying interleukin-2 receptor expression. J Immunol 135: 1819-1825

Ledbetter JA, Parsons M, Martin PJ, Hansen JA, Rabinovitch PS, and June CH. (1986). Antibody binding to CD5 (Tp67) and Tp44 T cell surface molecules: effects on cyclic nucleotides, cytoplasmic free calcium, and cAMP-mediated suppression. J Immunol 137: 3299- 3305

Ledbetter JA, Clark EA, Norris NA, Shu G, and Hellström I (1987) Expression of of a functional B-cell receptor CDw40 (Bp50) on carcinomas in Leukocyte Typing III. White cell differentiation antigens. A.J. McMichael ed. Oxford University Press. Oxford. p.432-435

Ledbetter JA, Shu G, Gallagher M & Clark EA (1987) Augmentation of normal and malignant B cell proliferation by monoclonal antibody to the B cell specific antigen Bp50 (CDw40). J Immunol 138: 788-794

Leibrock J, Lottspeich F, Hohn A, Hofer M, Hengerer B, Masiakowski P, Thoenen H & Barde Y-A (1989) Molecular cloning and expression of brain-derived neurotrophic factor. Nature 341: 149-152

Lesslauer W, Koning F, Ottenhof T, Giphart M, Goulmy E, and Van Rood JJ (1986) T90/44 (9.3 antigen) a cell surface molecule with a function in human T cell activation. Eur J Immunol 16: 1289-1296

Levi-Montalcini R (1987) The nerve growth factor 35 years later. Science 237: 1154-1162

Liu Y-J, Joshua DE, Williams GT, Smith CA, Gordon J & MacLennan ICM (1989) Mechanism of antigen-driven selection in germinal centers. Nature 342: 929-931

Loetscher, H, Pan YE, Lahm H, Gentz R, Brockhaus M, Tabuchi H, and Lesslauer W (1990) Molecular cloning and expression of the human 55 kd tumor necrosis factor receptor. Cell 61: 351-359

Lowenthal JW, Ballard DW, Bohnlein E & Greene WC (1989) Tumor necrosis factor α induces proteins that bind specifically to κB-like enhancer elements and regulate interleukin 2 receptor α-chain gene expression in primary T lymphocytes. PNAS 86: 2331-2335

Maisonpierre PC, Belluscio L, Squinto S, Ip NY, Furth ME, Lindsay RM & Yancopoulos GD (1990) Neurotrophin-3: a neurotrophic factor related to NGF and BDNF. Science 247: 1446-1451

Mallett S, Fossum S, and Barclay AN (1990). Characterization of the MRC OX40 antigen of activated CD4 positive T lymphocytes-a molecule related to nerve growth factor receptor. EMBO J 9: 1063-1068

Martorell J, Rojo I, Vilella R, Martinez-Caceres E, Vives J (1990) CD27 induction on thymocytes. J Immunol 145: 1356-1363

Meuer SC, Hussey RE, Fabbi M, Fox D, Acuto O, Fitzgerald KA, Hodgdon JC, Protentis JP, Schlossman SF, Reinherz EL (1984) An alternative pathway of T cell activation: a functional role for the 50 kD T11 sheep erythrocyte receptor protein. Cell 36: 897-906

Miller J, McLachlan, Klug A (1985) Repetitive zinc-binding domains in the protein transcription factor IIIA from Xenopus oocytes. EMBO J 4: 1609-1614

Mosley B, Beckmann P, March CJ, Idzerda RL, Gimpel SD, VandenBos T, Friend D, Alpert A, Anderson D, Jackson J, Wignall JM, Smith C, Gallis B, Sims JE, Urdal D, Widmer MB, Cosman D, Park LS (1989) The murine interleukin-4 receptor: Molecular cloning and characterization of secreted and membrane forms. Cell 59: 335-348

Nishizuka Y (1984) The role of protein kinase C in cell surface signal transduction and tumour promotion. Nature 308: 693-698

Otten U, Ehrhard P, Peck R (1989) Nerve growth factor induces growth and differentiation of human B lymphocytes. Proc Natl Acad Sci USA 86: 10059-10063

Paillard F, Sterkers G, Vaquero C (1990) Transcriptional and post-transcriptional regulation of TcR, CD4 and CD8 gene expression during activation of normal human T lymphocytes. EMBO J 9: 1867-1872

Patel MD, Samelson LE, and Klausner RD (1987) Multiple kinases and signal transduction. Phosphorylation of the T cell antigen receptor complex. J Biol Chem 262: 5831-5838

Paterson DJ, Jefferies WA, Green JR, Brandon MR Corthesy P Puklavec M, Williams AF (1987) Antigens of activated rat T lymphocytes including a molecule of 50,000 Mr detected only on CD4 positive blasts. Mol Immunol 24: 1281-1290

Paulie S, Ehlin-Henriksson B, Mellstedt H, Koho H, Ben-Aissa H, Perlmann P (1985) A p50 surface antigen restricted to human urinary bladder carcinomas and B lymphocytes. Cancer Immunol Immunother 20: 23-28

Paulie S, Rosen A, Ehlin-Henriksson B, Braesch-Andersen S, Jakobson E, Koho H & Perlmann P (1989) The human B lymphocyte and carcinoma antigen CDw40 is a phosphoprotein involved in growth signal transduction. J Immunol 142: 590-595

Perez C, Albert I, DeFay K, Zachariades N, Gooding L, Kriegler M (1990) A nonsecretable cell surface mutant of tumor necrosis factor (TNF) kills by cell-to-cell contact. Cell 63: 251-258

Radeke MJ, Misko TP, Hsu C, Herzenberg LA, Shooter EM (1987) Gene transfer and the molecular cloning of the rat nerve growth factor receptor. Nature 325: 593-597

Rogers S, Wells R, Rechsteiner M (1986) Amino acid sequences common to rapidly degraded proteins: the PEST hypothesis. Science 234: 364-368

Samelson LE, Phillips AF, Luong ET, Klausner RD (1990) Association of the fyn protein-tyrosine kinase with the T-cell antigen receptor. Proc Natl Acad Sci USA 87: 4358-4362

Schall TJ, Lewis M, Koller KJ, Lee A, Rice GC, Wong GHW, Gatanaga T, Granger GA, Lentz R, Raab H, Kohr WJ, Goeddel DV (1990) Molecular cloning and expression of a receptor for human tumor necrosis factor. Cell 61: 361-370

Shin J, Doyle C, Yang Z, Kappes D, Strominger JL (1990) Structural features of the cytoplasmic region of CD4 required for internalization. EMBO J 9 : 425- 434

Smith CA, Davis T, Anderson D, Solam L, Beckmanm MP, Jerzy R, Dower SK, Cosman D, Goodwin RG (1990) A receptor for tumor necrosis factor defines an unusual family of cellular and viral proteins. Science 248: 1019-1023

Smith KA (1988) Interleukin-2: inception, impact and implications. Science 240: 1169-1176

Stamenkovic I, Clark EA, and Seed B (1989) A B-lymphocyte activation molecule related to the nerve growth factor receptor and induced by cytokines in carcinomas. EMBO J 8: 1403-1410

Testi R, Phillips JH, Lanier LL (1989) Leu23 induction as an early marker of functional CD3/T cell antigen receptor triggering. J Immunol 142: 1854-1860

Thomas ML (1989) The leukocyte common antigen family. Ann Rev Immunol 7: 339-369

Upton C, DeLange AM, McFadden G (1987) Tumorigenic poxviruses: genomic organization and DNA sequence of the telomeric region of the Shope Fibroma Virus. Virology 160: 20-30

Van Lier RAW, Borst J, Vroom TM, Klein H, van Mourik P, Zeylemaker WP, Melief CJM (1987) Tissue distribution and biochemical and functional properties of Tp55 (CD27), a novel T cell differentiation antigen. J Immunol 139: 1589-1596

Van Lier, RAW, Oudkerk Pool, M, Kabel P, Mous S, Terpstra F, De Rie M, Melief CJM & Miedema F (1988) Anti-CD27 monoclonal antibodies identify two functionally distinct subpopulations within the $CD4^+$ T cell subset. Eur J Immunol 18: 811-816

Weiss A, Imboden J, Hardy K, Manger B, Terhorst C, Stobo J (1986) The role of the T3/antigen receptor complex in T cell activation. Ann Rev Immunol 4: 593-619

Yarden Y and Ullrich A (1988) Growth factor receptor tyrosine kinases. Ann Rev Biochem 57: 443-478

MEMBRANE RECEPTOR CLUSTERING AS A CELLULAR STIMULUS - THE MAST CELL CASE

I. Pecht, E. Ortega and R. Schweitzer-Stenner
Department of Chemical Immunology
The Weizmann Institute of Science
Rehovot 76100, Israel

Introduction

The secretory response of the mast cell to clustering of its surface membrane receptors for Fc_{ε} domains of IgE (Type 1 Fc_{ε} receptors - $Fc_{\varepsilon}RI$) has long been recognized as an instructive paradigm for stimulus-secretion coupling mechanisms (Froese 1984; Metzger et al 1986; Siraganian 1988). Moreover, since the initial triggering event is the specific antibody-antigen reaction taking place at the cell-surface, it provides an excellent model system for the diversity of cellular activation processes encountered in the immune system where the interaction between antigen receptors and their polyvalent ligands is the very first event (cf. De Lisi 1980). Mainly two features make mast cells useful and informative; a rather fast response which can be quantitated with relative ease and the availability of at least one cell-line (Barsumian et al 1981) accessible to biophysical and biochemical studies.

Considerable efforts have been made in order to define in quantitative terms the functional requirements for cellular triggering by receptor clustering. This is illustrated by the different approaches to define the minimal size of an $Fc_{\varepsilon}RI$ cluster that yields a cellular response and the dose response to such clusters per cell. Several experimental approaches were devised in order to answer these questions and some interesting insights have already emerged. In one approach, reagents which would cause $Fc_{\varepsilon}RI$ aggregation have been developed and employed. These included: a) the combination of chemically synthesized divalent haptens and specific IgE (Siraganian et al 1975; De Lisi and Siraganian 1979; Schweitzer-Stenner et al 1987); b) Chemically crosslinked oligomers of monoclonal, IgE class antibodies fractionated into discrete oligomers (i.e. dimers, trimers) (Segal et al 1977). c) Specific antibodies to surface epitopes domains of the $Fc_{\varepsilon}RI$ (Ishizaka et al 1977; Ishizaka and Ishizaka, 1978).

NATO ASI Series, Vol. H 52
Biological Signal Transduction
Edited by E. M. Ross and K. W. A. Wirtz

Results of experiments using these reagents have indeed established the notion that it is the clustering of $Fc_{\varepsilon}RI$ itself which is required for initiating secretory response (Ishizaka et al 1970). These studies also provided evidence that a minimal initial size of an effective $Fc_{\varepsilon}RI$ aggregate is a dimer (Segal et al 1977). More recently, this approach resolved possible orientational constraints on the dimers formed in terms of their signalling capacity (Ortega et al 1988; Pecht et al 1991).

Another experimental approach employed lipid haptens carried in liposomes as triggering agents for specific IgE mediated mast cells secretion (Balakrishnan et al 1982). The remarkable observation was that even freely mobile lipid haptens carried in liposome bilayers were found to cause mediator secretion from mast cells of the RBL-2H3 line. Haptens carried on lipids, in particular in "fluid" membranes cannot bring about IgE-$Fc_{\varepsilon}RI$ crosslinking in an analogous way to that attained by the above described reagents. Hence the conclusion proposed by McConnell and his associates (Balakrishnan et al 1982; Weiss et al 1982) was that, if indeed receptor clustering is a requirement for cell-activation, it does not necessarily involve molecular crosslinking. Hence, the simplest general model formulated by these authors, assumes the existence of a pre-equilibrium between monomeric and specific $Fc_{\varepsilon}RI$ complexes in the cell membrane. In the resting state, the concentration of such complexes is too low to yield any resolvable degranulation. Receptor crosslinking would then be one possible pathway to displace this equilibrium, increase the concentration of $Fc_{\varepsilon}RI$ complexes and cause secretion. It was further suggested that if one assumes the existence of limited regions of membrane-membrane contact between the mast cell and the hapten-lipid carrying liposome, then it is easy to imagine that IgE-$Fc_{\varepsilon}RI$ complexes would diffuse and concentrate into these contact regions. The enhanced $Fc_{\varepsilon}RI$ concentration would in turn also increase the concentration of specific molecular complexes above the critical local level, resulting in the observed degranulation (Balakrishnan et al 1982). The diffusion and concentration of $Fc_{\varepsilon}RI$ into such contact regions induced by IgE-hapten binding, has been observed already in the above studies and was recently further corroborated (McCloskey and M.m. Poo, 1986).

In recent years we have also been involved in developing reagents that would enable a quantitative and rigorous analysis of the $Fc_{\varepsilon}RI$ mediated triggering process. While earlier work employed monoclonal IgE-class antibodies and a family of synthetic divalent haptens as model antigens (Schweitzer-Stenner et al., 1987) recently we have concentrated on a different approach; several monoclonal antibodies (mAbs) specific for the $Fc_{\varepsilon}RI$ were raised and used as homogeneous and chemically defined crosslinking agents for this receptor on the rat mucosal-like mast cell line RBL-2H3 (Ortega et al 1988; Pecht et al., 1991). These mAbs were found to be rather useful agents in the quantitative analysis of the $Fc_{\varepsilon}RI$ mediated secretory response (Ortega et al 1988). In the first phase, intrinsic affinities of three such mAbs for the $Fc_{\varepsilon}RI$ on the RBL-2H3 were determined using their radiolabeled Fab fragments. In parallel, IgE binding to cells of the same batches were also carried out. As the $Fc_{\varepsilon}RI$ binds a single IgE molecule, these measurements yielded also information about the stoichiometry of Fab fragments binding to the $Fc_{\varepsilon}RI$. Indeed, though the affinities of the examined three different Fabs differed by over twenty fold, the stoichiometry of all three was 1 Fab: 1 $Fc_{\varepsilon}RI$. This result implied that these IgG class mAbs may crosslink the receptors into dimers only. We have further measured the binding of the labeled intact mAbs to the RBL-2H3 cells on identical batches to those on which IgE and the respective Fabs binding were done. These parallel measurements enabled to determine the extent of binding of the mAbs and, more important, the degree to which they cause $Fc_{\varepsilon}RI$ dimerization (Ortega et al., 1988).

The cellular dose response of the RBL-2H3 cell line was measured in terms of mediator secretion induced by each of these mAbs. The results showed that while two of the mAbs (H10 and J17) caused comparable and intermediate levels of secretion at their respective optimal concentration, the third one (F4) caused markedly higher response.

At this stage we assigned this difference in secretory response to the distinct affinities of the mAbs and hence their differing capacity to induce $Fc_{\varepsilon}RI$ dimerization. However upon combining the above data, i.e. the number of $Fc_{\varepsilon}RI$ per cell, the intrinsic affinity of each mAb and its binding as an intact divalent mAb to these receptors we could calculate the extent of $Fc_{\varepsilon}RI$ dimerization (Ortega et al 1988). The unexpected outcome of these calculations was that the highest secretory response was found to be produced by the mAb which caused the lowest extent of dimerization. Namely, mAb F4 which dimerized ~ 25% of the cell receptors causes the highest response (secretion of > 80% of mediators contents) while H10, which converts practically all the cell's $Fc_{\varepsilon}RI$ into dimers (already at ~

5×10^{-9} M) induces only 35% secretion of the cells' mediator contents. We tried to rationalize these intriguing results by the following two notions, both based on the distinct properties of the $Fc_\varepsilon RI$ dimers produced by the different mAbs.

1) The first rationale is based on dynamic considerations (cf. for example De Lisi, 1980) and assumes that only if receptors stay dimerized for times longer than a critical threshold period they would transduce a stimulatory signal to the cell. Therefore, only those $Fc_\varepsilon RI$ dimers which exist for that time-span would be productive. We have therefore studied the kinetics of association and dissociation of these three mAbs to and from the cell's $Fc_\varepsilon RI$, so as to assess the lifetimes of dimers produced by each mAb (Ortega et al, submitted). Results obtained to-date do not correlate with the amplitudes of the secretory response to the mAbs, and hence do not support this interpretation.

2) The second rationale assumes that each of the three mAbs brings the two $Fc_\varepsilon RI$'s together into a dimer with distinct spatial properties. In view of the asymetric multi-subunit structure of the $Fc_\varepsilon RI$, one would expect that the relative apposition of the $Fc_\varepsilon RI$ subunits in each dimer would depend on the individual epitope-mAb interactions and hence be distinct. Moreover, different configurations of $Fc_\varepsilon RI$ dimers would affect its signalling capacity. Therefore, it is not only the <u>number</u> of aggregated $Fc_\varepsilon RI$'s per cell that determines the amplitude of the secretory response but rather, other constraints imposed by each individual mAb are also decisive. We proceeded to suggest that these constraints are primarily of orientational nature implying an expected specificity in the interactions between the clustered receptors (Ortega et al, 1988).

<u>Receptor dimerization and biochemical coupling signals</u>

$Fc_\varepsilon RI$ aggregation initiates a cascade of biochemical reactions coupling it to the secretory process. Among these reactions, production of inositol phosphates and the rise in free cytosolic Ca^{2+} ions $[Ca^{2+}]_i$ received considerable attention. In order to try and resolve potential differences in the biochemical response to the $Fc_\varepsilon RI$ dimerization by the different mAbs, we also examined the extent of triggered coupling processes by measuring phosphatidyl inositides hydrolysis and the transient changes in $[Ca^{2+}]_i$ levels, and tried to relate these with extent of $Fc_\varepsilon RI$ dimerization and the secretory response it caused.

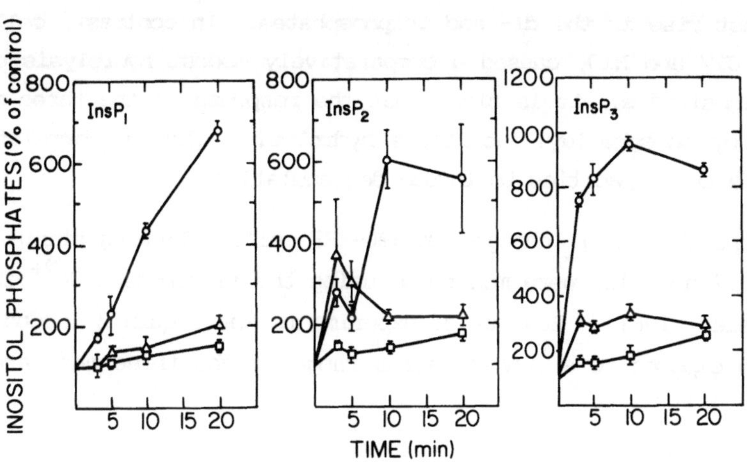

Fig. 1. Production of inositol phosphates in RBL-2H3 cells stimulated by Fc$_\varepsilon$RI-specific mAbs. RBL-2H3 cells were incubated with ^{3}H-myo-inositol (3 μCi/ml) for 16 hr. Following this incubation and washing, cells were incubated for 20 min in Tyrodes buffer containing 5 mM LiCl at 37°C. Aliquots of the cells were then incubated in suspension at 1.0-1.5 x 10^6/350 μl with the appropriate reagents. The cells were stimulated with mAb F4 (2x10^{-8}M) (o); mAb J17 (5x10^{-9}M) (Δ); or mAb H10 (5x10^{-8}M) ( ). At the indicated times, samples were withdrawn for determining release of β-hexosaminidase, and the reaction was immediately stopped by ice-cold chloroform-methanol. The inositol phosphates were extracted and separated by anion exchange chromatography. Results are expressed as the contents of each of the inositol phosphates relative to that in samples of unstimulated cells.

Hydrolysis of inositol phospholipids has been shown to be an integral part of the cellular mechanism of Ca^{2+} mobilization in response to diverse stimuli (Berridge and Irvine 1984; Majerus et al 1990). Production of inositol phosphates by RBL-2H3 cells in response to stimulation by mAbs F4, J17 and H10 was determined by employing reported protocols (Beaven et al 1984). In parallel, also the extent of degranulation was monitored by following the activity of secreted β-hexosaminidase. Figure 1 shows the results of such experiments. mAb F4 produced a significant hydrolysis of phosphatidyl inositides as measured by the production of three inositol phosphates ($InsP_1$, $InsP_2$, and $InsP_3$). This process lasted for at least 20 minutes, as reflected by the continuous accumulation of the mono-phosphate and the fast rise in the di- and triphosphates. In contrast, cell stimulation by J17 and H10, caused a comparatively modest hydrolysis of phosphatidyl-inositides. It is clear that the response to the three mAbs, as reflected by phosphatidyl-inositides hydrolysis, closely correlates with their different capacities to induce degranulation.

The temporal $[Ca^{2+}]_i$ changes in RBL-2H3 cells following stimulation by mAbs F4, J17 and H10, were monitored using the fluorescent Ca^{2+} indicator quin 2 (Tsien, 1981). Results of experiments where quin 2 loaded RBL-2H3 cells were exposed to a concentration range of the three mAbs are illus-

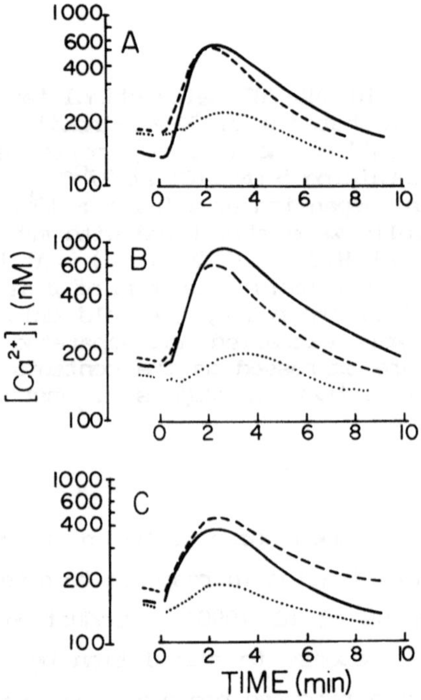

Fig. 2. Changes in free cytosolic Ca^{2+} ions concentrations $[Ca^{2+}]_i$ induced by binding of $Fc_\varepsilon RI$-specific mAbs. RBL-2H3 cells were loaded with the fluorescent Ca^{2+} indicator quin 2. After washing off the external indicator, the cells were suspended at a concentration of 10^6 cells/ml. Aliquots of the cell suspension were placed in a thermostated quartz cuvette and the fluorescence emitted at 492 nm (excitation at 329 nM) was monitored. Shown are the traces from the fluorimeter, along with the calibration scale of $[Ca^{2+}]_i$ concentrations. The following concentrations of mAbs were added at time 0: A) 2.5 x 10^{-7} M, B) 5 x 10^{-9} M. Solid lines: F4; dashed lines: J7; dotted lines: H10.

trated in Fig. 2. The scales showing the corresponding values of $[Ca^{2+}]_i$ were calculated as described elsewhere (Sagi-Eisenberg et al 1985). It is noteworthy that the $[Ca^{2+}]_i$ values obtained using the quin 2 as indicator, show some variability among experiments, a feature which is inherent to this method. Although this precludes the desired accuracy in the $[Ca^{2+}]_i$ determinations and hence its rigorous correlation, qualitative comparisons can be made from measurements done on samples derived from identical cell batches. Moreover, it should be stressed that the relative responses to each of the three mAbs always showed very similar patterns. From the results of several experiments of the kind illustrated in Fig. 2, (n=12) the following conclusions emerged: All three mAbs induced a clear increase in $[Ca^{2+}]_i$, albeit of different magnitudes. F4 induced the highest elevation, which at optimal F4 concentrations reached at least 600-700 nM. J17 raised the $[Ca^{2+}]_i$ levels to the range of 400 to 600 nM, only slightly lower than those induced by F4. In contrast, H10 produced considerably lower $[Ca^{2+}]_i$ levels, usually <200 nM. (The basal $[Ca^{2+}]_i$ in unstimulated cells were in the range of 100 to 150 nM). It is interesting to note that the rates of increase in $[Ca^{2+}]_i$ induced by the three mAbs were similar, reaching a maximum at ca. 2.5 minutes after addition of the mAbs, and then declining to basal levels at about 8-10 minutes.

The parallel examination of the RBL-2H3 cells secretory response showed that mAbs J17 and H10 induced similar maximal secretion whereas F4 yielded a considerably higher one. Thus it appears that no simple quantitative correlation can be made by this method between $[Ca^{2+}]_i$ levels, degranulation and the extent of $Fc_{\varepsilon}RI$ dimerization by these mAbs. Still, it is clear that qualitatively the elevation in $[Ca^{2+}]_i$ correlates well with the extent of degranulation and these two cellular responses are not comensurate with the extent of $Fc_{\varepsilon}RI$ dimerization. A better inverse correlation is however observed between the latter stimulus and the amounts of IP_3 formed. This may suggest a more direct linkage between $Fc_{\varepsilon}RI$ dimerization and phosphatidyl-inositides hydrolysis. In contrast, the $[Ca^{2+}]_i$ levels which reflect the composite contributions from more than a single coupling process do not show such a correlation.

<u>Receptor dimerization and desensitizing processes</u>

We began, at this point, to examine the possibility that at least part of the differences observed in cellular response to the different mAbs may be due to desensitization processes. Thus, we considered the possibility that inhibitory signals can be produced by the particular nature (structural or dynamic) of $Fc_{\varepsilon}RI$ dimers formed by the different mAbs employed for crosslinking. We first examined the possibility that the low level of secretion induced by H10 or J17 could be enhanced by further crosslinking the $Fc_{\varepsilon}RI$ bound mAbs with polyclonal antibodies specific to murine immunoglobulins. RBL-2H3 cells were stimulated with H10 or J17 for 30 min at

37°C. The cells were washed, and then further incubated with polyclonal antibodies specific for murine immunoglobulins for another 30 min at 37°C. Consistently no further secretion could be observed as the result of incubation with the anti-Ig antibodies. These results suggest that those $Fc_{\varepsilon}RIs$ crosslinked by the mAbs in the first phase of the experiment and which yielded only modest response, are unable to cause further secretion upon a more extensive crosslinking induced by the secondary polyclonal antibodies.

We then proceeded to examine whether the above observations are due to inhibitory signals which affect <u>only</u> those $Fc_{\varepsilon}RIs$ that were initially crosslinked and the cells would secrete in response to crosslinking of those $Fc_{\varepsilon}RIs$ which remained unaggregated. Thus, we investigated whether cells first stimulated by mAbs H10 or J17 could be induced to further secretion by F4. RBL-2H3 cells were treated (for 30 min at either 37 or 4°C) with doses of mAbs H10 or J17 that would occupy only a fraction of the cells' $Fc_{\varepsilon}RIs$ yet induce measurable secretion. Following this treatment, a range of ^{125}I-F4 concentrations was added and the cells were incubated at 37°C for further 30 minutes. Both ^{125}I-F4 binding and secretion of β-hexosamindase were then determined. The extent of secretion from cells treated with H10 or J17 at 37°C, (which was in the range of only 20-30%) was not increased by the treatment with mAb F4, eventhough significant binding of F4 was observed (Figure 3A). We have further ascertained that this result is not due to limited availability of free $Fc_{\varepsilon}RI$ in the H10 or J17 treated cells. This is shown by experiments where the density of free $Fc_{\varepsilon}RI$ was decreased by added IgE and this did not affect the maximal response (~90% degranulation) that F4 induced. Moreover, it did not affect the relationship between F4 binding and secretion (Fig 3). By contrast, when the cells were treated with H10 or J17 at 4°C, then F4 did cause an increase in the level of secretion as compared to that observed when the cells reacted with H10 or J17 only (Fig 3B). These data indicate that partial stimulation and secretion renders the cells unable to respond to further $Fc_{\varepsilon}RI$ crosslinking, even by mAb F4. In other words, the primary RBL-2H3 stimulation induces desensitization, probably along with the stimulatory signals, inhibitory ones are generated that limit the cells secretory response to a second, even more extensive $Fc_{\varepsilon}RI$ crosslinking.

In order to investigate whether such inhibitory signals are generated also in the absence of secretion, or whether desensitization would occur

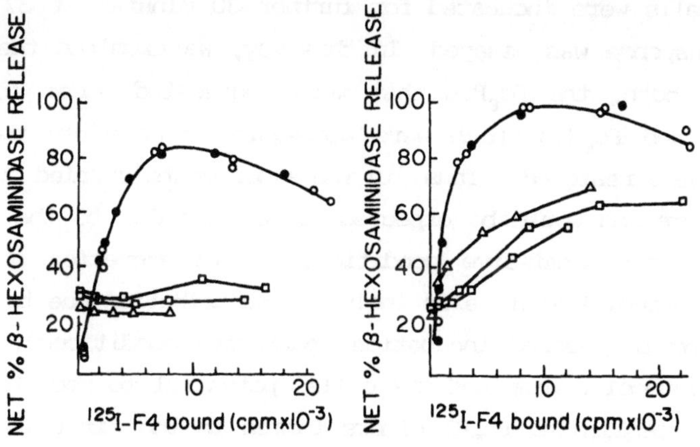

Fig. 3. Effect of preincubation of RBL-2H3 cells with IgE, H10 and J17 on the secretion induced by mAb F4. RBL-2H3 cells monolayers in microwells were incubated with unlabeled IgE (10 nM) (O), H10 (5 nM) (Δ), or J17 (4 nM and 1 nM) ( ), for 30 min at either 37°C (left panel) or at 4°C (right panel). After this incubation, ^{125}I-labeled mAb F4 was added, and all the plates were then incubated for 30 min at 37°C. Afterwards, secretion of β-hexosaminidase was determined in the supernatants. The plates were washed and the amount of ^{125}I-F4 bound was determined by lysing the cells and transferring the contents of each well to test tubes for gamma counting. Points are the mean values of triplicates. The employed concentrations of IgE, J17 and H10 were those that inhibited only partially the binding of F4. Higher concentrations fully inhibited F4 binding (and F4-induced secretion). Lower concentrations of J17 and H10 did not induce significant release by themselves, and did not cause resolvable inhibition of F4 binding (although some inhibition of F4-induced secretion was observed).

only when this final step takes place, we conducted the following experiments: First, RBL-2H3 cells were exposed, for 30 minutes, to a range of concentrations of each of the three $Fc_{\varepsilon}RI$ specific mAbs under conditions where <u>no</u> secretion took place (either at 4°C, at 25°C, or at 37°C in the absence of extracellular Ca^{2+} ions). Following this exposure for 30 minutes, the cells were treated by one of the following protocols (A) The cells were provided with secretion-permisive conditions by raising the temperature, or by adding Ca^{2+} ions (without washing off the free mAb). (B) The free mAb was washed away, and the cells were provided with secretion-permissive conditions. (C) The free mAb was washed away, secondary polyclonal mouse Ig specific antibodies were added and the cells were provided with secretion-permisive conditions. Following each of these protocols, the cells were incubated for further 30 minutes at 37°C and their secretory response was assayed. In this way, we examined the stimulatory capacity of both, the $Fc_{\varepsilon}RIs$ that were aggregated under non-permissive conditions, and $Fc_{\varepsilon}RIs$ which were aggregated only after the permissive conditions were restored. Thus, in the experiments carried out under protocol B, secretion would be expected to be only due to the $Fc_{\varepsilon}RI$ dimers formed under non-permissive conditions, while secretion observed as a result of protocol A would also include a contribution due to $Fc_{\varepsilon}RI$ dimers formed during the second incubation (permisive conditions). Experiments following protocol C examined the cells' potential to secrete as a result of further crosslinking $Fc_{\varepsilon}RI$ dimers formed in the first incubation done under non-permissive conditions, by the anti-immunoglobulin antibodies, under permissive conditions.

Figure 4 presents results of experiments where RBL-2H3 cells were first reacted with a concentration range of F4, J17 and H10 for 30 min at 4°C, and subsequently treated by one of the above described protocols. For comparison, Fig. 4A illustrates the secretory response of identical cell monolayers stimulated with the indicated concentrations of the different mAbs, for 30 min at 37°C. As expected, no secretion was observed as a result of the first 30 min of incubation at 4°C. Cells, which after this incubation, were placed at 37°C for 30 min (without washing away the free mAb), showed slightly lower levels of secretion than the control (cells incubated with the mAb for 30 min at 37°C). The respective dose-response curves showed essentially the same patterns as the control (Fig 4B). When the unbound mAb were washed away prior to transferring the cells to 37°C, then also the observed release was practically the same as that of the control (Fig 4C). If, however, following removal of unbound mAb, further crosslinking is induced by secondary polyclonal anti murine Ig antibodies, then the dose response patterns to all three mAbs became virtually identical (Fig. 4D).

As described above, similar experiments were performed where the first incubation was carried out at 4°C, at 25°C, or at 37°C but in the absence of extracellular Ca^{2+} ions (2mM EGTA). No secretion was observed in response to any of the mAbs after the first incubation phase under these conditions. The results obtained using mAb F4 are presented in Figure 5. Very similar results were obtained when the first incubation was performed at either 25°C (Fig 5B) or at 37°C in the absence of Ca^{2+} (Figure 5C), and

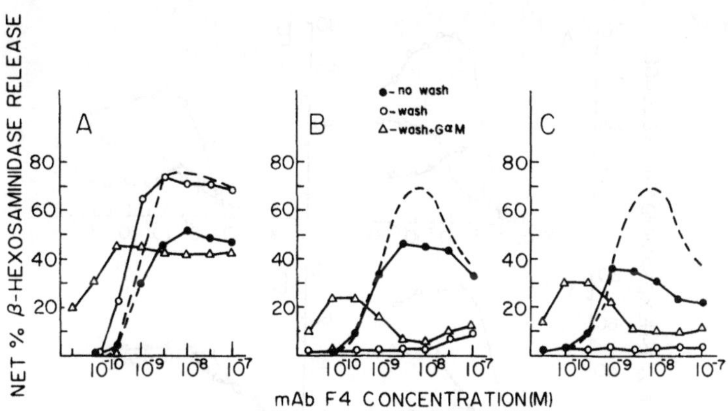

Fig. 4. RBL-2H3 cells secretion induced by $Fc_{\varepsilon}RI$-specific mAbs. Monolayers of RBL cells in microwells, were treated with the indicated concentrations of F4 (O), J17 ( ), or H10 (Δ), as follows: A) Immediately after addition of the mAbs, the cells were placed at 37°C. Secretion was assayed after 30 min.. B) The mAbs were added to cells preincubated at 4°C. After adding the mAbs, the cells were incubated for further 30 min at 4°C. Following this incubation, the plates were placed at 37°C (without washing), and secretion was assayed after 30 further minutes. C) The mAbs were added to cells preincubated at 4°C. After adding the mAbs, the cells were incubated for 30 further min at 4°C. After this incubation, the plates were washed once with cold Tyrode's buffer, to remove unbound mAb. 150 µl of fresh Tyrode's buffer was added, and the plates incubated for 30 min at 37°C before determining secreted β-hexosaminidase. D) Same as C, except that following washing to remove unbound mAbs, 150 µl of Tyrode's containing 15 µg/ml of goat anti-mouse Ig's were added, and the plates were incubated for 30 min at 37°C before determining the secreted β-hexosaminidase. Points are the mean values of sixtuplicates.

the response was significantly different from that observed when the first incubation was done at 4°C (Fig. 5A).

Following the first incubation performed at either 25°C or at 37°C in Ca^{2+}-free buffer, the results observed were: 1) Washing away the free mAb after the first incubation, cause a marked suppression of the secretion upon restoring permissive conditions (37°C and 2 mM Ca^{2+}). 2) When the free mAb was not washed away, considerable secretion is induced upon restoring permissive conditions. 3) When free mAb is washed away, and the cell-bound mAb are crosslinked by secondary polyclonal antibodies, secretion is observed only at those mAb concentrations that were too low to induce secretion without further crosslinking. We interpret these results as follows: $Fc_{\varepsilon}RI$ dimers formed during the first incubation at 25°C or at

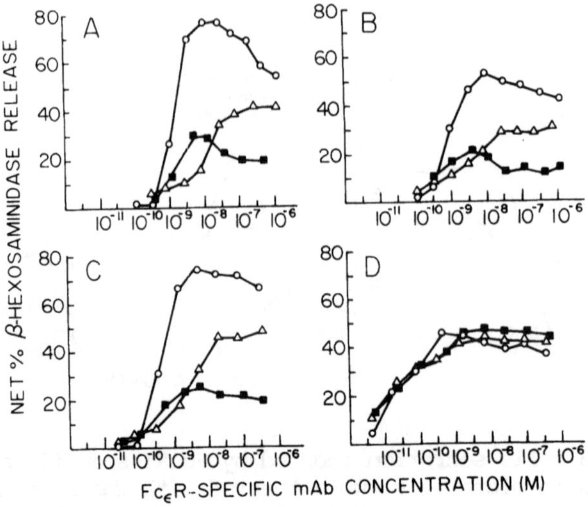

Fig. 5. RBL-2H3 desensitization induced by mAb F4. Monolayers of RBL-2H3 cells were incubated in microwells with the indicated concentrations of mAb F4, for 30 min under the following conditions: A) Tyrode's buffer, at 4°C, B) Tyrode's buffer, at 25°C, and C) Ca^{2+}-free Tyrode's buffer containing 2 mM EGTA, at 37°C. No secretion of β-hexosaminidase was detected following any of these treatments. Following 30 min, the cells were treated in one of these different ways: (i) The cells were placed under permissive conditions (without washing away unbound mAb) by raising the temperature to 37°C and/or by adding Ca^{2+}-ions (O). (ii) The cells were washed once with ice-cold Tyrode's buffer to remove unbound mAb, fresh buffer (containing Ca^{2+}) was added, and the plates incubated for 30 min at 37°C (O) or (iii). The cells were washed with ice cold buffer to remove unbound mAbs, and fresh buffer containing goat anti-mouse Ig antibodies was added, and the plates incubated for 30 min at 37°C (Δ). After each of these treatments, the cells were incubated for further 30 minutes at 37°C and secretion was assayed by determining β-hexosaminidase activity in the supernatants. Results are the mean of sixtuplicates. The dashed lines represent the control experiments were the release was induced by F4 added to cells for 30 min at 37°C, without any pre-incubation.

37°C in the absence of Ca^{2+}, undergo desensitization and lose their capacity to induce secretion when permissive conditions are restored. These Fc_{ε}RIs do not induce secretion even upon being further crosslinked by secondary antibodies. The cells, however, maintained their capacity to secrete in response to new Fc_{ε}RI dimers produced under secretion-permisive conditions by the mAb that was <u>not</u> washed away. The results obtained with

J17 and with H10 showed esentially the same patterns as those observed with F4, with the only difference being that the secretion levels were lower (compare data shown in Fig. 4).

The above results are consistent with the hypothesis that $Fc_{\varepsilon}RI$ dimerization caused by any of the three mAbs, undergo desensitization, i.e. inhibitory signals are produced which can affect both dimerized and/or undimerized $Fc_{\varepsilon}RIs$, depending on the conditions at which receptor dimerization takes place. Thus, if low extents of $Fc_{\varepsilon}RI$ dimerization are produced at 37°C in the presence of 2 mM Ca^{2+}, then the cells secrete proportionally but are unable to secrete further in response to newly formed $Fc_{\varepsilon}RI$ aggregates (Fig. 3) (even if the initial secretory response amounts to no more than 20-30% of the cell contents). If the initial dimerization of the $Fc_{\varepsilon}RI$ takes place at 25°C or at 37°C in Ca^{2+}-free medium, then those $Fc_{\varepsilon}RI$ that underwent dimerization are unable to induce secretion upon restoring permissive conditions. Still the cells can secrete in response to the newly formed $Fc_{\varepsilon}RI$ dimers under permissive conditions (Fig. 5). In contrast, if the initial $Fc_{\varepsilon}RI$ dimerization takes place at 4°C, these aggregates can still produce secretion upon restoring permissive conditions, and the formation of new $Fc_{\varepsilon}RI$ aggregates would also produce enhanced secretion (Figs. 4 and 5A).

It has been shown that aggregated $Fc_{\varepsilon}RIs$ are internalized within minutes of being crosslinked (Isersky et al, 1979, 1983; Furuichi et al 1984). Hence we examined the possibility that inability of $Fc_{\varepsilon}RI$ aggregates formed under non-permissive conditions, to induce secretion upon restoring permissive conditions, may be rationalized by internalization of the aggregated receptors. Early experiments used covalently linked oligomers of IgE to induce $Fc_{\varepsilon}RI$ aggregation and internalization, and it was shown that the internalization is incomplete, with 40% or more of the oligomeric IgE remainning on the cell surface (Isersky et al, 1983). Therefore, assuming that a similar extent of internalization takes place under our conditions, it is clear that the present results cannot be explained solely on the basis of internalization of aggregated receptors. Under the present conditions it can be expected that the amount of $Fc_{\varepsilon}RIs$ dimers remaining on the cell surface would sufice for inducing secretion, particularly in the case of mAb F4. Moreover, internalization of aggregated $Fc_{\varepsilon}RIs$ would not explain the inability of mAb F4 to induce secretion from cells previously stimulated by J17 or H10, eventhough significant binding of F4 was observed (Fig. 3).

The results of the experiments shown in Figure 4D, provide further support to the hypothesis that the distinct secretory patterns evoked by each $Fc_{\varepsilon}RI$ specific mAb are determined by the intrinsic properties of the dimers formed. This is demonstrated by the experiments where the $Fc_{\varepsilon}RI$-specific mAbs are reacted with the cells at low temperature (at which no secretion is induced), and then secretion is allowed to proceed in the absence of further $Fc_{\varepsilon}RI$ crosslinking. The secretory response patterns induced by each mAb were found to maintain their distinct dose response patterns. However, if the crosslinking configuration of the $Fc_{\varepsilon}RI$ dimers is modified as for example is done by crosslinking with the secondary antibodies, before secretion takes place, then the distinct secretory pattern characteristic of each mAb is lost, and identical dose response curves are observed.

Concluding Remarks

A system was established where quantitative analysis of a cellular response to receptor oligomerization can be carried out. Specifically, a well characterized mucosal mast cells line and monoclonal antibodies specific to its $Fc_{\varepsilon}RI$ were combined for studying the relation between the extent of receptor aggregation and the cells' secretory response. This analysis clearly established that already $Fc_{\varepsilon}RI$ dimer formation is a sufficient "unit signal" for initiating the secretory reaction of these mast cells. This provided the possibility to assess the actual number of $Fc_{\varepsilon}RI$ dimers required per cell for its activation. It also resolved marked differences in the secretory dose-response as the result of treatment by the different mAbs. Furthermore, it called at the first time for attention to the detailed properties of receptor-dimers produced by the different mAbs; Receptor complexes that may differ markedly in their structure (i.e. relative orientation of the $Fc_{\varepsilon}RI$ subunits in the dimer) and life-times. Therefore the secretory response to these dimers also differ considerably. We have now further employed these $Fc_{\varepsilon}RI$ specific mAbs in experiments where two biochemical processes that are assumed to couple $Fc_{\varepsilon}RI$ oligomerization to secretion were examined: The elevation in free cytosolic calcium ions concentrations is the result of several different processes where both influx and release from internal stores are involved. Therefore it became apparent that its dose response pattern to $Fc_{\varepsilon}RI$ dimerization is complex and cannot be simply correlated with the extent of $Fc_{\varepsilon}RI$ clustering. In contrast, the production of inositol phosphates is probably more directly coupled to $Fc_{\varepsilon}RI$ clustering and correlated better with the receptor dimerization by the different mAbs.

Finally, we employed these well defined $Fc_{\varepsilon}RI$ dimerizing mAbs for examination of the desensitization process. The biochemical basis for this phenomenon, where receptor clustering induced under conditions that are non-permissive for secretion leads to a lowered response upon restoring permissive conditions is still poorly understood (Fewtrell 1985). Hence it is rather important that protocols where desensitization is brought about in a mast cell system, such as the RBL-2H3 line, are established. This has now being attained using the above mAbs. The application of these reagents clearly illustrated their usefullness also for investigation of this aspect of signal transduction by receptor clustering. Thus a new potential approach is now available for the biophysical and biochemical characterization of the desensitization process.

ACKNOWLEDGEMENTS

The generous support by the Thyssen Foundation and the Council for Tobacco Research-USA Inc. (grant # 2799) is gratefully acknowledged.

REFERENCES

Balakrishnan K, Hsu FJ, Cooper AD, McConnell HM (1982) J Biol Chem 257:6427-6433
Barsumian EL, Isersky C, Petrino MG, Siraganian R (1981) Eur J Immunol 11:317-323
Beaven MA, Rogers J, Moore JP, Hesketh TR, Smith GA, Metcalfe JD (1984) J Biol Chem 259:7129-7136
Berridge MJ, Irvine RF (1984) Nature 341:197-205
De Lisi C, Siraganian RP (1979) J Immunol 122:2286-2292
De Lisi C (1980) Q Rev Biophy 13:201-230
Fewtrell C (1985) In: Rubin, R.P., Weiss, G.B. and Putney, J.W. Jr (eds) Calcium in Biological Systems. Plenum Publishing Corp. p 129
Froese A (1984) Prog Allergy 34:142-187
Furuichi K, Rivera J, Iserski C (1984) J Immunol 133:1513-1517
Iserski C, Rivera J, Segal D, Triche T (1983) J Immunol 131:388-394
Ishizaka K, Ishizaka T (1968) J Immunol 101:68-78
Ishizaka T, Chang TH, Taggart M, Ishizaka K (1977) J Immunol 119:1589-1596
Ishizaka T, Ishizaka K (1978) J Immunol 120:800-805
Ishizaka T, Ishizaka K (1984) Prog Allergy 34:188-205

Majerus PW, Ross TS, Cunnigham TW, Caldwell UK, Jefferson AM, Bansal VS (1990) Cell 63:459-465
McCloskey MA, Poo Mm (1986) J Cell Biol 102:2185-2196
Metzger H, Alcaraz G, Hohman R, Kineti JP, Pribluda K, Quarto R (1986) Annu Rev Immunol 4:419-470
Ortega E, Hazun B, Zor U, Pecht I (1989) Eur J Immunol 19:2251-2256
Ortega E, Schweitzer-Stenner R, Pecht I (1988) EMBO J 7:4101-4109
Pecht I, Ortega E, Jovin T (1991) Biochemistry, in press
Sagi-Eisenberg R, Lieman H, Pecht I (1985) Nature 313:59-60
Schweitzer-Stenner R, Licht A, Luescher E, Pecht I (1987) Biochemistry 26:3602-3612
Segal DM, Taurog JD, Metzger H (1977) Proc Nat Acad Sci USA 74:2993-2997
Siraganian RP, Hook WA, Levine BB (1975) Immunochemistry 12:149-157
Siraganian RP (1988) In: Galin JI, Goldstein IM, Snyderman R (eds) Inflammation, basic principles and clinical correlates. Raven Press New York p 513
Tsien RY (1980) Biochemistry 19:2396-2404
Weiss RM, Balakrishnan K, Smith BA, McConnell HM (1982) J Biol Chem 257:6440-6445

G PROTEIN-COUPLED RECEPTORS:
STRUCTURAL BASIS OF SELECTIVE SIGNALING

Elliott M. Ross

Department of Pharmacology
Southwestern Graduate School of Biomedical Science
University of Texas Southwestern Medical Center
5323 Harry Hines Boulevard
Dallas, Texas, 75235-9041, U.S.A.

Animal cells must respond appropriately to multiple hormonal signals. These signals, which may be mutually potentiative or antagonistic, must be integrated to yield appropriate intracellular signals in the form of second messengers, such as cyclic AMP, inositol phosphates, Ca^{2+}, *etc.* Although receptors are responsible for detecting extracellular signals, signal integration and sorting frequently depend on the activities of GTP-binding transducer proteins known as G proteins.

G protein-coupled receptors act by catalyzing the binding of GTP to appropriate G proteins. GTP binding activates the G protein, allowing it to bind and activate cellular effector proteins that generate second messengers, as is described in other chapters. Such three-protein relays sort messages both convergently and divergently. Several different receptors may trigger activation of a single G protein and different G proteins may all activate a single effector. One single receptor may also trigger activation of several different G proteins, however, and one G protein can activate multiple effectors. The pattern of a cell's responses to incoming information therefore reflects its complement of receptors, G proteins and effectors, as well as their relative selectivity for each other.

NATO ASI Series, Vol. H 52
Biological Signal Transduction
Edited by E. M. Ross and K. W. A. Wirtz

THE REGULATORY GTPase CYCLE

A G protein is activated by binding one molecule of GTP (Fig. 1). However, G proteins display intrinsic GTP-hydrolyzing activity, and bound GTP is rapidly ($t_{1/2}$ ~15 sec) converted to bound GDP, which does not cause activation. Reactivation involves the release of the tightly bound GDP and the binding of a second GTP molecule, both of which are far slower reactions. Thus, a steady-state GTPase cycle is defined in which the relative extent of G protein activation is the steady-state concentration of the G protein-GTP intermediate. Under resting conditions, GTP-activated G protein represents less than 1% of the total. However, an agonist-liganded receptor can promote both GDP release and GTP binding to increase the steady-state amount of G protein-GTP to greater than 60% of the total (Brandt and Ross, 1986). Agonist-liganded receptor acts as a formal catalyst of GDP/GTP exchange. A single receptor can interact sequentially with multiple G protein molecules in a process involving lateral diffusion in the plasma membrane. Under optimal conditions, receptor-stimulated GTP exchange is sufficiently rapid that one receptor can maintain the activation of multiple G proteins and thereby amplify the signal substantially. Receptor-catalyzed G protein activation is expressed as

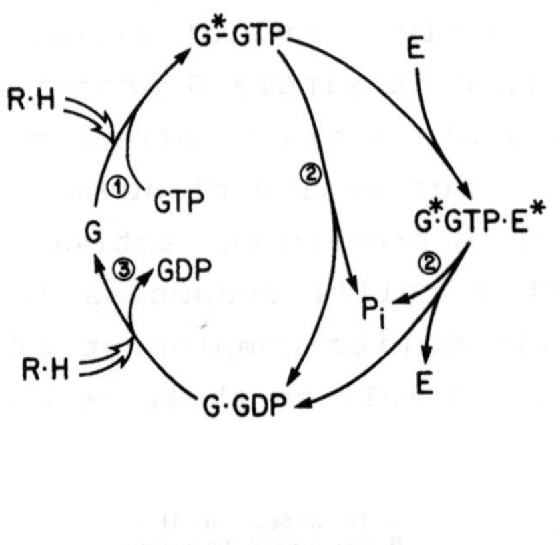

Figure 1. The regulatory GTPase cycle. Binding of GTP activates a G protein (G*-GTP) such that it can bind and activate an effector protein (G*·GTP·E*). Hydrolysis of bound GTP (reaction 2) terminates activation, leaving GDP-liganded, inactive G protein (G·GDP). Agonist-liganded receptor (R·H) catalyzes the release of GDP (reaction 3) and the binding of GTP (reaction 2). Because the basal rates of reactions 1 and 3 are slower than that of reaction 2, G proteins are largely inactive in the presence of GTP unless agonist is present. (From Ross, 1989.)

activation of downstream intracellular effector proteins. The detailed mechanism of these processes has been reviewed elsewhere (Ross, 1989; Gilman, 1987; Chabre, this volume).

G PROTEINS AND THEIR SUBUNITS

The last five years have seen a dramatic increase in what we know about the structural components of G protein-mediated signaling systems. Many G proteins have been purified and their sequences determined from cloned cDNA's (Lochrie and Simon, 1988; Strathman *et al.*, 1989). Numerous G protein-coupled receptors have also been purified and cloned (Parker and Ross, 1990).

G proteins are composed of three subunits, denoted α (largest), β and γ (smallest). The α subunit can be readily separated from the $\beta\gamma$ subunits, which form a stable complex that has not been resolved under non-denaturing conditions. The α subunit binds and hydrolyzes GTP. Furthermore, isolated GTP-liganded α subunit can activate its appropriate effector in the absence of $\beta\gamma$. Selectivity for receptor also seems to reside in the α subunit. The $\beta\gamma$ subunits to regulate the binding of nucleotides to α, help anchor α to the plasma membrane and mediate regulation of the α subunit by receptor (Ross, 1989; Gilman, 1987, for reviews).

G protein α subunits are a family of homologous proteins, ranging in size from about 39,000 Da to 45,000 Da. Homology is which are also conserved in other GTP-binding proteins. X-ray crystallography of the bacterial elongation factor Tu and the oncogene product $p21^{ras}$ indicates that these three regions surround the GTP binding site (Jurnak, 1985; Pai *et al.*, 1989). The α subunits, while homologous, are distinct for each G protein. Thus, α_s determines G_s, $\alpha_{i,2}$ determines $G_{i,2}$, and so on. Sequence homology among the α subunits and the similarity of their biochemical properties suggests that they have a similar tertiary structure. It is reasonable to assume that each α subunit will have a definable receptor-binding domain that is reasonably conserved with respect to structure and function, but which is adequately distinct such that selectivity of the G protein for receptor is maintained. A similar argument can be made for an effector-binding domain.

In contrast to the obvious functional individuality of the α subunits, the $\beta\gamma$ subunits are frequently considered as a common pool that is shared among the α's. This view is correct in part, but it also reflects the difficulty in separating distinct species of $\beta\gamma$. When the $\beta\gamma$ subunits are prepared from G protein trimers or are isolated from chromatographic fractions of plasma membrane extracts that do not include α subunits, these fractions appear able to regulate any of the α subunits. Recently, however, we found that different preparations of $\beta\gamma$ subunits differentially allow coupling of a specific receptor to different G proteins when receptor, α and $\beta\gamma$ are co-reconstituted into phospholipid vesicles. These $\beta\gamma$ fractions contain the products of up to four distinct β subunit genes (M. I. Simon, personal communication) and, based on electrophoretic analysis, three or more species of γ. At very least, a mixture of two β and three γ subunits yield a diverse group of regulatory $\beta\gamma$ dimers.

G PROTEIN-COUPLED RECEPTORS

Receptors that regulate G proteins are a large family of highly homologous, hydrophobic, integral plasma membrane proteins. Extensive biochemical studies of rhodopsin (Findlay and Pappin, 1986, for review) and more recent biochemical and genetic studies of the β-adrenergic receptor (see below) have contributed to a reasonably good understanding of the overall structure of these proteins. The G protein-coupled receptors are formed from a bundle of seven hydrophobic, membrane-spanning, α-helices that are connected by hydrophilic sequences on either side of the plasma membrane. The amino terminus lies on the extracellular side of the membrane and the carboxyl terminus is cytoplasmic.

Sequence similarity among the receptors is not uniformly distributed. The membrane-spanning helices display the most homology among receptors and there is little sequence similarity among the amino-terminal domains, the central portions of the large cytoplasmic loops that connect spans five and six and the carboxy terminal domains. Even the lengths of these non-homologous regions vary dramatically. The overall impression of the G protein-coupled receptors is therefore one of a highly conserved hydrophobic core of membrane spans, a glycosylated extracellular amino terminus, and two large cytoplasmic regions (Fig. 2).

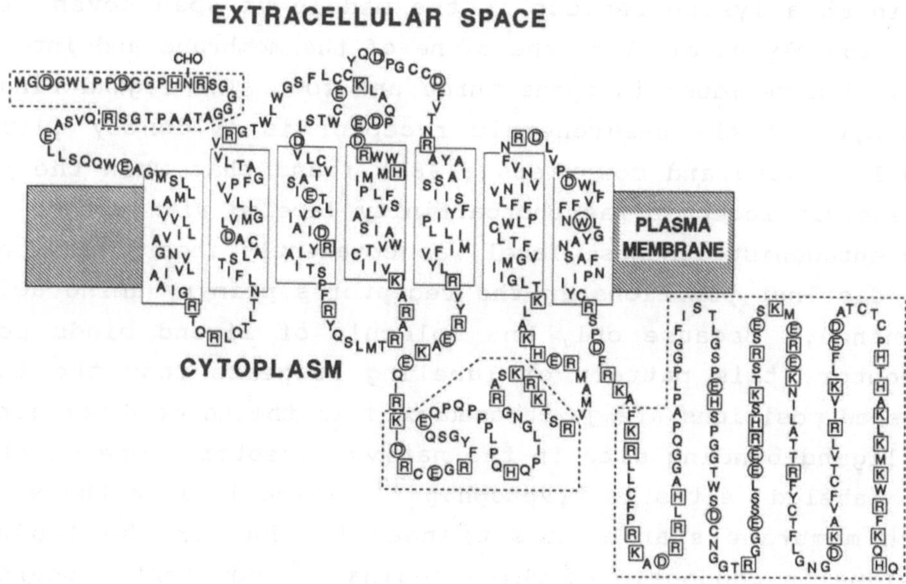

Figure 2. The amino acid sequence of the β-adrenergic receptor is shown in a pattern that displays its membrane-spanning helical regions and the associated cytoplasmic loops in relation to the plasma membrane bilayer. Basic residues are shown in squares and acidic residues are shown in light circles. Those portions of the receptor that can be proteolytically or genetically deleted without loss of function are outlined by the dashed lines. (From Ross, 1989).

The importance of the hydrophobic core of the receptors and of the short hydrophilic segments nearest to the membrane surface has been confirmed by the observation that the function of the β-adrenergic receptor is insensitive to proteolysis or genetic deletion of the non-conserved hydrophilic regions (Rubenstein *et al.*, 1987; Dixon *et al.*, 1987a,b). Removal of 29 of 40 N-terminal amino acid residues, more than half of the large third cytoplasmic loop, and almost all of the C terminal region does not diminish the ability of the β-adrenergic receptor to regulate G_s in response to agonists. Thus, both ligand binding and G protein regulation are subserved by the core structure.

In the case of the opsins and the receptors for biogenic amines, the ligand binding site lies within the bundle of membrane spanning helices. Retinal is covalently bound to opsin at a lysine residue in the middle of span seven. It lies roughly parallel to the plane of the membrane and interacts with residues in spans three and four. The ligand binding site of the β-adrenergic receptor is evidently quite similar. Wong and coworkers (1988) found that when the β-adrenergic receptor was photoaffinity labeled with either of two antagonist ligands, label was covalently incorporated at two distinct positions in the receptor's primary amino acid sequence. Because only one molecule of ligand binds per receptor, this pattern of labeling suggests that the two labeled positions are juxtaposed to form the three dimensional ligand-binding site in the native receptor. One of the two labeled sites is tryptophan^{330}, which lies in the seventh membrane span at a site near to that of the lysine residue in rhodopsin to which retinal is covalently bound. The second labeled site lies somewhere between spans two and four, suggesting that these spans form the opposite side of the catecholamine binding site. Other residues involved in catecholamine binding were identified by Strader, Dixon and coworkers (1989) by challenging site-directed mutants in the receptor with a panel of appropriately designed catecholamine analogs. They showed that an aspartate residue in span three acts as a counter-ion for the positive charge of the catecholamine, as is true for the Schiff's base in rhodopsin, and that two serine residues in span four confer recognition of the two catechol hydroxyl groups. The key question now is how the binding of a β-adrenergic agonist or *trans*-retinal within the bundle of helices can alter the structure of the receptor to convey a signal to the G protein bound on the cytoplasmic surface.

THE G PROTEIN-REGULATORY SITE OF RECEPTORS

Importance of Cationic Domains

Because much of the third cytoplasmic loop and the carboxyl terminal domain of receptors can be removed without loss of regulatory function, the more homologous regions near the bilayer surface must form the site that selectively binds and regulates G proteins. The importance of cationic sequences such as these for G protein activation has been supported by our observation that short, positively charged peptides can catalyze G protein activation through a receptor-like mechanism (Higashijima *et al.*, 1988, 1990). Mastoparans are tetra-cationic, amphiphilic, tetradecapeptides from wasp venoms that were first described as potent stimulators of secretion from a wide variety of secretory cells (Hirai *et al.*, 1979). The likely role of G proteins in triggering secretion (B. Gomperts, this volume) suggested that mastoparan might act directly on a G protein. We found that mastoparan stimulates the rate of nucleotide exchange by several G protein and this effect displays several characteristic properties of receptor-mediated regulation. The effect of mastoparan on G_i or G_o is blocked by pertussis toxin-catalyzed ADP-ribosylation of the α subunit. Such ADP-ribosylation generally does not alter the function of an isolated G protein but does block its ability to respond to receptor (van Dop *et al.*, 1984; Okajima *et al.*, 1985). Mastoparan also stimulates nucleotide exchange at low concentrations (below 100 nM) of Mg^{2+} without altering the intrinsic k_{cat} for hydrolysis of bound GTP. Mastoparan also appears to bind to G proteins competitively with respect to receptors. These data, which are summarized by Higashijima *et al.* (1990), suggest that the mastoparans may be structurally as well as functionally similar to the G protein-binding domain of cell surface receptors.

Although there is no significant similarity between the sequence of mastoparans and those of the cytoplasmic face of the G protein-coupled receptors, it is likely that there is overall similarity of tertiary structure. Membrane-bound mastoparan forms an α helix that is oriented such that its four positive charges are directed toward the aqueous medium (Wakamatsu *et al.*, 1983; Higashijima *et al.*, 1983). Each of the regions that are candidates for G protein binding sites -- the first and second intracellular loops, both origins of the large, third intracellular loop, and the beginning of the cytoplasmic carboxy terminal region -- are also strongly positively charged. We should now be able to use synthetic mastoparan analogs to determine what physical properties allow the activation of G proteins by positive charge clusters and what other properties allow selectivity.

Selectivity of mastoparans among G proteins is not as great as that displayed by most cell surface receptors, but it is still impressive. Mastoparan itself is relatively selective for G_o and G_i. We have synthesized other cationic peptides that selectively activate G_s and G_z. Other peptides and cationic compounds selectively antagonize the regulatory effects of "agonist" peptides. We should now be able to use synthetic mastoparan analogs and mutants of the β-adrenergic receptor to focus on what key features of these cationic arrays are important for G protein binding and activation.

Short Sequences Determine Selectivity for G Proteins

Defining the specific structures on receptors that comprise that selectively recognize and regulate G proteins is a particular challenge. Such understanding will allow us to predict which signaling pathway a specific receptor will utilize and also give us an immediate insight into the structural basis of the recognition process.

We have begun to identify G protein recognition signals by analyzing the selectivity among G proteins of genetically engineered chimeric receptors that contain putative selectivity sequences from one receptor grafted onto the core of another. The design of these chimeras is vastly simplified by two facts. First, because G proteins are fundamentally peripheral proteins bound to the inner face of the plasma membrane, the contact sites on a receptor must be on the cytoplasmic face. Second, the results of the proteolysis and deletion mutagenesis experiments discussed above rule out a role for the center of the large third intracellular loop and the extreme C-terminal domain. In essence, only five stretches of sequence need to be considered, as shown in Fig. 3.

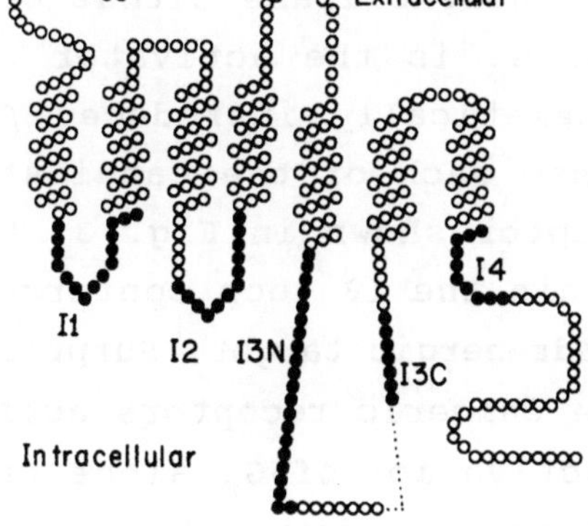

Figure 3. Potential G protein interaction sites (solid circles) are depicted in the context of the human m1 muscarinic cholinergic receptor. Chimeric receptors were constructed that have substituted sequence from one receptor into the homologous position on a second receptor. I1, I2 and I3 refer to the three intracellular loops. I3N and I3C refer to limited substitutions at either end of the I3 loop. Substitutions were made singly and together to detect functional interactions among the individual regions. The cDNA's were then expressed in A293 and COS cells to study their cellular signals (Wong *et al.*, 1989) and in Sf9 cells to allow their purification and reconstitution with isolated G proteins.

Initial studies of chimeric G protein-coupled receptors indicated the importance of the third intracellular loop in selectivity among G proteins. Kubo *et al.* (1988) showed that a chimeric m2 muscarinic cholinergic receptor that contained the third intracellular (I3) loop of the m1 receptor was able to elicit a characteristic m1 signal when the chimera was expressed in frog oocytes. The converse chimeric construct, an m1 receptor with the I3 loop of the m2 receptor, did not convey this signal. Similarly, (Kobilka *et al.* (1988) showed that a α_2-adrenergic receptor chimera that contained spans five and six and the I3 loop from the β-adrenergic receptor could stimulate G_s, the β-adrenergic receptor target.

In fact, the structural determination of G protein selectivity by receptors is both more simple and more complex than suggested by the first studies. Relatively short sequences within the I3 loop are sufficient to determine specificity in part, but two separate sequences must interact for faithful selection of G protein targets.

Wong *et al.* (1989) recently constructed a series of M1 muscarinic cholinergic:β-adrenergic chimeras and evaluated them for their ability to activate both G_p, the target of the m1 receptor, and G_s the target of the β-adrenergic receptor. G_p is a pertussis toxin-insensitive G protein that activates an inositol phospholipid-specific phospholipase C that has been only tentatively identified; G_s is the activator of adenylyl cyclase. When Wong systematically introduced β-adrenergic amino acid sequence into each of the candidate regions in the m1 muscarinic receptor shown in Fig. 3, he found that only substitutions within the I3 loop conferred the ability to activate G_s, the β-adrenergic target. Surprisingly, though, he found that these chimeric receptors activated both G_s and G_p. Selective activation of G_s alone required the introduction of β-adrenergic sequence both into the I2 and I3 loops, even though replacement of the I2 loop alone had little if any effect on the activity or selectivity of the m1 muscarinic receptor (Fig. 4).

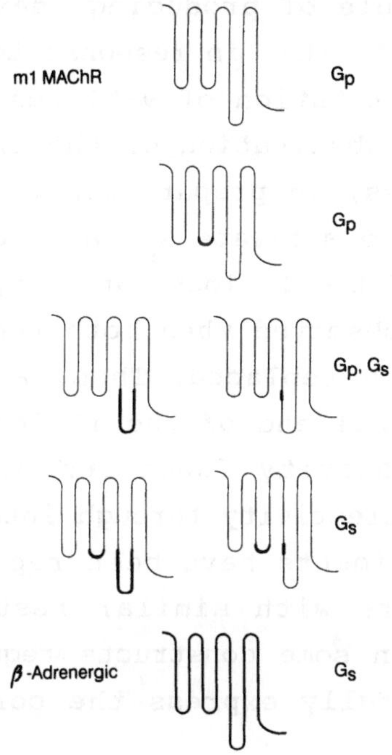

Figure 4. Phenotypes of chimeric m1 muscarinic cholinergic : β-adrenergic receptors. We constructed M1 muscarinic receptors (light lines) that contained substituent sequences from the avian β-adrenergic receptor (dark lines). The chimeras were assayed for their abilities to activate either G_p, which leads to IP_3 release, or G_s, which activates adenylyl cyclase. Substitution of either the entire I3 loop or its short I3N region led to activation of G_s, but replacement of I2 plus either I3 or I3N was required to eliminate activation of G_p. Substitution of I2 by itself had no effect on selectivity, nor did substitution of the I1 loop or the I4 region in any combination. Although the I3C region of the I3 loop is known to be important for the regulation of G proteins (Franke *et al.*, 1988; O'Dowd *et al.*, 1988; Cotecchia *et al.*, 1990), its replacement had no effect on selectivity among G proteins.

The requirement for sequences in both the I2 and I3 loops has now been confirmed in chimeras composed of sequences from several different donors and acceptors in the adrenergic and cholinergic families. The ability of I3 sequence to confer the ability to regulate novel G proteins and the requirement for I2 sequence to completely redirect coupling appears to be generally valid. It is not yet clear whether the two regions act independently, with I3 acting permissively and I2 acting restrictively, or whether there is important structural interaction between the loops to produce a larger and more strictly selective G protein binding site on the cytoplasmic face of the receptor.

Another surprising outcome of these studies was that a chimera that contained as few as four amino acid residues of

β-adrenergic receptor sequence near the N-terminal end of the I3 loop (I3N in Fig. 3) was capable of producing maximal stimulation of adenylate cyclase activity in response to the muscarinic agonist carbachol. Substitution of 9-11 residues was actually more effective than substitution of the entire I3 loop (S. K-F. Wong and E.M. Ross, in preparation). These I3N chimeras retained the ability to activate G_p, as did the chimeras with total replacements of the I3 loop, and complete redirection of coupling was only observed when both the I3N sequence and the I2 loop were both replaced. Thus, a very limited sequence near the N-terminal end of the I3 loop is responsible for all of the selectivity functions of the entire loop, and helps determine selectivity through interaction with I2. Again, these experiments have been repeated with multiple donor and acceptors with similar results, although the critical I3N region in some constructs required up to 20 amino acid residues to fully express the correct selectivity function.

The control of G protein selectivity by such limited regions on the receptors suggests the existence of a relatively small protein-protein interface that can potentially be mapped at a very detailed level. The relevant sequences are cationic, consistent with the structures of the mastoparans. It is particularly striking that all of these chimeric receptors retain the ability to regulate the activity of the hybrid G protein-binding domain in response to agonists. Our understanding of the important structures should lead to the synthesis of selective G protein-directed probes for studying *in vivo* regulation and, perhaps, the development of G protein-directed drugs.

ACKNOWLEDGEMENT

Studies from the author's laboratory were supported by NIH grant GM30355 and R. A. Welch Foundation Grant I-0982.

REFERENCES

Brandt, DR, and Ross, EM (1986) Catecholamine-stimulated GTPase cycle: Multiple sites of regulation by β-adrenergic receptor and Mg^{2+} studiedin reconstituted receptor-G_s vesicles. J Biol Chem 261:1656-1664

Cotecchia, S, Exum, S, Caron, MG, and Lefkowitz, RJ (1990) Regions of the α_1-adrenergic receptor involved in coupling to phosphatidylinositol hydrolysis and enhanced sensitivity of biological function. Proc Natl Acad Sci USA 87:2896-2900

Dixon, RAF, Sigal, IS, Candelore, MR, Register, RB, Scattergood, W, Rands, E, and Strader, CD (1987) Structural features required for ligand binding to the β-adrenergic receptor. EMBO J 6:3269-3275

Dixon, RAF, Sigal, IS, Rands, E, Register, RB, Candelore, MR, Blake, AD, and Strader, CD (1987) Ligand binding to the β-adrenergic receptor involves its rhodopsin-like core. Nature 326:73-77

Findlay, JBC, and Pappin, DJC (1986) The opsin family of proteins. Biochem J 238:625-642

Fong, HKW, Amatruda, I, T T, Birren, BW, and Simon, MI (1987) Distinct forms of the β subunit of GTP-binding regulatory proteins identified by molecular cloning. Proc Natl Acad Sci USA 84:3792-3796

Franke, RR, Sakmar, TP, Oprian, DD, and Khorana, HG (1988) A single amino acid substitution in rhodopsin (Lysine 248 Leucine) prevents activation of transducin. J Biol Chem 263:2119-2122

Gilman, AG (1987) G proteins: Transducers of receptor-generated signals. Ann Rev Biochem 56:615-649

Higashijima, T, Burnier, J, and Ross, EM (1990) Regulation of G_i and G_o by mastoparan, related amphiphilic peptides and hydrophobic amines: mechanism and structural determinants of activity. J Biol Chem 265:14176-14186.

Higashijima, T, Uzu, S, Nakajima, T, and Ross, EM (1988) Mastoparan, a peptide toxin from wasp venom, mimics receptors by activating GTP-binding regulatory proteins (G proteins). J Biol Chem 263:6491-6494

Higashijima, T, Wakamatsu, K, Takemitsu, M, Fujino, M, Nakajima, T, and Miyazawa, T (1983) Conformational change of mastoparan from wasp venom on binding with phospholipid membrane. FEBS Lett 152:227-230

Hirai, Y, Yasuhara, T, Yoshida, H, Nakajima, T, Fujino, M, and Kitada, C (1979) A new mast cell degranulating peptide "Mastoparan" in the venom of *Vespula lewisii*. Chem Pharm Bull 27:1942-1944

Jurnak, F (1985) Structure of the GDP domain of EF-Tu and location of the amino acids homologous to ras oncogene proteins. Science 230:32-36

Kobilka, BK, Kobilka, TS, Daniel, K, Regan, JW, Caron, MG, and Lefkowitz, RJ (1988) Chimeric α_2-, β_2-adrenergic receptors: Delineation of domains involved in effector coupling and ligand binding specificity. Science 240:1310-1316

Kubo, T, Bujo, H, Akiba, I, Nakai, J, Mishina, M, and Numa, S (1988) Location of a region of the muscarinic acetylcholine receptor involved in selective effector coupling. FEBS Lett 241:119-125

Lochrie, MA, and Simon, MI (1988) G protein multiplicity in eukaryotic signal transduction systems. Biochemistry 27:4957-4965

O'Dowd, BF, Hnatowich, M, Regan, JW, Leader, WM, Caron, MG, and Lefkowitz, RJ (1988) Site-directed mutagenesis of the cytoplasmic domains of the human β_2-adrenergic receptor. J Biol Chem 263:15985-15992

Okajima, F, Katada, T, and Ui, M (1985) Coupling of the guanine nucleotide regulatory protein to chemotactic peptide receptors in neutrophil membranes and its uncoupling by islet-activating protein, pertussis toxin. J Biol Chem 260:6761-6768

Pai, EF, Kabsch, W, Krengel, U, Holmes, KC, John, J, and Wittinghofer, A (1989) Structure of the guanine-nucleotide-binding domain of the Ha-ras oncogene product p21 in the triphosphate conformation. Nature 341:209-214

Parker, EM, and Ross, EM (1990) G protein-coupled receptors: Structure and function of signal transducing proteins. In: Current Topics in Membranes and Transport, v 36, T Claudio, Ed, Academic Press, NY pp 131-144

Ross, EM (1989) Signal sorting and amplification through G protein-coupled receptors. Neuron 3:141-152

Rubenstein, RC, Wong, SK-F, and Ross, EM (1987) The hydrophobic tryptic core of the β-adrenergic receptor retains G_s-regulatory activity in response to agonists and thiols. J Biol Chem 262:16655-16662

Strader, CD, Sigal, IS, and Dixon, RAF (1989) Structural basis of β-adrenergic receptor function. FASEB J 3:1825-1832

Strathmann, M, Wilkie, TM, and Simon, MI (1989) Diversity of the G-protein family: Sequences from five additional α subunits in the mouse. Proc Natl Acad Sci USA 86:7407-7409

Van Dop, C, Yamanaka, G, Steinberg, F, Sekura, R, Manclark, CR, Stryer, L, and Bourne, HR (1984) ADP-ribosylation of transducin by pertussis toxin blocks the light-stimulated hydrolysis of GTP and cGMP in retinal photoreceptors. J Biol Chem 259:23-25

Wakamatsu, K, Higashijima, T, Fujino, M, Nakajima, T, and Miyazawa, T (1983) Transferred NOE analyses of conformations of peptides as bound to membrane bilayer of phospholipid; mastoparan X. FEBS Lett 162:123-126

Wong, SK-F, Parker, EM, and Ross, EM (1990) Chimeric muscarinic cholinergic:β-adrenergic receptors that activate G_s in response to muscarinic agonists. J Biol Chem 265: 6219-6224

Wong, SK-F, Slaughter, C, Ruoho, AE, and Ross, EM (1988) The catecholamine binding site of the β-adrenergic receptor is formed by juxtaposed membrane-spanning domains. J Biol Chem 263:7925-7928

HORMONE RECEPTOR-EFFECTOR COUPLING IN PROLACTIN PRODUCING PITUITARY CELL LINES (GH CELLS)

Ruth H. Paulssen, Eyvind J. Paulssen, Kaare M. Gautvik
and Jan O. Gordeladze
Institute of Medical Biochemistry
University of Oslo
P.O.Box 1112 Blindern
0317 Oslo 3
Norway

INTRODUCTION

We have previously used clones of rat pituitary tumour cells (GH cells) to study cell membrane signal transduction in response to hormone stimulation (Gautvik et al. 1988). Recently, we have been interested in the coupling of pituitary peptide hormone receptors to specific GTP-binding regulatory proteins (G proteins) and how these signal proteins initiate intracellular messenger effector systems.

We have investigated the distribution of G protein mRNA and protein levels in these cells and delineated how vasoactive intestinal peptide (VIP), thyroliberin (TRH) and somatostatin (SRIF) couple to these proteins. In order to study these molecular mechanisms of hormone action, we have also used transient expression of G protein antisense RNA.

This review will thus briefly describe the GH cell system, some of the hormone regulated, G protein dependent signalling systems and the principle of using antisense techniques in the study of the function of these membrane proteins.

THE GH CELL SYSTEM

The GH cells are differentiated clonal pituitary cell strains that produce and spontaneously secrete prolactin (PRL) and/or growth hormone (GH) (Tashjian, 1979). GH cells respond to several hypothalamic and steroid hormones in a manner analogous to normal

NATO ASI Series, Vol. H 52
Biological Signal Transduction
Edited by E. M. Ross and K. W. A. Wirtz

anterior pituitary cells. GH_3 and GH_4C_1 cells produce both PRL and GH, while GH_12C_1 cells predominantly produce GH (table 1).

Table 1: Production of prolactin (PRL) and growth hormone (GH) in different rat pituitary tumour cell lines (GH_12C_1, GH_3 and GH_4C_1 cells).GH_4C_1 cells produce predominantly PRL while GH_3 cells produce both, PRL and GH. GH_12C_1 cells produce mainly GH.

Hormone production in GH cell lines

	PRL	GH
GH_12C_1	(+)	++
GH_3	++	+
GH_4C_1	+++	(+)

HORMONE RECEPTORS AND THE ACTIONS OF NEUROENDOCRINE HORMONES IN GH CELLS

GH_4C_1 and GH_3 cells contain receptors for the neuroendocrine peptide hormones thyroliberin (TRH), vasoactive intestinal peptide (VIP), and somatostatin (SRIF) (Hinkle et al. 1973; Gautvik et al. 1981; Bjøro et al. 1984; Bjøro et al. 1987), whereas the GH_12C_1 cells may have lost their functional receptors for TRH and SRIF (Gautvik et al., unpublished observations).
The action of TRH has been extensively studied in the clonal strains of the pituitary tumour cells (for review see Gautvik et al. 1983). The binding of [^{3}H]-TRH is stereospecific and shows existence of two independent binding sites (Gautvik et al. 1981), although only one polypeptide of 64 kDa on SDS-polyacrylamide gel electrophoresis has been demonstrated (Wright et al. 1988). The different binding affinities, and the fact that PRL release by TRH is biphasic (Gautvik et al. 1988) (see figure 1), has been suggested the evidence for the presence of TRH binding sites with the potential to activate several transmembrane signalling systems separately.
Vasoactive intestinal peptide (VIP) is also able to cause PRL

release in GH cells. The presence of specific VIP receptors in GH cells has been demonstrated (Bjøro et al. 1984). There is also a profound effect of VIP on PRL mRNA accumulation and hormone synthesis (Paulssen et al. 1990b).

TRH and VIP both stimulate PRL secretion, but their effects are different with regard to the onset of PRL release in response to

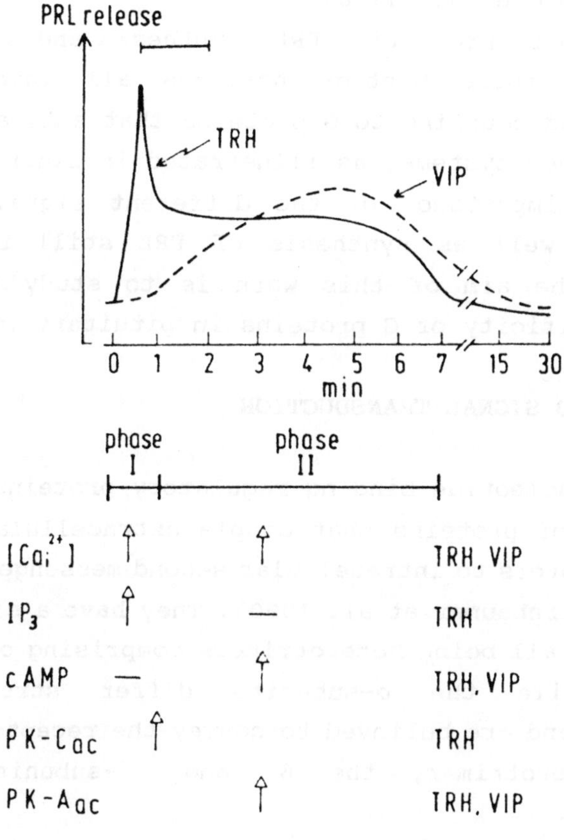

<u>Figure 1:</u> Schematical representation of the secretion profile of prolactin (PRL) from GH cells induced by thyroliberin (TRH) and vasoactive intestinal peptide (VIP). Also shown are the intracellular mediators known to be formed or activated by cellular treatment with TRH and/or VIP: Cytosolic Ca^{2+} concentrations ([Ca^{2+}]$_i$), inositol trisphosphate (IP_3), cyclic 3',5'-adenosine monophosphate (cAMP), protein kinase C activation (PK-C_{ac}) and protein kinase A activation (PK-A_{ac}).

maximum concentrations of the hormones (Gautvik et al. 1988) (see figure 1).

The phorbolester 12-0-tetradecanoyl phorbol 13-acetate (TPA) also gives a secretory effect which in onset resembles VIP, but may be related to the second phase of the TRH response (Negro-Vilar et al. 1985) and it also stimulates prolactin synthesis.

The hypothalamic and gastrointestinal polypeptide somatostatin (SRIF) inhibits basal as well as all kind of stimulated hormone secretion (Bjøro et al. 1988).

The regulation of prolactin (PRL) synthesis and secretion in GH_3 cells by these peptide hormones all involve receptor interaction and coupling to G proteins that activate or inhibit second messenger systems, as illustrated in figure 2.

The relative importance of the different signal systems for secretion as well as synthesis of PRL still remains to be elucidated. The aim of this work is to study the receptor-effector specificity of G proteins in pituitary (GH) cells.

G PROTEINS AND SIGNAL TRANSDUCTION

The guanine nucleotide binding regulatory proteins (G proteins) are a family of proteins that couple extracellularly activated membrane receptors to intracellular second messenger systems (for review, see Birnbaumer et al. 1990). They have a similar protein organization, all being heterotrimers comprising on α-, ß-and - subunits. While the α-subunits differ structurally and functionally and are believed to convey the receptor specificity of each heterotrimer, the ß- and -subunits show less heterogeneity.

The G_s protein is believed to be the sole stimulator of AC activity. The $G_s\alpha$-subunit is known as two different protein products (Northup et al. 1980), and four corresponding mRNAs that originate from the same gene have been described (Bray et al. 1986; Robishaw et al. 1986). G_s has also been shown to activate ligand regulated calcium channels (Yatani et al. 1988).

G_i was named for its inhibitory effect on adenylyl cyclase. Three different proteins ($G_{i-1}\alpha$, $G_{i-2}\alpha$ and $G_{i-3}\alpha$) from separate genes (Jones et al. 1987; Itoh et al. 1988; Goldsmith et al. 1988) are known. G_o has two distinct α-subunits ($G_o\alpha$) (Van Meurs et al. 1987; Hsu et al. 1990). Both $G_{i-3}\alpha$ and $G_o\alpha$ are believed to activate potassium channels (VanDongen et al. 1988; Mattera et al. 1989).

The recently cloned pertussis toxin (PTX)-insensitive α-subunit of G_x/G_z (Fong et al. 1988; Matsuoka et al. 1988) has yet no known function. The model for activation of GTP-binding proteins has been reviewed by several groups (Casey et al. 1988; Neer et al. 1988; Birnbaumer et al. 1987; Spiegel, 1987; Weiss et al. 1988; Allende, 1988).

Effector systems known to be regulated through G proteins are thus adenylyl cyclase (AC), phospholipase C (PL-C), phospholipase A_2 and receptor operated potassium and calcium channels.

DISTRIBUTION OF GUANINE NUCLEOTIDE BINDING REGULATORY PROTEINS IN RAT PITUITARY CELL LINES (GH CELLS)

We have recently studied the expression of different G-proteins in the rat pituitary cell lines (Paulssen et al. 1990). Northern blots of RNA from different cell lines as well as from rat pituitary have been hybridized to different cDNA probes for $G_s\alpha$, $G_o\alpha$, $G_{i-1}\alpha$, $G_{i-2}\alpha$, $G_{i-3}\alpha$ and $G_x\alpha$.

All cells contained mRNA for these G proteins, with the exception that $G_{i-1}\alpha$ mRNA was only found in normal pituitary cells. The mRNA distribution correlated well to the G protein pattern shown by immunoblotting. A summary of these results is outlined in table 2. The analyses showed surprising and characteristic differences in the expression of G protein mRNA between the various cell lines and tissues. For $G_x\alpha$ and $G_o\alpha$ we found several transcripts, some of which seemed to be preferentially expressed in the different tissues and cell lines.

The different distribution of G protein mRNAs is important when functional studies on signal transduction is performed.

We therefore also assayed basal and hormone stimulated adenylyl cyclase and phospholipase C activity in the same cell lines.

These studies could form the basis of functional implications of the observed differential expression of G proteins, but no firm conclusions could be drawn. $G_o\alpha$ and $G_x\alpha$ were considerably more

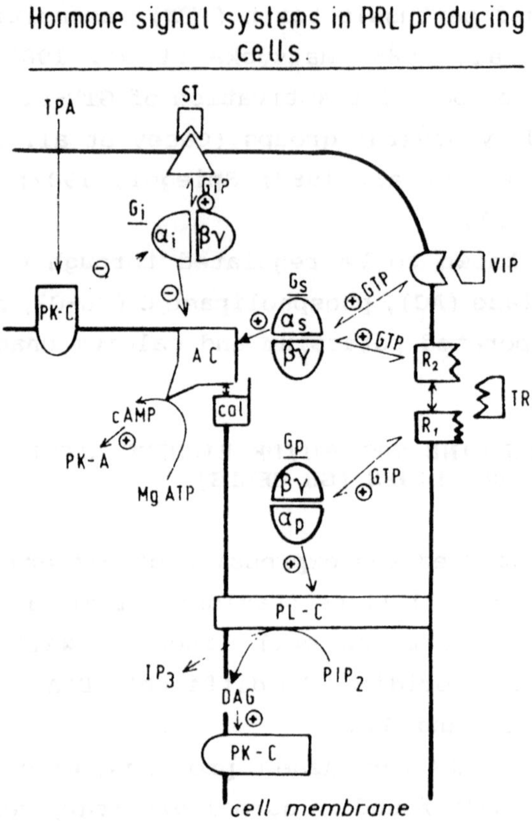

Figure 2: Hormone signal systems in GH cells. A schematic representation of the action of thyroliberin (TRH), vasoactive intestinal peptide (VIP), somatostatin (SRIF), and the phorbolester 12-o-tetradecanoyl phorbol-acetate (TPA). After peptide hormone receptor coupling, this complex interacts with GTP binding proteins which in presence of GTP is able to dissociate the heterotrimer structure into functional subunits that connect to an effector molecule. Thus α_p activates phospholipase C (PL-C) generating inositol trisphosphate (IP_3) and diacylglycerol (DAG) from phosphatidyl inositol 4,5-bisphosphate (PIP_2). Proteinkinase C (PK-C), which is activated by DAG or TPA, inactivates the G_i complex by inactivating its α_i subunit thereby releasing the inhibition of adenylyl cyclase (AC). In contrast, SRIF stimulates the inhibitory activity of α_i reducing the AC formation of cyclic AMP (cAMP) from its substrate (MgATP). VIP activates the G_s complex by dissociating α_s, thereby activating AC with formation of cAMP and activation of protein kinase A (PK-A).

Table 2: Distribution of guanine nucleotide-binding regulatory proteins (G proteins) in rat pituitary tumour cell lines (GH_12C_1, GH_3 and GH_4C_1 cells) and rat pituitary tissue (Pit.).

Distribution of G proteins in GH cells and pituitary

	$G_s\alpha$	$G_o\alpha$	$G_{i-1}\alpha$	$G_{i-2}\alpha$	$G_{i-3}\alpha$
GH_12C_1	+	++	−	+	+
GH_3	++	+	−	+	++
GH_4C_1	++	+	−	+	++
Pit.	+	+	+	+	(+)

generally expressed than $G_{i-1}\alpha$, opening the possibility that $G_{i-1}\alpha$ is not responsible for coupling to phospholipase C.

HORMONE SIGNALLING SYSTEMS IN GH CELLS

Both TRH and VIP generate more than one intracellular signal of importance for PRL secretion and/or synthesis.

In GH_3 cells, TRH regulates intracellular Ca^{2+} levels by rapidly increasing inositol polyphosphate (IP_3, IP_4?) production (Gershengorn et al. 1985). Apparently, this process and subsequent PRL release are mediated via a pertussis toxin (PTX) insensitive GTP-binding protein (Aub et al. 1986). In addition, TRH also enhances adenylyl cyclase (AC) activity in GH_3 and GH_4C_1 cells (Gordeladze et al. 1988a). It has been suggested that this is solely due to the effect of protein kinase C (PK-C) activated by diacylglycerols (DAG), since we have demonstrated that phorbol esters (TPA) which activate PK-C directly, inactivate the inhibitory $G_i\alpha$ protein coupling to AC (Gordeladze et al. 1988b). We could prove this notion to be incorrect by using a PKC inhibitor polymyxin B, which abolished TPA-enhanced AC but left all of the TRH-responsive AC intact.

These functional studies suggest that TRH is able to activate directly at least two different signalling systems; one which couples to a G_p-complex[1] activating phospholipase C (PL-C) (Straub et al. 1986; Martin et al. 1986), and the other which activate AC via a G_s-complex. The action mechanism of VIP, in relation to G proteins, appears to be monofunctional with its sole action on the G_s-complex, resulting in AC activation (Gordeladze et al. 1988a).

ANTISENSE TECHNIQUES

Transcriptional units that generate antisense RNAs have been introduced to many cells by DNA transfection and shown to inhibit expression of cellular genes (Izant et al. 1985; Kim et al. 1985; Hélène et al. 1990). The use of antisense RNA expression has been proposed as a general method for specific inhibition of gene expression (Izant et al. 1985; Hélène et al. 1990). Antisense RNA molecules contain sequences complementary to a portion or all of the target mRNA. The mechanism of inhibition may be due to the formation of RNA:RNA hybrids that might result in blocked RNA translation (Melton, 1985), the failure of RNA transport to the cytoplasm (Kim et al. 1985), an increased RNA turnover (Crowley et al. 1985) or a combination of these events.

After having worked out the optimal conditions for transfection and transient expression of foreign genes in GH cells and the effect this technique has on the endogenous hormone production itself (Paulssen et al. 1990b), we first used this approach to express antisense rat GH RNA to inhibit growth hormone production (Paulssen et al. 1990a). The rGH cDNA used in our experiments specified the entire coding sequence for the rat GH precursor and also contained part of the 5′ and all of the 3′ untranslated region of rGH mRNA. The results indicated that transfection of differentiated rat pituitary tumour cells with an antisense rGH expression plasmid resulted in specific decrease of levels of this RNA.

---

[1] G_p denotes the G protein complex that is coupled to phospholipase C. G_p may turn out to be more than one G protein.

ELUCIDATION OF RECEPTOR EFFECTOR SPECIFICITY BY THE USE OF ANTISENSE TECHNIQUES

To delineate which G protein TRH receptors couple to we likewise constructed an antisense RNA expression vector that carried part of the $G_s\alpha$ cDNA (Jones et al. 1987) (encoding the 52 kDa protein for $G_s\alpha$) in an inverted orientation relative to an inducible metallothionein promoter (Palmiter et al. 1983). Initial results indicated that transfected cells showed a specific inhibition of $G_s\alpha$ mRNA transcription that was also followed by decreased protein levels for $G_s\alpha$ after 24 hours. Induction of the metallothionein promoter with $CdCl_2$ was followed by a further decrease in $G_s\alpha$ mRNA expression. Transient expression of $G_s\alpha$ was followed by 80-90% inhibition of TRH activated AC and 40% reduced VIP stimulated adenylyl cyclase (AC) activity, while the phospholipase C (PL-C) activity remained unaffected under same conditions (R.H.Paulssen et al., manuscript in preparation). We therefore conclude that the TRH receptor in GH_3 cells is directly coupled to a G_s or a G_s-like protein to the adenylyl cyclase system.

CONCLUDING REMARKS

Hormone receptor-regulated second messenger signals features a network of interactions between receptors, G protein complexes and second messenger proteins. Regulation in signal conduction must be based upon altered levels of regulatory peptides or peptide modification by phosphorylation etc. A detailed knowledge of this interaction can be obtained by inhibition of or modification (mutation) of existing peptides or substitution of peptides that are constitutionally absent. We have, after studies of hormone receptors, existing G proteins and second messenger enzymes, chosen expression inhibition of G proteins as our main approach to elucidate the molecular mechanisms occurring at the membrane level. We believe this to be a generally suitable method in the study of cellular signal mechanisms.

ACKNOWLEDGEMENTS

The work upon which this review is based received economical support by the Norwegian Society against Cancer, the Anders Jahre foundation for the Promotion of Sciences, Oslo, and Insulinfondet, Copenhagen, Denmark.

REFERENCES

Allende JE (1988) GTP-mediated macromolecular interactions: the common features of different systems. FASEB J 2: 2356-2367

Aub DL, Frey EA, Sekura RD and Cote TE (1986) Coupling of the thyrotropin-releasing hormone receptor to phospholipase C by a GTP-binding protein distinct from the inhibitory or stimulatory protein GTP-binding protein. J Biol Chem 261: 9333-9340

Birnbaumer L, Codina J, Mattera R, Yatani A, Scherer N, Toro M-J and Brown AM (1987) Signal transduction by G proteins. Kidney Int 32: S14-S17

Birnbaumer L, Abramowitz J and Brown AM (1990) Receptor-effector coupling by G proteins. Biochem Biophys Acta 1031: 163-224

Bjøro T, Haug E, Sand O and Gautvik KM (1984) Vasoactive intestinal peptide causes a calcium dependent stimulation of prolactin secretion in GH4C1 cells. Mol Cell Endocrinol 37: 41-50

Bjøro T, Wiik P, Opstad PK, Gautvik KM and Haug E (1987) Binding and degradation of vasoactive intestinal peptide in prolactin-producing tumour cells (GH4C1). Acta Physiol Scand 130: 609-618

Bjøro T, Østberg BC, Sand O, Torjesen PA, Penman E, Gordeladze JO, Iversen J-G, Gautvik KM and Haug E (1988) Somatostatin inhibits prolactin secretion by multiple mechanisms involving a site of action distal to increased cyclic adenosine 3',5'-monophosphate and elevated cytosolic Ca^{2+} in rat lactotrophs. Acta Physiol Scand 133: 271-282

Bray P, Carter A, Simons C, Guo V, Puckett C, Kamholz J, Spiegel A and Nirenberg M (1986) Human cDNA clones for four species of $G_\alpha s$ signal transduction protein. Proc Natl Acad Sci USA 83: 8893-8897

Casey PJ and Gilman AG (1988) G protein involvement in receptor-effector coupling. J Biol Chem 263: 2577-2580

Crowley TE, Nellen W, Gomer RH and Firtel RA (1985) Phenocopy of discoidin I-minus mutants by antisense transformation in dictyostelium. Cell 43: 633-641

Fong HKW, Yoshimoto KK, Eversole-Cire P and Simon MI (1988) Identification of a GTP-binding protein α subunit that lacks an apparent ADP-ribosylation site for pertussis toxin. Proc Natl Acad Sci USA 85: 3066-3070

Gautvik KM and Lystad E (1981) Demonstration of heterogenous population of binding sites for thyroliberin in prolactin producing tumour cells and their possible functional significance. Eur J Biochem 116: 235-242

Gautvik KM, Gordeladze JO, Jahnsen T, Haug E, Hansson V and

Lystad E (1983) Thyroliberin receptor binding and adenylyl cyclase activation in cultured prolactin-producing rat pituitary tumor cells (GH Cells). J Biol Chem 258: 10304-10311

Gautvik KM, Bjøro T, Sletholt K, Østberg BC, Sand O, Torjesen P, Gordeladze JO, Iversen J-G and Haug E (1988) Regulation of prolactin secretion and synthesis by peptide hormones in cultured rat pituitary cells, in:Molecular Mechanisms in Secretion. Alfred Benzon Symposium 25. eds. N.A. Thorn, M. Treiman and O.H. Pedersen (Munksgaard, Copenhagen) pp.211-227

Gershengorn MC and Thaw C (1985) Thyrotropin-releasing hormone (TRH) stimulates biphasic elevation of cytoplasmic free calcium. Further evidence that TRH mobilizes cellular and extracellular Ca^{2+}. Endocrinology 116: 591-596

Goldsmith P, Rossiter K, Carter A, Simonds W, Unson CG, Vinitsky R and Spiegel AM (1988) Identification of the GTP-binding protein encoded by G_{i-3} complementary DNA. J Biol Chem 263: 6476-6479

Gordeladze JO, Bjøro T, Østberg BC, Sand O, Torjesen P, Haug E and Gautvik KM (1988a) Phorbol esters and thyroliberin have distinct actions regarding stimulation of prolactin secretion and activation of adenylate cyclase in rat pituitary tumour cells (GH_4C_1 cells). Biochem Pharmacol 37: 3133-3138

Gordeladze JO, Sletholt K, Thorn NA and Gautvik KM (1988b) Hormone-sensitive adenylate cyclase of prolactin-producing rat pituitary adenoma (GH_4C_1) cells: Molecular organization. Eur J Biochem 177: 665-672

Hinkle PM and Tashjian AHJr (1973) Receptor for thyroid-releasing hormone in prolactin producing rat pituitary cells in culture. J Biol Chem 248: 6180-6186

Hsu WH, Rudolph U, Sanford J, Bertrand P, Olate J, Nelson C, Moss LG, Boyd ABIII, Codina J and Birnbaumer L (1990) Molecular cloning of a novel splice variant of the α subunit of the mammalian G_o protein. J Biol Chem 265: 11220-11226

Hélène C and Toulmé J-J (1990) Specific regulation of gene expression by antisense, sense and antigene nucleic acids. Biochem Biophys Acta 1049: 99-125

Itoh H, Toyama R, Kozasa T, Tsukamoto T, Matsuoka M and Kaziro Y (1988) Precence of three distinct molecular species of Gi protein α subunit. J Biol Chem 263: 6656-6664

Izant JG and Weintraub H (1985) Constitutive and conditional suppression of exogenous and endogenous genes by anti-sense RNA. Science 229: 345-352

Jones DT and Reed RR (1987) Molecular cloning of five GTP-binding protein cDNA species from rat olfactory neuroepithelium. J Biol Chem 262: 14241-14249

Kim SK and Wold BJ (1985) Stable reduction of thymidine kinase activity in cells expressing high levels of anti-sense RNA. Cell 42: 129-138

Martin TF, Lucas DO, Bajjalieh SM and Kowalchyk JA (1986) Thyrotropin-releasing hormone activates a Ca^{2+}-dependent phosphoinositide phosphodiesterase in permeable GH_3 cells. GTP S potentiation by a cholera and pertussis toxin-

insensitive mechanism. J Biol Chem 261: 2918-2927
Matsuoka M, Itoh H, Kozasa T and Kaziro Y (1988) Sequence analysis of cDNA and genomic DNA for a putative pertussis toxin-insensitive guanine nucleotide-binding regulatory protein α subunit. Proc Natl Acad Sci USA 85: 5384-5388
Mattera R, Yatani A, Kirsch G, Graf R, Okabe K, Olate J, Codina J, Brown AM and Birnbaumer L (1989) Recombinant αi-3 subunit of G protein activates Gk-gated K+ channels. J Biol Chem 264: 465-471
Melton DA (1985) Injected anti-sense RNAs specifically block messenger RNA translation in vivo. Proc Natl Acad Sci USA 82: 144-148
Neer EJ and Clapham DE (1988) Roles of G protein subunits in transmembrane signalling. Nature 333: 129-134
Negro-Vilar A and Lapetina EG (1985) 1,2-didecanoylglycerol and phorbol 12,13-dibutyrate enhance anterior pituitary hormone sectretion in vitro. Endocrinology 117: 1559-1564
Northup JK, Sternweis PC, Smiegel MD, Schleifer LS, Ross EM and Gilman AG (1980) Purification of the regulatory component of adenylate cyclase. Proc Natl Acad Sci USA 77: 6516-6520
Palmiter RD, Norstedt G, Gelinas RE, Hammer RE and Brinster RL (1983) Metallothionein-human GH fusion genes stimulate growth of mice. Science 222: 809-814
Paulssen EJ, Paulssen RH, Haugen TB, Gautvik KM and Gordeladze JO (1990) Distribution of guanine nucleotide binding regulatory proteins in rat pituitary cell lines. Mol Cell Endocrinol (in press)
Paulssen RH, Paulssen EJ, Aleström P, Gordeladze JO and Gautvik KM (1990a) Specific antisense RNA inhibition of growth hormone production in differentiated rat pituitary tumour cells. Biochem Biophys Res Commun 171: 293-300
Paulssen RH, Paulssen EJ, Aleström P and Gautvik KM (1990b) Electroporation of rat pituitary (GH) cells lines: Optimal parameters and effects on endogenous hormone production. Biochem Biophys Res Commun 171: 1029-1036
Robishaw JD, Smigel MD and Gilman AG (1986) Molecular basis for two forms of the G protein that stimulates adenylate cyclase. J Biol Chem 21: 9587-9590
Spiegel AM (1987) Signal transduction by guanine nucleotide binding proteins. Mol Cell Endocrinol 49: 1-16
Straub GE and Gershengorn MC (1986) Thyrotropin-releasing hormone and GTP activate inositol trisphosphate formation in membranes isolated from rat pituitary cells. J Biol Chem 261: 2712-2717
Tashjian AHJr (1979) Clonal strains of hormone producing pituitary cells. Methods Enzymol 58: 527-535
Van Meurs KP, Angus CW, Lavu S, Kung H-F, Czarnecki SK, Moss J and Vaughan M (1987) Deduced amino acid sequence of bovine retinal $G_o\alpha$: Similarities to other guanine nucleotide-binding proteins. Proc Natl Acad Sci USA 84: 3107-3111
VanDongen AMJ, Codina J, Olate J, Mattera R, Joho R, Birnbaumer L and Brown AM (1988) Newly identified brain potassium channels gated by the guanine nucleotide binding protein Go. Science 242: 1433-1437
Weiss ER, Kelleher DJ, Woon CH, Sopakarkar S, Osawa S, Heasley

LE and Johnson GL (1988) Receptor activation of G proteins. FASEB J 2: 2841-2848

Wright M, Høgset A, Aleström P and Gautvik KM (1988) A 64 kDa proteinis a candidate for a thyrotropin-releasing hormone receptor in prolactin producing rat pituitary tumor cells (GH_4C_1 cells). Biochem Biophys Res Commun 157: 875-882

Yatani A, Imoto Y, Codina J, Hamilton SL, Brown AM and Birnbaumer L (1988) The stimulatory G protein of adenylyl cyclase, Gs, also stimulates dihydropuridine-sensitive Ca^{2+} channels. J Biol Chem 263: 9887-9895

EFFECT OF AGING ON THE ß-ADRENERGIC SIGNAL TRANSDUCTION SYSTEM IN RAT KIDNEY, LUNG, AND MONONUCLEAR LEUKOCYTES

N. Fraeyman, E. Van de Velde and P. Vanscheeuwijck
Heymans Institute of Pharmacology
De Pintelaan 185
B-9000 GHENT
BELGIUM

INTRODUCTION

Increasing evidence suggests that the process of aging is accompanied by, or results in, altered function of the adrenergic system (Pfeifer et al., 1982; Roberts and Tumer, 1987). In this overview recent data on the effect of aging on the properties and the function of the ß-adrenergic signal transduction (ßAST) system in kidneys, lungs and mononuclear leukocytes will be discussed. Earlier results have been summarized before (Ebstein et al., 1985; Krall, 1985; Feldman, 1986), while the effect of aging on the ßAST system in the heart has been reviewed recently (Weiss et al., 1984; Nielson and Vestal, 1989).

In a first part, some of the biochemical properties of the ßAST chain and the possibilities for its study will be outlined. In the second part, the effect of aging on the ßAST system in kidney, lungs and splenic mononuclear leukocytes from rats will be described. For mononuclear leukocytes, the results obtained with human cells will also be discussed.

For recent theories on the aging process itself, we refer to some recent publications (Hayflick, 1985; Meites et al., 1987; Bjorksten and Tenhu, 1990; Foresta et al., 1990; Geokas et al., 1990; Lockshin and Zakeri, 1990; Roa and Cohen, 1990).

NATO ASI Series, Vol. H 52
Biological Signal Transduction
Edited by E. M. Ross and K. W. A. Wirtz

1. THE β-ADRENERGIC TRANSDUCTION SYSTEM

The β-adrenergic transduction system is one of the best described neurotransmitter systems (Strosberg, 1987; Raymond et al., 1990). It is composed of three units : the agonist binding site or receptor molecule, the stimulatory G_s-protein and the adenylate cyclase.

Based on the relative potency of β-adrenergic agonists and antagonists, three subtypes of **β-adrenergic receptors** are distinguished: $β_1$, $β_2$ and $β_3$ (Lands et al., 1967; Arch et al., 1983; Granneman, 1990; Kaumann, 1989). Their structure and molecular properties have been elucidated using cloning techniques (Frielle et al., 1988; Emorine et al., 1989).

The **G_s-protein** is composed of three subunits : α_s, β and γ (Freissmuth et al., 1989; Birnbaumer et al., 1990). The α-component which possesses GTPase activity (Casey and Gilman, 1988), is considered to be the active subunit for the transmission of the signal while the β-γ dimer is probably necessary for membrane attachment (Jones et al., 1990; Mumby et al., 1990). Under favourable conditions, the β-adrenergic receptor can be associated with the G_s-protein, resulting in a higher affinity configuration for the agonist than the uncoupled receptor.

The third component of the βAST chain, the effector enzyme **adenylate cyclase** is located in the plasma membrane and converts ATP into cAMP, using Mg^{2+}-ions as co-factor (Limbird, 1981; Casperson and Bourne, 1987).

When the receptor is stimulated by agonists, an interaction of the three components of the βAST chain within the micro environment of the plasma membrane is required for transduction of the signal (Chabre, 1987; Gierschik et al., 1987; Levitzki, 1988). Binding of the agonist (first messenger) to the binding site leads to dissociation of α_s from G_s and formation of the receptor-α_s complex, while the β-γ complex remains in the membrane. Via the exchange of GDP for GTP, the activated receptor transmits the signal to the G_s-protein which activates the adenylate cyclase. The final step in the transduction is the conversion of ATP into cyclic AMP which is considered as the biochemical response upon receptor stimulation (second messenger). The original situation for the G-protein is restored by the hydrolysis of GTP into GDP by the GTPase activity of the α_S-subunit.

It has to be stressed that after the transformation of the extracellular signal into a second, cytoplasmatic signal, the generation of the physiological effect (contraction, relaxation, secretion, glycogenolysis....) requires another set of enzyme systems such as kinases. The interactions and regulatory pathways of this highly complicated system will not be discussed here but it is clear that some age-related changes in physiological effects might well be caused by changes in this part of the ßAST system.

There are two major types of experiments for the study of the properties of the ßAST system. Either the properties of the binding site itself are measured or the activity of the ßAST system, in terms of cAMP synthesis, is assayed. For methodological details see Vanscheeuwijck et al. (1989a) and Yamamura et al. (1990).

Ligand binding experiments yield data on receptor density and equilibrium dissociation constant for the ligand. With the appropriate antagonists, data on the presence of ß-adrenergic receptor subtypes can be obtained, while the use of agonists may give an estimate of the percentage of receptors in high affinity configuration i.e. coupled with the G_S-protein.

For measurement of cAMP formation, the ßAST chain can be stimulated at different levels providing information on the activity and interaction of the different components of the chain. If the tissue preparation is stimulated at the level of adenylate cyclase using forskolin, information on the reactivity of the enzyme is obtained. Stimulation at the level of the G_S-protein, using F^--ions, GTP or a non-hydrolysable analogue of GTP, yields information on the interaction of the G_S-protein and the adenylate cyclase and on the activity of the enzyme itself. Stimulation at the level of the receptor with ß-agonists in the presence of optimal concentrations of co-factors for the G-protein (GTP or its analogue GppNHp) and for adenylate cyclase (ATP, an ATP-regenerating system and Mg^{2+}) requires coupling between the different components and yields data on the functionality of the whole ßAST system.

2. EFFECT OF AGING ON THE ßAST SYSTEM IN KIDNEY, LUNG, AND MONONUCLEAR LEUKOCYTES

2.1. Kidney

Aging is accompanied by the progressive deterioration of renal function (Anderson and Brenner, 1986; Vega et al., 1990). There are some recent data in the literature on the effect of age on the ßAST chain in the kidney of the rat. Galbusera et al. (1988) found that total ß-receptor density remained unchanged with age, while the ratio β_1/β_2 changed from approximately 3/4 to 4/2 ; synthesis of cAMP was not measured. In contrast to these results we found that the receptor density increased with aging and that the ratio between the subtypes was slightly shifted in favour of the β_2-subtype (Vanscheeuwijck et al., 1990a). The discrepancies between the two studies may be related to different experimental conditions : rat strain, preparation of the tissue (including the absence of EGTA in the experiments of Galbusera which might favour proteolysis, Benovic et al., 1983; Stiles et al., 1984), ligand and competitors. We further found a lowered percentage of receptors coupled with the G-protein upon aging Vanscheeuwijck et al. (1990a). The stimulation of cAMP formation by guanine nucleotides and forskolin was more pronounced in kidneys from old rats. This result could be suggestive for an increased amount of G-proteins or adenylate cyclase and/or a more effective coupling between these two compounds. The cAMP-synthesis upon stimulation with the agonist isoproterenol, however, was unaffected by age in spite of the observed changes in receptor density and coupling between receptor and

G_s-protein. We conclude from these results that a highly efficient mechanism of compensation within the ßAST chain exists; hence any impairment of the ßAST system mediated physiological function in kidney should be located at the post-ßAST site.

2.2. Lung

Beta-adrenergic receptors in the lungs of adult rats mediate a number of physiological functions, including relaxation of airway smooth muscle, tracheal fluid production and surfactant release into the alveoli from type II pneumocytes (Frossard and Landry, 1985). There is evidence for a decrease in physiological response to ß-adrenergic agonist stimulation with aging (Duncan et al., 1982; O'Donnel and Wanstall, 1983; Duncan and Douglas, 1985).

The results related to the effect of aging on the properties of the ßAST chain are contradictory. Abrass et al. (1982) reported that there was no change in ß-receptor density and equilibrium dissociation constant for the antagonist dihydroalprenolol in lung membranes from Fisher 344 rats of 3, 12 and 24 months of age. Later, it was shown that isoproterenol bound with low and high affinity to the receptor, and that upon aging, the affinity of the high affinity binding site decreased significantly (Scarpace and Abrass, 1983). In the same study, adenylate cyclase activity was shown to be reduced in the older animals in all stimulatory conditions. In further experiments, Scarpace and Yu (1987), demonstrated that both receptor density and adenylate cyclase activity upon agonist stimulation were lowered in older animals. The authors explained the contradiction with their previous results, as far as receptor density is concerned, by the fact that another strain of rats was used. In another paper of the same group, a slight increase of the receptor density was observed (Scarpace and Baresi, 1988). In this case the use of purified membrane preparations could explain the discrepancy with previous results.

It has to be mentioned that at least one other paper has shown an increase in receptor density in lung tissue from rats (Vanscheeuwijck et al. 1989b). We investigated several other parameters of the ßAST chain and found that the percentage of receptors with high affinity configuration for the agonist was lower in older animals, and that in all stimulatory conditions, except for forskolin, a similar stimulation of the production of cAMP was obtained in lungs from young and old rats. Most interestingly, similar to what was found for kidneys, isoproterenol stimulation under optimal conditions, resulted in a comparable increase in the activity of adenylate cyclase. We concluded that, as for kidney, higly efficient homeostatic mechanisms are present, which may compensate the age-related changes at the different levels of the ßAST system.

2.3. Mononuclear leukocytes

Since the discovery of ß-adrenoceptors on human mononuclear leukocytes (Williams et al., 1976), these easily obtainable blood cells have been frequently used as a model for less accessible tissues. This raises the question in how far mononuclear leukocytes are representative for other tissues, but it is beyond the scope of this short overview to discuss this problem (Brodde et al., 1989). Up until now, however, no definite physiological function for ß-receptors in human or animal mononuclear leukocytes has been found, although the interaction between the neuroendocrine and the immune system has been discussed recently (Cavagnaro et al., 1988). From a biochemical point of view, the ßAST system in mononuclear leukocytes is functional : the presence of G_s-protein and of adenylate cyclase have been demonstrated and stimulation of the receptors with isoproterenol leads to an increase of the intracellular cAMP concentration.

The effect of aging on receptor density in a membrane preparation of human mononuclear leukocytes has been reported first by Schocken and Roth (1977) who found a lowered receptor density. However, numerous studies using membrane preparations (Abrass and Scarpace, 1981; Kraft and Castleden, 1981; Landmann et al., 1981; Doyle et al., 1982; Brodde, 1983; Landmann et al., 1983; Halper et al., 1984; Zahniser et al., 1988) or intact lymphocytes (Middeke at al., 1984) failed to confirm their findings. It has to be mentioned that the methodology for the purification of the cells and the preparation of membranes is important (Marinetti et al., 1983; Meurs et al., 1982; Sandnes et al., 1987). It is nowadays generally accepted that aging has no effect on the receptor density of ß-receptors in mononuclear leukocytes.

The effect of aging on the coupling between receptor and G-protein has been investigated by Feldman et al., (1984). In their elegant study, they found that the percentage of receptors coupled with the G-protein decreased significantly upon aging. These findings were in contrast to those of Landmann et al. (1983) but were confirmed later by Zahniser et al. (1988).

The activity of adenylate cyclase under a number of stimulatory conditions has been investigated repeatedly. In most studies, the cAMP-response was slightly decreased (Abrass and Scarpace, 1982; Krall et al., 1983; O'Hara et al., 1985; Krall et al., 1987), while in one case no change in activity was found (Mader and Davis, 1989).

From this overview, one might conclude that in human mononuclear leukocytes there is no evidence for an altered receptor density ; the coupling between receptor and G-protein is impaired while the basal or forskolin stimulated adenylate cyclase activity is affected only to a very small extent. The production of cAMP upon isoproterenol stimulation in the presence of guanine nucleotides is unaffected or at most slightly reduced.

We recently studied the ßAST chain in splenic leukocytes of rats. Very few age-dependent differences in the properties of the ßAST chain were found : receptor density and ligand affinity were both unaltered with aging (Vanscheeuwijck et al., 1990b). As for lung and kidney (Vanscheeuwijck et al., 1989b, 1990a), cAMP production upon stimulation of the receptors with isoproterenol in the presence of GTP was unaltered with age, although a significant decrease in percentage high affinity binding sites was observed.

We believe that in rat splenic leukocytes, the age-related effect on the ßAST system is comparable to that in lung and kidney : the decrease in percentage high affinity binding sites for agonist is compensated for, since a similar production of cAMP is obtained for a similar stimulus with agonist. As far as human lymphocytes are concerned, Oppenheim et al. (1984) similarly concluded from the measurement of isoproterenol induced synthesis of cAMP in lymphocytes from young and old individuals, that the important characteristic of lymphocyte ß-adrenergic function is that the self-regulation remains intact in aged man.

3. FINAL REMARKS AND CONCLUSIONS

From this overview it is clear that there are many contradictory results on the effect of aging on the function of the ß-adrenergic system, and only part of the discrepancies can be explained by variable experimental conditions or manipulations.

A second conclusion is that, even when the coupling between binding site and G-protein is impaired, or when other properties of the ßAST system, such as receptor density or the activity of adenylate cyclase, are changed, this has few consequences at the level of the biochemical response. Indeed, the synthesis of cAMP is often found to be unaffected by age, which is suggestive for the presence of highly efficient homeostatic systems. If this is true one could argue that the observed alterations in physiological responses may originate from changes at the post-ßAST system level, rather than at the level of the ßAST chain itself.

An important parameter is probably the fluidity of the plasma membrane. As mentioned earlier, the interaction between the three components of the ßAST chain occurs in the plasma membrane (Hanski et al., 1979; Peters, 1988). Since it is known that, upon aging, the fluidity of the cell membrane tends to be reduced (Nokubo, 1985), this could explain the lowered degree of coupling between the binding site and the G-protein, which is almost always found in tissues from older animals. It remains to be established, both in vitro and in vivo, whether the age-induced modulation of the plasma membrane fluidity is really the cause for the decrease in coupling.

4. REFERENCES

Abrass IB, Scarpace PJ (1981) Human lymphocyte beta-adrenergic receptors are unaltered with age. J Gerontol 36:298-301

Abrass IB, Scarpace PJ (1982) Catalytic unit of adenylate cyclase : reduced activity in aged-human lymphocytes. J Clin Endocrinol Metabol 55:1026-1028

Abrass IB, Davis JL, Scarpace PJ (1982) Isoproterenol responsiveness and myocardial ß-adrenergic receptors in young and old rats. J Gerontol 37:156-160

Anderson S, Brenner BM (1986) Effects of aging on the renal glomerulus. Am J Med 80:435-442

Arch JRS, Ainsworth AT, Cawthorne MA, Piercy V, Sennitt MV, Thody VE, Wilson C, Wilson S (1984) Atypical ß-receptor on brown adipocytes as target for anti-obesity drugs. Nature 309:163-165

Benovic JL, Stiles GL, Lefkowitz RJ, Caron MG (1983) Photoaffinity labelling of mammalian beta-adrenergic receptors : metal-dependent proteolysis explains apparent heterogeneity. Biochem Biophys Res Comm 110:504-511

Birnbaumer L, Abramowitz J, Brown AM (1990) Receptor-effector coupling by G proteins. Biochim Biophys Acta 1031:163-224

Bjorksten J, Tenhu H (1990) The crosslinking theory of aging -added evidence. Expl Gerontol 25:91-95

Brodde O-E (1983) Endogenous and exogenous regulation of human α-and ß-adrenergic receptors. J Recept Res 3:151-162

Brodde O-E, Michel MC, Gordon EP, Sandoval A, Gilbert EM, Bristow MR (1989) ß-adrenoceptor regulation in the human heart : can it be monitored in circulating lymphocytes ? Eur Heart J 10(suppl B):2-10

Casperson GF, Bourne HR (1987) Biochemical and molecular genetic analysis of hormone-sensitive adenylyl cyclase. Ann Rev Pharmacol Toxicol 27:371-384

Casey PF, Gilman AG (1988) G-protein involvement in receptor-effector coupling. J Biol Chem 263:2577-2580

Chabre M (1987) The G protein connection : is it in the membrane or the cytoplasm ? TIBS 12:213-215

Cavagnaro J, Waterhouse GAW, Lewis RM (1988) Neuroendocrine-immune interactions: immunoregulatory signals mediated by neurohumoral agents. In: The year in immunology, (Eds Cruse JM, Lewis RE, Jackson), Karger, Basel; Vol 3:228-246

Doyle V, O'Malley K, Kelly JG (1982) Human lymphocyte ß-adrenoceptor density in relation to age and hypertension. J Cardiovasc Pharmacol 4:738-740

Duncan PG, Douglas JS (1985) Age-related changes in guinea pig respiratory tissues: considerations for assessment of bronchodilators. Eur J Pharmacol 108:39-48

Duncan PG, Brink C, Douglas JS (1982) ß-receptors during aging in respiratory tissues. Eur J Pharmacol 78:45-52

Ebstein RP, Stessman J, Eliakim R, Menczel J (1985) The effect of age on ß-adrenergic function in man : a review. Isr J Med Sci 21:302-311

Emorine LJ, Marullo S, Briend-Sutren M-M, Patey G, Tate K, Delavier-Klutchko C, Strosberg D (1989) Molecular characterisation of the human β_3-adrenergic receptor. Science 245:1118-1121

Feldman RD, Limbird LE, Nadeau J, Robertson D, Wood AJJ (1984) Alterations in leukocyte ß-receptor affinity with aging. New Engl J Med 310:815-819

Feldman RD (1986) Physiological and molecular correlates of age-related changes in the human ß-adrenergic receptor system. Fed Proc 45:48-50

Foresta P, Ramacci MT, Giuliani A, Angelucci L (1990) Aging and immune response : a multivariate approach. Expl Gerontol 25:117-126

Freissmuth M, Casey PJ, Gilman AG (1989) G proteins control diverse pathways of transmembrane signaling. FASEB J 3:2125-2131

Frielle T, Daniel KW, Caron MG, Lefkowitz RJ (1988) Structural basis of ß-adrenergic receptor subtype specificity studied with chimeric β_1/β_2-adrenergic receptors. Proc Natl Acad Sci 85:9494-9498

Frossard N, Landry Y (1985) Physiological approach of beta receptor coupling to adenylate cyclase in rat airways : ontogenical modification and functional antagonism. J Pharmacol Expl Ther 233:168-175

Galbusera M, Garattini S, Remuzzi G, Mennini T (1988) Catecholamine receptor binding in rat kidney : effect of aging. Kidney Internat 33:1073-1077

Geokas MC, Lakatta EG, Makinodan T, Timiras PS (1990) The aging Process. Ann Int Med 113:455-466

Gierschik P, Grandt R, Marquetant R, Jakobs KH (1987) Role of G-proteins in signal transduction. J Cardiovasc Pharmacol 10 (suppl 4):S6-S10

Granneman JG (1990) Norepinephrine and BRL 37344 stimulate adenylate cyclase by different receptors in rat brown adipose tissue. J Pharmacol Expl Ther 254:508-513

Halper JP, Mann JJ, Weksler ME, Bilezikian JP, Sweeney JA, Brown RP, Golbourne T (1984) Beta adrenergic receptors and cyclic AMP levels in intact human lymphocytes : effects of age and gender. Life Sci 35:855-863

Hanski E, Rimon G, Levitzki A (1979) Adenylate cyclase activation by the ß-adrenergic receptors as a diffusion-controlled process. Biochem 18:846-853

Hayflick L (1985) Theories of biological aging. Expl Gerontol 20:145-159

Jones TLZ, Simonds WF, Merendino JJ, Brann MR, Spiegel AM (1990) Myristoylation of an inhibitory GTP-binding protein α-subunit is essential for its membrane attachment. Proc Natl Acad Sci USA 87:568-572

Kaumann AJ (1989) Is there a third heart ß-adrenoceptor ? TIPS 10:316-320

Kraft CA, Castleden CM (1981) The effect of aging on ß-adrenoceptor-stimulated cyclic AMP formation in human lymphocytes. Clin Sci 60:587-589

Krall JF, Connelly-Fittingoff M, Tuck ML (1983) Lymphocyte adenylate cyclase and human aging. Proc Soc Expl Biol Med 173:475-480

Krall JF (1985) Aging: effect on the biochemistry of beta-adrenergic catecholamine action. Rev Clin Bas Pharmacol 5:159-174

Krall JF, Fernandez EI, Connelly-Fittingoff M (1987) Human aging: effect on the activation of lymphocyte cyclic AMP-dependent protein kinase by Forskolin. Proc Soc Expl Biol Med 184:396-402

Lands AM, Arnold A, McAuliff JP, Luduena FP, Brown TG (1967) Differentiation of receptor systems activated by sympathomimetic amines. Nature 214:597-598

Landmann R, Bittiger H, Bühler FR (1981) High affinity beta-2-adrenergic receptors in mononuclear leukocytes : similar density in young and old normal subjects. Life Sci 29:1761-1771

Landmann R, Bürgisser E, Bühler FR (1983) Human lymphocytes as a model for beta-adrenergic receptors in clinical investigation. J Recept Res 3:71-88

Levitzki A (1988) From epinephrine to cyclic AMP. Science 241:800-806

Limbird LE (1981) Activation and attenuation of adenylate cyclase. Biochem J 195:1-13

Lockshin RA, Zakeri ZF (1990) Programmed cell death : new thoughts and relevance to aging. J Gerontol : Biol Sci 45:B135-140

Mader SL, Davis PB (1989) Effect of age on acute regulation of beta-adrenergic responses in mononuclear leukocytes. J Gerontol 44:M168-173

Marinetti GV, Rosenfeld SI, Thiem PA, Condemi JJ, Leddy JP (1983) Beta-adrenergic receptors of human leukocytes. Biochem Pharmacol 31:2033-2043

Meites J, Goya Y, Takahashi S (1987) Why the neuroendocrine system is important in aging processes. Expl Gerontol 22:1-15

Meurs H, Van Den Bogaard W, Kauffman HF, Bruynzeel PLB (1982) Characterization of (-)-[^{3}H]dihydroalprenolol binding to intact and broken cell preparations of human peripheral blood lymphocytes. Eur J Pharmacol 85:185-194

Middeke M, Remien J, Holzgreve H (1984) The influence of sex, age, blood pressure and physical stress on β_2-adrenoceptor density of mononuclear cells. J Hypert 2:261-264

Mumby SM, Heukeroth RO, Gordon JI, Gilman AG (1990) G-protein α-subunit expression, myristoylation, and membrane association in COS cells. Proc Natl Acad Sci USA 87:728-732

Nielson CP, Vestal RE (1989) α-adrenoceptors, ß-adrenoceptors and aging. In Handbook of Hypertension vol 12: Hypertension in the Elderly. Eds Amery A, Staessen J. Elsevier Sci

Nokubo M (1985) Physical-chemical and biochemical differences in liver plasma membranes in aging F-344 rats. J Geront 40:409-414

O'Hara N, Daul, AE, Fesel R, Siekmann U, Brodde O-E (1985) Different mechanisms underlying reduced β_2-adrenoceptor responsiveness in lymphocytes from neonates and old subjects. Mech Ageing Devel 31:115-122

O'Donnell SR, Wanstall JC (1984) Beta-1 and beta-2 adrenoceptor-mediated responses in pulmonary artery and aorta from young and aged rats. J Pharmacol Expl Ther 228:733-738

Oppenheim G, Mintzer J, Halperin Y, Eliakim R, Stessman J, Ebstein RP (1984) Acute desensitization of lymphocyte beta-adrenergic-stimulated adenylate cyclase in old age and Alzheimer's disease. Life Sci 35:1795-1802

Peters R (1988) Lateral mobility of proteins and lipids in the red cell membrane and the activation of adenylate cycclase by ß-adrenergic receptors. FEBS letters 234:1-7

Pfeifer MA, Weinberg CR, Cook D, Best JD, Reenan A, Halter JB (1983) Differential changes of autonomic nervous system function with age in man. Am J Med 75:249-258

Raymond JR, Hnatowich M, Lefkowitz RJ, Caron MG (1990) Adrenergic receptors. Models for regulation of signal transduction processes. Hypert 15:119-131

Roa KM, Cohen HJ (1990) The role of the cytoskeleton in aging. Expl Geront 24:7-22.

Roberts J, Tumer N (1987) Age-related changes in autonomic function of catecholamines. Rev Biol Res Aging 3:257-298

Sandnes D, Jacobsen FW, Jacobsen S (1987) Modes of determining ß-adrenoceptor number in human mononuclear leukocytes. Pharmacol Toxicol 61:265-270

Scarpace PJ, Yu PB (1987) Diet restriction retards the age-related loss of beta-adrenergic receptors and adenylate cyclase activity in rat lung. J Gerontol 42:442-446

Scarpace PJ, Baresi LA (1988) Increased beta-adrenergic receptors in the light-density membrane fraction in lungs from senescent rats. J Gerontol 43:B163-167

Scarpace PJ, Abrass IB (1983) Decreased beta-adrenergic agonist affinity and adenylate cyclase activity in senescent rat lung. J Gerontol 38:143-147

Schocken DD, Roth GS (1977) Reduced beta-adrenergic receptor concentrations in ageing man. Nature 267:856-859

Stiles GL, Strasser RH, Kilpatrick BF, Taylor SR, Lefkowitz RJ (1984) Endogenous proteinases modulate the function of the ß-adrenergic receptor-adenylate cyclase system. Biochim Biophys Acta 802:390-398

Strosberg AD (1987) Molecular and functional properties of beta-adrenergic receptors. Am J Cardiol 59:3F-9F

Vanscheeuwyck P, Fraeyman N (1989a) Evaluation of a semiautomatic cell harvester filtration for the determination of ß-adrenoceptors in human mononuclear leukocytes. J Pharmac Methods 21:299-307

Vanscheeuwijck P, Van de Velde E, Fraeyman N (1989b) Effect of aging on properties and function of ß-adrenoceptors in rat lung. Eur J Pharmacol (Mol Pharmacol Sect) 172:373-380

Vanscheeuwijck P, Van de Velde E, Fraeyman N (1990a) The ß-adrenergic transduction system in kidneys from young and senescent rats. Eur J Pharmacol (Mol Pharmacol Sect) 188:129-137

Vanscheeuwijck P, Van de Velde E, Fraeyman N (1990b) Characterization of the ß-adrenergic transduction system in spleen mononuclear leukocyte membranes of young and senescent rats. Biochem Pharamcol 39:2035-2040

Vega JA, Ricci A, Amenta F (1990) Age-dependent changes of the sympathetic innervation of the rat kidney. Mech Ageing Devel 54:185-196

Weiss B, Clarck MB, Greenberg LH (1984) Modulation of catecholaminergic receptors during development and aging in Handbook of Neurochemistry, vol 6: receptors in the nervous system; Ed Lajtha, Plenum Press

Williams LT, Snyderman R, Lefkowitz RJ (1976) Identification of ß-adrenergic receptors in human lymphocytes by (-)[^{3}H]alprenolol binding. J Clin Invest 57:149-157

Yamamura HI, Enna SJ, Kuhar MJ (Eds) (1990) Methods in neurotransmitter receptor analysis. Raven Press New York

Zahniser NR, Parker DC, Bier-Laning CM, Miller JA, Gerber JG, Nies AS (1988) Comparison between the effects of aging on antagonist and agonist interactions with beta-adrenergic receptors on human mononuclear and polymorphonuclear leukocyte membranes. J Gerontol 43:M151-157.

THE TRANSDUCIN CYCLE IN THE PHOTOTRANSDUCTION CASCADE

Marc CHABRE, Bruno ANTONNY and T.Minh VUONG
Institut de Pharmacologie Moléculaire et Cellulaire du CNRS
660, route des Lucioles, Sophia Antipolis, F-06560 Valbonne

Among the numerous heterotrimeric G-proteins that have now been characterized, transducin, the G-protein that conveys the visual transduction signal in vertebrate retinal rods, was often considered as a special case: It is enormously abondant in the highly specialised rod outer segment (10 to 20% of total protein content of the retinal rod outer segment) and it is soluble: its extraction from ROS membranes does not require any detergent. Indeed, in the physiological cycle, the TαGTP subunit is released from the membrane after the rhodopsin-catalysed exchange of GDP for GTP, and it takes a cytoplasmic route to reach its membrane bound effector. This solubility might be uniquely related to the very peculiar morphology of the ROS that are filled up with tightly stacked discs. The cytoplasm is divided in extremely thin layers, (150 Å thickness) in which transducin remains confined at a high concentration (500 μM). Thus a soluble TαGTP cannot drift very far while shuttling between rhodopsin and PDE molecules present on the two membrane surfaces that limit the cytoplasmic layer. Other G-proteins, in more usual cellular environments, would dilute in the cytoplasm if they were released from the cell membrane. However transducin does not seem to differ drastically from the other G-proteins, none of which is an intrinsic membrane protein. The degrees of membrane attachments are related to acylations or farnesylations of subunits which are otherwise very analogous. Indeed, the β subunit of transducin is strictly identical to the 36 kDa subunit of Gs and Gi. Tα has no acylation or farnelysation, but limited sequence differences with α subunits of other G-protein, as expected from their different receptor and effector specificities. The GTP induced dissociation of α from βγ, first observed on transducin, is now recognised as a general phenomenon for all activated heterotrimeric G-proteins.

NATO ASI Series, Vol. H 52
Biological Signal Transduction
Edited by E. M. Ross and K. W. A. Wirtz

Thus the visual transduction cascade provides a very instructive model for analysing the cycle (Fig 1) of an heterotrimeric G-protein that flip-flops between an active, dissociated GTP-bound state and an inactive, reassociated GDP-bound state, and shuttles between a membrane receptor, rhodopsin, and an effector of intracellular second messenger, the cGMP-phosphodiesterase (Chabre and Deterre, 1989).

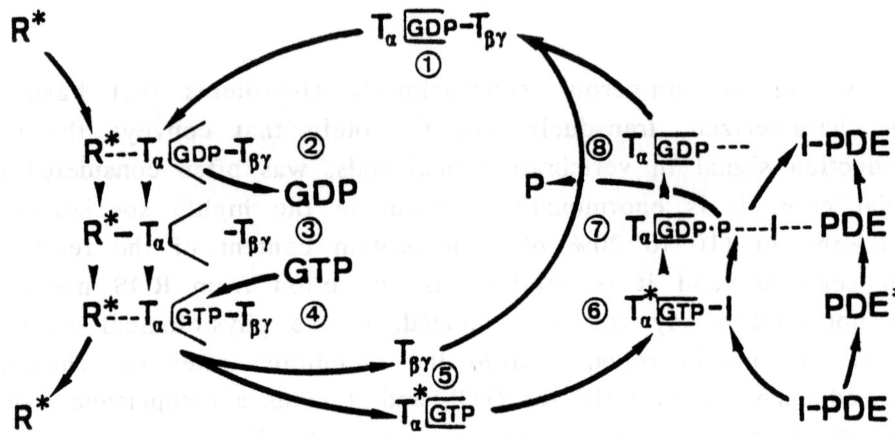

Fig 1. The transducin cycle

On the receptor side, rhodopsin stands as the archetype of the now very large class of "7 α helix" transmembrane receptors. Rhodopsin is exceptionally abundant in the disc membrane. As compared with the total of 10^4 to 10^5 receptors found in the plasma membrane of an hormone responsive cell, there are nearly 10^8 rhodopsin molecules in a mammalian retinal rod, and more than 10^9 in the large rods of batrachians. These enormous numbers are made necessary by the special physical character of the signal: Unlike chemical transmitters, photons do not diffuse in the extracellular medium around the cell until they are eventually captured by a receptor. Whereas a limited number of receptors are sufficient to confer to a cell an efficiency close to the theoretical maximum for the capture of a diffusing molecule (Berg, 1983), a much larger number of pigmented receptors is required to get a significant probability of absorption for photons that cross the cell only once in a straight line. Thus in visual cells, the number of receptors is in large excess over the number of G-proteins to which these receptors can couple. This anomalous stoichiometry has major implications. Conceptually it excludes *a priori* any "precoupling" between inactivated receptors and the G-proteins: As there are 10 times more rhodopsins than transducins, if the latter were bound to rhodopsin before illumination, only 10% of the receptor would be precoupled, and the remaining 90% would not be able to couple when they are activated. Even a lucky precoupled rhodopsin would activate only the single transducin to which it is already bound, as no free pool of transducin would be left to allow subsequent amplification. The very rapid activation of hundreds of transducin by a single photoexcited rhodopsin calls for fast sequential catalytic R*-T interactions.

The effector, cGMP phosphodiesterase, is peripherally bound to the disc membrane in substoichiometric amount with respect to transducin. Unlike the fast catalytic activation of transducin by rhodopsin (+ GTP), activation of PDE by transducin requires the continued interaction of one (or 2) TαGTP with each PDE, and terminates when the bound Tα inactivates upon hydrolysis of its GTP. The size of the PDE pool therefore sets the amplitude of the GMPc response, and the rate of GTP hydrolysis in Tα controls its termination.

HOLOTRANSDUCIN-GDP DIFFUSES LATERALLY ON THE MEMBRANE AND DOES NOT COUPLE TO RHODOPSIN

When no rhodopsin is activated, in the dark, if the medium ionic composition is kept close to that of the cytoplasm, transducin remains bound to the membrane as an heterotrimeric complex TαGDP-Tβγ. However the binding is weak: if the membrane suspension is too diluted (less than 0.1 mg/ml rhodopsin), a significant proportion of the transducin pool can be eluted from the membrane (Bruckert *et al.*, 1988). The membrane attachment is also very sensitive to the ionic strength and composition of the solution, and to the surface charge of the membrane, of which rhodopsin is a major component. But transducin is not specifically bound to a particular rhodopsin, and must diffuse rather freely on the membrane surface. However its lateral mobility, which may be critical for the rate of collisions with rhodopsin, has never been measured directly and it is often implicitly assumed that rhodopsin mobility, which is well known, is the sole factor influencing the collision rate. Recent kinetic measurements by a fringe photobleaching technique indicate that the relative diffusion of transducin with respect to rhodopsin is not the main rate limiting factor for the activation of transducin by R* (Bruckert *et al.*, in preparation). This suggests that the lateral mobility of transducin is higher than that of rhodopsin, as could be expected from the fact that the G-protein is in the cytoplasm on the membrane, while rhodopsin is embedded in the less fluid lipid layer. When discussing collision coupling models for other G-protein systems, the emphasis is also generally put on the receptor mobility. It is however very likely that, as in the case of the visual system, the most mobile component is the G-protein. Even if attached to the membrane by acyl or farnesyl chains that dip into the lipids, a G-protein whose entire polypeptide chain is in the cytoplasm diffuses faster than a receptor whose polypeptide chain crosses many times the lipid bilayer.

T-GDP INTERACTS WITH R*: THE "HIGH AFFINITY COMPLEX" AND NUCLEOTIDE EXCHANGE

Inactive rhodopsin contains a permanently bound retinal which, in its native "11-cis" conformation, behaved as a perfect antagonist. Light converts this antagonist into an agonist: the photoisomerization causes a displacement of the protonated Schiff-base bond between the retinal and a lysine ((Lys 277 on helix 7), which was in tight interaction with the negative charge of an Aspartic (Asp 113, on helix 3). Interestingly, in adrenergic receptors, an aspartic residue located at a similar position on helix 3 interacts with the charged $-NH_3^+$ of the agonist that binds. The pertubation on this aspartic seems to be the trigger for the transconformation of rhodopsin, and other 7 helix receptors, to the activated state R*. By contrast with R, R* has now an affinity for holotransducin TαGDP-Tβγ which upon collision binds to the cytoplasmic domain of R*. The binding specificity seems conferred mainly by the third cytoplasmic loop that connects helices 5 and 6, but other cytoplasmic loops contribute to the binding energy. The binding to R* induces in transducin a transconformation which results in "opening" the nucleotide site: Tα keeps an affinity in the 10^{-5} M range for GDP but the k_{off} rate increases by more than 6 orders of magnitude, from less than 10^{-3} to about 10^{3} sec^{-1}. That is the only effect of R*. The k_{on} rates of GDP and GTP in the empty site of a R*-bound Tα are comparable, R* does not increase the affinity of Tα for GTP. But under physiological conditions the fast exchange of GDP for GTP is favored over the simple GDP/GDP exchange because GTP is present at high concentration in the cell. The binding of GTP is quasi irreversible: it induces a structural change in Tα which locks the new nucleotide into the site; TαGTP then dissociates from R* as well as from Tβγ.

In situ, with about 1 mM GTP in the cytoplasm, an R*-catalyzed GDP/GTP exchange reaction is completed in the order of a millisecond (Vuong *et al.*, 1984). But *in vitro*, if all nucleotides are suppressed from the medium, T cannot bind GTP and the reaction stops at the transition state in which the R*-bound Tα has an open and empty nucleotide site. The R^*_{ret}-$T\alpha_{empty}$-$T\beta\gamma$ complex remains tight and stable quasi indefinitely unless one re-adds nucleotide. In this complex, Tα keeps permanently its binding capacity for GTP, and R* keeps its bound all-trans retinal, blocked in the "Meta II" spectroscopic state, whereas the retinal of a free R* would be rapidly released (Bornancin *et al.*, 1989). Thus the R*-T interaction which keeps the nucleotide site open in the G-protein, conversely maintains the (agonist) retinal blocked in the activated receptor R*. This effect is the equivalent of the "high affinity binding" of agonists observed on all receptors that couple to G-proteins, when nucleotides are suppressed from the medium. This increase of affinity is a consequence of the coupling to the G-protein which occurs after the agonist induced receptor activation, but not an evidence for a putative "precoupling" that would precede the binding of the agonist.

Retinal can be removed from R* in stabilized R^*_{ret}-T_{empty} complexes, by attacking its Schiff-base bond to Lys 277 with concentrated hydroxylamine. But even after the retinal is gone, the nucleotide free R^*_{empty}-T_{empty} complex remains stable. Adding GTP after a delay will liberate instantly a functionnal TαGTP, but then the released R^*_{empty} denaturates rapidly. A very long incubation in the absence of nucleotide will yield some dissociation, but free T_{empty} will then rapidly denaturate: the bound nucleotide seems required for the stability of the α subunit of an heterotrimeric G-protein, except for the very short time when it is interacting with an agonist-activated receptor.

THE ACTIVE STATE TαGTP. ACTIVATION BY FLUOROMETALLIC COMPLEXES

The changes induced in Tα by the binding of a GTP are responsible for its dissociation from R* and from Tβγ, and for its solubility. Other indications for a significant structural change are the decreased proteolytic susceptibility of TαGTP versus TαGDP, and a large increase of intrinsic tryptophane fluorescence.

Physiologically, the active TαGTP state is reached by a R* catalyzed GDP/GTP exchange in TGDP, rather than by a phospho-transfer reaction on the bound GDP. But it has recently been understood that the active state can also be artificially reached by a condensation reaction with metallofluoride complexes that bypasses the catalytic action of the receptor. This is the basis of the long known activating effect of fluoride salts (NaF or KF) on transducin, as well as on other heterotrimeric G-proteins (Bigay *et al.*, 1985; Bigay *et al.*, 1987). The reactive species are fluorometallic complexes generally formed with aluminium often present in traces amount in solutions, or rapidly extracted from the glassware by fluoride: in millimolar fluoride, aluminium as well as beryllium form ionic complexes such as AlF_4^- or $BeF_3(OH_2)^-$, which are stereo-analogs of a phosphate group. But these complexes contrast with phosphate by the ionic nature of their bond: the P-O bond is covalent but the Al-F and the Be-F bonds are ionic. This implies that in aquous solution the fluoride anions bond to the metal cation in these complexes are slowly exchangeable. Owing to this property, these complexes can bind to GDP in the nucleotide site of Tα by exchanging one of their fluorides for the terminal oxygen of the GDP β-phosphate, thus mimicking a GTP and promoting the activation of the α subunit.

While monitoring by intrinsic tryptophane fluorescence the kinetics of TαGDP activation by these complexes, we observed (Antonny *et al.*, 90) that the requirement of aluminium or beryllium for transducin activation is not strict when the NaF concentration is increased above 3 mM and disappears when it reaches 15 mM (Fig 2). The rate of activation of Tα by fluoride "alone" contrasts with those formed with ALF_4^- or BeF_3^- by a monotonous dependance of the activation kinetics with respect to fluoride concentration and the very fast deactivation kinetics (> 1 sec^{-1}) after fluoride dilution. This argues

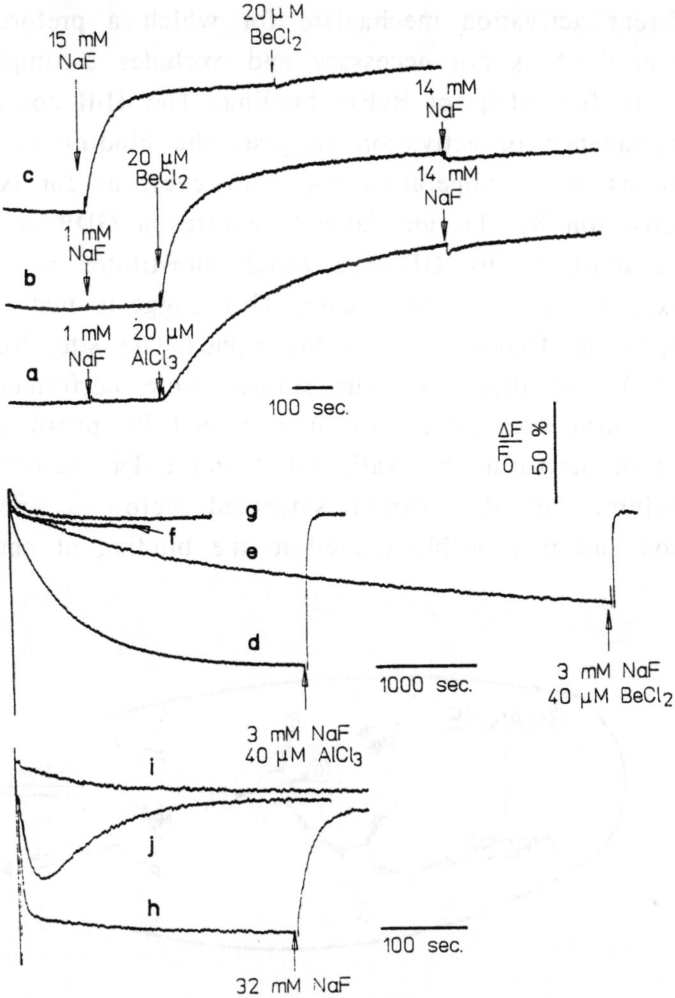

Fig 2. Kinetics of fluorescences changes correlated with transducin activation or deactivation. Traces **a, b, c** : Activation kinetics; 50 nM solutions of TαGDP in TKM (Tris, KCl, $MgCl_2$) buffer are activated by injections of NaF, $AlCl_3$ or $BeCl_2$ to the indicated concentrations. **d,e,f,g**: deactivation kinetics and controls; concentrated solutions of TαGDP that have been activated at saturation by incubating for 25 minutes at 25°C in TKM buffer with 3 mM NaF and 40 µM $AlCl_3$ (d), or 3 mM NaF and 40 µM $BeCl_2$ (e) are diluted 200 folds (50 nM final) in TKM buffer or (f,g) diluted in their respective incubation medium for controls. **h,i**: deactivation kinetics and control for transducin activated by incubation (25 minutes, 25°C) in TKM buffer with 32 mM NaF; (h) dilution (100 folds, 50 mM final) in TKM buffer or (i) in TKM buffer and 32 mM NaF. **j:** Evolution of transducin upon its dilution (100 folds, 50 nM final) from TKM buffer with 32 mM NaF alone into TKM buffer with 1 mM NaF and 10 µM $AlCl_3$.

for a different activation mechanism for which a preformed complexe with Al^{3+} or Be^{2+} is not necessary and excludes a simple bimolecular mechanism as for AlF_4^- or BeF_3^- binding. The Hill coefficient for the fluoride dependence of activation suggests the binding of at least three fluoride anions in a cooperative way. However, as for AlF_4^- or BeF_3^- binding, activation by fluoride "alone" requires a GDP in the nucleotide site and is inhibited by GDPβS, which substitutes a sulfure for the terminal oxygen on the β phosphate. This suggests that the targets for the binding of the fluorides is still the γ-phosphate site. But what is the counterion? As all these previous studies were performed in a buffer containing 2 mM magnesium, we investigated its possible influence on the kinetics of activation by NaF. AlF_4^- and BeF_3^- activation depended on magnesium, but this effect saturated below 2 mM magnesium concentration and is probably related to the binding of one magnesium

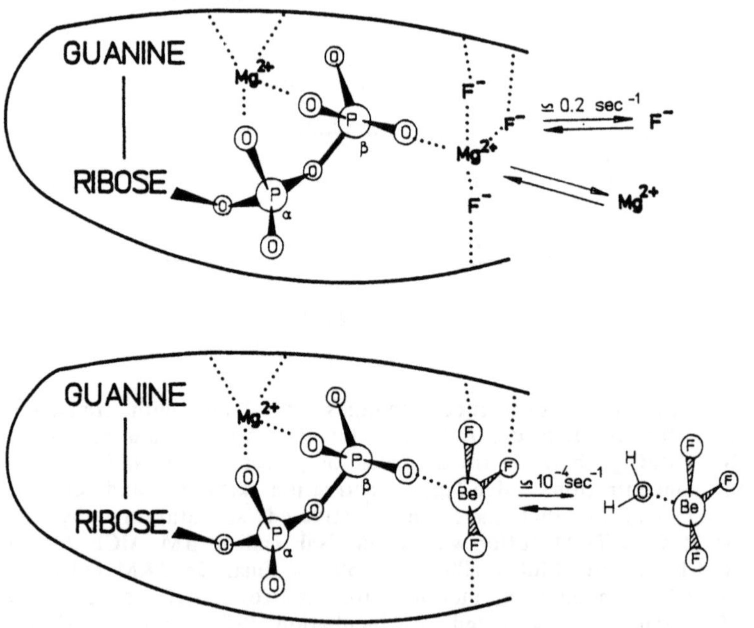

Fig 3. Models for insertion of fluorometallic complexes in the nucleotide site of TαGDP. Preformed AlF_x or BeF_x tetrahedral complexes interact by a bimolecular reaction with the terminal oxygen of the GDP β-phosphate. Fluoride and magnesium ions, wich do not form stable commplexes in solution, can also bind separately in the site to form a γ-phosphate analogue. The dotted lines represent hydrogen bonds or weak ionic bonds.

to the α and β phosphate of GDP in the site. By contrast the magnesium dependence of the activation rate by fluoride "alone" does not saturate but becomes linear for magnesium concentrations above 2 mM. This suggests the binding of one magnesium with the 3 fluorides. Our hypothesis is that although magnesium and fluoride do not form complexes in solution, they can cooperate in the nucleotide site to form a γ-phosphate analog, albeit with a much lower affinity than AlF_4^- or BeF_3^- (Fig 3).

The phosphate-analog property of metallofluoride complexes extend to many different enzymes such as actin, tubulin and various ATPase/ATPsynthase, which are certainly not related to the heterotrimeric G-protein family (Chabre, 1990). But surprisingly these complexes seem not to have any effect on other GTP binding proteins, such as the "small G" of the Ras family, or the bacterial elongation factor EF-Tu which, on the basis of sequence conservations in four regions known to constitute the nucleotide binding site of EFTu, are generally considered as providing good models for the nucleotide sites of Tα or the Gα's (Bourne *et al.*, 1988). The fluorometallic complexes demonstrate that the γ-phosphate site of transducin and other heterotrimeric G-proteins are different from that of Ras or EF-Tu.

TαGTP INTERACTS WITH THE INHIBITORY SUBUNITS OF THE PDE

The cGMP phosphodiesterase is constituted of two nearly identical catalytic subunits α (88 kD) and β (84 kD) which, in the basal state are inhibited by two copies of a small inhibitory subunit γ (13 kD) (Deterre *et al.*, 1988). The active TαGTP diffusing in the cytoplasm couple by collision with the PDE complex which, under physiological ionic conditions, is peripherically bound to the membrane by carboxy-methylation and possibly farnesylation of its main subunits terminals. TαGTP binds to one of the γ subunits and changes its conformation so that it is released from the cGMP hydrolytic site. In a low ionic strength medium TαGTP-PDEγ complexes are eluted separately from the activated PDE$\alpha\beta$ complexes. Thus the TαGTP-PDEγ interaction is stronger than the residual interaction of PDEγ with PDE$\alpha\beta$, but there are indications that under normal ionic conditions the whole ensemble

$(T\alpha GTP\text{-}PDE\gamma)_2$-PDE$\alpha\beta$ remains associated and membrane bound. The possibility of exchange of inhibitor molecules between fully activated and native PDE, led to the proposal of a rapid inactivation mechanism of the PDE (Deterre *et al.*, 1988; Chabre and Deterre, 1989) by exchange of inhibitory subunits between native PDE$\alpha\beta\gamma_2$ and fully activated PDE$\alpha\beta$. But a fast exchange kinetics implies a low binding constant for one of the inhibitor on native PDE, which is not observed. The overall dimeric structure of the PDE does not seem to allow significant cooperativity between the two catalytic sites: The K_m of the two catalytic subunits are identical, and the V_{max} of the enzyme which has lost one inhibitor is just half of that of the fully activated enzyme with no inhibitor left: the enzyme activity is inversely linear with the number of interacting TαGTP.

THE GTP HYDROLYSIS IN TαGTP TERMINATES THE PDE ACTIVITY

The physiological response of a retinal rod to a flash is fast and transient: it is terminated in less than one second. As the cGMP dependent channels do not show any desensitization, their reopening requires the fast recovery of the basal cGMP concentration. This depends on the activity of the guanylate cyclase as well as on that of the PDE.

The guanylate cyclase is regulated by a 26 kD calcium dependent protein: the decrease in intracellular calcium subsequent to the closure of cGMP gated channels removes the negative control by this protein. This feed-back reaction of the guanylate cyclase helps regenerate the basal cGMP level. But the PDE activity must also turn off rapidly for the recovery of the rod sensitivity: if after a flash the PDE remained fully active for a long time, the regeneration of the basal cGMP level being only due to the enhancement of guanylate cyclase activity, the rod would remain unsensitive to a new flash long after recovering its basal depolarization.

The PDE activity is directly controlled by the level of Tα in the GTP state. This level depends on two parameters: the rate of TαGTP formation catalyzed by R*, and the rate of its decay by GTP hydrolysis. A fast termination of the PDE activity implies both a fast inhibition of the activatory coupling of R* to the pool of TαGDP-Tβγ, and a fast decay of the pool of TαGTP. The uncoupling of R* is insured by the consecutive actions of rhodopsin kinase, which phosphorylates specifically in R* serine and threonine residues located near the C terminal end, and of arrestine which then binds to phosphorylated R* and hinders its further coupling to transducin. This process can occur in 200 mS under physiological conditions (Schleicher *et al.*, 1989). Therefore, as supported by electrophysiological studies of rods injected by non hydrolizable GTP analogs (Sather and Detwiler, 1987), the hydrolysis of GTP in Tα remains the rate limiting step in the deactivation of the PDE. The transition back to the TαGDP conformation loosens the interaction with PDEγ which can return to its inhibitory site.

As with other heterotrimeric G-proteins, the GTPase activity of transducin appears slow. The first *in vitro* GTPase measurement yielded turnover rates on the order of 1 to 3 phosphates released/min/transducin. These early studies were carried out with reconstituted systems of lipid vesicles, purified rhodopsin and transducin, or with diluted suspensions of illuminated Rod outer segments (ROS), in the presence of 1-10 mM GTP. More recently, rates of up to 20-30 GTP hydrolyzed/min/transducin were reported in more concentrated ROS suspensions, and with higher GTP concentrations (Sitaramayya *et al.*, 1988). But these high rates included a large contribution from an uncontrolled "dark" GTPase activity in the ROS that is probably not due to transducin. Only the GTPase rate increment induced by illumination is relevant.

In the above experiments, the lifetime of active TαGTP is deduced from steady state measurements of the GTPase rate under saturating illumination of the sample, to prevent the GTP loading step catalyzed by R* to be rate limiting. Such measurements assay the duration of a complete cycle that include the collision coupling of TαGDP-Tβγ to R*, the R* catalyzed GDP/GTP exchange on Tα, the dissociation of TαGTP from R* and Tβγ, the diffusion of TαGTP toward the PDE and its interaction with PDEγ, the GTP hydrolysis and finally the reassociation of TαGDP with Tβγ before a new cycle can begin. The activation steps are known to be rapid, but the hydrolytic step per-se might be followed by slower steps, such as a long-lived, inactive TαGDP-Pi conformation (as for tubuline or actine) or a slow return of TαGDP to the disc membrane. Steady state measurements give a sum total of all the times in the cycle.

Suspecting that the lifetime of TαGTP is short compared to the duration of the whole cycle, a time-resolved technique has been developped to analyse the contributions of the various steps in a transiently activated sample (Vuong and Chabre, 1990). The lifetime of TαGTP was measured by monitoring the heat coming from the hydrolysis of GTP from a pool of transducin that had been activated synchronously by flash-activated rhodopsin. The lifetime of R* could be reduced to less than a second by the presence of hydroxylamine, to prevent the turn-over of transducin after it has loaded and hydrolyzed its first GTP. The decay of TαGTP over a single cycle was monitored from the heat of GTP hydrolysis. The sample was a thin (0.4 mM) layer of ROS suspension sandwiched between the thermal sensor, a 20 micron thick pyroelectric film of poled PVDF film, and a transparent cover milar window. The cell is thermally insulated and thus heats up adiabatically as the reaction proceeds. A high impedance preamplifier picks up the displacement current resulting from the dilation of the heated PVDF poled film. Upon a weak flash in the presence of GTP in the ROS suspension, a characteristic heat pulse rises and decays within a second (Fig 4). All controls indicate that it is related to the hydrolysis of GTP. If hydroxylamine is reduced or removed from the suspension, the pulse is followed by a long tail, characteristic of a slow turn-over of GTP hydrolysis. This suggests that a transducin can load and hydolyses fast a GTP but then, even if R* is still active, this TαGDP cannot recycle as fast. This suggests that a "dead time" may exist, during which a transducin that has just hydrolyzed its GTP is inactive but cannot instantly be reactivated by R*.

The in vitro kinetics of the transducin GTPase cycle thus appear to include at least 4 steps:

1 - Activation	TαGDP-Tβγ + GTP --> TαGTP + Tβγ + GDP	taking about 0.1 s
2 - Deactivation	TαGTP --> $T\alpha_{refractory}$	completed within 1 s
3 - Regeneration	$T\alpha_{refractory}$ --> TαGDP + free Pi, and/or	} lasting many seconds
4 - Reassociation	TαGDP + Tβγ -->TαGDP-Tβγ	

The slow "resetting" steps during which deactivated transducin cannot reload GTP would account for the previously reported low steady-state GTPase rates *in-vitro*. These steps could be largely artifactual. Indeed, the solubilization and dilution of TαGTP into the bulk solution might have as an unwanted side-effect a slow rebinding of the resulting Tα-GDP to the disc membrane before reassociation with Tβγ

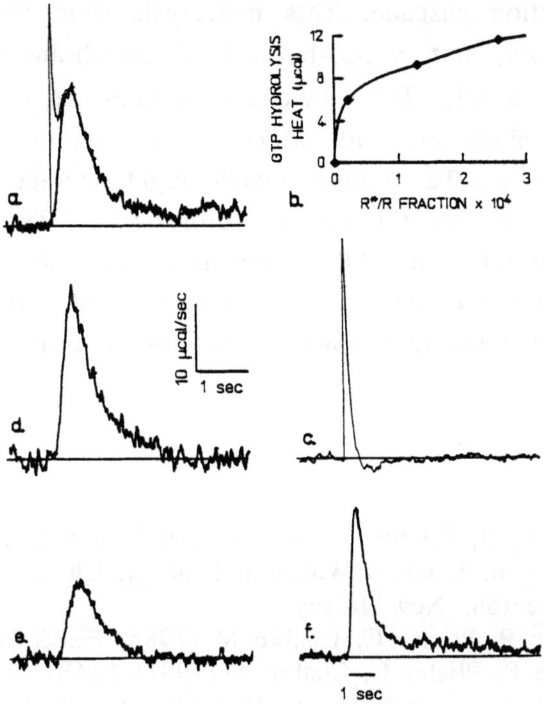

Fig 4. Time resolved microcalorimetric measurement of GTP hydrolysis. *a*) Light-induced GTP hydrolysis signal at low NH_2OH: 63 µM rhodopsin, 2 mM GTP, 5 mM NH_2OH, R*/R= $1.3x10^{-4}$ (thin trace). The spike preceding the broader signal is due to direct heating of the PVDF film by the light flash. *b*) Integrated areas (in µcal) of GTP hydrolysis signals as a function of photolysis levels: 75 µM rhodopsin, 2 mM GTP, 50 mM NH_2OH, R*/R=$2.2x10^{-5}$, $1.3x10^{4}$, and $2.6x10^{-4}$. *c*) Average of eight flash spikes collected after GTP exhaustion; its subtraction from the raw signal reveals the heat generated by the sample itself (*a*, thick trace). The slight negative component is due to cooling of the PVDF film by the exhausted sample. *d*, *e*) GTP hydrolysis signals at higher NH_2OH from: 75 µM (*d*), 33 µM (*e*) rhodopsin, 2 mM GTP, 25 mM NH_2OH, R*/R= $1.3x10^{-4}$. *f*) Average of two experiments at low NH_2OH levels (same composition and illumination as in *a*) shown on a x2 compressed time scale. These GTP hydrolysis signals are repeatable up to 50 times. The signals in *a*, *d* and *e* come from aliquots of the same preparation. (The significant light-independent GTPase activity of these ROS samples adds only a constant to the baseline.)

can occur. A characteristics of steady-state GTPase measurements seems to be that the more diluted the ROS membrane suspension or reconstituted vesicles, the lower the steady-state GTPase rate.

Occurring in less than a second, endogenous GTP hydrolysis by transducin is fast enough to qualify as a shut-off mechanism for the visual transduction cascade. This hydrolysis time does not depend on soluble cofactors, but yet-to-be-defined membrane-bound components could well play a role. Two obvious candidates are the PDE inhibitor, to which TαGTP binds and with which it stays on the membrane (Deterre *et al.*, 1986) and the Tβγ subunit which might interact with the Tα-PDEγ complex before or after the actual hydrolysis of GTP. It thus remains to be seen whether this fast GTPase time is intrinsic to the TαGTP subunit or whether it is conditioned by the interactions of TαGTP with the other components of the cGMP cascade on the rod disc membrane.

REFERENCES

Antonny B, Bigay J, Chabre M (1990) FEBS Letters 268:277-280
Berg HC (1983) In: Random Walks in Biology. Ch. 2. Princeton University Press, Princeton, New Jersey
Bigay J, Deterre P, Pfister C, Chabre M (1985) FEBS Lett 191:181-185
Bigay J, Deterre P, Pfister C, Chabre M (1987) EMBO J 6:2907-2913
Bornancin F, Pfister C and Chabre M (1989) Eur J Biochem 184:687-698
Bourne HR, Masters SB, Miller RT, Sullivana KA, Heideman W (1988) Cold Spring Harbor Symp Quant Biol LIII:221-229
Bruckert F, Vuong TM, Chabre M (1988) Eur Biophys J 16:207-218
Chabre M (1990) Trends Bioch Sci 15:6-10
Chabre M, Deterre P (1989) Eur J Biochem 179:255-266
Deterre P, Bigay J, Robert M, Pfister C, Kühn H, Chabre M (1986) Proteins Struct Funct Genet 1:188-193
Deterre P, Bigay J, Forquet F, Robert M, Chabre M (1988) Proc Natl Acad Sci USA 85:2424-2428
Sather WA, Detwiler P (1987) Proc Natl Acad Sci USA 84:9290-9294
Schleicher A, Kühn H, Hofmann KP (1989) Biochemistry 28:1770-1775
Sitaramayya A (1988) Biochemistry 27:4880-4887
Vuong TM, Chabre M, Stryer L (1984) Nature (London) 311:659-661
Vuong TM, Chabre M (1990) Nature 346:71-74

EXPRESSION OF TWO ISOFORMS OF THE GUANINE NUCLEOTIDE BINDING PROTEIN, G_o, IN NG108-15 NEUROGLIOMA CELLS.

Ian Mullaney+ and Graeme Milligan+=
Molecular Pharmacology Group,
Departments of Biochemistry+ and Pharmacology=,
University of Glasgow, Glasgow G12 8QQ,
Scotland, U.K.

Introduction

Members of a family of guanine nucleotide binding proteins (G-proteins) have been demonstrated to be neccessary to allow transduction of information from hormone-activated cell-surface receptors to effector systems which are either ion channels or enzymes which alter the rate of synthesis of intracellular second messengers [Gilman, 1987, Milligan, 1989]. Each of these G-proteins is a heterotrimer consisting of distinct α, β and γ subunits. A degree of genetic diversity has been noted for both β and γ subunits [Gao *et al.*, 1987, Gautam *et al.*, 1989], but it is at the level of the α subunits where the greatest diversity occurs. Whilst a high degree of homology exists between the individual α subunits, it is believed that the selectivity of transmembrane signalling is defined by the differences between the individual α subunits.
The traditional means of identifying G-proteins has relied upon the ability of either cholera toxin or pertussis toxin to catalyse an NAD^+-dependent ADP-ribosylation on the α subunits of particular G-proteins [Milligan, 1988]. Pertussis toxin was originally believed to ADP-ribosylate but a single G-protein, which, based on the functional effects of the modification, was defined as the inhibitory G-protein of the adenylate cyclase cascade (G_i) [Katada and Ui, 1982]. The site of pertussis toxin-catalysed ADP-ribosylation is a cysteine residue located 4 amino acids from the C-terminus of the α subunit. To date, six distinct cDNA clones corresponding to G-protein α subunits have been identified (TD1, TD2, Gi1, Gi2, Gi3 and Go) with a cysteine residue at this position and each of these proteins has been demonstrated to be a substrate for pertussis toxin. As these gene products are all between 39 and 41 kDa, then more selective

NATO ASI Series, Vol. H 52
Biological Signal Transduction
Edited by E. M. Ross and K. W. A. Wirtz

means of identification of the individual polypeptides have been required and a series of antisera, generated either against purified proteins or against synthetic peptides corresponding to parts of the primary sequence of these polypeptides, have been produced [Pines *et al.* 1985, Gierschik *et al.* 1986, Mumby *et al.* 1986, Goldsmith *et al.* 1987].

G_o was first identified as a pertussis toxin sensitive G-protein which essentially copurified with G_i1 from bovine brain [Strenwies and Robishaw, 1984, Neer *et al.* 1984, Milligan and Klee, 1985]. Although, the function of this protein has been a matter of debate, considerable evidence now suggests that it acts to transduce receptor-mediated inhibition of Ca^{2+} channels [Hescheler *et al.*, 1987, Harris-Warrick *et al.*, 1988, Ewald *et al.*, 1988, McFadzean *et al.*, 1989]. Initial evidence for multiple forms of $G_o\alpha$ was provided by the purification from bovine brain of two polypeptides which had immunological characteristics of $G_o\alpha$ (G_o and G_o^*) [Goldsmith *et al.* 1988]. More recent studies have augmented this work and demonstrated the separation of four G_o-like polypeptides [Inanobe *et al.*, 1990]. It is still unclear if these are the products of distinct genes, stem from differential splicing of nuclear RNA transcribed from a single gene or represent differentially covalently modified forms of the same polypeptide. However, the recent isolation of two cDNA clones for $G_o\alpha$ from a hamster insulin secreting tumour [Hsu *et al.*, 1990] which differ only in the last third of the polypeptide and may arise from differential splicing of the $G_o\alpha$ gene has provided a basis for the identification of multiple forms of $G_o\alpha$.

We have reported the ocurrence of two forms of $G_o\alpha$ in membranes of neuroblastoma x glioma, NG108-15, hybrid cells and demonstrated that cyclic AMP-induced increases in the levels of G_o, which are associated with the 'differentiation' of these cells [Mullaney and Milligan, 1989], are essentially restricted to the more acidic of these isoforms which is equivalent to brain G_o rather than G_o^* [Mullaney and Milligan, 1990]. Three distinct antipeptide antisera, generated against regions of the primary sequence of $G_o\alpha$, and which have all been demonstrated to identify selectively $G_o\alpha$ [Mullaney and Milligan, 1989] were unable to discriminate between these two isoforms [Mullaney and Milligan, 1990]. Each of these isoforms is a substrate for pertussis toxin-catalysed ADP-ribosylation and neither form appears to be phosphorylated or to contain endogenously added ADP-ribose. These isoforms may correspond to the cDNA's for G_o1 and G_o2 [Hsu *et al.*, 1990, Strathmann *et al.*, 1990].

Brabet *et al.,* have also noted recently the appearance of a distinct isoform of $G_o\alpha$ consequent upon DMSO-induced differentiation of murine neuroblastoma N1E-115 cells [Brabet *et al.*, 1990] but again it is unclear at what level these polypeptides differ from one another.

Results

The anti-$G_o\alpha$ antiserum, ON1, [Mullaney and Milligan, 1989] which was generated against a synthetic peptide corresponding to amino acids 2-17 predicted from a cDNA corresponding to $G_o\alpha$ (amino acids 1-16 of the mature protein), identified an apparently single polypeptide of some 39kDa in membranes of neuroblastoma x glioma hybrid, NG108-15, cells following resolution of the membrane proteins in 1-dimensional SDS-PAGE (10% (w/v) acrylamide) and transfer to nitrocellulose. Differentiation of the cells by treatment for 6 days with agents which increase levels of intracellular cAMP produced a distinct increase in immunologically detected levels of this polypeptide [Mullaney and Milligan, 1989].

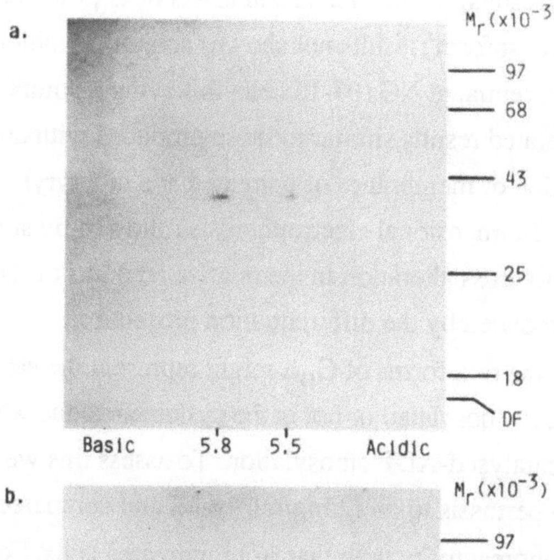

Figure 1.
Two polypeptides which have the immunochemical characteristics of $G_o\alpha$ are present in membranes of NG108-15 cells. Effects of cAMP-induced differentiation. Membranes (100 µg) of untreated (a) and dibutyryl cAMP-differentiated (1mM, 6 days) (b) NG108-15 cells were resolved in 2-dimensional SDS-PAGE . Following transfer to nitrocellulose. Antiserum ON1 (1: 200 dilution) was used as the primary reagent in an immunoblot.

In contrast to the 1-dimensional gels, immunoblotting of 2-dimensional gels with antiserum ON1 identified two distinct polypeptides of some 39 kDa with isoelectric points (pI) of some 5.5 and 5.8 (Figure 1). Greater immunological reactivity was observed routinely for the more basic form in membranes from the undifferentiated cells. Immunoblots of membranes of the dibutyryl cAMP-differentiated cells with antisera ON1 also demonstrated the expression of both forms of $G_o\alpha$. However, levels of the basic form of $G_o\alpha$ were virtually unchanged from those of the undifferentiated cells whilst markedly elevated levels of the more acidic form were seen (Figure 1). Indeed, the acidic form was now the predominant species in the membranes of the dibutyryl cAMP-differentiated cells. Differentiation of NG108-15 cells with each of 8-bromo cAMP (1mM), prostaglandin E1 (10 μM) or forskolin (10 μM), produced qualitatively similar results, ie. the increase in levels of G_o was due to an increase in amounts of the acidic isoform (results not shown) and immunoblotting membranes of both control and differentiated NG108-15 cells following resolution by 2-D electrophoresis generated results similar to those produced with antiserum ON1.
By contrast, resolution of membranes of untreated and dibutyryl cAMP-differentiated NG108-15 cells by 2-dimensional electrophoresis followed by silver staining of the gels indicated that no gross alteration in the relative amounts of the most prevelant polypeptides was produced by the differentiation procedure.
It was possible that the two forms of $G_o\alpha$ might represent the same polypeptide which differed in being ADP-ribosylated or not at the cysteine residue which is the acceptor for pertussis toxin-catalysed-ADP-ribosylation. To assess this we treated NG108-15 cells in culture with pertussis toxin (25ng/ml, 16 h.) and compared the mobility of the forms of $G_o\alpha$ immunoreactivity with that from untreated cells. Following pertussis toxin-treatment, two polypeptides of some 39kDa were still identified by antiserum ON1, but both migrated as more acidic forms in comparison to the polypeptides in membranes from untreated cells. Attempts to incorporate radioactivity, using [^{32}P]NAD$^+$ and thiol-activated pertussis toxin, into polypeptides in membranes of the pertussis toxin-treated NG108-15 cells was unsuccessful (results not shown), demonstrating that the entire pool of pertussis toxin-sensitive G-proteins had been ADP-ribosylated by the pretreatment. To further assess this question we separated membranes which were derived from undifferentiated NG108-15 cells which had been pretreated with pertussis toxin by 1D-SDS-PAGE (10% acrylamide) followed by immunoblotting with anti-$G_o\alpha$ antisera. Such experiments demonstrated that forms of G_o in these membranes migrated more slowly through the gel than $G_o\alpha$ from NG108-15 cells which had not been treated with the toxin (Figure 2), presumably due to the

extra mass provided by the addition of ADP-ribose. Following incubation of membranes from pertussis toxin-pretreated cells with mercuric acetate, a treatment known to remove ADP-ribose from cysteine residues [Jacobson *et al.*, 1990], the $G_o\alpha$ immmunoreactivity again migrated identically with that from membranes of untreated cells (Figure 2). Mercuric acetate treatment of membranes of untreated NG108-15 cells had no effect on the mobility in such gels of $G_o\alpha$ immunoreactivity, indicating that $G_o\alpha$ did not contain endogenous ADP-ribose on a cysteine residue (Figure 2).

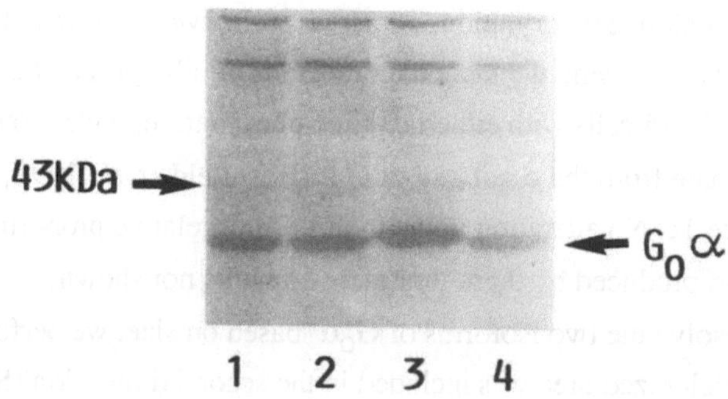

Figure 2.
Pertussis toxin-catalysed ADP-ribosylation of 'G_o' from NG108-15 cells is reversed by treatment with mercuric acetate.
Membranes (40μg) of NG108-15 cells which had been treated with either vehicle (1, 2) or pertussis toxin (25ng/ml, 16 h) (3, 4) were treated with (2, 4) or without (1, 3) mercuric acetate (10mM) in 50% formic acid for 15 minutes at 37°C. Following the treatments protein was precipitated with TCA (20% (w/v)final concentration) and recovered by centrifugation. The protein was dissolved in "Laemmli" sample buffer, the samples resolved by SDS-PAGE (10% acrylamide) and then immunoblotted for total $G_o\alpha$ using antiserum OC1 (directed against the C-terminal decapeptide of G_o) as primary reagent. $G_o\alpha$ in membranes derived from cells pretreated with pertussis toxin migrated more slowly through the gel than in membranes from untreated cells (compare lane 3 with lane 1). However, treatment of the membranes of pertussis toxin-pretreated cells with mercuric acetate restored the mobility of $G_o\alpha$ to that of untreated membranes (compare lanes 4 and 1).

Incubation of membranes of NG108-15 cells for an extended period with neutral hydroxylamine, a treatment which will remove ADP-ribose from arginine or asparagine residues [Jacobsen *et al.* 1990], also had no effect on the mobility through 1D-SDS-PAGE of immunoreactive $G_o\alpha$ derived from either untreated or pertussis toxin pretreated cells, indicating also the lack of endogenous ADP-ribose on such amino acids.

We have demonstrated recently that the pertussis toxin-sensitive G-protein, G_i2, can act as a substrate for phosphorylation by protein kinase C in hepatocytes [Bushfield *et al.*, 1990]. Further, this protein, which appears to be the true G_i of the adenylate cyclase cascade in hepatocytes [Bushfield *et al.*, 1990] and in NG108-15 cells [McKenzie and Milligan, 1990], is no longer able to mediate guanine nucleotide regulation of adenylate cyclase following such phosphorylation. To assess if the two isoforms of $G_o\alpha$ differed only in their phosphate content, we treated membranes of dibutyryl cAMP-differentiated NG108-15 cells with either alkaline phosphatase, under conditions which can remove phosphate from the α subunit of G_i2 [Bushfield *et al.*, 1990], or with protein phosphatase 1b. No alteration in the mobilities or relative proportions of the two $G_o\alpha$ isoforms were produced by these treatments (results not shown).

In an attempt to resolve the two isoforms of $G_o\alpha$ based on size, we performed 2-D gels in which 4M deionized urea was included in the second dimension (SDS-urea-PAGE). Noticable resolution of the two forms of G_o was achieved in immunoblots of such gels, particularly when the SDS-urea-PAGE was performed in gels produced using lowered bis-acrylamide concentrations. The more acidic isoform migrated slightly more slowly than the more basic isoform.

Whilst 2-dimensional electrophoresis allowed the resolution of the two forms of $G_o\alpha$, in many experiments we noted that a distinct fraction of the total $G_o\alpha$ immunoreactivity appeared not to penetrate adequately into the isoelectic focussing tube gel and following the 2nd dimension separation, was thus detected at the extreme basic edge of the immunoblot. We thus employed 1-dimensional SDS-urea-PAGE gels to attempt to resolve the two forms of $G_o\alpha$ under conditions in which all the available protein would enter the gel. Under these conditions two distinct polypeptides were resolved which were both identified by each of the three separate anti-$G_o\alpha$ anti-peptide antisera (Figure 3 and data not shown). The form of G_o with lower mobility through this type of gel (which we have called G_o1 [Mullaney and Milligan, 1990]) was present in greatly increased levels in membranes of cAMP-differentiated NG108-15 cells compared to the undifferentiated cells, whilst the form with greater mobility (which we have called G_o2) was unaltered in amount (Figure 3, compare lanes 2 and 3). The more slowly migrating form of $G_o\alpha$ in NG108-15 cells (G_o1) migrated in such gels exactly as did the major form of $G_o\alpha$ from rat brain. However no polypeptide corresponding to the more rapidly migrating form of $G_o\alpha$ in NG108-15 cells (G_o2) was detected in rat brain (Figure 3). We are currently attempting to assess if the two isoforms of $G_o\alpha$ in NG108-15 cells correspond to the two $G_o\alpha$ cDNA species isolated by Hsu *et al.*[1990] by immunoblotting with antisera generated against diverging sections of the primary sequence of these two forms.

We confirmed that the two G_o-like polypeptides present in NG108-15 cells were not simply rat and mouse variants of $G_o\alpha$ produced by distinct chromosomes derived from the parental cells of the hybrid by demonstrating that mouse brain $G_o\alpha$ also migrated identically with the more slowly migrating polypeptide of NG108-15 cells in the SDS-urea-PAGE system (Figure 3). As with rat brain, no immunoreactive polypeptide was detected in mouse brain which corresponded to the more rapidly migrating form of $G_o\alpha$ of NG108-15 cells.

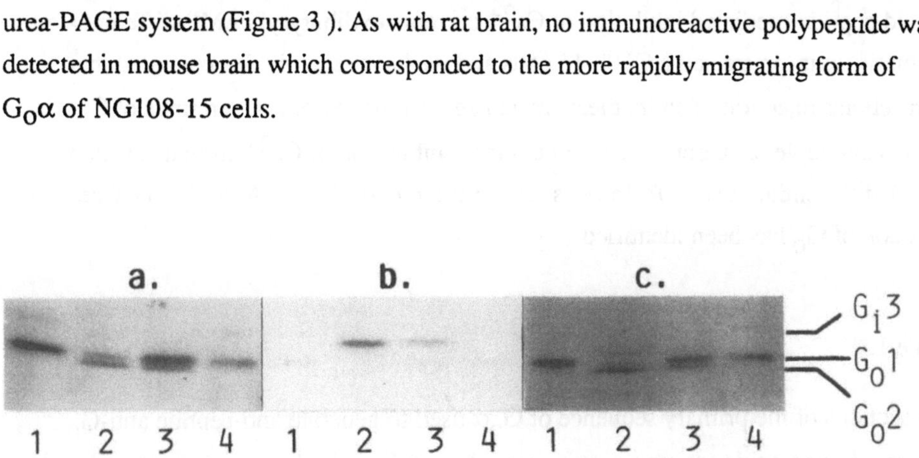

Figure 3.
1-dimensional SDS-urea-PAGE analysis of G_o1 and G_o2.
Membranes of (1) rat brain (5 µg), (2) untreated NG108-15 cells (50 µg), (3) dibutyryl cAMP-differentiated (1 mM, 6 days) NG108-15 cells (50 µg) and (4) mouse brain (5 µg) were resolved by 1-dimensional SDS-urea-PAGE and immunoblotted with antisera IM1 (1 : 200 dilution (**a**), I3B (anti-$G_i3\alpha$)(1: 200 dilution) (**b**) or OC1 (1 : 1,000 dilution).

Each of the anti-$G_o\alpha$ antisera also identified a second polypeptide in brain. However this polypeptide ($G_o{}^*\alpha$ in the terminology of Goldsmith *et al.* (1988)) migrated considerably more slowly in the 1-dimensional SDS-urea-PAGE system than did either of the $G_o\alpha$ forms in NG108-15 cells.

Discussion

Members of a highly homologous family of heterotrimeric G-proteins function to allow transduction of hormonal and neurotransmitter encoded information from plasma membrane spanning receptor proteins to effector systems [Gilman, 1987, Milligan,

1989]. Whilst the roles of a number of the G-proteins have been well defined, the functions of a number of others have been less well characterized. G_o was first identified as a polypeptide of 39 kDa which co-purified through a number of chromatographic steps with the major form of Gi (G_i1) from bovine brain [Sternweis and Robishaw, 1984, Neer *et al.* 1984, Milligan and Klee, 1985]. Although G_o may represent as much as 1% of brain membrane protein [Sternweis and Robishaw, 1984, Gierschik *et al.* 1986], it has been difficult to define its function. Hescheler *et al.* (1987) were able to demonstrate that $G_o\alpha$ from bovine brain was able to reconstitute opioid peptide-mediated inhibition of Ca^{2+} currents to dibutyryl cAMP-differentiated, pertussis toxin-pretreated NG108-15 cells and we have recently demonstrated that intracellular injection of antibodies directed against the extreme C-terminus of $G_o\alpha$ is able to uncouple $\alpha2$ adrenergic receptors from inhibition of Ca^{2+} currents in such PGE1-differentiated NG108-15 cells [McFadzean *et al.* 1989]. As such, one (the) function of G_o has been identified.

Table 1.

The sections of the primary sequence of $G_o\alpha$ used to generate anti-peptide anti-G_o antisera. Amino acids are represented using the one letter code.

Antiserum	Peptide Used	G-protein sequence
OC1	ANNLRGCGLY	$G_o\alpha$ 345-354
ON1	GCTLSAEERAALERSK	$G_o\alpha$ 1-16
IM1	NLKEDGISAAKDV	$G_o\alpha$ 22-35

We have recently demonstrated that a number of agents which, as a common mode of action, increase intracellular levels of cAMP are able to cause a morphological differentiation of neuroblastoma x glioma hybrid cells [Mullaney *et al.*, 1988] and concomitant with this process, marked increases in the levels of immunologically detectable $G_o\alpha$ are noted in the membranes of these cells [Mullaney and Milligan, 1989].

Detailed examination of fractions containing pertussis toxin-sensitive G-proteins during purifications from bovine brain have indicated the presence of two distinct populations of $G_o\alpha$ [Goldsmith *et al.*, 1988]. These two forms differed in isoelectric point and as such could be resolved on high performance ion exchange resins [Goldsmith *et al.*, 1988]. A series of different anti-G_O-antisera were unable to discriminate between the two polypeptides and as such it is unclear if the two forms were derived from distinct but highly homologous genes or if they were generated by either the differential splicing of pre-mRNA derived from a single gene or by alterations in covalent modification subsequent to synthesis. Both of the potential explanations which would be consistent with the two forms being derived from a single gene have parallels in other G-proteins. Firstly, it is well established that up to 4 forms of $G_s\alpha$ can be generated via differential splicing mechanisms [Bray *et al.* 1986, Robishaw *et al.* 1986] and that multiple forms of G_s α so generated can be detected in a single cell. Furthermore, the recent demonstration of the production of alternatively spliced forms of $G_O\alpha$ in drosophila [De Sousa *et al.* 1989, Thambi *et al.* 1989, Yoon *et al.* 1989] offers a further potential explanation for the identification of distinct isoforms of $G_O\alpha$ in NG108-15 cells as demonstrated herein. It has also recently been recorded that alternate forms of $G_O\alpha$ cDNA can be isolated from a hamster cell line [Hsu *et al.*, 1990], and that these forms may be produced by differential splicing of mRNA. Furthermore multiple transcripts have been identified on northern blots using a mammalian $G_O\alpha$ cDNA as probe [Jones and Reed, 1987]. Recent studies on the purification of G_O-like polypeptides from brain have led to the isolation of 4 distinct fractions [Inanobe *et al.*, 1990] which based on peptide mapping could be divided into two distinct families [Inanobe *et al.*, 1990].

It has also been noted that members of the "G_i-like" gene products have been demonstrated to be substrates for each of protein kinase C [Katada *et al.* 1985, Pyne *et al.* 1989, Bushfield *et al.* 1990], protein kinase A [Watanabe *et al.* 1988] and the insulin receptor tyrosyl kinase activity [O'Brien *et al.* 1987, Krupinski *et al.* 1988] under various conditions and that $G_s\alpha$ can be endogenously ADP-ribosylated [Jacquemin *et al.*, 1986]. It was thus possible that the two isoforms of $G_O\alpha$ in NG108-15 cells differed only by some form of post-translational modification.

In an attempt to assess whether the two forms of $G_O\alpha$ were identical save for the post-translational addition of a phosphate group, we incubated membranes of dibutyryl cAMP-differentiated NG108-15 cells, which contain high levels of the acidic form of $G_O\alpha$ (G_O1), with alkaline phosphatase or protein phosphatase 1b, resolved the membrane proteins by 2-dimensional electrophoresis and then immunoblotted using antiserum ON1. No alterations in the relative amounts or mobilities of the two forms were noted. This is despite the fact that treatment of hepatocyte membranes under these conditions causes the dephosphorylation of phosphorylated $G_i2\alpha$ [Bushfield *et al.*, 1990].

To assess if the more acidic form of $G_o\alpha$ had become ADP-ribosylated by an endogenous ADP-ribosyltransferase activity we performed a series of experiments. In one set of these we demonstrated that both isoforms of $G_o\alpha$ in these cells were substrates for pertussis toxin-catalysed ADP-ribosylation. Following pretreatment of the cells with pertussis toxin, immunoblotting subsequent to 2-D electrophoresis showed that both isoforms now migrated as more acidic proteins. As such, neither form could have endogenous ADP-ribose on the cysteine residue which is the acceptor site for pertussis toxin-catalysed ADP-ribosylation. Further, ADP-ribosylated $G_o\alpha$ migrates more slowly through 1-D SDS-PAGE than non ADP-ribosylated $G_o\alpha$. Treatment of membranes with mercuric acetate causes the removal of ADP-ribose from cysteine groups. Such treatment of $G_o\alpha$ which had been ADP-ribosylated by pertussis toxin restored the mobility of the polypeptide to that of the untreated polypeptide, defining that the G_o did not contain endogenous ADP-ribose attached to a cysteine residue. We cannot demonstrate definitively that the more acidic isoform has not been ADP-ribosylated by an endogenous ADP-ribosyltransferase activity at a site other than a cysteine residue, but treatment of membranes of NG108-15 cells with neutral hydroxylamine under conditions which remove ADP-ribose which has been incorporated onto an arginine residue into the α subunits of each of G_s G_i2 and G_i3 failed to modify the mobility of $G_o\alpha$ in gels.

As the α subunits of forms of G_i and G_o are myristoylated at their N-terminus, then it might be suggested that the two G_o isoforms differ in that one isoform does not contain covalently linked myristic acid and that this could account for the differences in electrophoretic mobility. Such an arguement is not consistent with the fact that myristoylation is essential for membrane attachment of such G-proteins [Jones *et al.* 1990] and that both of the $G_o\alpha$ isoforms are membrane associated polypeptides as membrane preparations were the starting material for these studies. Furthermore, Brabet *et al.*, [1990] have indicated that the two forms of $G_o\alpha$ immunoreactivity in DMSO-differentiated N1E-115 cells both contain myristic acid. It has also recently been noted that $G_o\alpha$ from NG108-15 cells contain covalently attached palmitic acid, presumably at a cysteine residue [Parenti *et al.*, 1990]. It is currently unknown if both isoforms of $G_o\alpha$ maintain this fatty acid.

Resolution of the forms of $G_o\alpha$ in a 1-dimensional system would obviously be a major advantage as entry of G-protein α subunits into the isoelectric focussing dimension of 2-dimensional gels is poor. We have achieved this aim by the inclusion of 4M deionized urea in the resolving phase of SDS-PAGE. The results obtained by this approach are entirely consistent with those obtained using the 2-dimensional system. Each of three distinct anti-G_o antisera identified each of the two polypeptides and each antiserum demonstrated that cAMP-induced differentiation of NG108-15 cells produced elevated levels of only one of these isoforms (G_o1). Because the polypeptide whose expression was increased by the differentiation procedure was the acidic form and because this polypeptide migrated identically with rat brain G_o in the SDS-urea-PAGE system then this can be identified as G_o rather than G_o^* [Goldsmith *et al.* 1988]. The more basic $G_o\alpha$-like polypeptide of NG108-15 cells (G_o2) is not, however, identical with G_o^*. This conclusion is based on the differences in mobility of G_o^* and G_o2 from NG108-15 cells in 1-dimensional SDS-urea-PAGE . Go* migrates in this system considerably more slowly than G_o1 whilst G_o2 actually migrated slightly more rapidly than G_o1.

The possibility that the more rapidly migrating form (G_o2) represents a proteolytic product of G_o1 is rendered unlikely by the fact that antisera directed against both the extreme N-terminal and C-terminal region of G_o1 identify both forms. Further evidence to suggest the integrity of the C-terminal region of each polypeptide is that, as noted above, both isoforms are substrates for pertussis toxin-catalysed ADP-ribosylation. As this modification occurs only 4 amino acids from the C-terminus then any modification at the C-terminus would have to be extremely minor to be consistent with the immunological detection of this polypeptide. Indeed, detailed examinations of the cross-reactivity or otherwise between antisera raised against the C-terminal decapeptides of G_i2, G_i3 and G_o indicate a key role for the C-terminal tyrosine residue in recognition of $G_o\alpha$ by antiserum OC1 [Milligan, 1990].

As none of the antisera we currently have available provide a selective probe to discriminate between the two isoforms of $G_o\alpha$ in NG108-15 cells then we cannot currently assess which of these polypeptides plays the key role in receptor regulation of voltage-sensitive Ca^{2+} channels in NG108-15 cells [Hescheler *et al.* 1987, McFadzean *et al.* 1989]. As cAMP-induced differentiation is necessary to record neurotransmitter-mediated inhibition of such currents, it is tempting to speculate that the acidic isoform (G_o1) is of primary importance, as this is the form which is increased markedly in levels during the differentiation process. Assessment of the validity of this hypothesis will require further experimentation.

Acknowledgement

These studies were supported by Medical Research Council project grants to G.M.

References

Brabet, P., Pantaloni, C., Rodriguez, R. Martinez, J., Bockaert, J. and Homburger, V. (1990) Neuroblastoma differentiation involves the expression of two isoforms of the α subunit of G_o. J. Neurochem. 54, 1310-1320.

Bray, P., Carter, A., Simons, C., Guo, V., Puckett, C., Kamholz, J., Spiegel, A. and Nirenberg, M. (1986) Proc. Natl. Acad. Sci. (U.S.A.) Human cDNA clones for four species of $G\alpha_s$ signal transduction protein. 83, 8893-8897.

Bushfield, M., Murphy, G.J., Lavan, B.E., Parker, P.J., Hruby, V.J., Milligan, G. and Houslay, M.D. (1990) Hormonal regulation of $G_i2\alpha$ subunit phosphorylation in intact hepatocytes. Biochem. J. 268, 449-457.

de Sousa, S.M., Hoveland, L.L., Yarfitz, S. and Hurley, J.B. (1989) The *drosophila* $G_o\alpha$-like G-protein produces multiple transcripts and is expressed in the nervous system and in ovaries. J. Biol. Chem. 264, 18544-18551.

Ewald, D.A., Sternweis, P.C. and Miller, R.J. (1988) Guanine nucleotide binding protein G_o induced coupling of neuropeptide Y receptors to Ca^{2+} channels in sensory neurons. Proc. Natl. Acad. Sci (U.S.A.) 85, 3633-3637.

Gao, B., Gilman, A.G., and Robishaw, J.D. (1987). A second form of the β subunit of signal transducing G-proteins. Proc. Natl. Acad. Sci. (U.S.A.) 84, 6122-6125.

Gautam, N., Baetscher, M., Aebersold, R. and Simon, M.I. (1989) A G-protein gamma subunit shares homology with ras proteins. Science 244, 971-974.

Gierschik, P., Milligan, G., Pines, M., Goldsmith, P., Codina, J., Klee, W. and Spiegel, A. (1986). Proc. Natl. Acad. Sci. (U.S.A.) Use of specific antibodies to quantitate the guanine nucleotide binding protein G_o in brain. 83, 2258-2262.

Gilman, A.G. (1987) G-proteins: transducers of receptor generated signals. Ann. Rev. Biochem. 56, 615-649.

Goldsmith, P., Gierschik, P., Milligan, G., Unson, C.G., Vinitsky, R., Malech, H. and Spiegel, A. (1987) Antibodies directed against synthetic peptides distinguish between GTP binding proteins in neutrophil and brain. J. Biol. Chem. 262, 14683-14688.

Goldsmith, P., Backlund, P.S., Rossiter, K., Carter, A., Milligan, G., Unson, C.G. and Spiegel, A. (1988) Purification of heterotrimeric GTP binding proteins from brain: Identification of a novel form of G_o. Biochemistry 27, 7085-7090.

Harris-Warrick, R., Hammond, C., Paupardin-Tritsch, D., Homberger, V., Rouot, B., Bockaert, J. and Gershenfeld, H.M. (1988) An α40 subunit of a GTP binding protein immunologically related to G_o mediates a dopamine-induced decrease of Ca^{2+} current in snain neurons Neuron 1, 27-32.

Hescheler, J., Rosenthal, W., Trautwein, W. and Schultz, G. (1987) The GTP binding protein, G_o, regulates neuronal calcium channels. Nature (London) 325, 445-447.

Hsu, W.H., Rudolph, U., Sanford, J., Bertrand, P., Olate, J., Nelson, C., Moss, L.G., Boyd, A.E.III, Codina, J. and Birnbaumer, L. (1990) Molecular cloning of a novel splice variant of the α subunit of the mammalian G_o protein. J. Biol. Chem. 265, 11220-11226.

Inanobe, A., Shibasaki, H., Takahashi, K., Kobayashi, I., Tomita, U., Ui, M. and Katada, T. (1990) Characterization of four G_o-type proteins purified from bovine brain membranes. FEBS Lett. 263, 369-372.

Jacobson, M.K., Loflin, P.T., Aboul-Ela, N., Mingmuang, M., Moss, J. and Jobson, E.L. (1990) Modification of plasma membrane protein cysteine residues by ADP-ribose *in vivo*. J. Biol. Chem. 265, 10825-10828.

Jacquemin, C., Thibout, M., Lambert, B. and Correze, C. (1986) Endogenous ADP-ribosylation of G_s subunit and autonomous regulation of adenylate cyclase. Nature 323, 182-184.

Jones, D.T. and Reed, R.R. (1987) Molecular cloning of five GTP-binding protein cDNA species from the rat olfactory neuroepithelium. J. Biol. Chem. 262, 14241-14249.

Jones, L.Z., Simonds, W.F., Merendino, J.J., Brann, M.R. and Spiegel, A.M. (1990) Myristoylation of an inhibitory GTP-binding protein α subunit is essential for its membrane attachment. Proc. Natl. Acad. Sci. (U.S.A.) 87, 568-572.

Katada, T. and Ui, M. (1982) ADP-ribosylation of the specific membrane protein of C6 cells by islet activating protein associated with modification of adenylate cyclase. J. Biol. Chem. 257, 7210-7216.

Katada, T., Gilman, A.G., Watanabe, Y., Bauer, S. and Jakobs, K.H. (1985) Protein kinase C phosphorylates the inhibitory guanine nucleotide binding regulatory component and apparently suppressess its function in hormonal inhibition of adenylate cyclase. Eur. J. Biochem. 151, 431-437.

Krupinski, J., Rajaram, R., Lakonuhok, M., Benovic, J.L. and Cerione, R.A. (1988) Insulin dependent phosphorylation of GTP binding proteins in phospholipid vesicles. J. Biol. Chem. 263, 12333-12341.

McFadzean, I., Mullaney, I., Brown, D.A. and Milligan, G. (1989) Antibodies to the GTP binding protein, G_o, antagonize noradrenaline-induced calcium current inhibition in NG108-15 hybrid cells. Neuron, 3, 177-182.

McKenzie, F.R. and Milligan, G. (1990) δ opioid receptor mediated inhibition of adenylate cyclase is transduced specifically by the guanine nucleotide binding protein Gi2. Biochem. J. 267, 391-398.

Milligan, G. (1988) Techniques used in the identification and analysis of function of pertussis toxin sensitive guanine nucleotide binding proteins. Biochem. J. 255, 1-13.

Milligan, G. (1989) Guanine nucleotide-binding proteins and cellular control. Curr. Opinion Cell Biol. 1, 196-200.
Milligan, G. (1990) Immunological probes and the identification of guanine nucleotide binding proteins. In: G-proteins as mediators of cellular signalling processes. Houslay, M.D. and Milligan, G (eds), pp 31-46. Wiley and Sons Ltd, Chichester.
Milligan, G. and Klee, W.A. (1985) The inhibitory guanine nucleotide binding protein (Ni) purified from bovine brain is a high affinity GTPase. J. Biol. Chem. 260, 2057-2063.
Mullaney, I., Magee, A.I., Unson, C.G. and Milligan, G. (1988) Differential regulation of amounts of the guanine nucleotide binding proteins G_i and G_o in neuroblastoma x glioma hybrid cells in response to dibutyryl cyclic AMP. Biochem. J. 256, 649-656.
Mullaney, I. and Milligan, G. (1989) Elevated levels of the guanine nucleotide binding protein, G_o, are associated with differentiation of neuroblastoma x glioma hybrid cells. FEBS Lett. 244, 113-118.
Mullaney, I. and Milligan, G. (1990) Identification of two distinct isoforms of the guanine nucleotide binding protein in neuroblastoma x glioma hybrid cells: Independent regulation during cyclic AMP-induced differentiation. J. Neurochem. 55, in press.
Mumby, S.M., Kahn, R.A., Manning, D.R. and Gilman, A.G. (1986) Antisera of defined specificity for subunits of guanine nucleotide binding proteins. Proc. Natl. Acad. Sci. (U.S.A.) 83, 265-269.
Neer, E.J., Lok, J.M. and Wolf, L.G. (1984) Purification and properties of the inhibitory guanine nucleotide regulatory unit of brain adenylate cyclase. J. Biol. Chem. 259, 14222-14229.
O'Brien, R.M., Houslay, M.D., Milligan, G. and Siddle, K. (1987) The insulin receptor tyrosyl kinase phosphorylates holomeric forms of the guanine nucleotide regulatory proteins G_i and G_o.FEBS Lett. 212, 281-288.
Parenti, M., Milligan, G. and Magee, A.I. (1990) The α subunit of G_o is palmitoylated but not isoprenylated FEBS Lett. (in press).
Pines, M., Gierschik, P., Milligan, G., Klee, W. and Spiegel, A. (1985) Antibodies against the carboxyl-terminal 5kDa peptide of the α subunit of transducin crossreact with the 40 but not the 39kDa guanine nucleotide binding protein from brain. Proc. Natl. Acad. Sci. (U.S.A.) 82, 4095-4099.
Pyne, N.J., Murphy, G.J., Milligan, G. and Houslay, M.D. (1989) Treatment of intact hepatocytes with either the phorbol ester TPA or glucagon elicits the phosphorylation and functional inactivation of the inhibitory guanine nucleotide regulatory protein G_i. FEBS Lett. 243, 77-82.

Robishaw, J.D., Smigel, M.D. and Gilman, A.G. (1986) Molecular basis for two forms of the G-protein that stimulates adenylate cyclase. J. Biol. Chem. 261, 9587-9590.

Sternweis, P.C. and Robishaw, J.D. (1984) Isolation of two proteins with high affinity for guanine nucleotides from membranes of bovine brain. J. Biol. Chem. 259, 13806-13813.

Strathmann, M., Wilkie, T.M. and Simon, M.I. (1990) Alternative splicing produces transcripts encoding two forms of the α subunit of the GTP-binding protein G_O. Proc. Natl. Acad. Sci. (U.S.A.) 87, 6477-6481.

Thambi, N.C., Quan, F., Wolfgang, W.J., Spiegel, A. and Forte, M. (1989) Immunological and molecular characterization of $G_O\alpha$-like proteins in *drosophila* central nervous system. J. Biol. Chem. 264, 18552-18560.

Watanabe, Y., Imaizumi, T., Masaki, N., Iwakura, K. and Yoshida, H. (1988) Effects of phosphorylation on inhibitory GTP-binding protein by cyclic AMP-dependent protein kinase on its ADP-ribosylation by pertussis toxin, islet activating protein. FEBS Lett. 236, 372-374.

Yoon, J., Shortridge, R.D., Bloomquist, B.T., Schneuwly, S., Perdew, M.H. and Pak, W.L. (1989) Molecular characterization of *drosophila* gene encoding $G_O\alpha$ subunit homolog. J. Biol. Chem. 264, 18536-18543.

MECHANISMS UNDERLYING HORMONAL MODIFICATION OF DOPAMINE RECEPTOR SENSITIVITY

Paul B. Foley, Donald L. Cameron and Ann D. Crocker
Department of Clinical Pharmacology and Centre for Neuroscience,
School of Medicine,
The Flinders University of South Australia
Bedford Park, South Australia, 5042
Australia

Introduction and Background

The identification of factors that affect responsiveness to dopamine agonists (dopamine agonist sensitivity) has important implications for the understanding and treatment of disorders of central dopamine transmission such as schizophrenia and Parkinson's disease. Increases in agonist sensitivity, supersensitivity phenomena, are well known in the peripheral nervous system where changes in receptor concentration or coupling of receptors to second messenger systems have been identified as mechanisms underlying these phenomena (Fleming & Westfall, 1988). The aim of the present report is to describe experiments in which the importance of these mechanisms in determining dopamine agonist sensitivity in the central nervous system has been investigated. In these experiments we have used an animal model, the hypothyroid rat, which exhibits increased behavioural sensitivity to dopamine receptor agonists compared to euthyroid controls (Cameron & Crocker, 1990b).

Dopamine agonists acting on receptors in the striatum (comprising Caudate-Putamen and nucleus Accumbens) induce characteristic behavioural stereotypies that can be quantified (Cameron *et al*, 1988) and used as an index of agonist action in the central nervous system. We have chosen to utilise the dopamine-agonist induced stereotyped head-down sniffing (SHDS) response to investigate agonist action *in vivo* as it follows a clear, monophasic dose-response relationship.

NATO ASI Series, Vol. H 52
Biological Signal Transduction
Edited by E. M. Ross and K. W. A. Wirtz

Using this paradigm we have demonstrated that the SHDS dose-response curve was shifted to the left in hypothyroid rats compared to euthyroid controls. This occurred in the absence of any change in baseline or maximal responses. This phenomenon was observed using either the mixed dopamine D1/D2 receptor agonist apomorphine, or the selective D2 agonist, quinpirole. In addition, this increased sensitivity could be reversed by replacement with either of the thyroid hormones, tri-iodothyronine (T3) or tetra-iodothyronine (T4) (Cameron & Crocker, 1990b).

The SHDS response to quinpirole was competitively antagonised by the selective D2 receptor antagonist, raclopride. Schild analysis of the raclopride induced parallel shifts in the quinpirole-SHDS dose-response curves demonstrated a higher *apparent* pA_2 in hypothyroid animals indicating a decrease in antagonist-sensitivity (Cameron & Crocker, 1990b). These data suggest that a change in D2 receptor function is associated with the hypothyroid state, making it a suitable model in which to study facilitated dopamine receptor-mediated signal transduction.

Attempts to define the cellular substrates of this phenomenon initially focussed on determining if there was a change in either receptor concentration or affinity, since several reports indicated that increased agonist sensitivity may be associated with increased receptor concentration in other systems (Maura *et al*, 1988). However, binding studies carried out in striatal homogenates using the D1 receptor ligand $[^3H]$SCH23390 and the D2 receptor ligand $[^3H]$spiperone failed to show any significant difference in the concentration or affinity of either D1 or D2 dopamine receptor subtype in hypothyroid rats compared to controls.

Nevertheless, the possibility remained that a change in receptor concentration in a discrete part of the striatum in the hypothyroid rat could be masked in an assay conducted in a homogenate of whole striatum. Parallel work in the laboratory employing intracerebral injections of irreversible antagonists and quantitative autoradiography enabled us to test this idea, and localised the receptors responsible for mediating the dopamine agonist-induced SHDS to a small region of the ventromedial striatum (Cameron & Crocker, 1989). Subsequent examination of the dopamine receptor binding properties of this region in eu- and hypothyroid rats using the more sensitive technique of quantitative autoradiography showed a small (20%) but significant decrease in D2 receptor concentration in hypothyroid rats (Cameron & Crocker, 1990a). The most parsimonious explanation of this observation is that this decrease in dopamine receptor concentration represents a compensatory response to an increase in sensitivity somewhere in the chain of neuronal events that are involved in the agonist effect.

It has been established that D1 receptors are positively coupled to adenylate cyclase while D2 receptors are negatively coupled to this enzyme (Stoof & Kebabian, 1981). We therefore investigated adenylate cyclase activity in eu- and hypothyroid rats. No difference was detected in cAMP levels between hypothyroid rats and controls in response to selective stimulation of either receptor subtype. Further, there were no differences in the actions of sodium fluoride, which activates G_s, or forskolin, which stimulates adenylate cyclase directly, between groups (Table 1.). The finding that there was no alteration in striatal dopamine-mediated adenylate-cyclase activity in hypothyroid conditions points to an alteration in some other second messenger for dopamine. Certainly other work from our laboratory has since questioned the role of adenylate cyclase in the SHDS response. For example, we have demonstrated that although a reduction in dopamine receptor concentration by an irreversible antagonist, leads to an attenuation of the agonist-induced SHDS response, this is not paralleled by a change in dopamine-stimulated adenylate cyclase activity (Cameron & Crocker, 1988). Thus it appears that cAMP is not the second messenger involved in D2-mediated induction of SHDS.

The focus of our current research, therefore, has been on the possible actions of the thyroid hormones at a point distal to the dopamine receptor. Mesostriatal dopamine neurons synapse on the terminals of cholinergic interneurons in the striatum (Lehmann & Langer, 1983) and a known action of agonists at striatal D2 receptors is the inhibition of acetylcholine (ACh) release (Stoof & Kebabian, 1982).

Evidence that the D2-mediated inhibition of ACh release is necessary for the expression of the SHDS response has been provided by the observation that

TABLE 1.
BASELINE AND STIMULATED cAMP LEVELS
Values for cAMP production ($pmol.min^{-1}.mg^{-1}$ protein) are expressed as the mean ± s.e.m. for striata from 5 to 8 animals.

Condition	Euthyroid	Hypothyroid
Basal	3.13 ± 0.23	2.55 ± 0.31
10μM Dopamine	4.11 ± 0.43	4.01 ± 0.40
100μM Dopamine	4.70 ± 0.56	4.07 ± 0.31
Basal	3.40 ± 0.15	3.50 ± 0.06
4mM NaF	4.90 ± 0.40	4.80 ± 0.19
8mM NaF	5.00 ± 0.21	5.10 ± 0.10
Basal	2.70 ± 0.40	2.35 ± 0.40
0.1μM Forskolin	6.30 ± 0.95	5.35 ± 1.10
100μM Forskolin	16.80 ± 2.80	16.90 ± 3.10
Basal*	6.11 ± 0.66	5.63 ± 0.49
10μM Dopamine*	5.57 ± 0.36	3.76 ± 0.96
100μM Dopamine*	4.83 ± 0.16	3.62 ± 0.89

***0.1μM Forskolin was added to elevate basal adenylate cyclase activity in order for the inhibitory effect of dopamine at the D2 receptor to be observed more readily. In addition, 1μM SCH23390 was also added to remove the stimulatory effect of dopamine at the D1 receptor. cAMP levels were determined by the method of Neilsen *et al* (1984).**

muscarinic agonists reduce the effects of dopamine agonists on this response (Crocker & Russell, 1990). This is consistent with a decrease in muscarinic stimulation being required for the manifestation of dopamine agonist-induced SHDS. We have recently provided further support for the hypothesis that an inhibition of muscarinic receptor function in the striatum is involved in the SHDS response by showing that microinjection of an irreversible muscarinic receptor agonist into the region of the striatum identified as mediating the SHDS response also attenuates the effects of dopamine agonists. (Crocker & Cameron, 1990). The conclusion drawn from these experiments is that the dopamine and muscarinic receptors involved in the SHDS response are in close anatomical association in the ventromedial striatum.

Consequently, it is likely that the next link in the chain of events involved in the dopamine agonist-induced SHDS involves the D2 mediated modulation of ACh release.

In this paper we report some preliminary results of our investigations into the dopamine agonist-induced inhibition of stimulated ACh release in hypothyroid rats *in vitro*.

Methods and Results

Male Sprague Dawley rats were rendered hypothyroid by the administration of propylthiouracil (PTU) via their drinking water, available *ad libitum*, for a period of 3 to 7 weeks, at which time they were between 100 and 130 days of age. Their hypothyroid status was confirmed by measuring a 1 to 2 degree (Celsius) drop in core body temperature, and also by observing an increased behavioural (SHDS) response to apomorphine challenge (0.65μmol.kg^{-1} s.c.), as described above.

Following sacrifice by decapitation, the brains were quickly removed and placed in ice-cold oxygenated Krebs buffer (118mM NaCl, 5mM KCl, 25mM $NaCO_3$, 1mM $MgCl_2$, 1.2mM $CaCl_2$, 5.6mM glucose). The striata from three rats were then dissected out, pooled and cut into 2 x 0.3 x 0.3mm blocks using a McIlwaine tissue chopper at 4C. These blocks were then incubated with 10nM ^{3}H-labelled choline for 20 minutes at 37C in a shaking water bath under 95% O_2 : 5% CO_2. The [^{3}H]choline is taken up by the neurons in the tissue block, metabolised to acetyl-[^{3}H]choline and incorporated into the vesicular pool of ACh that is released by membrane depolarisation (Szerb, 1975). Depolarisation-induced [^{3}H]ACh efflux has been established as a reliable index of total ACh release (Beani *et al*, 1984). A control set of tissue blocks was also incubated with the selective choline uptake inhibitor hemicholinium-3 (10μM) in order to assess the degree of non-specific [^{3}H]choline uptake into the tissue. The tissue was then washed three times with fresh Krebs buffer, aliquotted into 1.5mL Eppendorf tubes (approximately 5mg wet weight tissue/tube), centrifuged (10000g, 4 minutes) and the supernatant discarded. The tissue was resuspended in 1mL of control buffer, high potassium buffer (77mM KCl, 46mM NaCl) or calcium free/high potassium buffer (which included 10^{-4}M EGTA) and incubated at 37C for 4 minutes. Depolarisation-induced ACh release was elicited by incubation in the high potassium containing buffer. Each of these buffers contained varying concentrations of a test drug, such as apomorphine. Following incubation the tubes were centrifuged (10000g, 4 minutes), and the supernatant collected and assayed for tritium in a Beckman LS3801 scintillation counter. The tissue was dissolved in 0.1M HCl and the radioactivity also counted.

Initially this technique was used to compare the inhibition, by the mixed D1/D2 agonist apomorphine, of acetyl[^{3}H]choline release by striatal blocks from control and PTU-treated rats. A dose-dependent inhibition of release was observed at doses between 10^{-9} and 10^{-6}M, beyond which the level of inhibition increased no further, and in some experiments even dropped markedly. With PTU-treated rats, this inhibition curve was shifted to the left, suggesting an increased sensitivity of the dopamine receptors involved in ACh release (Fig 1).

ACh release was not observed with incubation in Ca^{2+}-free high K^+ buffer, indicating the physiological validity of our system. Inclusion of hemicholinium-3 in the incubation with [^{3}H]choline reduced both tissue and supernatant levels to less than 3% observed in the control group, indicating that a specific uptake was occurring, and not non-specific binding and subsequent release of labelled choline from tissue membranes.

The effect of preincubation of the [^{3}H]choline loaded tissue blocks with either 1μM haloperidol (mixed D1/D2 antagonist) or 1μM raclopride (D2 antagonist) on the agonist-induced attenuation of ACh release was investigated. Neither antagonist reduced the apomorphine-induced inhibition of ACh release. In fact, raclopride produced a dose dependent inhibition by itself or in combination with the mixed agonist, an effect that was Ca^{2+} dependent. Clearly this result was unexpected and raised questions about the method we were using to investigate changes in ACh release. In particular, our method involves the examination of striatal responses in

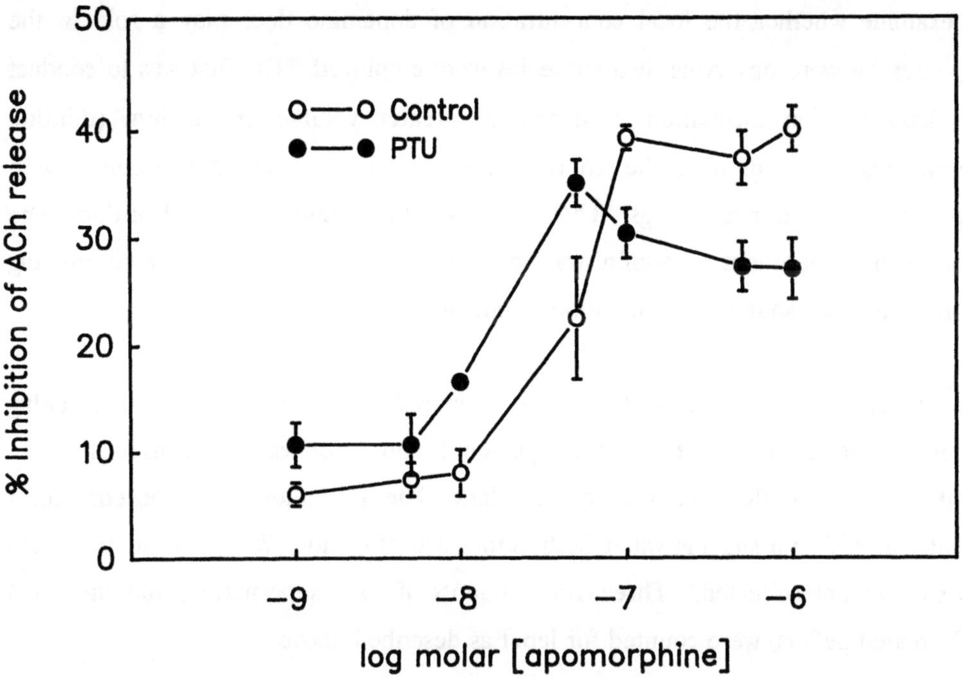

Figure 1. Apomorphine-induced inhibition of depolarisation-induced acetylcholine release in striatal blocks from euthyroid (control) and hypothyroid (PTU) rats. Each point is the mean ± s.e.m. of 3 determinations.

stationary medium, in contrast with the perfusion of the tissue blocks described initially by Stoof and Kebabian (Stoof & Kebabian, 1982). One disadvantage of our system is that it allows the accumulation of substances, other than the ACh released by K^+ induced depolarisation, including endogenous dopamine. While this may more closely mimic the physiological situation than the perfusion system in which the tissue is, in effect, being continually and rapidly washed, the multiple interactions it makes possible in the tissue render it difficult to interpret the actions of specific agents accurately. At the very least, the effects of undefined levels of endogenous dopamine acting upon both dopaminergic (via autoreceptors) and cholinergic neurons (via post-synaptic receptors) may modify the findings. Interestingly, several laboratories have found it difficult to demonstrate *in vivo* the inhibitory effects of apomorphine on ACh release observed *in vitro*, perhaps reflecting similar difficulties (Bertorelli & Consolo, 1990).

To examine whether the local concentration of dopamine does play a role in the responses we were observing, two strategies were employed. The first was to conduct the depolarisation incubation in a tenfold greater volume of medium (dilution experiments), thus diluting the concentrations of other released substances with respect to that of our test drugs. The second was to deplete the animal of dopamine prior to the experiment by administration of the drug reserpine, thereby eliminating the most obvious source of obfuscation in our model.

For the dilution experiments, tissue was prepared as above, but instead of being aliquotted into Eppendorf tubes, the equivalent amount of tissue was dispensed into 10mL of the test buffer in a 50mL conical flask. The incubation then proceeded for 4 minutes at 37C in a shaking water bath before centrifugation (7000g, 10 minutes) and the supernatant collected. Three 1mL aliquots of each supernatant, and the 0.1M HCl treated pellets, were counted for label as described above.

For the reserpine experiments, the original procedure was pursued, except that reserpine ($10mg.kg^{-1}$ s.c.) was administered to the animals 18 hours prior to sacrifice. The effect of the drug was confirmed in the rats by observing a 2 degree drop in body temperature, and a marked behavioural catalepsy.

In both experimental systems no inhibition of ACh release was observed with either haloperidol or raclopride alone, while dose-dependent inhibition of release by apomorphine was evident that could be blocked by 1μM raclopride. Thus the results from both the dilution and reserpine experiments supported our hypothesis that dopamine accumulation in our stationary system contributed to the apparently paradoxical results obtained with antagonists described. It is proposed therefore, that under the original assay conditions, depolarisation-induced dopamine release may actually have been enhanced by the action of the antagonists at pre-synaptic dopamine autoreceptors.

To confirm the validity of the previous observation that the dose-response curve to apomorphine was shifted to the left in hypothyroid rats, the experiment was repeated using the dilution experiment protocol. The results of this experiment confirmed that apomorphine (10^{-8} - 10^{-7}M) induced a greater inhibition of ACh release in tissue from PTU-treated rats (39 - 41%) compared to that from controls (23 - 31%).

Discussion

These results confirm that hypothyroid rats do exhibit heightened responsiveness at the D2 receptor with respect to the inhibition of cholinergic function and therefore the possible mechanisms underlying this need to be considered.

Regulation of D2 agonist sensitivity may occur at any of three major levels: the receptor itself, the second messenger whose activity it stimulates, or the G protein which mediates the interaction between the receptor and its effector. As discussed above, our laboratory has shown that in the hypothyroid rat exhibiting elevated D2 responsiveness, D2 receptor numbers are actually reduced by about 20%, while D1 receptor numbers are unchanged (Cameron & Crocker, 1990a). Affinities at both receptors are also unaltered (Cameron & Crocker, 1987). Therefore, changes in the status of the membrane-bound receptor do not explain the increased agonist efficacy observed.

Several intracellular second messenger systems have been associated with the D2 receptor. The best described is the inhibition of adenylate cyclase activity via G_i, opposing D1 activation via G_s. However, Stoof and Kebabian (1982) demonstrated that the inhibitory action of dopaminergic agents at the D2 receptor upon adenylate cyclase was independent of their inhibitory action upon ACh release. In addition, our laboratory has shown that adenylate cyclase activity is not altered in the hypothyroid rat, so that reduced cAMP synthesis does not explain the enhanced agonist-response we observed.

In contrast, a high correlation has been described between the inhibition of the electrical- and K^+-evoked release of ACh from guinea-pig striatal slices by the D2 agonist quinpirole, and the inhibition by this agent of the rise in intracellular calcium concentration ($[Ca^{2+}]_i$) (Fujiwara *et al*, 1987). This relationship is important, as neurosecretion is known to be calcium dependent, although the precise role of Ca^{2+} in the process is yet to be firmly defined (Llinás, 1982). This attenuated rise in $[Ca^{2+}]_i$ was ascribed to an inhibition of Ca^{2+} release from intracellular stores rather than a decrease in Ca^{2+} influx, and was independent of changes in adenylate cyclase activity. It was proposed that the inhibition was enacted through depression of phosphatidylinositol turnover (Fujiwara *et al*, 1987), but subsequent investigations have failed to substantiate this finding (Kelly *et al*, 1988; Vallar & Meldolesi, 1989; Vallar *et al*, 1988).

It is possible that $[Ca^{2+}]_i$ might also be modified by a change in ion channel conductivity. Recent studies have established that D2 receptors activate K^+ channels in the pituitary (Castelletti *et al*, 1989), the substantia nigra (Lacey *et al*, 1987) and the striatum (Freedman & Weight, 1988). Vallar *et al* (1990) have recently linked this activation with a decrease in Ca^{2+} conductance in a pituitary cell line expressing the D2 receptor. This activation is independent of adenylate cyclase activity, and is pertussis toxin sensitive, indicating coupling through a member of the G_i/G_o family, designated G_K (Codina *et al*, 1988). Experiments examining reconstitution of D2 receptors with G_i and G_o suggest that the protein involved in this effect is not homologous with that involved in the inhibition of adenylate cyclase (Ohara *et al*, 1988).

However, there is some debate as to whether the inhibition of the depolarisation induced rise in $[Ca^{2+}]_i$ is due solely to changes in release of Ca^{2+} from intracellular stores, or whether a modification of Ca^{2+} influx is also involved. It has been suggested that D2 receptors are negatively coupled to Ca^{2+} channels via a G protein as has been described for other inhibitory receptors, including the M_2-muscarinic and the α_2-adrenergic receptors (Vallar & Meldolesi, 1989). Certainly, in sensory neurons dopamine blocks the low threshold T-type Ca^{2+} channel and delays activation of the high threshold L-type Ca^{2+} channel (Marchetti *et al*, 1986). Such an effect of D2 activation upon Ca^{2+} conductance would provide a satisfactory explanation for the inhibition of depolarisation-induced $[Ca^{2+}]_i$ elevation.

Thus the effector system involved in dopaminergic inhibition of ACh release in the striatum is probably the G protein-mediated regulation of ion conductance in the cholinergic interneuron. However, the mechanism by which the system is rendered supersensitive in hypothyroidism remains to be elucidated. Analogies can be drawn here with the described influence of thyroid status upon adrenergic regulation of adipose tissue and liver adenylate cyclase (for review see Malbon *et al* (1988)). Thyroid hormones play a so-called "permissive" role in modulating the response to agonists by affecting the availability (and perhaps activity) of the relevant G proteins. In particular, hypothyroidism is associated with an upregulation of the α and β subunits of G_i, and a slight decrease or no change in G_s levels (Ros *et al*, 1989). Interestingly, glucocorticoid status, also shown by our laboratory to affect D2 responsiveness in the absence of a change in receptor concentration, is suggested to play a similar "permissive" role in regulating G protein availability (Watkins *et al*, 1987).

It would therefore be of interest to examine G protein status in the cholinergic neurons from the striata of hypothyroid rats. The specific G protein mediating the action of the D2 receptor upon $[Ca^{2+}]_i$ however, has not yet been definitely identified. In fact, it is not yet clear whether the G protein involved interacts directly with the Ca^{2+} channel, or if generation of a second messenger (e.g. diacylglycerol (Rane & Dunlap, 1986)) is necessary. Hescheler *et al* (1987) have demonstrated that the α-subunit of G_o is sufficient to overcome pertussis toxin blockade of the inhibitory effect of dopamine, noradrenaline and GABA upon Ca^{2+} conductance in cultured neurons, but this does not establish conclusively that G_o is the normal coupling involved. The next step, then, in the investigation of increased D2 agonist response is the identification of the G protein mediating the inhibition by apomorphine of the depolarisation-induced rises of both $[Ca^{2+}]_i$ and ACh release, and to compare its status in the euthyroid and hypothyroid states.

In addition, work from some laboratories suggests the involvement of a D1 receptor-linked mechanism in the release of ACh in the striatum independent of, and opposing, D2-initiated events (Saller & Salama, 1986; Consolo *et al*, 1987; Bertorelli

& Consolo, 1990). Thus, hormonal modification of D1-associated cellular responses may need to be considered. Also deserving attention is the activity of the enzyme phosphodiesterase; while adenylate cyclase has been discounted in the D2-mediated inhibition of ACh release, a decline in cAMP levels brought about by an increase in the activity of the catabolic enzyme may be involved.

Conclusion

Summarising the evidence reviewed above, it is possible that the D2 agonist inhibition of depolarisation induced ACh release is the result of the receptor acting via an as yet unidentified G protein/s to reduce Ca^{2+} flux into the cholinergic neuron. Thus it can be hypothesised that in the hypothyroid rat there is an increased availability of the G protein involved, upregulating the D2- initiated response. Whether this increased availability of G protein is due to reduced catabolism of the protein, accelerated cycling of the protein, or to stimulation of transcription remains to be explored.

In conclusion, the determination of the mechanisms by which D2 agonist sensitivity is modified by hormonal status is of significance in understanding the basis of supersensitivity phenomena in the central nervous system. In addition, the elucidation of such mechanisms has important implications for comprehending the neuro-biochemical alterations underlying pathologies of striatal dopamine transmission.

Acknowledgements

This work was supported by a grant from the NH&MRC of Australia to A.D.C. and P.B.F. is in receipt of a Flinders University Postgraduate Research Award. Gifts of raclopride (Astra) and SCH23390 (Schering-Plough) are gratefully acknowledged.

References

Beani L, Bianchi C, Siniscalchi A, Sivilotti L, Tanganelli S, Veratti E (1984) Different approaches to study acetylcholine release: endogenous ACh versus tritium efflux. Naunyn-Schmiedeberg's Arch Pharmacol 328:119-126

Bertorelli R, Consolo S (1990) D1 and D2 dopaminergic regulation of acetylcholine release from striata of freely moving rats. J Neurochem 54:2145-2148

Cameron DL, Crocker AD (1987) Post-synaptic changes and increased dopamine receptor sensitivity in hypothyroid rats. Pharmacol Biochem Behav 28(2):193-196

Cameron DL, Crocker AD (1988) Alkylation of striatal dopamine receptors abolishes stereotyped behaviour but has no effect on dopamine stimulated adenylate cyclase activity. Neurosci Lett 90(1):165-171

Cameron DL, Crocker AD (1990a) Increased dopamine agonist sensitivity is accompanied by a decrease in striatal D2 receptor concentration in the hypothyroid rat. European J Pharmacol 184(4):1410-1411

Cameron DL, Crocker AD (1990b) The hypothyroid rat as a model of increased sensitivity to dopamine receptor agonists. Pharmacol Biochem Behav 37:(in press)

Cameron DL, Crosbie J, Crocker AD (1988) A fixed interval momentary sampling method for assessing on-going behaviours induced by dopamine receptor agonists. Prog Neuro-Psychopharmacol & Biol Psychiat 12:595-606

Cameron DL, Crocker AD (1989) Localization of striatal dopamine receptor function by central injection of an irreversible dopamine receptor antagonist. Neuroscience 32:769-778

Castelletti L, Memo M, Missale C, Spano PF, Valerio A (1989) Potassium channels involoved in the transduction mechanism of dopamine D_2 receptors in rat lactotrophs. J Physiol 410:251-265

Codina J, Olate J, Abramowitz J, Mattera R, Cook RG, Birnbaumer L (1988) Alpha c3 cDNA encodes the alpha subunit of Gk the stimulatory G protein of receptor regulated K^+ channels. J Biol Chem 263:6746-6750

Consolo S, Wu CF, Fusi R (1987) D-1 receptor-linked mechanism modulates cholinergic neurotransmission in rat striatum. J Pharmacol Exp Ther 242:300-305

Crocker AD, Cameron DL (1990) Investigation of striatal dopamine-acetylcholine interactions using the muscarinic receptor alkylating agent BM 123. CINP, Proc. XVIIth Congress, Kyoto 2:253

Crocker AD, Russell RW (1990) Pretreatment with an irreversible muscarinic agonist affects responses to apomorphine. Pharmacol Biochem Behav 35:511-516

Fleming WW, Westfall DP (1988) Adaptive supersensitivity. In: Handbook of experimental pharmacology, vol 90/I, Trendelenburg U, Weiner N (eds) Berlin: Springer-Verlag p. 509-559.

Freedman JE, Weight FF (1988) Single K^+ channels activated by D_2 dopamine receptors in acutely dissociated neurons from rat corpus striatum. Proc Natl Acad Sci USA 85:3618-3622

Fujiwara H, Kato N, Shuntoh H, Tanaka C (1987) D_2-dopamine receptor-mediated inhibition of intracellular Ca^{2+} mobilization and release of acetylcholine from guinea-pig neostriatal slices. Br J Pharmac 91:287-297

Hescheler J, Rosenthal W, Trautwein W, Schultz G (1987) The GTP-binding protein, G_o, regulates neuronal calcium channels. Nature 325:445-447

Kelly E, Batty I, Nahorski SR (1988) Dopamine receptor stimulation does not affect phosphoinositide hydrolysis in slices of rat striatum. J Neurochem 51(3):918-925

Lacey MG, Mercuri NB, North RA (1987) Dopamine acts on D_2 receptors to increase potassium conductance in neurones of the rat substantia nigra zona compacta. J Physiol 392:397-416

Lehmann J, Langer SZ (1983) The striatal cholinergic interneuron: Synaptic target of dopaminergic terminals?. Neuroscience 10:1105-1120

Llinás RR (1982) Calcium in synaptic transmission. Sci American 247(4):56-65

Malbon CC, Rapiejko PJ, Watkins DC (1988) Permissive hormone regulation of hormone-sensitive effector systems. Trends Pharmacol Sci 9:33-36

Marchetti C, Carbone E, Lux HD (1986) Effects of dopamine and noradrenaline on Ca channels of cultured sensory and sympathetic neurons of chick. Pflüg Arch 406:104-111

Maura G, Giardi A, Raiteri M (1988) Release regulating D-2 dopamine receptors are located on striatal glutamatergic nerve terminals. J Pharmacol Exp Ther 247:680-684

Nielsen M, Klimek V, Hyttel J (1984) Distinct target size analysis of dopamine D-1 and D-2 receptors in rat striatum. Life Sciences 35:325-332

Ohara K, Haga K, Bernstein G, Haga T, Ichiyama A (1988) The interaction between D-2 dopamine receptors and GTP-binding proteins. Molec Pharmacol 33:290-296

Rane SG, Dunlap K (1986) Kinase C activator 1,2-oleoylacetylglycerol attenuates voltage dependent calcium current in sensory neurons. Proc Natl Acad Sci USA 83:184-188

Ros M, Northrup JK, Malbon CC (1989) Steady state levels of G proteins and β-adrenergic receptors in rat fat cells. J Biol Chem 263:4362-4368

Saller CF, Salama AI (1986) D-1 and D-2 dopamine receptor blockade: interactive effects *in vitro* and *in vivo*. J Pharmacol Exp Ther 236(3):714-720

Stoof JC, Kebabian JW (1981) Opposing roles for D-1 and D-2 dopamine receptors in efflux of cyclic AMP from rat neostriatum. Nature 294:366-368

Stoof JC, Kebabian JW (1982) Independent in vitro regulation by the D-2 dopamine receptor of dopamine stimulated efflux of cyclic AMP and K^+-stimulated release of acetylcholine from rat neostriatum. Brain Res 250:263-270

Szerb JC (1975) Endogenous acetylcholine relaees and labelled acetylcholine formation from [^{3}H]choline in the myenteric plexus of the guinea-pig ileum. Can J Phys Pharm 53:566-574

Vallar L, Vicenti LM, Meldolesi J (1988) Inhibition of inositophosphate production is a late Ca^{2+}-dependent effect of D2 dopaminergic receptor activation in rat lactotroph cells. J Biol Chem 263:10127-10134

Vallar L, Meldolesi J (1989) Mechanisms of signal transductuction at the D_2 receptor. Trends Pharmacol Sci 10:74-77

Vallar L, Muca C, Magni M, Albert P, Bunzow J, Meldolesi J, Civelli O (1990) Differential coupling of dopaminergic D2 receptors expressed in different cell types - Stimulation of phosphatidylinositol 4,5-bisphosphate hydrolysis in LtK-fibroblasts, hyperpolarization, and cytosolic-free Ca2+ concentration decrease in GH4C1 cells. J Biol Chem 265(18):10320-10326

Watkins DC, Northrup JK, Malbon CC (1987) Regulation of G proteins in differentiation. J Biol Chem 262:10651-10657

PHOSPHORYLATION OF THE GROWTH-FACTOR ACTIVATABLE Na^+/H^+ ANTIPORTER: A MECHANISM OF REGULATION.

Sardet, C., Wakayabashi, S., Fafournoux, P., Pouysségur, J.
CNRS, Centre de biochimie,
Université de Nice, Parc Valrose,
Nice 06034
FRANCE

INTRODUCTION

Na^+/H^+ exchange occurs across the plasma membrane in prokaryotic and eukaryotic cells. This process has a central role in cell homeostasis : regulation of intracellular pH (pHi), cell Na^+ content, cell volume, transcellular Na^+ and HCO_3^- absorption in epithelial cells are some of the parameters regulated by Na^+/H^+ exchange (reviewed by Seifert and Aronson, 1986; Grinstein, 1988). In vertebrate cells, the antiporter operates as a H^+ extruding system driven by the inwardly directed Na^+ gradient. It is activated cooperatively by intracellular H^+ through a proton regulatory site whereby H^+ binding exerts an allosteric activation on H^+ extrusion (reviewed by Aronson, 1985). Therefore, H^+ efflux can be considered as a self controlled mechanism where the pHi sensitivity of the Na^+/H^+ antiporter contributes to the determination of the resting pHi. In fact, the Na^+/H^+ antiporter is virtually turned off at neutral and basic pHi and is maximally active at acidic pHi below 6. This characteristic clearly defines the antiporter as a system adapted to protect cells against intracellular acidification. In addition to the Na^+/H^+ antiporter, their exists at least two other membrane components that regulate intracellular pH: a Na^+ dependant- and a Na^+ independant- Cl^-/HCO_3^- antiporter. This duplication of pHi regulating systems (reviewed by Madshus *et al.*, 1988) is probably due to the fact that cytoplasmic pH must be finely regulated to permit optimal metabolic conditions for the development and growth of multicellular organism. This has been demonstrated in eukaryotic cells in culture which are totally devoid of their pH regulating systems. Chinese Hamster Lung Fibroblasts mutants ("Proton-Suicide" Mutants (PS)) disrupted for Na^+/H^+ antiport activity (Pouysségur *et al.*, 1984) and grown in a bicarbonate free medium, buffered at neutral pH, stop proliferating and enter a reversible quiescent state (Pouysségur et al, 1984).

NATO ASI Series, Vol. H 52
Biological Signal Transduction
Edited by E. M. Ross and K. W. A. Wirtz

In these conditions, their intracellular pH is decreased by 0.2-0.3 pH units. This drop in pHi is the sole limiting parameter for growth since this effect can be completely reversed simply by restoring an alkalinic pH in the culture medium by buffering the extracellular pH above 7.5. An important feature that could be related to this interdependance between growth control and pHi is that the Na^+/H^+ antiporter is activated in response to growth factors and various cell activating agents such as sperm, neurotransmitters, hormones, chemotactic peptides, phorbol esters and mitogenic lectins (reviewed by, Grinstein *et al.*, 1989). In absence of bicarbonate, this activation results in persistent cytoplasmic alkalinisation of 0.1 to 0.3 pH units which is suppressed by the specific inhibitors of the Na^+/H^+ antiporter, or in mutants deficients in Na^+/H^+ antiport activity. Kinetic data suggest that activating agents trigger their effect on the Na^+/H^+ exchange by increasing the affinity of the antiporter protein for H^+ at the internal H^+ regulatory site (Moolenaar *et al*, 1983; Paris and Pouysségur, 1984). The means by which external signals are propagated from the cell surface to specific membrane, cytoplasmic and nuclear targets is still poorly understood. In that context, the Na^+/H^+ antiporter, a ubiquitous plasma membrane protein which is activated by all types of growth factors is an interesting molecule. From the available data one can raise the important question as to whether the antiporter serves as an integrator of the diverse output signals coordinating the mitogenic events. As growth factors increase protein phosphorylation of cellular effectors on tyrosyl as well as on seryl and threonyl residues, it is strongly suggested that phosphorylation cascades are involved in mediating early cell growth events. Therefore, our favored hypothesis predicts an activation of the Na^+/H^+ exchange *via* a phosphorylation-induced conformational change of the antiporter leading to an increase affinity of "the proton regulatory site". In this report we present arguments that strongly support this hypothesis.

THE Na^+/H^+ ANTIPORTER IS A PHOSPHOPROTEIN

First, in order to determine if the Na^+/H^+ is a phosphoprotein, a specific antibody was developed to gain direct access to the antiporter protein. As decribed below, the cloning of the Na+/H+ antiporter gene eventually allowed construction of a fusion protein and the generation of an antibody. As previously reported (Pouyssegur *et al.*, 1984, Franchi *et al.*, 1986b, Sardet *et al.*, 1989) the power of reverse genetics was exploited to clone the gene. The isolated cDNA codes for a protein of 815 amino-acids (Mw = 90704) with two potential N-linked glycosylation sites. The hydropathy profiles reveals the existence of 10 to

12 putative transmembrane segments followed by a long hydrophilic stretch of 315 amino-acids, presumably cytoplasmic (Sardet *et al.*, 1989 and 1990) (Fig 1).

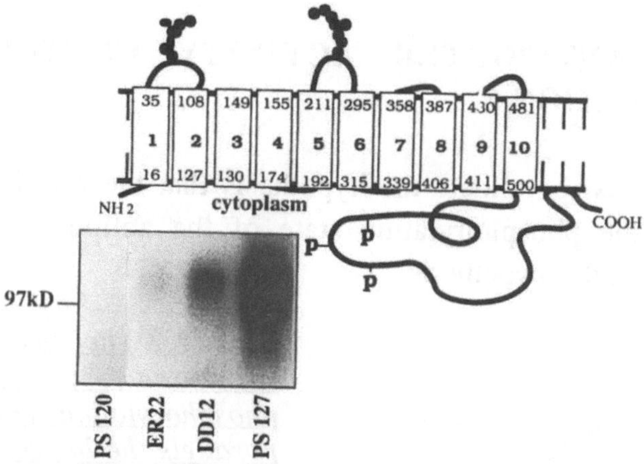

***Figure 1** <u>Putative transmembrane organization of the Na+/H+ antiporter and Immunoprecipitation of ^{32}P-labelled protein.</u>*
This putative model of the Human amiloride-sensitive Na+/H+ antiporter was established from the c-DNA deduced amino-acids sequence. The antiporter is a glycosylated protein of 815 amino-acids with at least ten transmembrane domains and several phosphorylation sites (P) in the large COOH-terminal cytoplasmic domain (Sardet et al., 1989). Exponentionally growing cells were labelled with ^{32}(P)orthophosphate and immunoprecipitated with polyclonal antibodiy RP-c28 (Sardet et al., 1990). A single labelled polypeptide of 105-110 kD is immunoprecipitated in ER22 cells (a CCL39-derived clone that expresses 800,000 human EGF receptors (L'Allemain et al., 1989), DD12 (a CCL39-derivated clone that overexpresses its endogenous Na+/H+ activity 10 to15 times (Franchi et al., 1986a) and PS127A (a PS120 derivate that overexpresses Na^+/H^+ activity 10-15 times the Human Na^+/H^+ antiporter c-DNA (Sardet et al., 1990). This phosphoprotein is not detectable in PS120 (a CCL-39 mutant deficient for Na^+/H^+ activity (Pouysségur et al., 1984)).

From the known nucleotide sequence, a fusion protein was constructed by expressing in *E.coli* the last 157 amino-acids of the hydrophilic carboxyterminus region linked to bacterial β-galactosidase. A large production of this fusion protein was used to raise specific polyclonal antibodies to the antiporter (RP-c28) (Sardet *et al.*, 1990). Using this antibody, immunoprecipitation of membrane proteins from [^{32}P] labelled growing cells revealed a single glycosylated phosphoprotein of 105-110 kD (Fig.2). Today with the same antibody, phosphorylated antiporter has been demonstrated in several cell types:

Chinese Hamster Lung Fibroblast (CCL39, ER22, DD12) (Sardet *et al.*, 1990), Human Carcinoma cells (A431) (Sardet *et al*, 1990), Mouse embryonic EC cells (P19), and Human Platelets (non published).

IS NA+/H+ ANTIPORTER ACTIVITY CONTROLLED BY PHOSPHORYLATION ?

The next step was to validate the hypothesis that the Na^+/H^+ exchange is controlled by the phosphorylation state of the antiport protein. Several arguments support this hypothesis.

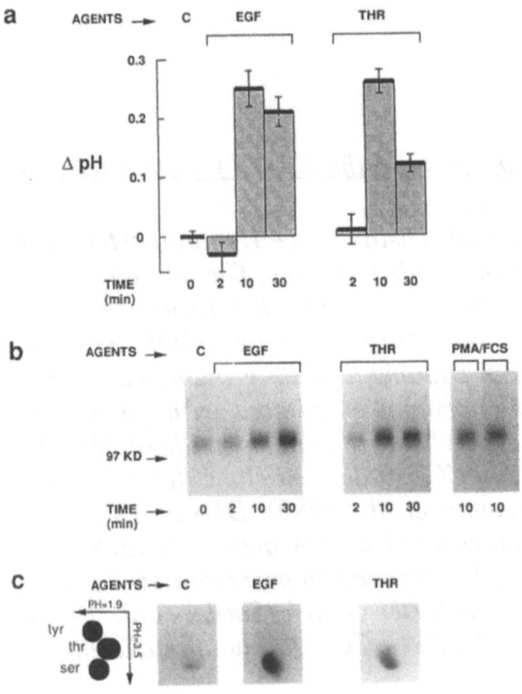

Figure.2 *<u>The time course of the increase in antiporter phosphorylation level by mitogens parallels the increase in Na+/H+ exchange activity. Mitogens induced phosphorylation occurs on serine residues.</u>* *Quiescent ER22 cells were stimulated by either 40nM EGF, 10 nM thrombin (THR), 100 ng/ml phorbol dibutyrate (PMA) or 10% fetal calf serum (FCS). The intracellular pH increase (ΔpH) (a) and in vivo phosphorylation (b) were measured on parallel cultures and under the same conditions of stimulations, Phospho amino-acid analysis (c) was performed under the same conditions of 30 min stimulation with EGF and 30 min stimulation with thrombin (Cooper et al, 1983). For pH determinations, quiescent cells were cultured in bicarbonate-free culture medium buffered with 20mM Hepes, pH 7.4 and then stimulated by mitogens. Intracellular pH was measured with (^{14}C)-benzoic acid as previously reported (L'Allemain et al., 1984). For immunoprecipitation, quiescent ER22 cells were labelled to equilibrium in bicarbonate- and phosphate-free culture medium containing (^{32}P)-orthophosphate and immunoprepreciptated with RP-c28 as previously described. Amino-acids analysis were performed on this immunoprecipitated Na^+/H^+ antiporter (Sardet et al., 1990).*

(i) Growth-factors induce a parallel increase in Na^+/H^+ exchange activity and Na^+/H^+ antiporter phosphorylation.

We analyzed the capacity of growth factors to modify the phosphorylation state of the exchange protein in secondary cultures of CHL fibroblast (ER22) (Sardet *et al.*, 1990). ER22 is a CCL39-derived clone that expresses high levels of human EGF receptors and responds equally well to mitogenic stimulation with either EGF or α-thrombin (THR) (L'Allemain *et al.*, 1989).

As shown in Fig.2a, both EGF and thrombin induce a rise in intracelluiar pH after a lag of 2 min, that peaks around 10 min and declines slightly thereafter, persisting as long as the stimulus is maintained (Van Obberghen-Schilling *et al.*, 1985). The *in vivo* state of phosphorylation of the Na^+/H^+ antiporter, was analyzed by immunoprecipitation with polyclonal antibodies RP-c28, under the same conditions of mitogenic stimulation, on ER22 cells prelabelled with (^{32}P) to equilibrium. Fig.2b shows clearly that the Na^+/H^+ antiporter is already phosphorylated in G0 arrested ER22 cells (time 0), then after a lag of at least 2 min, EGF and thrombin stimulate the phosphorylation of the antiporter. Fetal calf serum and phorbol esters, agents known to activate the antiporter in CHL cells also stimulate its phosphorylation (Fig.2b). The activation of the Na^+/H^+ exchange by EGF, thrombin or serum occurs in a time-responsive manner which correlates with the increased level of phosphorylation in the antiporter.

(ii) Phosphatases inhibition induces a parallel increase in Na^+/H^+ exchange activity and Na^+/H^+ antiporter phosphorylation.

An other striking evidence of the relationship between protein phosphorylation and Na^+/H^+ antiporter activity comes from the use of okadaic acid on living cells. Okadaic acid rapidly stimulates protein phosphorylation in intact cells (Haystead *et al.*, 1989), and behaves like a specific phosphatase inhibitor in a variety of metabolic processes. *In vitro*, it inhibits phosphatases 1 and 2A which are specific for serine/threonine phosphoryl residues (For review, Cohen *et al.*, 1990). Alone, okadaic acid mimics the effect of thrombin on Na^+/H^+ exchange activation in hamster fibroblasts ER22; namely increased levels of phosphorylation of the antiporter protein (Fig.3b) and production of intracellular alkalinization (Fig.3a), thus suggesting that the latter process is stimulated by a serine/threonine phosphorylation. Morever, this type of experiment outlines the major role of phosphatases in the regulation of the Na^+/H^+ exchange. The inhibition of PP1/PP2A by okadaic acid can only increase the phosphorylation state of a protein if the opposite protein kinase is at least partially active. It means that there exists at least one protein kinase which phosphorylates the Na^+/H^+ antiporter in the resting state. This raises the following questions: how the level of antiporter phosphorylation is connected to Na^+/H^+ exchange kinetic properties, and as a consequence what is the role of the phosphorylation in the resting state ?

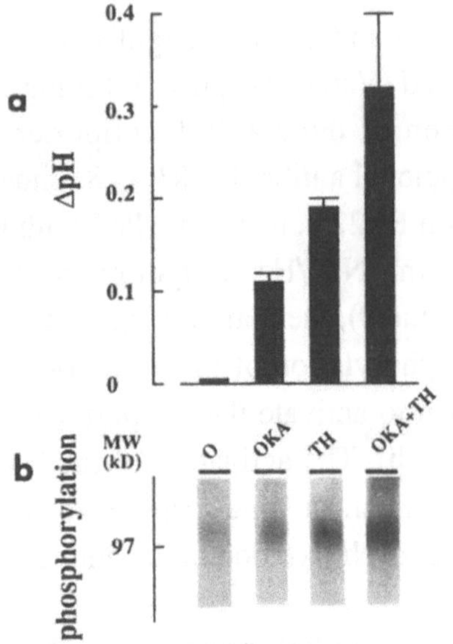

***Figure 3** <u>Na+/H+ antiporter activation by okadaic acid.</u> Quiescent ER22 cells were treated with 1 μM okadaic acid for 15 min in the presence or absence of 10 nM Thrombin. This phosphatase inhibitor provokes an amiloride sensitive intracellular alkalinisation (ΔpH) (a) and a parallel increase in Na+/H+ antiporter phosphorylation level (b) (Sardet et al., in preparation). Intracellular pH and immunoprecipitation of the Na+/H+ antiporter were performed as described in Fig.2.*

Does the addition of phosphate modulate the Na^+/H^+ antiporter activity ?

It was previously observed in A431 carcinoma cells (Cassel *et al.*, 1986) and in spontaneously beating cardiac cells (Weissberg *et al.*,1988) that ATP depletion by incubation with metabolic inhibitors (this would decrease the antiporter resting phosphorylation) modifies reversibly the kinetic properties of the Na^+/H^+ exchange. In these cells, an inhibition of Na^+/H^+ exchange activity correlates with the extent of the reduction of the ATP pool. The apparent mechanism consists of an increased dependency for intracellular protons in ATP-depleted cells. This would suggest that resting phosphorylation could be responsible for the antiporter proton affinity value in absence of growth factors. On the opposite, in presence of growth factors, an increase of phosphate groups on the same site would explain the observed decreased dependency for intracellular proton. It is generally assumed that protein phosphorylation stabilizes different conformational states of proteins to enhance or inhibit their activity, however direct evidences about molecular mechanisms which allow phosphoryl transfert to trigger these changes are poorly understood (Sprang *et al.*, 1988). We have recently observed that the Na^+/H^+ antiporter is an oligomer (Fafournoux *et al.*, in preparation). Thus, phosphorylation of the antiporter could strength interactions between subunits that could in turn alter the binding site for allosteric effector and substrate, a mechanism operating for the activation of glycogen phosphorylase (Sprang *et al.*, 1988). Alternatively the phosphorylation could modify directly the pKa of the allosteric site through addition of negative charges around this site.

DIFFERENT CLASSES OF GROWTH-FACTORS STIMULATES NA+/H+ ANTIPORTER PHOSPHORYLATION ON SERINE RESIDUES : A COMMON MECHANISM OF ACTIVATION ?

To investigate the enzymatic cascade initiated either from receptor tyrosine kinases (EGF receptor) or from G-protein coupled receptors, it appears essential to analyse the key phosphorylation site(s) on the antiporter molecule

As described on ER22 cell line (Fig.2), EGF and thrombin (THR) have an equal potency to activate the Na^+/H^+ exchange. (L'Allemain *et al.*, 1989; Sardet *et al.*, 1990) and Na+/H+ antiporter phosphorylation. Whatever the nature of the stimulus (EGF or THR), all the phosphates are incorporated in serine residues (Fig. 2c) (Sardet *et al.*, 1990) regardless of the mitogen used as is the case for ribosomal protein S6 phosphorylation (Martin-perez *et al.*,1984). This simple observation raises the question whether there exists (a) common Na+/H+ antiporter serine-kinase(s) which is (are) activated by different growth-factors that use separate signalling pathways (Chambard *et al.*, 1987, L'Allemain *et al.*, 1986). The transduction pathways involved in CCL39/ER22's Na^+/H^+ exchange regulation have been summarized in Fig.4.

One or several Na+/H+ antiporter kinases ?

Most growth factor receptors that are tyrosine kinases (EGF, Insulin, IGF-1, FGF, PDGF, CSF-1) (reviewed by Yarden and Ulrich, 1988), have been described to be able to activate the Na^+/H^+ antiporter in various cell lines (for example: L'Allemain and Pouysségur, 1986; Hartmann *et al.*, 1990). The fact that these receptors are tyrosine kinases suggested that the initial event following growth factor binding may be stimulation of receptor tyrosine kinase activity. However most of the changes in protein phosphorylation in response to growth factors occur on serine and threonine residues. Thus, as has recently been described for EGF, there is growing evidence for the existence of a protein kinase cascade (Ahn *et al.*, 1990). Several serine /threonine protein kinases may be activated by a cascade of events starting with the tyrosine phosphorylation of at least one of these kinases, for example, Raf-1 (Morrison *et al.*, 1988) and finishing with the activation of cellular targets like the ribosomal protein S6 (for review, Kozma *et al.*, 1989) or the Na^+/H^+ antiporter. The last kinase in the cascade can phosphorylate one or several cellular effectors. Thus, for the growth factor dependent phosphorylation of the antiporter, the question remains whether the last kinase(s) is(are) a specific Na^+/H^+ antiporter kinase(s) activated by growth factors, like the S6-kinase (Ballou *et al.*, 1988) or a serine kinase(s) which is (are) known to phosphorylate a large number of targets, for example, A-kinase, C-kinase or Ca^{++}/calmodulin dependent kinase.

Previously it has been suggested in the litterature that these latter kinases are involved in Na+/H+ exchange activation by another class of mitogens that bind to G-protein coupled receptors, including thrombin, vasopressin, bombesin, serotonin, ß and α-adrenergic agonists, *etc*... These cell activating agents initially trigger their intracellular effect via specific coupling of G-protein family members with either ions channels or enzymatic effectors such as phospholipase-C (PLC), phospholipase-A_2 (PLA_2) and adenylate-cyclase (AC) (reviewed by Pouysségur, 1989; Birnbaumer *et al.*, 1990).

Growth factors, like thrombin, which act through a stimulation of phospholipase C catalyse the hydrolysis of phosphatydylinositol 4,5 biphosphate (PIP_2) into diacylglycerol (DAG), a potent activator of protein kinase-C, and inositol 1,4,5 triphosphate (IP_3), a mediator of calcium release from internal stores that in turn activates various Ca^{++} binding protein (reviewed by Berridge,1987) such as calmodulin dependent kinases (MacNicol *et al.*, 1990). In a variety of systems, principally on non polarized-cells like fibroblasts, lymphocytes or smooth muscle cells, it has been demonstrated that the stimulation of the Na^+/H^+ antiporter is mediated at least in part *via* protein kinase C activation.

Summarizing the results of a large number of publications, this has been proved by three different approaches in intact cells: i) direct stimulation of kinase C by DAG or phorbol esters activates the Na^+/H^+ exchange (Moolenar *et al.*, 1984 Paris and Pouysségur, 1984); ii) potential protein kinase-C inhibitors like sphingosine (Lowe *et al.*, 1990) block Na^+/H^+ exchange activation by phorbol esters. iii) phorbol esters fail to activate the Na+/H+ exchange on protein kinase-C depleted cells (Huang *et al.*,1987). In contrast, it should be mentioned that

kinase-C activation has also been reported to inhibit apical Na^+/H^+ exchange in epithelial cells from small intestine and kidney (Rood *et al.*,1988).
However, it has been demonstrated on certain cells that some growth-factors like thrombin (Huang *et al.*, 1987) are still able to activate the Na^+/H^+ exchange after proteine kinase-C depletion. Therefore it has been suggested that protein kinase-C independent transduction pathways are also able to regulate the Na^+/H^+ antiporter.
The transient intracellular Ca^{++} ($[Ca^{++}]_i$) elevation induced by several mitogenic agents precedes the increase in intracellular pH (Hesketh et al,1985). Therefore it has been proposed that a relationship exist between the two responses. Thus, the pH response could be mediated in part *via* the modulation of protein kinase-C activity, and/or *via* the activation of the Ca^{++}/calmodulin kinase as it has been recently demonstrated that phosphatidylinositol signaling pathway activates this latter kinase (MacNicol *et al.*, 1990). However the role of $[Ca^{++}]_i$ rise in mitogen-induced alkalinisation remains controversial, and seems different from one cell type to another. The prevention of mitogen-induced $[Ca^{++}]_i$ rise inhibits the activation of the Na^+/H^+ exchange in smooth muscle cells (Huang *et al.*, 1987) and human fibroblast of fetal skin origin (WS-1) (Hendey et al, 1989). In contrast, inhibiting Ca^{++} transients has no effect on Swiss 3T3 cells (Ives and Daniel,1987). The direct elevation of $[Ca^{++}]_i$ by calcium ionophores gives the same type of contradictory results. It activates by itself the Na^+/H^+ exchange in WS-1 fibroblasts (Hendey *et al.*, 1989) but fails to produce this effect in 3T3 cells (Ives and Daniel, 1987), HF human fibroblasts (Moolenaar et al, 1983), smooth muscle cells (Huang et al, 1987), and Chinese hamster lung fibroblast (CCL-39) (data non published). Lastly, Ca^{++}/calmodulin has been shown to inhibit the Na^+/H^+ exchange on ileal brush-border vesicles (Emmer *et al.*, 1989).
Also controversial is the role of intracellular cAMP variation on the Na^+/H^+ exchange. Receptors coupled to the regulation of adenylate cyclase through G-proteins, such as ß-adrenergic, α_2-adrenergic, or prostaglandin E_1 receptors have been shown to activate the Na^+/H^+ exchange in various non polarized cells (for example, Ganz *et al.*, 1990). However it has been suggested that this effect could be mediated independently of changes in cAMP levels and independently of a cholera toxin sensitive G protein (Limbird, 1988, Ganz *et al.*, 1990). Alternatively, in several studies on epithelial brush border membranes, an increase in intracellular cAMP concentration has been reported to inhibit the Na^+/H^+ exchange in membrane vesicles (Kahn *et al.*, 1985). This effect could be mediated *via* the activation of cAMP dependent kinase (Weinman *et al.*, 1988).

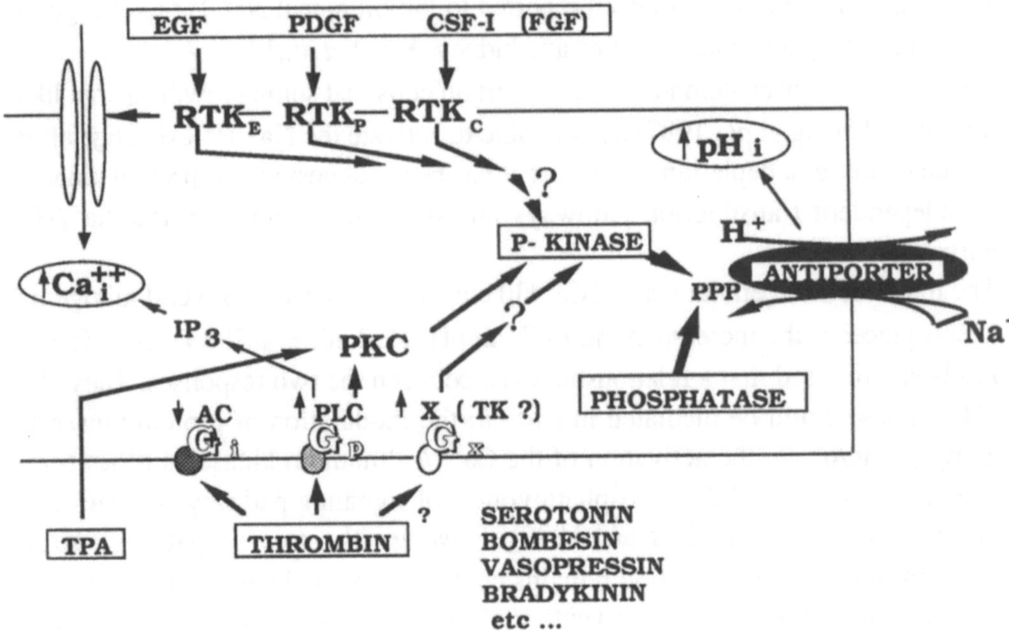

Figure 4: *Growth-signaling pathways and Na^+/H^+ antiport activation in CCL39 cells.*
This sheme outlines the hypothetical mitogenic pathways leading to activation of the Na^+/H^+ antiporter in CCL39 cells and subsequent rise in cytoplasmic pH. Tyrosine kinase activating growth factors as EGF, PDGF or FGF activate a transmembrane receptor (RTK). A second family of growth promoting agents acting through G protein-coupled receptors known to inhibit adenyl cyclase (AC), activate polyphosphoinositide-specific phospholipase C (PLC), and as a working hypothesis by activating X, other unknown effectors. As is illustrated, protein kinase C (pKC) also activated by phorbol esters (TPA) does not represent the only way to activate the antiporter. $[Ca]_i$ modification does not regulate the antiporter in these cells. A common protein kinase (P-kinase) activated by different pathways could be a specific Na^+/H^+ antiporter kinase. This scheme also points out the potential major role of a phosphatase in Na^+/H^+ antiport activity regulation.

One or several Na^+/H^+ antiporters ?

The observed differences in Na^+/H^+ exchange regulatory mechanisms from cell type to cell type could reflect (i) different transduction pathways depending on the cell and/or (ii) the presence of different isoforms of the Na^+/H^+ antiporter. In fact, two spatially and pharmacologically distinct antiporters have been recently described in kidney (Haggarty *et al.*, 1988) and intestinal (Knickelbein *et al.*, 1988) epithelial cells. The apical brush border membrane Na^+/H^+ exchange is less sensitive to inhibition by amiloride or its derivativees than the basolateral exchange. The latter presents similar pharmacological characteristics to the previously described plasma membrane Na^+/H^+ exchange from non-polarized cells. Regarding the previous data on Na^+/H^+ exchange regulation, the apical isoform would be inhibited by C-kinase and A-kinase. On the contrary, the basolateral isoform would be activated by C-kinase. This hypothesis of at least two distinct Na^+/H^+ antiporters would not be surprising as there also exists a growing family of genes for several transmembrane transport protein like anion exchangers (Kopito *et al.*, 1989), Na^+/K^+-ATPases (reviewed by Sweader, 1989) as well as glucose transporters (reviewed by Gould and Bell, 1990). In these examples, different protein isoforms are expressed in different cell types, where they contribute to specialized properties with non identical regulation; however the functional differences between isoforms are still poorly understood.

Na^+/H^+ antiporter phosphorylation site(s)

In order to definitively answer the questions exposed in the titles of the last two paragraphs: "does phosphorylation control the antiporter activity ? and is there a common mechanism of antiporter activation shared by different signalling pathways ?", the Na^+/H^+ antiporter phosphorylation sites have to be precisely identified and specifically mutagenized to obtain functional antiporter (sensitive to amiloride and activatable by cytoplasmic activation) resistant to activation by growth-factors. There already exists informations on the localization and characteristics of these phosphorylations sites:

(i) The first indication comes from recent studies on the $Na+/H+$ antiporter fonctionnal domains. Various deleted or mutated human antiporter c-DNAs have been transfected in the fibroblast mutant cell line PS which is devoid of Na^+/H^+ exhange activity. These constructions can be rapidly tested for any functional expression (Sardet *et al.*,1989; Wakayabashi *et al.*, in preparation). A set of deletions within the cytoplasmic domain has revealed several interesting features (Fig. 5). Complete deletion of the cytoplasmic domain (Δ515) produce a truncated protein which is still able to carry out an amiloride sensitive Na^+/H^+ exchange but has lost growth factor regulation (Wakabayabashi *et al.*, in preparation). The fact that ion transport requires only the transmembrane domain has also been observed with the anion transporters (Kopito *et al*, 1989). We suggest that the Na^+/H^+ antiporter can be separated into at least two distinct functional domains: the transmembrane domain that has all the features required to carry an amiloride-sensitive Na^+/H^+ exchange with a proton regulatory site which confers an allosteric activation; and a cytoplasmic domain (kinetic data not shown) which determine the "set point" value. As growth-factors modifie this set-point value, we suggest that growth-factor dependent phosphorylations would take place in this latter domain.

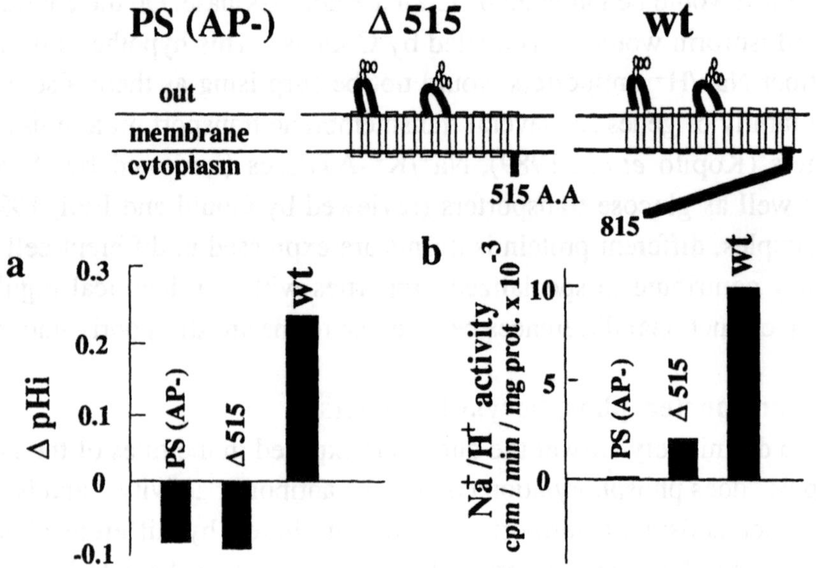

Figure 5: Expressions of various deleted forms of Na+/H+ antiporter PS antiporter deficient mutants.

PS (AP-) cells, a CCL-39 mutant deficient in Na^+/H^+ exchange activity, was stably transfected by Human Na^+/H^+ antiporter c-DNA coding for the complete protein (wt) (815 amino-acids), or coding only for the NH_2 transmembrane domain (Δ515) (515 amino-acids) (Wakabayabashi et al, in preparation). Quiescent cells PS, Δ515 and wt were stimulated by 7.5% fetal calf serum. The increase in intracellular pH (ΔpH) was measured as decribed in Fig3. Amiloride sensitive Na^+/H^+ exchange activity (Vmax) was measured by $^{22}Na^+$ uptake as previously described (Franchi et al., 1986b). Δ515 deletion is still able to carry out a Na^+/H^+ exchange but is no longer regulated by growth factors.

(ii) The analysis of phosphopeptides tryptic map shows an another important information. Tryptic peptides of *in vivo* ^{32}P-labeled Na^+/H^+ antiporter from ER22 cells treated with growth factors (serum, EGF, thrombin) were analysed by two dimensional thin-layer electrophoresis/chromatography (Fig.6)(Fafournoux *et al.*, in preparation). These maps suggest the existence of several sites of phosphorylation since the relative amount of phosphate incorporated differ in some peptides in the absence and presence of growth factors (data not shown); and more striking was the appearance in the map of a new labeled peptide (⇒) due to growth-factor stimulation.

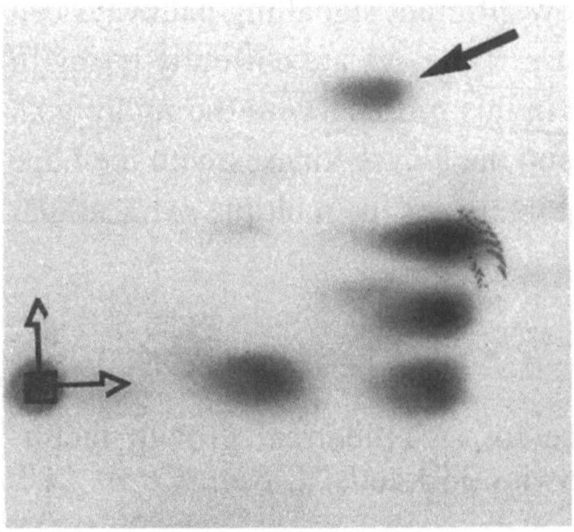

Figure 6: <u>Tryptic phosphorylation map of the Human Na+/H+ antiporter.</u> *PS 127A exponentionally growing cells were labelled to equilibrium with (^{32}P) orthophosphate and subsequently immunoprecipitated with the polyclonal antibody RP-c28. Trypsin digestion of the immunoprecipitated Na^+/H^+ antiporter was then analysed by two dimensional thin iayer elctrophoresis / Chromatography. In response to mitogenic stimulation (EGF or thrombin),this map shows the occurence of a new phosphorylation site (⇒)*

We now know that most phosphoproteins contain multiple phosphorylation sites and that these sites are often located in clusters. From the study of the enzyme glycogen synthetase (for review, Roach, 1990), it has been shown that multiple phosphorylations could also occur in a hierarchal fashion; one kinase introduces a phosphate group that is a requirement for the action of an another. Hierarchal phosphorylations involving several kinases whose activation depends on different intracellular signals could explain the contradictory results on the Na^+/H^+ exchange activation noted above.

(iii) Finally, the recent progress in recognizing specific phosphorylation site motifs (Kemp and Pearson, 1990) has allowed the definition of potential "consensus sites". Several of these sites (C-Kinase, calmodulin-dependent kinase II for example,) have been identified in the antiporter cytoplasmic domain, and thus constitute good candidate sites to mutagenize in future experiments. However, these "consensus"sites have to be considered with considerable caution as there are many exceptions for the moment. In addition, if the antiporter is phosphorylated by a specific antiporter-kinase, it will probably be impossible to identify the phosphorylation sites simply by looking at the amino-acid sequence.

CONCLUSION

The development in the last two years of specific antibodies against the growth-factor activatable Na^+/H^+ antiporter has allowed the demonstration that (i) this protein is phosphorylated *in vivo* , (ii) this phosphorylation increases rapidly in response to growth factor suggesting this is the mechanism by which cell activating agents modify the Na^+/H^+ exchange in the plasma membrane. As all types of growth factors stimulate this phosphorylation on serine residues this may serve as a good model to understand how different signalling pathways can converge to trigger the same cellular response. Thus, we are currently trying to map the key phosphorylation sites involved in this process, with the future goal to identify the Na^+/H^+ antiporter kinase(s) and the kinase-kinases with the hope to identify the "swich- or master kinase" that integrates the multiple extracellular stimuli.

References

Ahn, N. and Krebs, E.G. (1990) Evidence for an epidermal growth factor-stimulated protein kinase cascade in Swiss 3T3 cells. *J. Biol. Chem.* **265**: 11495

Aronson, P.S., Nee, J. and Suhm, M.A. (1982) Modifier role of internal Na^+ inactivating the Na^+/H^+ exchanger in renal microvillus membrane vesicles. *Nature* **299**: 161.

Aronson, P.S. (1985) Kinetic properties of the plasma membrane Na^+/H^+ exchanger. *Annu. Rev. Physiol.* **47**: 545.

Ballou, L.M., Siegmann, M. and Thomas, G. (1988) S6 kinase in quiescent Swiss mouse 3T3 cells is activated by phosphorylation in response to serum treatment. *Proc. Natl. Acad. Sci. USA* **85**: 7154.

Berridge,M.J. (1987) Inositol triphosphate and diacylglycerol : two interacting second messengers. *Annu. Rev. Biochem.* **56**: 159

Birnbaumer, L., Abramowitz, J. and Brown, A.M. (1990) Receptor-effector coupling by G proteins. *Biochim. Biophys. Acta* **1031**: 163

Cassel, D., Katz, M. and Rotman, M. (1986) Depletion of cellular ATP inhibits Na^+/H^+ antiport in cultured human cells. *J. Biol. Chem.* **261**: 5460.

Chambard, J.C., Paris, S., L'Allemain, G. and Pouysségur, J. (1987) Two growth factor signalling pathways in fibroblasts distinguished by pertussis toxin. *Nature* **326**: 800.

Cohen, P., Holmes, C.F.B. and Tsukitani, Y. (1990) Okadaic acid : a new probe for the study of cellular regulation. *TIBS* **15**: 98.

Cooper, J., Hunter, T. & Sefton, B. (1983) Detection and quantification of phosphotyrosine in proteins. *Meth. Enzymol.* **99**, 387

Emmer, E., Rood, R.P., Wesolek, J.H., Cohen, M.E., Barithwaite, R.S., Sharp, G.W.G. and Donowitz, M. (1989) Role of calcium and calmodulin in the regulation of the rabbit ileal brush-border membrane Na^+/H^+ antiporter. *J. Memb. Biol.* **108**: 207.

Franchi, A., Cragoe, E.Jr and Pouysségur, J. (1986a) Isolation and properties of fibroblast mutants overexpressing an altered Na^+/H^+ antiporter. *J. Biol. Chem.* **261**: 14614.

Franchi, A., Perucca-Lostanlen, D. and Pouysségur (1986b) Functional expression of a transfected Na^+/H^+ antiporter human gene into antiporter-deficient mouse L cells. *Proc. Natl. Acad. Sci. USA* **83**: 9388.

Ganz, M.B., Pachter, J.A. and Barber, D.L. (1990) Multiple receptors coupled to adenylate cyclase regulate Na+/H+ exchange independent of cAMP. *J. Biol. Chem.* **265**: 8989.

Gould,G.W., and Bell, I.G., (1990) Facilitative glucose transporters: an expanding family, *TIBS* **15**:18

Grinstein, S.(1988) *"Na^+/H^+ Exchange"* (eds S. Grinstein) CRC press, Boca Raton

Grinstein, S., Rotin, D. and Mason, M.J. (1989) Na^+/H^+ exchange and growth factor-induced cytosolic pH changes. Role in cellular proliferation. *Biochim. Biophys. Acta* **988**: 73.

Haggerty, J.G., Agarwal, N., Reilly, R.F., Adelberg, E.A. and Slayman, C.W. (1988) Pharmacologically different Na/H antiporters on the apical and basolateral surfaces of cultured porcine kidney cells (LLC-PK1). *Proc. Natl. Acad. Sci. USA* **85**: 6797.

Hartmann, T., Seuwen, K., Roussel, M.F., Sherr, C.J. and Pouysségur, J. (1990) Functional expression of the human receptor for colony-stimulating factor 1 (CSF-1) in hamster fibroblasts : CSF-1 stimulates Na^+/H^+ exchange and DNA-synthesis in the absence of phosphoinositide breakdown. *Growth Factors* **2**: 289.

Haystead, T.A.J., Sim, A.T.R., Carling, D., Honnor, R.C., Tsukitani, Y., Cohen, P. and Hardie, D.G. (1989) Effects of the tumour promoter okadaic acid on intracellular protein phosphorylation and metabolism. *Nature* **337**: 78.

Hendey, B., Mamrack, M.D. and Putnam, R.W. (1989) Thrombin induces a calcium transient that mediates an activation of the Na^+/H^+ exchanger in human fibroblasts. *J. Biol. Chem.* **264**: 19540.

Hesketh, T.R., Moore, J.P., Morris, J.D.H., Taylor, M.V., Rogers, J., Smith, G.A. and Metcalfe, J.C. (1985) A common sequence of calcium and pH signals in the mitogenic stimulation of eukaryotic cells. *Nature* **313**: 481.

Huang, C.L., Cogan, M.G., Cragoe, E.J. and Ives, H.E. (1987) Thrombin activation of the Na^+/H^+ exchanger in vascular smooth muscle cells. *J. Biol.Chem.* **262**: 14134.

Ives, H.E. and Daniel, T.O. (1987) Interrelationship between growth factor-induced pH changes and intracellular Ca^{2+}. *Proc. Natl. Acad. Sci. USA* **84**: 1950.

Kahn, A.M., Dolson, G., Hise, M., Bennett, S., and Weinman, E. , (1985) Parathyroid hormone and dibutyryl cyclic AMP inhibit Na+/H+ countertransport in brush border membrane vesicles from a suspension of rabbit proximal tubule. *Am. J. Physiol.* **248**: F212.

Kemp, B.E. and Pearson, R.B. (1990) Protein kinase-recognition sequence motifs. *TIBS* **15**: 342.

Knickelbein, R.G., Aronson, P.S. and Dobbins, J.W. (1988) Properties of distinct luminal and basolateral Na:H exchangers in rabbit ileum. *FASEB J.* **2**: A941.

Kopito, R.R., Lee, B.S., Simmons, D.M., Lindsey, A.E., Morgans, C.W. and Schneider, K. (1989) Regulation of intracellular pH by a neuronal homolog of the erythrocyte anion exchanger. *Cell* **59**: 927.

Kozma, S.C., Ferrari, S. and Thomas, G. (1989) Unmasking a growth factor/oncogene-activated S6 phosphorylation cascade. *Cellular Signalling* **1**: 219.

L'Allemain, G., Paris, S. & Pouysségur, J. (1984). Growth factors action and intracellular pH regulation in fibroblasts. Evidence for a major role of the Na^+/H^+ antiporter*J. Biol. Chem.* **259**: 5809

L'Allemain, G. & Pouysségur, J. (1986) EGF and Insulin action in fibroblasts. Evidence that phosphoinositide hydrolysis is not an essential mitogenic signalling pathway *FEBS Letters* **197**: 344 .

L'Allemain, G., Seuwen, K., Velu, T. and Pouysségur, J. (1989) Signal transduction in hamster fibroblasts overexpressing the human EGF receptor. *Growth Factors* **1**: 311.

Limbird, L.E. (1988) Receptors linked to inhibition of adenylate cyclase : additional signaling mechanisms. *FASEB J.* **2**: 2686.

Lowe, J.H.N., Huang, C.H. and Ives, H.E. (1990) Sphingosine differentially inhibits activation of the Na^+/H^+ exchanger by phorbol esters and growth factors. *J. Biol. Chem.* **265**: 7188-7194.

MacNicol, M., Jefferson, A.B. and Schulman, H. (1990) Ca^{2+}/calmodulin kinase is activated by the phosphatidylinositol signaling pathway and becomes Ca^{2+}-independent in PC12 cells. *J. Biol. Chem.* **265**: 18055.

Madshus, I.H. (1988) Regulation of intracellular pH in eukaryotic cells. *Biochem. J.* **250**: 1

Martin-Pérez, J., Siegmann, M. and Thomas, G. (1984) EGF, PGF2α, and insulin induce the phosphorylation of identical S6 peptides in Swiss mouse 3T3 cells : effect of cAMP on early sites of phosphorylation. *Cell* **36**: 287.

Moolenaar, W.H., Tertoolen, L.G.J. and de Laat, S.W. (1984) Phorbol ester and diacylglycerol mimic growth factors in raising cytoplasmic pH. *Nature* **312**: 371.

Moolenaar, W.H., Tsien, R.Y., van der Saag, P.T. and de Laat, S.W. (1983) Na^+/H^+ exchange and cytoplasmic pH in the action of growth factors in human fibroblasts. *Nature* **304**: 645.

Morrisson, D. Kaplan, D., Rapp, U. and Roberts, T. (1988) Signal transduction from membrane to cytoplasm: Growth factors and membrane-bound oncogene products increase Raf-1 phosphorylation and associated protein kinase activity. *Proc.Natl.Acad.Sci. USA*, **85**, 8855.

Paris, S., and Pouysségur, J. (1984) Growth factors activate the Na^+/H^+ antiporter in quiescent fibroblasts by increasing its affinity for intracellular H^+. *J. Biol. Chem.* **259**: 10989.

Pouysségur, J., Sardet, C., Franchi, A., L'Allemain, G., and Paris, S. (1984) A specific mutation abolishing Na^+/H^+ antiport activity in hamster fibroblasts precludes growth at neutral and acidic pH., *Proc.Natl.Acad.Sci. USA* **81**: 4833.

Pouysségur, J., Franchi, A., L'Allemain, G. and Paris, S. (1985) Cytoplasmic pH, a key determinant of growth factor-induced DNA synthesis in quiescent fibroblasts. *FEBS Lett.* **190**: 115.

Pouysségur,J. (1989) G-proteins in growth factor action, in *"G-Proteins"*, edited by Birnbaumer,L. and R. Iyengar, Academic press , **23**:555

Roach, P.J. (1990) Control of glycogen synthase by hierarchal protein phosphorylation. *FASEB J.* **4**: 2961.

Rood, R.P., Emmer, E., Wesolek, J., McCullen, J., Hussain, Z., Cohen, M.E., Braithwaite, R.S., Murer, H., Sharp, G.W.G. and Donowitz, M. (1988) Regulation of the rabbit ileal brush-border Na^+/H^+ exchanger by an ATP-requiring Ca^{2+}/calmodulin-mediated process. *J. Clin. Invest.* **82**: 1091.

Sardet, C., Franchi, A., and Pouysségur, J. (1989) Molecular cloning, primary structure and expression of the human growth factor-activatable Na^+/H^+ antiporter.*Cell* **56**: 271.

Sardet, C., Counillon,L., Franchi, A., and Pouysségur, J. (1990).Growth factors induce phosphorylation of the Na^+/H^+ antiporter, a glycoprotein of 110-kD. *Science* **247**: 723

Seifter, J.L. and Aronson, P.S. (1986) Properties and physiologic roles of the plasma membrane sodium-hydrogen exchanger. *J. Clin. Invest.* **78**: 859. .

Sprang, S.R. Acharya, K.R., Goldsmith, E.J., Stuart, D.I;, Varvill, K., Fletterick, R.J., Madsen, N.B. and Johnson, L.N. (1988) Structural changes in glycogen phosphorylase induced by phosphorylation. *Nature* **336**: 215.

Sweadner, K.J. (1989) Isozymes of the Na^+/K^+-ATPase. *Biochim. Biophys. Acta* **988**: 185.

Van Obberghen-Schilling, E., Chambard, J.C., Paris, S., L'Allemain, G. & Pouysségur, J. (1985) α-Thrombin-induced early mitogenic signalling events and G0 to S-phase transition of fibroblasts require continaual external stimulation. *EMBO J.* **4**, 2927

Weinman, E.J., Dubinsky, W.P. and Shenolikar, S. (1988) Reconstitution of cAMP-dependent protein kinase regulated renal Na^+-H^+ exchanger.*J. Memb. Biol.* **101**: 11.

Weissberg, P.L., Little, P.J., Cragoe, E.J. and Bobik, A. (1988) The pH of spontaneously beating cultured rat heart cells is regulated by an ATP-calmodulin-dependent Na^+/H^+ antiport. *Circ.Res.* **64**: 676.

Yarden,Y., and Ulrich,A. (1988) Growth factor receptor tyrosine kinases. *Ann.Rev.Biochem.*, **57**: 443.

Overexpression of Protein Kinase C α in Swiss/3T3 Cells Reduces The Number of Both High and Low Affinity of the Epidermal Growth Factor Receptors.

Hagit Eldar-Finkelman, Yaffa Zismann,
Axell Ullrich* and Etta Livneh.
Department of Chemical Immunology
The Weizmann Institute of Science
Rehovot 76100, Israel.

The signaling pathways induced by growth factors and hormones can be divided into two major classes according to the second messangers they produce. In one pathway activation of G-proteins and adenylate cyclase produces the second massenger cAMP, while in the other pathway a rapid turnover of phospholipids generates two second messangers IP_3 and diacylglycerol (for review Berrdige, 1987).

Protein Kinase C (PKC), a ca^{+2} and phospholipid dependent serine/threonine kinase is thaught to play an important role in transducing mitogenic signals (for review, Nishizuka, 1986, 1988). This is mainly due to the fact the PKC is activated by diacylglycerol and ca^{+2} under physiological conditions. In addition PKC was shown to be activated and to represent the major receptor for the tumor promoting phorbol esters (Castagna et al. 1982, Ashendel et al. 1983, Leach et al. 1983). The involvment of the phorbol ester molecules in cell transformat-

*Max-Plank-Institute fur Biochem,
8033 Martinsried bei Munchen, West Germany.

NATO ASI Series, Vol. H 52
Biological Signal Transduction
Edited by E. M. Ross and K. W. A. Wirtz

mation and proliferation may suggest that PKC is a key regulatory enzyme in transformation processes and neoplasia.

Characterization of cDNA clones for PKC have revealled a gene family coding for closely related structural proteins. Four members of the gene family were characterized α, βI, βII and γ (Parker et al. 1986, Housey et al. 1987, Coussens et al. 1986), where βI and βII are derived from alternative splicing of a single mRNA transcript. These isoenzymes exhibit subtle different modes of activation, kinetic properties and tissue specific distribution which may suggest that PKC isoenzymes activate distinct cellular pathways in the cell. To elucidate further the functional role of specific forms of PKC we have introduced PKCα into Swiss/3T3 cells, thus generating cell lines stably producing elevated levels of PKCα. Here we describe the biological properties of Swiss/3T3 cells overexpressing PKCα and show its role in regulating the expression of EGF receptor molecules in these cells.

RESULTS

The full lengh cDNA coding for PKCα was subcloned into a retroviral expression vector (pZSF2, Eldar et al. 1990), which consisted two LTR promotors and a G418 resistence gene. The plasmid pZSF2 was used to transfect -2 cells by the calcium phosphate precipitation method. The medium collected from the Ψ-2 transfected cells was farther used to infect Swiss/3T3 cells. After two weeks of growth in selection medium containing G418, several clones were isolated designated SF1.1, SF6.1, SF6.3, SF1.3 and SF3.2. Control Swiss/3T3 cells were generated by transfection of the plasmid pZSF2 lacking the PKCα cDNA insert into Ψ-2 cells. Follwing infection of Swiss/3T3 cells and G418 selection resulted in the generation of control cell lines SC1 and SC2.

Characterization of generated cell lines- PKC activity was measured using partially purified cell extracts and histone

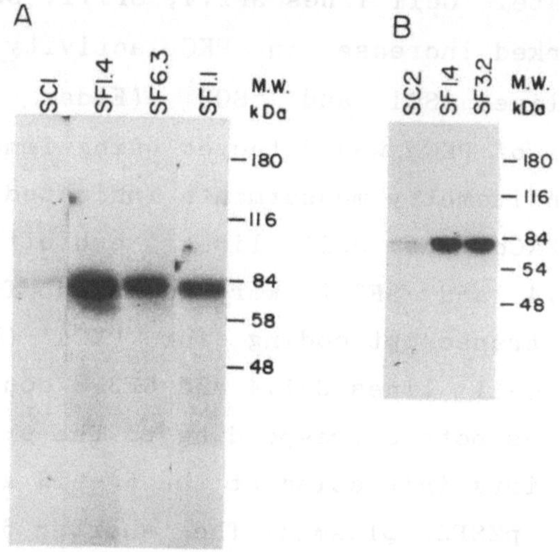

Fig.1 Expression of PKCα in Swiss/3T3 cells. Immunoblot analysis of control (SC1, SC2) and overexpressing PKCα cells. Two antibodies against PKCα were used: monoclonal antibody MC-5 (Amersham) (A) and polyclonal antibody R105 (B).

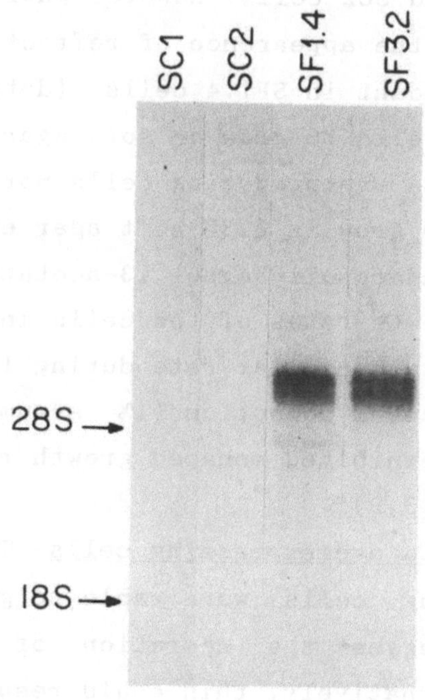

Fig.2 Nortern blot analysis of PKCα transcripts. mRNA transcripts was detected using ^{32}P-labeled 1.2-kb cDNA of PKCα. SC1, SC2 control cells, SF1.4, SF3.2 overexpressing PKCα cells.

III-S as substrate. Cell lines SF1.4, SF1.1, SF6.1, SF6.3, and SF1.3 showed marked increase in PKC activity compared to control cell lines SC1 and SC2 (Eldar et al. 1990). Increased level of PKCα was detected using immunoblot analysis (Fig.1 A,B). Densitometry measurments indicated 40 to 10 fold increase for PKCα. The cell lines exhibiting high PKC expression SF1.4 and SF3.2 were assayed for the size and abundant of RNA transcript coding for PKCα (Fig.2). The RNA isolated from cell lines SF1.4 and SF3.2 contained 6.45- and 6.0-kb RNA species both corresponding to the predicted size of the mRNA transcripts initiating at the 5'-LTR and terminates at the 3'-LTR of pZSF2 plasmid. The shorter 6.0-kb transcript corresponds to a spliced transcript, utilizing the 5' and 3' M-MulV splice sites (Mann et al. 1983).

<u>Growth properties of PKCα transfected cells</u>- PKCα expressing cell lines SF1.4 and SF3.2 displayed similar morphology of that of control SC1 and SC2 cells. however phorbol ester treatment (100ng/ml) induced the appearence of refractile and round cells that were more abundant in SF1.4 cells (data not shown). The ability of the cells to grow in soft agar was also examined. However neither PKCα overproducing cells nor the control cell lines were able to grow in 0.3% soft agar even in the presence of TPA (12-O tetradecanoylphorbol 13-acetate, 100ng/ml).

Examining the growth rates of the cells indicated that both cell types grew at similar rate during log phase (Fig.3,A). However under low serum condotion (1% serum) both SF1.4 and SF3.2 cell lines exhibited enhaced growth rate compared to the control cells (Fig.3,B).

<u>EGF Rceptor and PKCα overexpressing cells</u>- The fact that the PKCα overexpressing cells were able to grow under low serum conditions could suggest the secretion of factors enhancing their growth, alternatively, this could result from alterations in growth factor receptor responses. The second possibility was initially examined. EGF receptor was initially chosen for our studies since PKC was previously implicated to downregulate

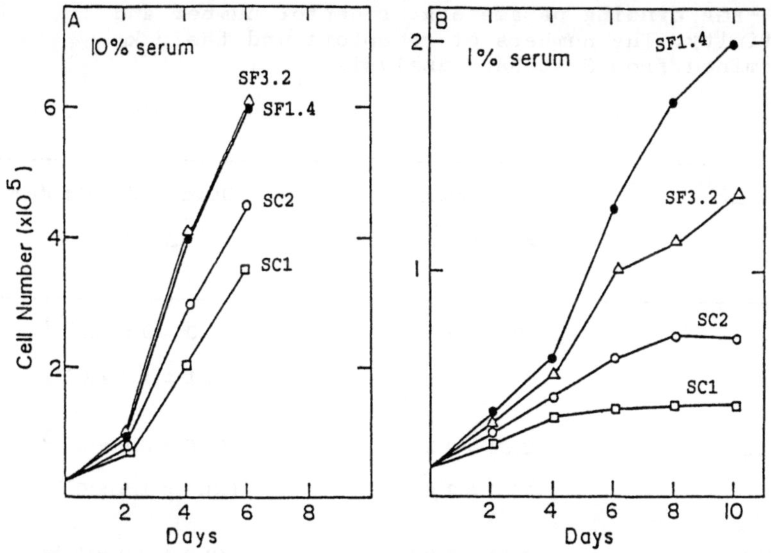

Fig.3 Growth curves of PKCα overexpressing cells. cells were plated at 2×10^4 cells/ 35 mm plate. Every two days cells were counted. The experiments were performed with DMEM containing 10% or 1% calf serum (A or B respectively).

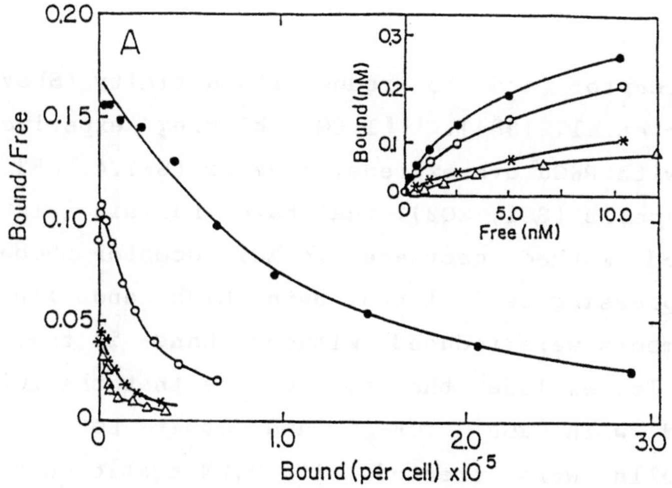

Fig.4 ^{125}I EGF binding experiments. EGF binding was determined for each of the cell lines as described under "Experimental Procedures (Eldar et al. 1990).○SF1.1, △SF1.4, ✱ SF3.2, ● SC2.

Table.1 ^{125}EGF binding parameters: receptor number and ligand affinity. The numbers of receptors and their Kd were determined from Scatchard analysis.

Cell line	Receptor number/ cell (x 10^3)	Dissociation Constants Kd (M)
SC2	9.3 ± 0.8	$(6.0 \pm 0.5)x10^{-10}$
	62.5 ± 2.4	$(13.1 \pm 1.7)x10^{-9}$
SF 1.1	2.32 ± 0.2	$(8.0 \pm 0.8)x10^{-10}$
	18.0 ± 2.6	$(33.1 \pm 1.5)x10^{-9}$
SF 1.4	0.99 ± 0.2	$(8.2 \pm 0.2)x10^{-10}$
	11.8 ± 1.8	$(39.7 \pm 4)x10^{-9}$
SF 3.2	1.59 ± 0.6	$(7.4 \pm 0.3)x10^{-10}$
	13.9 ± 2.3	$(17.0 + 0.5)x10^{-9}$

the EGF receptor and to reduce its affinity (Shoyab et al. 1979, Hunter et al. 1984). ^{125}I EGF binding experiments were performed with PKCα overexpressing cells (SF1.4, SF3.2, SF1.1) and control cells (SC1, SC2). Scatchard analysis (Fig.4, Table 1.) revealed marked decrease in EGF receptor numbers in all PKCα overexpressing cell lines. Both high and low affinity receptor numbers were reduced, without change in their apparent affinities. To exclude the possibility that the EGF receptor was occupied with EGF, TGF-α and other EGF like growth factors, cells were treated with 0.2N acetic acid prior the performance of binding experiments. Same binding data were obtained following acetic acid treatmnet. In order to confirm these results obtained from the ^{125}I EGF binding experiments cells were metabolically labeled with ^{35}S- metionine. Time course analysis showed a decrease in EGF receptor molecules syntesized in PKCα overexpressing cells (Fig.5,A). Moreover, reduced steady state levels of EGF receptor molecules in SF1.4 and SF3.2, determined by immunoblot analysis (fig.5,B) were also consistent with these results. Northern blot analysis of RNA isolated from SF1.4 and SF3.2 cells and from control cells SC1 and SC2, also showed constitutive depression in EGF receptor transcripts in the PKCα overexpressing cells when probed for the steady state mRNA levels of EGF receptors (Fig.6). Thus overexpression of PKCα in Swiss/3T3 cells regulated the expression of the EGF receptor molecules. This reduction was due to the reduction in the mRNA transcripts coding for the EGF receptor.

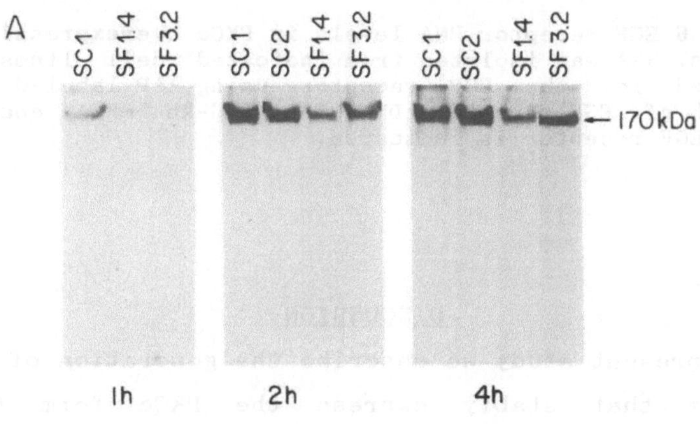

Fig.5 Immunoprecipitation and immunoblotting of EGF receptors. Cells were labeled with ^{35}S methionine for indicated times (A),detailed experimental procedures are found elsewhere (Eldar et al.1990). The detection of the steady state level of the EGF receptor in cells (B), an immunoblot analysis using antibodies against the EGF receptor.

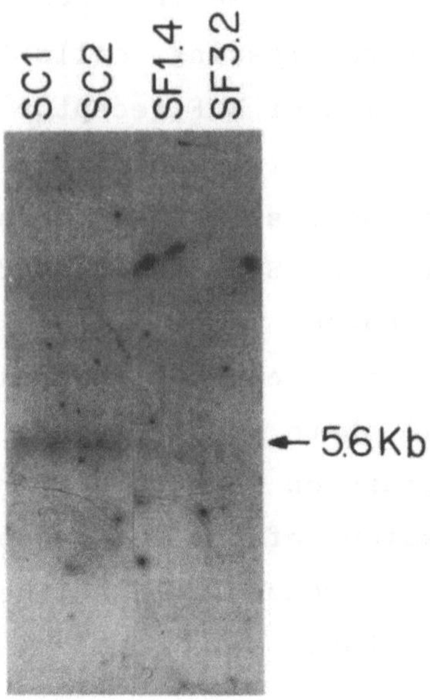

Fig. 6 EGF receptor RNA levels in PKCα overexpressing cells. RNA was isolated from indicated cell lines and probed for the EGF receptor using ^{32}P-labeled full lengh of EGF receptor cDNA. The 5.6-kb mRNA encoding the EGF receptor is indicated.

DISCUSSION

In the present study we describe the generation of Swiss/3T3 cell lines that stably express the PKCα form. Among the different isoenzyme of the PKC family, the α form is the most ubiquitous. Swiss/3T3 cells endogenously express this subtype only, in very low amount, thus providing an ideal cell system for our studies.

The generated Swiss/3T3 cell lines, expressing high levels of PKCα (about forty fold increase) were not morphologically altered relative to control cell lines and were not able to form colonies in soft agar. Previous reports describing the overexpression of PKCβI and PKC γ forms in Rat1 and NIH/3T3 cells indicated morphological changes, growth in soft agar and tumorogenicity when inculated into nude mice (Housey et al.

1988, Persons et al. 1988). However, recent reports showed that overexpression of PKCβI in a different cell type C3H/10T1/2 cells could not increase their tumorogenicity or their ability to grow in soft agar (Krauss et l. 1989). Morover the same PKCβI overexpressed in HT29 Colon cancer cells inhibited their growth and supressed their tumorogenicity (Choi et al. 1990). These different results togethr with our data suggest that the induction of transformation probably depends on the different cell types. Distinct cellular pathways induced by PKC in these cells may explain the diversity obtained in the various studies presented here.

Our studies revealed decreased levels of EGF receptor molecules expressed in PKCα overexpressing cells and suggested the involvment of PKCα in a cellular mechanism regulating the expression of the EGF receptor mRNA transcripts molecules in Swiss/3T3 cells. Several studies indicated that the expression of the EGF receptor is a regulated process. EGF itself, TPA, retinoic acid and thyroid hormone regulate the expression of the EGF receptor (Clark et al. 1985, Hudson et al. 1990). The regulation of the mRNA transcripts coding for the EGF receptor reported here could result from instability of these mRNA transcripts, aternatively it could be at the transcriptional level. Recently it was shown that TPA induced the repression of EGF receptor gene transcription (Gardner et al. 1990). These results correlates with our data and indicate that the reduction in the EGF receptor mRNA transcripts may be atthe transcriptional level.

The expression of the EGF receptor may play an important role in tumor formation or its progression in cells. The apparent loss of EGF binding sites was one of the earliest markers noted on certain virally or chemically transformed fibroblasts (Todaro et al. 1976). It was also shown that Chinese hamster embryo cells lost their cell surface EGF receptors gradually during their neoplastic progression (Wakshoull et al. 1985). On the other hand amplification of the EGF receptor was found in various human tumors and human cell lines including the epidermoid carcinoma A431 cell line (Merlino et al. 1984, Lin et al. 1984, Liberman et al. 1985). In addition overexpression of the EGF receptor in NIH/3T3 cells showed a transformed phenotype that was strictly dependent on the number of the EGF receptor and on the presence of EGF (Velu et al. 1987, Riedel et al. 1988). Thus two different mechanism controlling the regulation of the EGF receptor may play a role in tumor formation. Our studies shows that deregulation of PKC expression may be involved in at least one of these mechanism causing the reduction of the EGF receptor number, although altered PKC expression by itself is not sufficient to induce the transformed phenotype for certain cells.

References

C. L. Ashendel, J. M. Staller and R. K. Boutwell, Cncer Res. 43, 4333-4337, 1983.

M. J. Berridge, Biochemica et Biophysica Acta 907 33-45, 1987.

M. Castanga, Y. Takai, T. Mori, K. Kaibuchi, K. Sano,U. Kikkawa and Y. Nishizuka, J. Biol. Chem. 257, 7847-7851, 1982.

P. M. choi, K. Tchou-Wong and I. B. Weinstein, Mol. Cell. Biol. 4650-4657, 1990.

A. J. L. Clark, S. Ishii, N. Richert, G. T. Merlino and I. Pastan, Pro. Natl. Acad. Sci. u.S.A. 82, 8374-8378, 1985.

L. Coussens, P. J. Parker, L. Rhee, T. Yank-Feng, E. Chen, M. D. Waterfield, U. Francke and A. Ullrich, Science, 233, 853-859, 1986.

H. Eldar, Y. Zisman, A. Ullrich and E. Livneh, J. Biol. Chem. 265, 22, 13290-13296, 1990.

D. P. Gardner and N. Shimizu, FEBS 269, 288-291, 1990.

G. M. Housey, C. A. O'Brian, M. D. Johnson, P. Kirschmeir and I. B. Weinstein, Proc. Natl. Acad. Sci. U.S.A. 84, 1065-1069, 1987.

G. M. Housey, M. D. Johnson, W. L. W. Hsiao, C. A. O'Brian, J. P. Merphy, P. Kirschmeier and I. B. Weinstein, Cell, 52, 343-354, 1987.

L. G. Hudson, J. B. Santon C. K. Glass and G. N. Gill, Cell, 62, 1165-1175, 1990.

T. Hunter, N. Ling and J. A. Cooper, Nature, 279, 387-391, 1979.

R. S. Krauss, G. M. Housey, M. D. Johnson and I. B. Weinstein, Oncogene, 4, 991-998, 1989.

K. L. Leach, M. L. James and P. M. Blumberg, Proc. Natl. Acad. Sci. U.S.A, 80, 4208-4212, 1983.

D. A. Persons, W. O. Wilkison, R. M. Bell and O. J. Finn, Cell 52, 447-458, 1988.
T. A. Liberman, H. R. Nusbaum, N. Rason, R. Kris, I. Lax, H. Soreq, N. Whittle, M. D. Waterfield, A. Ullrich and J. Schlessinger, Nature, 313, 144-147, 1985.
C. R. Lin, W. S. Chen, W. Kruiger, L. S. Stolarski, W. Weber, R. M. Evans, I. M. Verma, G. N. Gill, and M. G. Rosenfeld, Science, 224, 843-848, 1984.
R. Mann, R. C. Mulligan and D. Baltimore, Cell, 33, 153-159, 1983.
G. T. Merlino, Y. H. Xu, S. Ishii, A. J. Clark, K. Semba, K. Toyoshima, T. Yamamoto and I. Pastan, Science, 224, 417-419, 1984.
Y. Nishizuka, Science, 233, 305-312, 1986.
Y. Nishizuka, Nature, 334, 661-665, 1988.
P. J. Parker, L. Coussens, N. Totty, L. Rhee, S. Young, L. Chen, L. Stabel, M. D. Waterfield and A. Ullrich, Science, 233, 853-859, 1986.
Shoyab, J. E. Delarco and G. T. Todaro, Nature, 279, 387-391, 1979.
G. J. Todaro, J. E. DeLarco and S. Cohen, Nature, 264, 26-31, 1976.
E. Wakshull, P. M. Kraemer and W. Wharton, cancer, Res. 45, 2070-2075, 1985.
T. J. Velu, L. Beguinot, W. C. vass, M. C. Willingham, G. T. Merlino, I. Pastan and D. R. Lowy, Science, 238, 1408-1410, 1987.

CHARACTERIZATION OF PROTEIN KINASE C ISOFORMS IN A SMALL CELL LUNG CARCINOMA CELL LINE: REGULATION OF PKCε AND ITS CATALYTIC FRAGMENT

Gregory Baxter and Berta Strulovici
Department of Biochemistry, Institute of Cancer and Developmental Biology
Syntex Research
3401 Hillview Avenue
Palo Alto, California 94304
U.S.A.

INTRODUCTION

The human small cell lung carcinoma (SCLC) is characterized by cells closely related to the neuroendocrine components of the respiratory epithelium (Bonikos and Bensch, 1977). Therefore, unlike other types of lung cancer, SCLC cells secrete numerous peptide hormones (Woll and Rozengurt, 1989). For example, bombesin and its mammalian homologue gastrin-releasing peptide (GRP) have been proposed as possible autocrine growth factors for SCLC (Cuttitta et al., 1985). In fact, specific GRP-receptor binding activity was found in some tumor-derived cell lines (Krist et al., 1987) and a monoclonal antibody raised against a C-terminal heptapeptide sequence common to amphibian bombesin and human GRP blocked the binding to cellular receptors and inhibited the clonal growth of SCLC in vitro and the growth of SCLC xenografts *in vivo* (Cuttitta et al., 1985).

Many hormone and growth factor responses are mediated through the stimulation of PI turnover (Berridge and Irvine, 1984). This bifurcating signal transduction pathway involves the elevation of intracellular calcium concentrations and the activation of protein kinase C (PKC) via a rise in endogenous diacylglycerol levels (Nishizuka, 1989). Recently, it has been shown that many malignant cells, including some SCLC cell lines, possess high levels of PKC activity (Hirai et al., 1989). In addition, overexpression of PKC in fibroblast cells results in abnormal growth regulation (Housey et al., 1988; Persons et al., 1988). These findings suggest that PKC plays a critical role in the regulation of cellular growth.

Until relatively recently, all biological responses involving PKC were attributed to one molecular species. Molecular cloning analysis has shown that PKC is a family of multiple isoforms having closely related, but different

NATO ASI Series, Vol. H 52
Biological Signal Transduction
Edited by E. M. Ross and K. W. A. Wirtz

structures (Nishizuka, 1989; Parker et al., 1989). Thus far, the sequences of seven isoforms-α, βI, βII, γ, δ, ζ and ε have been deduced from cDNAs obtained from brain and spleen libraries (Nishizuka, 1989; Ono et al., 1988). Some cells express one of the so far identified PKC isoforms (Heidenreich et al., 1990; Pelosin et al., 1987) while the majority coexpress multiple isoforms (Kiley et al., 1990; Pfeffer et al., 1990; Strulovici et al., 1990). It is anticipated that rather than act interchangeably, each of these isoforms mediate distinct cellular responses. Biochemically, PKCα, β and γ exhibit subtle differences and are known as Ca^{2+} and phospholipid dependent kinases (Kiley and Jaken, 1987), whereas PKCδ (Leibersperger et al., 1990) and ε (Schaap and Parker, 1990) are Ca^{2+} independent and display quite different substrate specificity (Schaap et al., 1989).

A large body of literature shows that bombesin acts as a potent mitogen on cultured Swiss 3T3 fibroblasts (Zachary et al., 1987; Zachary et al., 1986), where its addition elicits a set of early biological responses including PIP_2 hydrolysis and PKC activation(Zachary et al., 1987) and induction of the cellular oncogenes c-fos and c-myc (Sausville et al., 1987). However, little is known about the mechanism by which bombesin or its mammalian homologue GRP stimulate growth in SCLC cells.

In our study, we used specific antisera to PKCα, β, δ, γ and ε to assess the abundance, distribution and regulation of PKC isoforms in several established SCLC cell lines. Our data shows that PKCε is largely overexpressed in the SCLC cell line N417D and that it exhibits unusual subcellular distribution, the majority of it being in particulate form, where it is constitutively active. Moreover, we were able to characterize a 40 kDa protein in these cells which we identified as the PKC catalytic domain. Both PKCε and its catalytic fragment are regulated by GRP or phorbol esters. These data shed some light on the regulation of a recently identified PKC isoform in a relevant cellular system. The data may have broader implications regarding the role of PKC in carcinogenesis.

RESULTS

Protein kinase C isozymes in SCLC cell lines. To identify the types of PKC isoforms present in SCLC cells, whole cell extracts from four established SCLC cell lines were analyzed by immunoblot analysis using purified antibodies directed against synthetic peptides unique to PKCα, β, δ, γ and ε (Strulovici et al., 1990).

As shown in Fig. 1 immunoreactive PKCα, δ and ε were present to varying degrees in the SCLC cell lines H69, H82, H661 and N417D, while PKCβ and γ were undetectable (not shown). At longer exposure to film, immunoreactive PKCε was noted in all four SCLC cell lines with approximately a 10 fold amplification in the N417D cell line (not shown). A similar result was obtained when the N417D SCLC cell line was compared with human bronchial lung cells.

In the N417D SCLC cell line, the PKCε antibody detected a protein with an apparent molecular weight of 90 kDa, which agrees well with the reported apparent molecular weight of PKCε (Schaap et al., 1989), and a 40 kDa protein (Fig. 2, lane 1). In the human bronchial lung cells, only the 90 kDa protein was apparent.

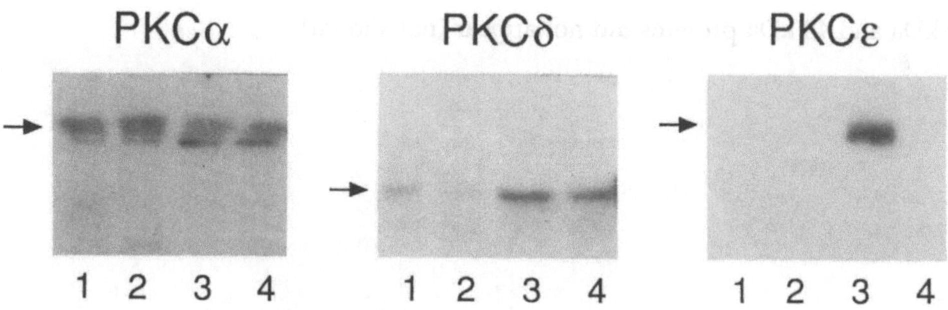

Fig. 1. Immunoblot analysis of four SCLC cell lines with antisera against PKCα, PKCδ and PKCε.

SCLC total cell homogenates (100μg protein) were subjected to SDS/PAGE, transferred to nitrocellulose membranes and immunoblotted with antisera against PKCα, PKCδ and PKCε (dil. 1:100) as described (Strulovici et al., 1990) [125I]-Protein A labeled immunoblots were exposed to Kodak XAR film (10 days for PKCα and PKCδ, 2 days for PKCε) at -70°C. The SCLC cell lines examined were: lane 1, H82; lane 2, H661; lane 3, N417D; lane 4, H69. Immunoreactive PKC isoforms are identified by arrows.

PKCε in N417D cells: identification of the 40 kDa protein. To determine the identity of the 40 kDa protein, we focused our attention on the N417D cell line. These cells were described before (Sausville et al., 1987) as a variant SCLC cell line. One of its major characteristics is its rapid growth rate. Fig. 2 shows the abundance and subcellular distribution of the 90 kDa and 40 kDa proteins in the N417D cells detected by its immunoreactivity with the purified PKCε antibody. When the cells were directly lysed by the addition of hot SDS/PAGE sample buffer and analyzed for immunoreactive PKCε, both proteins, the 90kDa and the 40 kDa, were detected (Fig. 2, lane 1). Upon subcellular fractionation, the 90 kDa protein appeared mostly in the particulate fraction (Fig. 2, lane 3) while the 40 kDa protein was cytosolic (Fig. 2, lane 2).

To provide evidence for the specific reactivity of the PKCε antibody, we assessed the ability of the synthetic substrate used as the antigen to compete for antibody binding. Fig. 2, lanes 4 and 5 show the result of one such experiment. The reactivity with the 90 kDa and 40 kDa bands was blocked by the peptide antigen, thus demonstrating that the 40 kDa protein is a fragment of the PKCε. Since our polyclonal antibody is directed against the C-terminus of PKCε (Strulovici et al., 1990), the possibility exists that it represents the catalytic domain of PKCε. When preimmune serum was used instead of the antiserum, both the 90 kDa and 40 kDa proteins did not appear (not shown).

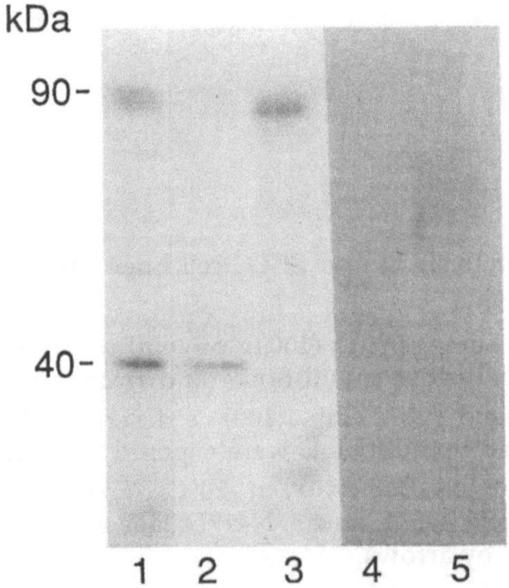

Fig. 2. Subcellular distribution of PKCε in the SCLC cell line N417D.

Cells were either lysed directly in SDS/PAGE sample buffer (lane 1) or in hypotonic lysis buffer (Strulovici et al., 1990). Cytosolic (lane 2) and particulate (lane 3) fractions were prepared as described. 50μg protein were separated by SDS/PAGE, transferred to nitrocellulose membranes and immunoblotted with antiserum against PKCε. Lanes 4 (cytosolic fraction) and 5 (particulate fraction): specific blockade with 0.1 mg/ml peptide antigen. The 90 kDa and 40 kDa PKCε immunoreactive proteins are indicated.

PKCε is regulated by GRP and TPA. We analyzed the effects of the phorbol diester 12-O-tetradecanoylphorbol 13-acetate (TPA), which binds directly to and activates PKC and GRP, a known mitogen for SCLC cells, on the fate of PKCε and its catalytic fragment. Serum starved cells were treated with either 20nM TPA or 50nM GRP for up to 8 hours and cytosol and particulate fractions were then analyzed for PKCε by Western blotting. We did not detect any short term effects (<30 min) on the abundance and distribution of immunoreactive 90 kDA or 40 kDa PKCε. However, upon prolonged treatment with either TPA or GRP (>1 hour - 8 hours), we found a significant increase in the amount of the PKCε catalytic fragment. This corresponded with a slight loss of particulate PKCε and a minor increase in the cytosolic PKCε (Fig. 3). In the presence of GRP, the increase in both cytosolic PKCε and the catalytic fragment was even more pronounced and more rapid than with TPA (Fig. 3). The PKCε catalytic fragment was detected by 1 hr and continued to increase up to 8 hr of treatment.

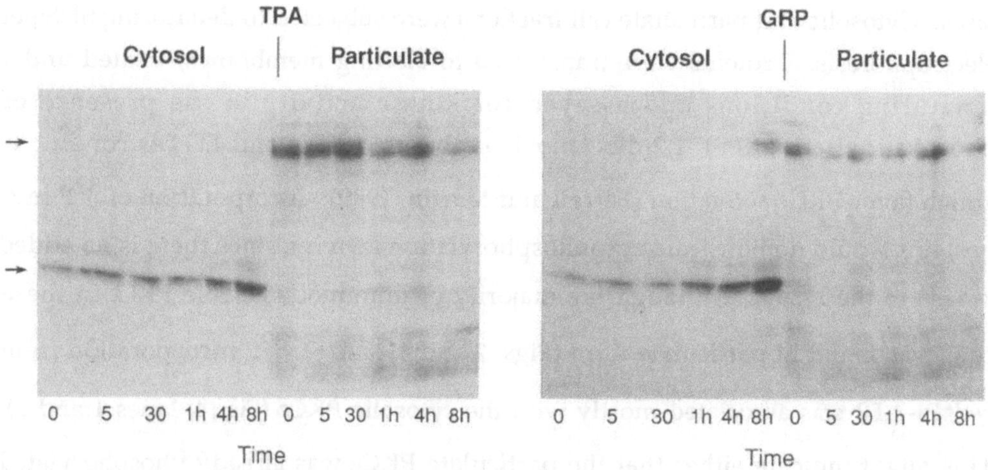

Fig. 3. Immunoblot analysis of PKCε in SCLC N417D cells: Fate of cytosolic and particulate PKCε after treatment with TPA or GRP.

Serum starved cells (18 hr) were treated with either 20nM TPA (left) or 50nM GRP (right) for the indicated times. Particulate and cytosolic fractions were prepared. 50μg protein were subjected to SDS/PAGE, transferred to nitrocellulose and immunoblotted with antiserum to PKCε. [125I]-Protein A labeled membranes were exposed to Kodak XAR film for 4 days at -70°C. The 90 kDa and 40 kDa PKCε immunoreactive proteins are indicated by arrows.

This increase in the 40 kDa catalytic fragment was blocked by the presence of cycloheximide, indicating that its appearance requires *de novo* protein synthesis (data not shown). The TPA stimulated increase in cytosolic PKCε was also slightly decreased in the presence of cycloheximide. In control cells (no TPA), cycloheximide slightly increased the 40 kDa protein by 8 hr.

<u>PKCε is autophosphorylated in SCLC cells</u>. One possible mechanism by which the 40 kDa PKCε catalytic fragment might be generated is via proteolytic cleavage of PKCε. Earlier reports (Kishimoto et al., 1983; Tapley and Murray, 1985) have shown that PKC is cleaved by a Ca^{2+}-dependent neutral protease (calpain) to produce a catalytically active fragment (protein kinase M) and that calpain I, which is active in the micromolar range of calcium concentrations, may react preferentially with the active form of PKC that is associated with the particulate cell fraction (Kishimoto et al., 1989). To test the hypothesis that the 40 kDa PKCε catalytic fragment might be derived from a constitutively active PKCε, we tested its autocatalytic activity by a method of renaturation of proteins on nitrocellulose blots. Cytosolic and particulate cell fractions were subjected to denaturing SDS gel electrophoresis. Proteins were transferred to blotting membranes, treated under renaturing conditions and assayed for kinase activity in the presence of phosphatidylserine (PS), 1,2-dioctanoyl-*sn*-glycerol (diC_8) and EGTA, conditions which favor PKCε activation (Ferrell and Martin, 1989). Incorporation of ^{32}P into proteins would demonstrate autophosphorylation activity, since there is no added kinase to the assay. Although the majority of immunodetectable PKCε in these cells was found in particulate form (Figs. 2 and 4, lane 6), ^{32}P incorporation from [γ-^{32}P]-ATP was associated mostly with the cytosolic PKCε (Fig. 4, lanes 1 and 3). This might indicate either that the particulate PKCε was already phosphorylated *in situ* or that it was unable to undergo autophosphorylation under our assay conditions. Upon prolonged exposure to film (Fig. 4, lanes 3 and 4), additional kinases could be noticed. There was very little ^{32}P incorporation into the PKCε catalytic fragment (Fig. 4, lane 3). When the cell cytosol was subjected to two-dimensional SDS gel electrophoresis followed by immunoblotting with the PKCε antibody, the 40 kDa PKCε catalytic fragment appeared as multiple species representing distinct pI values, suggesting multiple phosphorylation states (Fig 7).

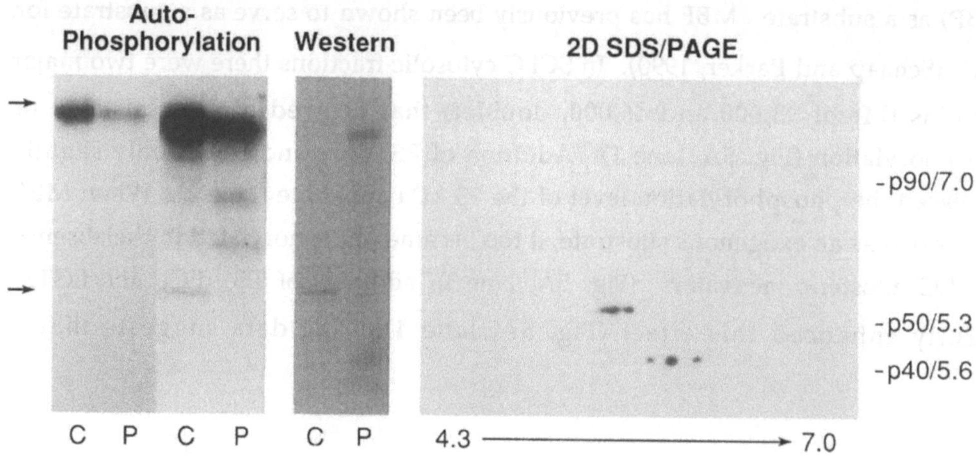

Fig. 4. *In vitro* activity of renaturable PKCε from SCLC N417D cells.

Left, cytosolic and particulate fractions were prepared and 50μg protein separated by SDS/PAGE and transferred to Nytran membranes. Lanes 1, 3 and 5 are cytosolic fractions (C) and lanes 2, 4 and 6 are particulate fractions (P). Lanes 1-4, membranes were allowed to renature and were incubated with [γ-32P]-ATP for 30 min and washed as described (Ferrell and Martin, 1989). Lanes 1 and 2 were exposed to film for 6 hr at -70°C while lanes 3 and 4 were exposed to film for 36 hr at -70°C. Lanes 5 and 6, membranes were immunoblotted with antiserum to PKCε, labeled with [125I]-Protein A and exposed to film for 2 days at -70°C. The 90 kDa and 40 kDa PKCε immunoreactive proteins are indicated by arrows. Right, cytosolic proteins (25μg) were subjected to two-dimensional SDS/PAGE, transferred to nitrocellulose membranes and immunoblotted with antiserum to PKCε Immunoblots were labeled with [125I]-Protein A and exposed to film for 4 days at -70°C. The apparent molecular mass and average pI values of the PKCε immunoreactive proteins are indicated.

Along with the 40 kDa catalytic fragment, we also observed a 50 kDa protein which crossreacts with our PKCε antibody and is apparently phosphorylated. This protein could be a precursor proteolytic cleavage product to the stable 40 kDa fragment, which we sometimes notice on our Western blots (not shown).

PKCε and its catalytic fragment are constitutively active. The unusual subcellular localization of PKCε in the SCLC cell (particulate vs. cytosolic, Figs. 2 and 4) and the presence of a soluble catalytic fragment suggested that the enzyme might already be in an active form in the membrane (90 kDa) and cytosol (40 kDa). To test this notion, the activity of PKCε from SCLC N417D cells was assessed by performing cell free kinase assays using either cytosolic or particulate fractions as the source of enzyme and substrate or exogenous myelin basic protein

(MBP) as a substrate. MBP has previously been shown to serve as a substrate for PKCε (Schaap and Parker, 1990). In SCLC cytosolic fractions there were two major proteins (Mr of 93,000 and 46,000, doublet) that showed high basal level of phosphorylation (Fig. 5A, lane 1). Addition of PS, diC_8 and EGTA only slightly increased the phosphorylation level of the 93 kDa substrate (lane 2). When MBP was added as an exogenous substrate, it too became phosphorylated in the absence of PKC allosteric activators (Fig. 5A, lane 3); addition of PS, diC_8 and EGTA slightly enhanced this effect (Fig. 5A, lane 4). The data suggests that a

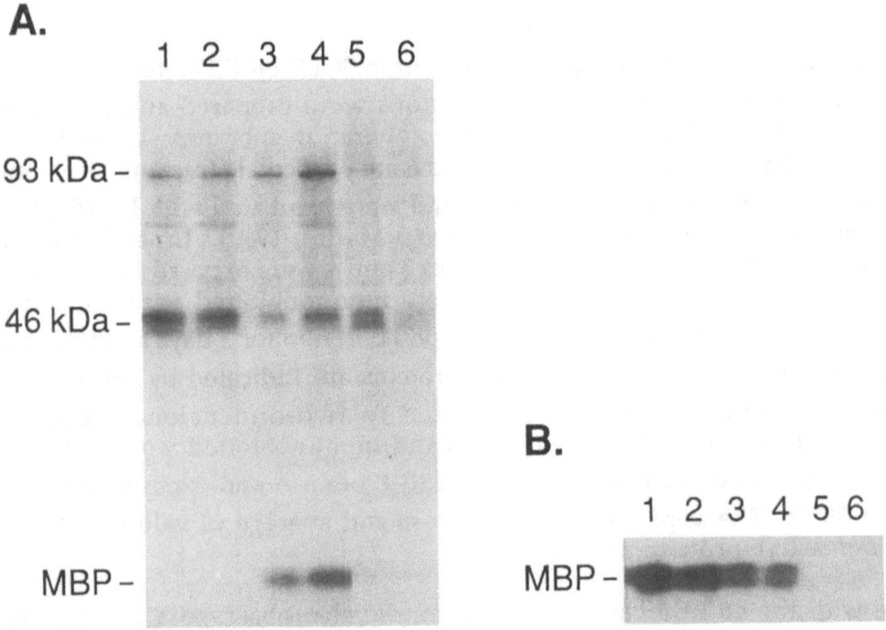

Fig. 5. *In vitro* phosphorylation of MBP by SCLC cytosol: Effects of PKC inhibitors. Left: Cytosolic fractions were prepared. Kinase assays were performed using 25μg total protein in the absence (lanes 1 and 4) and presence (lanes 2, 4 and 6) of 1μg/ml PS, 10μM diC8 and 1mM EGTA. Lanes 3 and 4 contained 10μg MBP. Lanes 5 and 6 contained MBP and 1mg/ml PKCε pseudosubstrate peptide. (B): Selective effect of two classes of PKC antagonists. Lanes 1, 3 and 5 had 1mM EGTA only; lanes 2, 4 and 6 contained 1μg/ml PS, 10μM diC8 and 1mM EGTA. Lanes 3 and 4 contained 50μM sphingosine and lanes 5 and 6 contained 1nM staurosporine. Reactions were terminated by the addition of SDS/PAGE sample buffer and the samples were analyzed by polyacrylamide gel electrophoresis followed by autoradiography for 1 day. Arrows indicate the location of MBP, the 93 and 46 kDa endogenous protein substrates.

constitutively active kinase, independent of PS and diC_8 is present in the SCLC cell cytosol. To further verify that the activity measured in our assays represented the activity of the catalytic fragment of PKCε, kinase assays were carried out in the presence of a synthetic peptide inhibitor modeled after the pseudosubstrate region of PKCε, ERMRPRKRQGAVRRRV. In the presence of the peptide inhibitor, both the basal and PS, diC_8 and EGTA-stimulated phosphorylation of MBP and of the endogenous 93 kDa substrate was dramatically reduced (Fig. 5A, lanes 5 and 6). There was only a slight inhibition of MBP phosphorylation by sphingosine (50μM) (Fig. 5B, lanes 3 and 4), while staurosporine (1nM) blocked both basal (Fig. 5B, lane 5) and PS, diC_8 and EGTA-stimulated phosphorylation of MBP (Fig. 5B, lane 6). This data is consistent with the notion that the SCLC cell cytosol contains a constitutively active PKCε catalytic fragment. SCLC cytosolic kinase activity was also measured by phosphorylation of pep-ε (Shaap et al., 1989), a synthetic peptide modeled after the PKCε pseudosubstrate site (ERMRPRKRQGSVRRRV). We found a very high level of kinase activity in the absence of added PKC allosteric activators (PS and diC_8). This activity was not inhibited by the synthetic peptide modeled after the PKCγ pseudosubstrate site or by the cyclic AMP-dependent protein kinase inhibitor, PKI. Phosphorylation of pep-ε was significantly inhibited by staurosporine yet was not inhibited by sphingosine, similar to the results observed using MBP as substrate.

To examine whether the particulate PKCε is constitutively active, membrane extracts were used as a source of enzyme, and exogenous MBP as the substrate in a cell-free kinase assay (Fig. 6). MBP was phosphorylated in the presence of EGTA alone (lane 3) or PS, diC_8 and EGTA (lane 4). The PKCε pseudosubstrate peptide inhibitor reduced significantly this phosphorylation (lanes 5 and 6). This data supports the notion of a constitutively active particulate PKCε. Autophosphorylation of PKCε was slightly enhanced in the presence of MBP (Fig. 6, lane 4), and further increased in the presence of PS, diC_8 and EGTA. The peptide antagonist was inhibitory in this respect (Fig. 6, lane 5).

Partial purification of the 40 kDa protein. We purified the 40 kDa PKCε immunoreactive protein from SCLC N417D cytosol by ion exchange chromatography (Fig. 7). Basal and PS/diC_8/EGTA-stimulated activity was measured using pep-ε (Shaap et al., 1989) as the substrate. Activity was retained

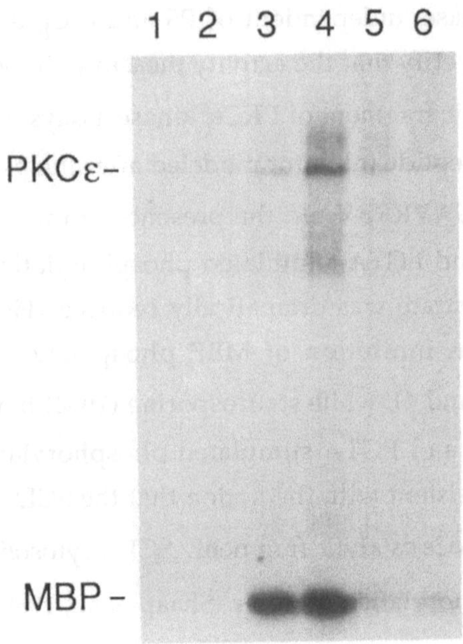

Fig. 6. *In vitro* phosphorylation of MBP in the presence of SCLC particulate fraction: Inhibition by PKCε pseudosubstrate peptide.

Particulate fractions were prepared as described. PKC assays were performed exactly as described in Legend to Fig. 5. See Legend to Fig. 5 also for the explanation of lanes 1-6. Reactions were terminated by the addition of SDS/PAGE sample buffer and the samples were analyzed by polyacrylamide gel electrophoresis followed by autoradiography for 1 day. Arrows indicate the location of MBP and endogenous autophosphorylated PKCε.

on a Mono Q (Pharmacia LKB) column and eluted by a linear gradient of NaCl. Fraction 26 contained a significant level of basal kinase activity (Fig. 7, left panel), which was neither further increased by the addition of PS/diC_8 (Fig. 7, right panel), nor did it show any detectable phorbol ester binding activity (not shown). By Western blotting with the PKCε antibody, we identified fraction 26 as the fraction containing the majority of the 40 kDa PKCε immunoreactive protein (Fig. 7, inset). These results thus establish the presence of a catalytically active PKCε fragment in the cytosol of SCLC N417D cells.

DISCUSSION:

The major conclusions of this investigation relate to the abundance, distribution and regulation of PKCε in an established SCLC cell line. Our data

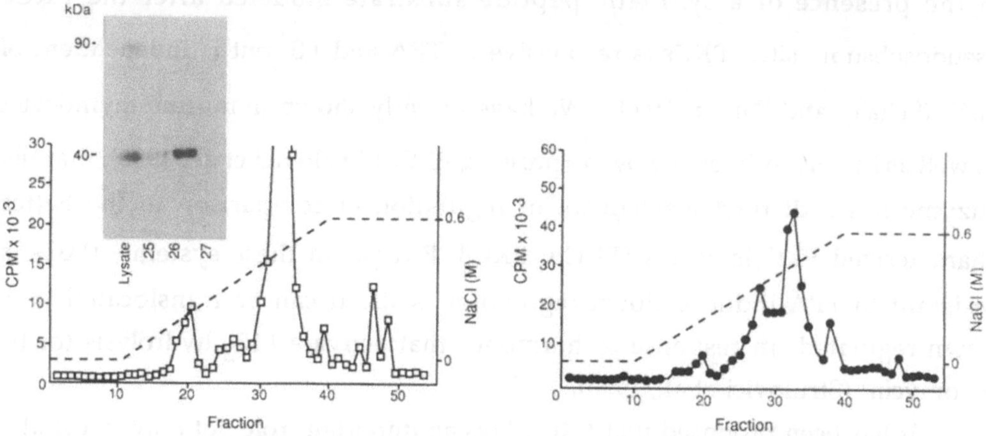

Fig. 7. Partial purification of PKCε catalytic fragment from SCLC N417D cell cytosol by ion exchange chromatography.

Cells were lysed in hypotonic lysis buffer and the cytosolic fraction was prepared as described. 40mg total protein was applied to a Mono Q HR5/5 (Pharmacia LKB Biotechnology Inc.) column which was connected to a Pharmacia FPLC system. Proteins were eluted by a linear concentration gradient of NaCl 0-0.6 M (as indicated by the dashed line) and 1ml fractions were collected. Aliquots of each fraction were assayed for kinase activity or PKCε immunoreactivity (inset). A. Kinase activity was determined in the presence of 1mM EGTA only (left graph) or 1μg/ml PS, 10μM diC8 and 1mM EGTA (right graph) using 20μM pep-ε as substrate. Inset. 50μl of cell lysate of fractions 25-27 were subjected to SDS/PAGE, transferred to nitrocellulose and immunoblotted with antiserum to PKCε. Immunoblots were labeled with [125I]-Protein A and subjected to autoradiography for 2 days at -70°C.

shows that the SCLC cell line N417D greatly overexpresses PKCε when compared with normal human bronchial lung cells. Moreover, PKCε displayed an unusual subcellular distribution in these cells with over 90% of the immunoreactive protein associated with the particulate fraction. In addition, we found a 40 kDa PKCε immunoreactive protein associated with the cytosolic fraction and identified it as a stable constitutively active catalytic fragment of PKCε, whose levels are regulated by TPA and the SCLC cell mitogen GRP.

PKCε belongs to a recently discovered family of related enzyme subspecies δ, ζ and ε (Parker et al., 1989). Recent studies with PKCε transiently overexpressed in COS-1 cells (Schaap et al., 1989) or using the baculovirus expression system (Schaap and Parker, 1990) have shown that the kinase activity was detected only in the presence of a synthetic peptide substrate modeled after the PKCε pseudosubstrate site. PKCε is responsive to TPA and PS, but is independent of Ca^{2+} (Schaap and Parker, 1990). We have recently shown in murine thymocytes as well as in CHO cells stably overexpressing PKCε (Strulovici et al., 1990), that the enzyme has a distinct mechanism of regulation in comparison to the better characterized PKC isoforms (PKCα, PKCβ, PKCγ): in these systems, PKCε is resistant to TPA-induced down-regulation, while it can be translocated (and down-regulated) in response to hormones that activate PIP_2 hydrolysis (or by ionomycin) (Strulovici et al., 1990).

It has been proposed that PKC plays an important role not only in cellular proliferation but also in cellular transformation. Recent reports, specifically addressing phenotypic effects of PKCα (Eldar et al., 1990), βI (Housey et al., 1988) or γ (Persons et al., 1988) isoforms expressed in Swiss 3T3 fibroblasts, NIH 3T3 and rat 1 cells, respectively, indicated enhanced morphological changes following exposure to TPA (Housey et al., 1988), decreased EGF receptor number associated with enhanced growth (Eldar et al., 1990) and tumorigenicity when inoculated into nude mice (Persons et al., 1988). Interestingly, another report (Choi et al., 1990) indicated that overexpression of PKC βI in the human colon cancer cell line HT29 actually decreased their tumorigenicity in nude mice, suggesting that in some instances, PKC might act as a growth-suppressor gene. A recent report (Hirai et al., 1989) has indicated that lung cancer cells, including one SCLC cell line, often exhibit significantly higher PKC activity than other types of cancer cells. PKCα was identified by these studies as the only PKC isoform present. Here we show that PKCε, the major isoform present in the SCLC N417D, is largely overexpressed in comparison with normal human bronchial lung cells or other cell lines. Together, these findings indicate that PKC plays a critical role in growth control and tumor promotion.

Most PKC activity in normal, nontransformed cells is found in the soluble fraction that can be extracted with a divalent cation-chelator and translocates to a particulate cellular compartment upon activation (Anderson et al., 1985). By contrast, transformed fibroblasts, for example, show reversed PKC distribution (Megidish and Mazurek, 1989; Persons et al., 1988). In our present study, PKCε is found mostly in particulate form (Fig. 2) with less than 10% in the cellular cytosol. Moreover, we have shown that the enzyme is constitutively active in a cell-free kinase assay, where an exogenous substrate was phosphorylated in the absence of TPA. The effect was substantially diminished by the specific PKCε peptide antagonist (Fig. 6).

In addition to the particulate 90 kDa PKCε, we discovered a 40 kDa PKCε immunoreactive protein. Our data show that the 40 kDa protein is constitutively present in these cells. Its appearance is not an artifact of post-extraction proteolysis since it is present also when cells are directly lysed in hot SDS/PAGE sample buffer (Fig. 2, lane 1). A number of criteria suggest that this 40 kDa protein is the catalytic fragment of PKCε. First, it was detected by a specific antiserum to PKCε. This signal was blocked by preincubation of the antibody with the peptide antigen, and did not appear when preimmune serum was used instead of the antiserum. Second, its activity was independent of phospholipid/diacylglycerol (Fig. 5) and was blocked by the PKCε peptide inhibitor but not by the synthetic peptide modeled after the PKCγ pseudosubstrate or by the cyclic AMP-dependent kinase inhibitor PKI further suggesting that the 40 kDa protein is the PKCε catalytic fragment. Additional evidence is based on the fact that staurosporine, a PKC antagonist which interacts with the catalytic domain of PKC (Tamaoki et al., 1986), inhibits the kinase activity, while sphingosine, which interacts with the regulatory lipid binding domain of PKC (Hannun and Bell, 1987), is inactive in this respect (Fig. 5). Thus, we have demonstrated the identity of the 40 kDa protein in the cytosol of the SCLC cells as a constitutively active fragment of PKCε.

A possible role for PKC and its catalytic domain in the development of Moloney murine leukemia virus-induced lymphoma in BALB/c mice (Wolfson et al., 1986) or in HMBA-mediated murine erythroleukemia differentiation (Melloni et al., 1987) was reported. In our study, although the 40 kDa kinase is constitutively present, its expression is enhanced by treatment of the cells with either TPA or GRP for up to 8 hr. This effect required *de novo* protein synthesis. It is not clear as yet whether the appearance of the kinase is the result of proteolytic cleavage of the particulate 90 kDa PKCε, although 8 hr TPA treatment reduces the amount of immunoreactive particulate 90 kDa, with a concomitant increase in the 40 kDa protein. GRP treatment however, induced a larger increase in the 40 kDa protein (and also of cytosolic 90 kDa PKCε), which could not be accounted for completely by the loss in particulate 90 kDa PKCε. GRP, thus, might induce the appearance of the PKCε catalytic fragment by a different mechanism than TPA.

Evidence presented by us in this report suggests that both the particulate form of PKCε and the active catalytic fragment may be phosphorylated. Though we were unable to prove directly that these enzymes were autophosphorylated endogenously, we demonstrated this by several indirect methods. We showed that the migration pattern of the PKCε catalytic fragment on two-dimensional gel electrophoresis displayed charge heterogeneity, suggesting multiple phosphorylation states. We also showed, by a method of renaturation, that though the cytosolic PKCε could be autophosphorylated, both the particulate form and the catalytic fragment possessed very low levels of autophosphorylation activity, possibly due to them having been already phosphorylated in situ, thus preventing further phosphorylation at this same site(s).

Autophosphorylation has been proposed to alter the association of PKC with membranes (Wolf et al., 1985) and to increase the enzyme's sensitivity to proteolytic down-regulation (Ohno et al., 1990). Proteolysis of brain PKCε was shown to occur through tryptic attack at multiple sites within the PKCε V2/V3 domain (Schaap et al., 1990). The brain catalytic fragment retained the ability to phosphorylate pep-ε in a lipid/phorbol ester independent manner (Schaap et al., 1990), as did the 40 kDa SCLC kinase (present study). However, several differences between brain and SCLC PKCε could be noticed: a) cytosolic vs. particulate location; b) the size of the catalytic fragment (50 kDa for the brain enzyme vs. 40 kDa for the SCLC enzyme); c) susceptibility to proteolysis: the 50 kDa catalytic fragment of brain PKCε could be further proteolysed into non-active fragments, while the 40 kDa SCLC PKCε was stable and was not susceptible to further proteolytic degradation. Thus, the SCLC PKCε might differ slightly from the brain enzyme.

Evidence for the existence of several PKCε splice variants was reported in brain tissue (Ono et al., 1988) and lung (Schaap et al., 1990). This mechanism has been shown to be operative in generating the two PKCβ isoforms, β1 and β2 from a single gene (Coussens et al., 1987). Thus, the PKCε from SCLC cells might be such a splice variant, which behaves differently in terms of subcellular localization and spontaneous generation of a stable catalytic fragment.

Alternatively, one or several point mutations could have occurred in the PKCε gene from SCLC cells, resulting in the expression of an enzyme that is constitutively active and behaves like an oncogene. The idea that the PKC genes may comprise a set of proto-oncogenes has been suggested previously. In a recent study, a mutant PKCα was found to be associated with an UV-induced fibrosarcoma. It contained four point mutations and it was predominantly in particulate form, in contrast to the wild type fibroblast PKCα. The mutant PKCα was shown to be responsible for the malignant transformation of these fibroblasts (Megidish and Mazurek, 1989). In another study, a primary melanoma cell line was found to have a tumor-specific deletion within the gene for PKCα (Linnenbach et al., 1988).

At this time, there is no direct explanation for the mechanism of generation of the PKCε catalytic fragment in the SCLC cells. An alternative explanation to proteolytic cleavage could be through gene rearrangement. It has been shown that amino-terminal deletions of several cellular proto-oncogenes results in constitutively active species and that these proteins are produced by gene rearrangement (Franz et al., 1989; Jackson and Baltimore, 1989; Khazaie et al., 1988). The constitutive activity of these oncogene proteins is thought to be responsible for their transforming abilities.

In summary, we have shown that the SCLC N417D cell line expresses high levels of PKCε. The enzyme is found in a particulate cell fraction and seems to be constitutively active. In addition to the intact 90 kDa PKCε, a 40 kDa PKCε immunoreactive protein was noted. It was identified as the PKCε catalytic domain and exhibited phospholipid/phorbol ester independent kinase activity which could be inhibited by the PKCε pseudosubstrate peptide and by staurosporine, but not by sphingosine. The levels of PKCε catalytic fragment were increased by treatment of the cells with phorbol esters or GRP. Further studies are required to determine the role of PKCε and its catalytic fragment in the control of proliferation of SCLC and other carcinoma cells.

REFERENCES

Anderson, W. B., Estival, A., Tapiovaara, H. and Gopalakrishna, R. (1985) Adv. Cyclic Nucleotide Protein Phosphorylation Res. **19:** 287-306

Berridge, M. J. and Irvine, R. F. (1984) Nature **312:** 315-321

Bonikos, D. S. and Bensch, K. G. (1977) Am. J. Medicine **63:** 765-771

Choi, P. M., Tchou-Wong, K.-M. and Weinstein, I. B. (1990) Mol. Cell. Biol. **10:** 4650-4657

Coussens, L., Rhee, L., Parker, P. J. and Ulrich, A. (1987) DNA **6:** 389-394

Cuttitta, F., Carney, D. N., Mulshine, J., Moody, T. W., Fedorko, J., Fischler, A. and Minna, J. D. (1985) Nature **316:** 823-826

Eldar, H., Zisman, Y., Ullrich, A. and Livneh, E. (1990) J. Biol. Chem. **265:** 13290-13296

Ferrell, J. E. and Martin, G. S. (1989) J. Biol. Chem. **264:** 20723-20729

Franz, W. M., Berger, P. and Wang, J. Y. J. (1989) EMBO J. **8:** 137-147

Hannun, Y. A. and Bell, R. M. (1987) Science **235:** 670-674

Heidenreich, K. A., Toledo, S. P., Watson, M. J., Brunton, L. L., Daniel-Issakani, S. and Strulovici, B. (1990) J. Biol. Chem. **265:** 15076-15082

Hirai, M., Gamou, S., Kobayashi, M. and Shimizu, N. (1989). Jpn. J. Cancer Res. **80:** 204-208

Housey, G. M., Johnson, M. K., Hsiao, W. L. W., O'Brian, C. A., Murphy, J. P., Kirschmeier, P. and Weinstein, I. B. (1988) Cell **52:** 343-354

Jackson, P. and Baltimore, D. (1989) EMBO J. **8:** 449-456

Khazaie, K., Dull, T. J., Graf, T., Schlessinger, J., Ullrich, A., Beug, H. and Vennstrom, B. (1988) EMBO J. **7:** 3061-3071

Kiley, S. and Jaken, S. (1887) Proc. Natl. Acad. Sci. U.S.A. **84:** 4418-4422

Kiley, S., Schaap, D., Parker, P., Hsieh, L. L. and Jaken, S. (1990) J. Biol. Chem. **265:** 15704-15712

Kishimoto, A., Kajikawa, N., Shiota, M. and Nishizuka, Y. (1983) J. Biol. Chem. **258:** 1156-1164

Kishimoto, A., Mikawa, K., Hashimoto, K., Yasuda, I., Tanaka, S., Tominaga, M., Kuroda, T. and Nishizuka, Y. (1989) J. Biol. Chem. **264:** 4088-4092

Krist, R. M., Hazan, R., Villines, J., Moody, T. W. and Schessinger, J. (1987) J. Biol. Chem. **262:** 11215-11220

Leibersperger, H., Gschwendt, M. and Marks, F. (1990) J. Biol. Chem. **265:** 16108-16115

Linnenbach, A. J., Huebner, K., Reddy, E. P., Herlyn, M., Parmiter, A. H., Nowell, P. C. and Koprowski, H. (1988) Proc. Natl. Acad. Sci. U.S.A. **85:** 74-78

Megidish, T. and Mazurek, N. (1989) Nature **342:** 807-811

Melloni, E., Pontremoli, L., Mechetti, M., Sacco, O., Cakiroglu, A. G., Jackson, J. F., Rifkind, R. A. and Marks, P. A. (1987) Proc. Natl. Acad. Sci. U.S.A. **84:** 5282-5286

Nishizuka, Y. (1989) J. Amer. Med. Assoc. **262:** 1826-1833

Ohno, S., Akita, Y., Konno, Y., Imajoh, S. and Suzuki, K. (1988) Cell **53:** 731-741

Ohno, S., Konno, Y., Akita, Y., Yano, A. and Suzuki, K. (1990). J. Biol. Chem. **265:** 6296-6300

Ono, Y., Fujii, T., Ogita, K., Kikkawa, U., Igarashi, K. and Nishizuka, Y. (1988) J. Biol. Chem. **263:** 6927-6932

Parker, P. J., Kour, G., Marais, R. M., Mitchell, F., Pears, C., Schaap, D., Stabel, S. and Webster, C. (1989) Mol. Cell. Endo. **65:** 1-11
Pelosin, J.-M., Vilgrain, I. and Chambaz, E. M. (1987) Biochem. Biophys. Res. Comm. **147:** 382-391
Persons, D. A., Wilkinson, W. O., Bell, R. M. and Finn, O. J. (1988) Cell **52:** 447-458.
Pfeffer, L. M., Strulovici, B. and Saltiel, A. R. (1990) Proc. Natl. Acad. Sci. U.S.A. **87:** 6537-6541
Sausville, E. A., Moyer, J. D., Heikkila, R., Neckers, L. M. and Trepel, J. B. (1988) Annals New York Acad. Sci. **547:** 310-321
Schaap, D., Hsuan, J. and Parker, P. J. (1990) Eur. J. Biochem. **191:** 431-435
Schaap, D. and Parker, P. J. (1990) J. Biol. Chem. **265:** 7301-7307
Schaap, D., Parker, P. J., Bristol, A., Kriz, R. and Knopf, J. (1989) FEBS Lett. **243:** 351-357
Strulovici, B., Daniel-Issakani, S., Baxter, G., Knopf, J., Sultzman, L., Cherwinski, H., Nestor, J., Webb, D. R. and Ransom, J. (1990) J. Biol. Chem. **In Press**
Strulovici, B., Daniel-Issakani, S., Oto, E., Nestor, J., Chan, H. and Tsou, A.-P. (1989) Biochemistry **28:** 3569-3576
Tamaoki, T., Nomoto, H., Takahashi, I., Kato, Y., Morimoto, M. and Tomita, F. (1986) Biochem. Biophys. Res. Comm. **135:** 397-402
Tapley, P. M. and Murray, A. W. (1985) Eur. J. Biochem. **151:** 419-423
Wolf, M., Cuatrecasas, P. I. and Sahyoun, N. (1985) J. Biol. Chem. **260:** 15718-15724
Wolfson, M., Aboud, M., Ofir, R., Weinstein, Y. and Segal, S. (1986) Int. J. Cancer **37:** 589-593
Woll, P. J. and Rozengurt, E. (1989) Biochem. Biophys. Res. Comm. **164:** 66-73
Zachary, I., Millar, J., Nanberg, E., Higgins, T. and Rozengurt, E. (1987) Biochem. Biophys. Res. Comm. **146:** 456-463

THE HUMAN TYPE I cGMP-DEPENDENT PROTEIN KINASE GENE

M. Sandberg[1], V. Natarajan[2], S. Ørstavik[1], S.M. Lohmann[2] and T. Jahnsen[1]
Laboratory for Gene Technology
Institute of Pathology
Rikshospitalet
University of Oslo
N-0027 Oslo 1
Norway

SUMMARY

The type I cGMP-dependent protein kinase has been shown to play a crucial role in relaxation of vascular smooth muscle and is believed to mediate its effects via regulation of the calcium levels in the cell. Two isoforms of type I cGMP-dependent protein kinase, designated type Iα and type Iß have been shown to exist. With exception of the amino terminal part of the molecules the two isoforms are identical. To elucidate which genetic mechanism is responsible for the existence of two isoforms we have undertaken the task of characterizing the human gene(s) for these proteins. Using cDNA probes encoding the type Iß cGMP-dependent protein kinase (Sandberg et al 1989), 8 exons covering approximately half the coding region have been isolated so far. Several exon/intron splice junctions have been conserved between Drosophila *and human cGMP-dependent protein kinase.*

INTRODUCTION

cGMP was isolated in 1963 (Ashman et al 1963), and 7 years later a protein kinase selectively activated by cGMP was discovered (Kuo and Greengard 1970). A few years later the existence of a specific mammalian cGMP-dependent protein kinase was demonstrated (Hofmann and Sold 1972) which could be separated from the cAMP-dependent protein kinase in extracts from several guinea pig tissues (Kuo 1974). The first purification of a cGMP-dependent protein kinase was accomplished from bovine lung in 1976 (Gill et al 1976, Lincoln et al 1977) and in 1984 the complete amino acid sequence of this cGMP-dependent protein kinase was elucidated (Takio et al 1984). Based on the amino acid sequence it was possible to divide the cGMP-dependent protein kinase into six functional domains (see fig. 1). It was shown that this cGMP-dependent protein kinase, designated type I, was very similar to the much better studied cAMP-dependent protein kinase. However, while the cAMP-dependent

1: Institute of Medical Biochemistry, University of Oslo, P.O.Box 1112, Blindern, N-0317 Oslo 3, Norway. 2: Labor für Klinische Biochemie, Medizinische Univ.-Klinik, University of Würzburg, Josef Schneider Strasse 2, D-8700 Würzburg, FRG

NATO ASI Series, Vol. H 52
Biological Signal Transduction
Edited by E. M. Ross and K. W. A. Wirtz

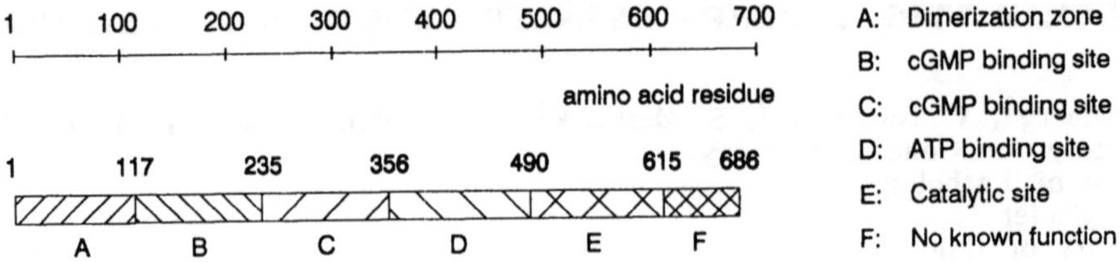

Figure 1. Schematic drawing showing the human type Iβ cGMP-dependent protein kinase. This kinase can be divided into 6 domains, of which 5 have been assigned specific functions (Takio et al 1984).

protein kinase was a tetramer consisting of two regulatory and two catalytic subunits, the type I cGMP-dependent protein kinase was a dimer, and each monomer had both regulatory and catalytic domains. Another distinct form of cGMP-dependent protein kinase has been characterized from rat and pig intestinal brush borders (deJonge 1981). This enzyme, which has been designated type II cGMP-dependent protein kinase, is membrane-bound in contrast to the soluble type I.

The type I cGMP-dependent protein kinase showed tissue specific expression, as can be seen from studies that have been carried out in the rat (Walter 1981). The richest sources of this kinase were cerebellar Purkinje cells, smooth muscle cells and blood platelets (Walter 1981, Waldmann et al 1986) suggesting that it plays an important functional role in these systems. The type I cGMP-dependent protein kinase is believed to be primarily regulated by the level of cGMP which is elevated by agents like nitrovasodilators (Arnold et al 1977, Katsuki et al 1977, Schultz et al 1977), endothelium derived relaxing factor (EDRF) (Rapoport et al 1983) and atrial natriuretic peptides (Walter 1989). There exists substantial evidence that the regulation of the cytosolic calcium level in vascular smooth muscle cells and platelets is probably the most important target for cGMP and the type I cGMP-dependent protein kinase (Felbel et al 1988, Cornwell and Lincoln 1989). It is, however, not clear which steps of the calcium mobilizing or calcium removal pathways are regulated by type I cGMP-dependent protein kinase phosphorylation. For a more detailed account of the physiological role of the type I cGMP-dependent protein kinase, see Walter (1989).

Drosophila was the first species from which the cGMP-dependent protein kinase was cloned (Foster et al 1988, Kalderon and Rubin 1989). Shortly thereafter the first sequences for mammalian type I cGMP-dependent protein kinases were reported (Wernet et al 1989, Sandberg et al 1989). These reports, together with studies performed by Wolfe (Wolfe et al 1989) proved the existence of two isoforms of the mammalian type I cGMP-dependent protein kinase, designated type Iα and type Iβ, respectively. The two isoforms appeared to differ only in the amino terminal part of the monomers. These two isoforms may be the result of alternative splicing of a single gene or may be the gene products of two similar but distinct genes. Based on amino acid and nucleotide similarities found throughout the Iα and Iβ isoforms, the most likely explanation would be alternative splicing of one single gene. To further investigate this, we have undertaken the task of cloning and characterizing the gene(s) for the human type I cGMP-dependent protein kinase. Some preliminary results from this study will be presented here.

MATERIALS AND METHODS

Library

A genomic leukocyte library in EMBL3 from Clontech, Palo Alto, CA (Cat.no. HL 1006d) was used. For the cDNA screening, a cerebellar library in lambda zap was used (Cat.no. 935201, Stratagene, La Jolla, CA).

Probes

The cDNA of the human type Iβ cGMP-dependent protein kinase (Sandberg et al 1989) was digested with various restriction enzymes and some of the resulting fragments were used as probes. Probes were labelled using the random priming method (Feinberg and Vogelstein 1983).

Screening

Screening of both the cDNA and the genomic libraries, was carried out using high stringency. Phage plaques were transferred to nitrocellulose filters, and after denaturation (in 1.5 M NaCl/0.5 M NaOH), the filters were baked at 68 °C. After prehybridization in 6 X SSC*/2 X Denhardt's solution at the same temperature, the filters were hybridized over night in the same solution with the radioactively labelled probe also at 68 °C. The filters were washed three times at 68 °C in 1 X SSC/0.5 % SDS before autoradiography was performed.

Sequencing

Nucleotide sequencing was performed by the dideoxy chain termination method under

---

*<u>The abbreviations used are:</u> SSC, standard saline citrate (0.15 M NaCl, 15 mM sodium citrate, pH 7.0)

standard conditions (Sanger et al 1977) using the modified T7 DNA polymerase (Tabor and Richardson 1987). Nucleotide and amino acid sequence data were analysed using the GCG program package (Devereux et al 1984).

RESULTS AND DISCUSSION

The gene for the type I cGMP-dependent protein kinase

In addition to the reported isolation of cDNA clones encoding the type Iβ cGMP-dependent protein kinase (Sandberg et al 1989), a cDNA clone encoding a part of the type Iα was isolated from a human cerebellar library, thus proving the existence of two type I cGMP-dependent protein kinase isoforms also in man (Natarajan et al, unpublished results). As mentioned in the introduction, these two isoforms can be the products of two distinct, but similar genes, or the products of alternative splicing of one single gene (fig. 2). To determine which genetic mechanism is used to generate these isoforms, genomic studies were undertaken. Various restriction fragments from

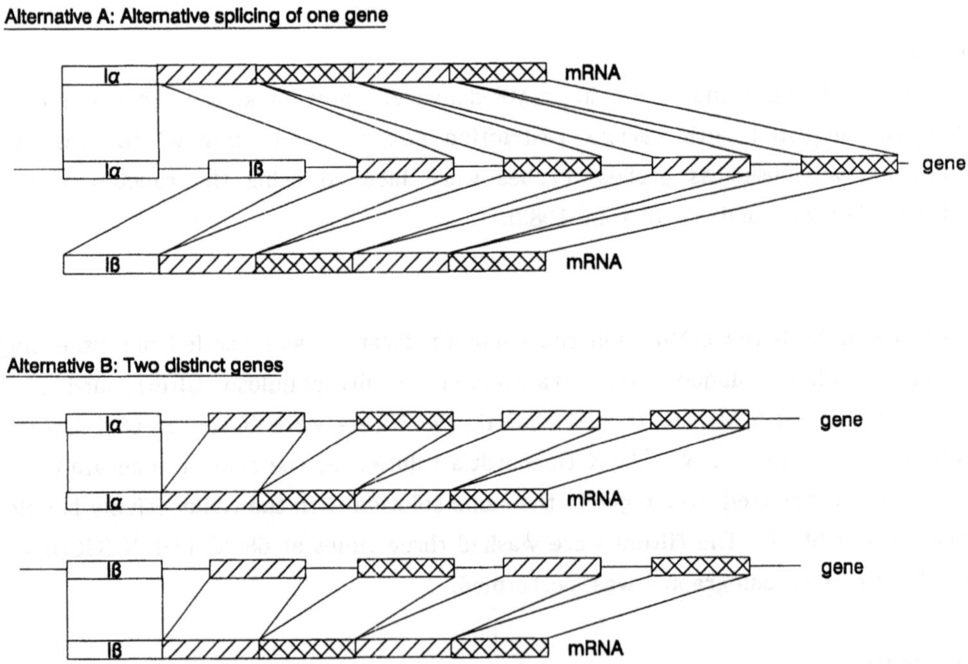

Figure 2. The two isoforms of the type I cGMP-dependent protein kinase can be the product of alternative splicing of one gene as outlined in the upper part of the figure. Transcription of two related genes may also explain the existence of the two isoforms as depicted in the lower part of the figure.

Exon no. (nuc. no)	5' intron	Exon	3' intron
H1 (? - 311)		ACA	GTAAGCAGG
H2 (479- 592)	TCCTTTTAG	ATG-108 nuc.-AGA	GTAAGACTA
H3 (593- 698)	TTTCTCCAG	CTC-100 nuc.-AAG	GTAGGATGC
H4 (699- 762)	GTTTTTCAG	CGT- 58 nuc.-GAG	GTAATTGTT
H5 (1173-1313)	TTCTTTAAG	GTC-135 nuc.-GAG	GTAAAGGCT
H6 (1314-1403)	AATTTACAG	ACT- 84 nuc.-TAG	GTAGGAGAT
H7 (1896-1962)	CTTTTCTAG	ATG- 61 nuc.-AGT	GTAAGTAGA
H8 (1963- ?)	TTTTTACAG	GTT	
Consensus sequence	CAG		GTAAGT

Table 1. The table shows the intron/exon and exon/intron junctions for the isolated exons for the type I cGMP-dependent protein kinase. The consensus sequences are also shown. The enumeration system is identical to the system used when the sequence of the type Iβ cDNA was reported (Sandberg et al 1989).

the cDNA clones encoding the type Iβ cGMP-dependent protein kinase were used as probes to screen a human genomic library. More than ten different clones have been isolated so far, most of them non-overlapping. Each clone is approximately 15 kb in length, indicating that the gene(s) for the type I cGMP-dependent protein kinase may be more than 100 kb long. As indicated in table 1, a total of 8 exons have been isolated so far from different regions of the gene. The exons that are exclusively in the reading frame vary from 64 to 141 nucleotides in length and all boundaries follow the usual rules for intron/exon and exon/intron junction consensus sequences (Shapiro and Senepathy 1987). As has been the case for several genes including the human genes encoding regulatory subunits of the cAMP-dependent protein kinase (Solberg et al, Øyen et al, unpublished data), the exon containing the last part of the reading frame as well as the 3' non-coding region is substantially longer than the other exons. For the type I cGMP-dependent protein kinase, this exon is more than 560 nucleotides in length. Seven of the isolated exons are from the region that is common to type Iα and type Iβ. The last exon encodes the entire type Iβ specific region, as well as at least 160 nucleotides from the transcribed region 5' of the start codon of this isoform. One exon encoding part of each cGMP binding site has been isolated (refer to fig. 1), as well as an exon encoding parts of each of the two binding sites. Two exons giving rise to the major part of the ATP binding site have been characterized as well as a short exon encoding part of the last domain.

Evolutionary considerations

It has been shown for Drosophila that two distinct genes, designated DG1 and DG2, encode the cGMP-dependent protein kinase (Kalderon and Rubin 1989), of which the

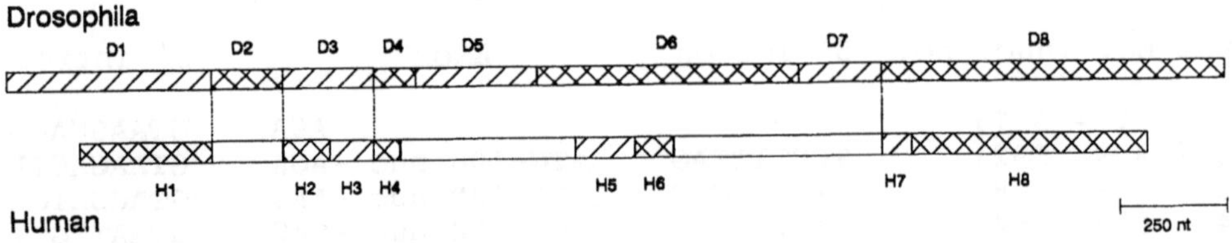

Figure 3. The upper part of the figure shows the organization of the exons in the reading frame of the Drosophila cGMP-dependent protein kinase (DG2) (Kalderon et al 1989). The exons are designated D1 to D8. The lower part depicts the human gene encoding the type Iβ cGMP-dependent protein kinase, and here the hatched areas designate the characterized exons. The vertical lines indicate the positions of the four corresponding splice junctions in the cDNAs from the two species.

DG2-gene may give rise to several polypeptides. In our study we have used the revised sequence with accesion number J04817 from GeneBank to compare the Drosophila and human cGMP-dependent protein kinase. This part of the Drosophila gene contains 7 introns (fig. 3). Fig. 3 also shows the localization of the splice junctions in the human type I cGMP-dependent protein kinase. It is interesting to note that 4 of the 7 splice junctions in Drosophila are also present in human. The regions corresponding to the remaining three splice junctions have so far not been characterized in human. These results indicate that in the human gene for the type I cGMP-dependent protein kinase all splice sites present in the Drosophila will be found. However, the converse is not the case. The region of the human gene corresponding to the Drosophila D3 exon, is for instance composed of 2 exons, while the part of the human gene corresponding to the Drosophila exon D6 must be composed of no less than 4 exons (fig. 3).

The cAMP- and cGMP-dependent protein kinases are closely related. Comparison of the characterized exons of the regulatory subunits RIα and RIIβ of the cAMP-dependent protein kinase (Solberg et al, Øyen et al, unpublished data) to the exons of the type I cGMP-dependent protein kinase indicates that the relative localization of the splice junctions are not conserved between the two different families of cyclic nucleotide dependent protein kinases. Based on this and the fact that splice junctions between Drosophila and human cGMP-dependent protein kinase are conserved, it is tempting to speculate that the cAMP- and cGMP-dependent protein kinases arose as distinct genes at an early point in evolution.

Conclusion

Presently, all available information about the genomic organization of the type I cGMP-dependent protein kinase points in the direction of alternative splicing of one single gene as the mechanism giving rise to the two isoforms. Our cloning and mapping studies indicate the existence of only one copy of each exon in the human genome. In addition, a relatively high degree of similarity between the specific parts of the two isoforms indicates a common evolutionary origin. This has also been shown to be the case for other known instances of alternative splicing after the "mutually exclusive" model (Breitbart et al 1987).

ACKNOWLEDGEMENTS

We gratefully acknowledge the technical assistance of Inger Ronander and Anne Keiserud. This work was supported by the Norwegian Cancer Society, the Norwegian Research Council, Anders Jahres Foundation for the Promotion of Science, Torsteds grant, Nordic Insulin Foundation, the Norwegian Council for Cardiovascular Diseases, and a NATO Collaborative Research Grant. Vasanti Natarajan was supported by a fellowship from the Deutsche Akademische Austauschdienst (DAAD).

REFERENCES

Arnold WP, Mittal C, Katsuki S, Murad F (1977) Nitric oxide activates guanylate cyclase and increases guanosine 3':5'-cyclic monophosphate levels in various tissue preparations. Proc Natl Acad Sci USA 74:3203-3207

Ashman DF, Lipton R, Melicow MM, Price TD (1963) Isolation of adenosine 3',5'-monophosphate and guanosine 3',5'-monophosphate from rat urine. Biochem Biophys Res Commun 11:330-334

Breitbart RE, Andreadis A, Nadal-Ginard B (1987) Alternative splicing: A ubiquitous mechanism for the generation of multiple protein isoforms from single genes. Ann Rev Biochem 56:467-495

Cornwell TL, Lincoln TM (1989) Regulation of intracellular Ca^{2+} levels in cultured vascular smooth muscle cells: reduction of Ca^{2+} by atriopeptin and 8-Br-cGMP is mediated by cGMP-dependent protein kinase. J Biol Chem 264:1146-1155

deJonge H (1981) Cyclic GMP-dependent protein kinase in intestinal brushborders. Adv Cycl Res 14:315-333

Devereux J, Haeberli P, Smithies O (1984) A comprehensive set of sequence analysis programs for the VAX. Nucleic Acids Res 12:387-395

Feinberg AP, Vogelstein B (1983) A technique for radiolabeling DNA restriction nuclease fragments to high specific activity. Anal Biochem 132:6-10

Felbel J, Trockur B, Ecker T, Landgraf W, Hofmann F (1988) Regulation of cytosolic calcium by cAMP and cGMP in freshly isolated smooth muscle cells from bovine trachea. J Biol Chem 263:16764-16771

Foster JL, Higgins GC, Jackson FR (1988) Cloning, sequence and expression of *Drosophila* cAMP-dependent protein kinase catalytic subunit gene. J Biol Chem 263:1676-1681

Gill GN, Holdy KE, Walton GM, Kanstein CB (1976) Purification and characterization of 3':5'-cyclic GMP-dependent protein kinase. Proc Natl Acad Sci USA 73:3918-3922

Hofmann F, Sold G (1972) A protein kinase activity from rat cerebellum stimulated by guanosine 3',5'-monophosphate. Biochem Biophys Res Commun 49:1100-1107

Kalderon D, Rubin GM (1989) cGMP-dependent protein kinase genes in Drosophila. J Biol Chem 264:10738-10748

Katsuki S, Arnold W, Mittal C, Murad, F (1977) Stimulation of guanylate cyclase by sodium nitroprusside, nitroglycerin, and nitric oxide in various tissue preparations and comparison to the effects of sodium azide and hydroxylamine. J Cyclic Nucleotide Res 3:23-35

Kuo JF (1974) Guanosine 3':5'-monophosphate-dependent protein kinases in mammalian tissues. Proc Natl Acad Sci USA 71:4037-4041

Kuo JF, Greengard P (1970) Cyclic nucleotide-dependent protein kinase VI. Isolation and partial purification of a protein kinase activated by guanosine 3',5'-monophosphate. J Biol Chem 245:2493-2498

Lincoln TM, Dills jr WL, Corbin J (1977) Purification and subunit composition of guanosine 3':5'-monophosphate-dependent protein kinase from bovine lung. J Biol Chem 252:4269-4275

Rapoport RM, Draznin MB, Murad F (1983) Endothelium-dependent relaxation in rat aorta may be mediated through cGMP-dependent protein phosphorylation. Nature 306:174-176

Sandberg M, Natarajan V, Ronander I, Kalderon D, Walter U, Lohmann SM, Jahnsen T (1989) Molecular cloning and predicted full-length amino acid sequence of the type Iβ isozyme of cGMP-dependent protein kinase from human placenta. FEBS Letters 255:321-329

Sanger F, Nicklen S, Coulson AR (1977) DNA sequencing with chain-terminating inhibitors. Proc Natl Acad Sci USA 74:5463-5467

Schultz KD, Schultz K, Schultz G (1977) Sodium nitroprusside and other smooth muscle-relaxants increase cyclic GMP levels in rat ductus deferens. Nature 265:750-751

Shapiro MB, Senepathy P (1987) RNA splice junctions of different classes of eukaryotes: sequence statistics and functional implications in gene expression. Nucleic Acids Res 15(17):7155-7174

Tabor S, Richardson CC (1987) DNA sequence analysis with a modified bacteriophage T7 DNA polymerase. Proc Natl Acad Sci USA 84:4767-4771

Takio K, Wade RD, Smith SB, Krebs EG, Walsh KA, Titani K (1984) Guanosine cyclic 3',5'-phosphate-dependent protein kinase, a chimeric protein homologous with two separate protein families. Biochemistry 23:4207-4218

Waldmann R, Bauer S, Göbel C, Hofmann F, Jakobs KH, Walter U (1986) Demonstration of cGMP-dependent protein kinase and cGMP-dependent phosphorylation in cell-free extracts of platelets. Eur J Biochem 158:203-210

Walter U (1981) Distribution of cyclic-GMP-dependent kinase in various rat tissues and cell lines determined by a sensitive and specific radioimmunoassay. Eur J Biochem 167:441-448

Walter U (1989) Physiological role of cGMP and cGMP-dependent protein kinase in the cardiovascular system. Rev of Physiology 113:41-88

Wernet W, Flockeri V, Hofmann F (1989) The cDNA of the two isoforms of bovine cGMP-dependent protein kinase. FEBS Letters 251:191-196

Wolfe L, Francis SH, Corbin J (1989) Properties of a cGMP-dependent monomeric protein kinase from bovine aorta. J Biol Chem 264:4157-4162

CONTROL OF CELLULAR ACTIVITY BY PROTEIN PHOSPHORYLATION-DEPHOSPHORYLATION: PHOSPHORYLASE KINASE FROM BOVINE STOMACH SMOOTH MUSCLE

T.G. Sotiroudis, V.G. Zevgolis, and A.E. Evangelopoulos
Institute of Biological Research
The National Hellenic Research Foundation
48 Vassileos Constantinou Avenue
Athens 116 35
Greece

Protein phosphorylation process appears to be the most important type of covalent modification involved in cellular functions (Krebs, 1985). Basically, the system involves a minimum of three proteins and two reactions:

$$\text{Protein} + n\text{NTP} \longrightarrow \text{Protein-P}n + n\text{NDP} \quad (1)$$

$$\text{Protein-P}n + nH_2O \longrightarrow \text{Protein} + n\text{Pi} \quad (2)$$

Reaction (1) is catalyzed by protein kinase(s) and reaction (2) by phosphoprotein phosphatase(s). In general NTP is ATP but several protein kinases are today known in which GTP is almost as effective as ATP (Krebs and Beavo, 1979).

A large number of hormones, neurotransmitters and other extracellular signals have been found to produce many of their metabolic responses through the action of protein kinases while phosphorylation of plasma membrane receptors has been recently shown to play a key role in the regulation of receptor function (Sibley et al, 1987). Second messengers involved in the regulation of protein kinase activities include cAMP, cGMP, Ca^{2+} and diacylglycerol. Most of the above second messenger-activated protein kinases show a broad substrate specificity and are serine/threonine kinases with striking sequence similarity in their catalytic domains. Tyrosine-specific protein kinases represent another important group of protein kinases because their enzymatic activity has appeared so far to be intrinsic for two classes of proteins: the transforming proteins of certain retrovival oncogens and the membrane receptors for certain cellular growth factors (Hunter and Cooper, 1985). In this latter family, the protein-tyrosine kinase is activated upon binding of the growth factor to its receptor and phosphorylates both the receptor itself and other cell target proteins (Yarden and Ullrich, 1988).

NATO ASI Series, Vol. H 52
Biological Signal Transduction
Edited by E. M. Ross and K. W. A. Wirtz

Protein serine/threonine phosphatases show broad substrate specificity in vitro and their classification requires the use of specific inhibitors and activators (Cohen, 1989). Four major classes of protein phosphatase catalytic subunits have been identified. Type 1 protein phosphatases selectively dephosphorylate the β subunit of phosphorylase kinase from rabbit skeletal muscle and are inhibited by the thermostable proteins inhibitor-1 and inhibitor-2. The enzymatic activity of protein phosphatase-1 is controlled by many extracellular stimuli and its catalytic subunit exists in several stable inactive or active conformations (Ballou and Fischer, 1986). Type 2 protein phosphatases have been separated into three enzymatic species (2A, 2B, 2C) differing in their relative specificity to phosphoprotein substrates and their regulation by divalent cations (Cohen, 1989). Very recently, the analysis of phosphorylation systems was revolutionized by the discovery that the toxin okadaic acid is a tumour promoter and at the same time a potent and specific inhibitor of protein phosphatases 1 and 2A (Cohen et al, 1990). In contrast to our extensive knowledge of protein-serine phosphatases little is known about the structure and function of protein tyrosine phosphatases, although substantial evidence is accumulating to indicate that these specific phosphatases constitute a family of important enzymes (Hunter, 1989). In this respect, the identification of integral membrane proteins with stuctural features of receptor molecules and intrinsic protein tyrosine phosphatase activity raises the exciting possibility of a novel mechanism of signal transduction (Tonks and Charbonneau, 1989).

It is widely recognized that Ca^{2+} is a major second messenger mediating the action of neurotransmitters, hormones, growth factors and mitogens (Rasmussen, 1986). Although the actions of Ca^{2+} do not appear to be mediated by a universal biochemical mechanism there is experimental evidence that certain of the physiological effects of Ca^{2+} are mediated or modulated by a protein phosphorylation cascade via stimulation of a number of Ca^{2+}-dependent protein kinases. Biochemical studies of the signalling systems that utilize Ca^{2+} as a second messenger in several tissues have revealed the presence of at least six protein kinases that can be regulated by Ca^{2+}. Ca^{2+}/calmodulin dependent protein kinase II and protein kinase C appear to be general or multifuntional protein kinases, while myosin light chain kinase and phosphorylase kinase are considered to be dedicated protein kinases. Two additional Ca^{2+}-regulated protein kinases,

Ca^{2+}/calmodulin-dependent protein kinase I and III, may also have a very limited substrate specificity but have not been extensively studied (Schulman, 1988).

In the last few years there has been a surge of interest in smooth muscle biochemistry and especially in the regulation of contractile process by protein phosphorylation (Kamm and Stull, 1989). Glycogen, which is the most conspicuous source of endogenous carbohydrate metabolism observed in the smooth muscle cells of different organs has been considered as one of the main available energy stores, which takes part in maintaining the vascular tone under stimulated conditions in the absence of glucose (Sotiroudis et al, 1986). In addition, it has been suggested that metabolism in smooth muscle was functionally compartmentalized and that this compartmentation reflected a particular enzymic localization, with a glycolytic enzyme cascade associated with the plasma membrane and glycogenolysis linked to contractile filaments or cytosolic elements (Paul, 1989). Thus, it is obvious that for a better understanding of the coordination of metabolic and contractile activity in smooth muscle, the purification of the enzymes involved in the corresponding regulatory mechanisms is needed.

Phosphorylase kinase is one of the regulatory enzymes involved in the cascade of reactions associated with glycogenolysis and it has been studied in a number of tissues and species (Carlson et al, 1979; Chan and Graves, 1984; Pickett-Gies and Walsh, 1986). However, only the rabbit skeletal muscle enzyme has been thoroughly characterized in terms of its physicochemical, enzymatic and regulatory properties. It is a large oligomeric enzyme (1.3 MDa) with the subunit structure $(\alpha\beta\gamma\delta)_4$. It is regulated in a complex way by phosphorylation and Ca^{2+} and it sits at the crossroad of glycogenolysis linking glycogen breakdown to both nervous and endocrine stimulation (Carlson et al, 1979; Chan and Graves, 1984; Pickett-Gies and Walsh, 1986). δ-subunit is identical to calmodulin and confers Ca^{2+} sensitivity to the enzyme (Cohen et al, 1978). γ-subunit has catalytic activity (Skuster et al, 1980) and is similar to other protein kinases (Reimann et al, 1984). The two large subunits, α and β, are homologous proteins (Zander et al, 1988; Kilimann et al, 1988), they cary all phosphorylation sites and at least one of their function is regulatory. The primary structures of γ- and δ-subunits have been determined and cDNAs encoding α-, β- and γ- subunits have been cloned (Zander et al, 1988; Kilimann et al, 1988; daCruz Silva and Cohen, 1987; Bender and Emerson, 1987).

In our effort to better understand the hormonal control of glycogen metabolism in smooth muscle we have purified and characterized phosphorylase kinase from chicken gizzard smooth muscle (Nikolaropoulos and Sotiroudis, 1985). It was interesting as well as puzzling that gizzard phosphorylase kinase showed a completely different subunit pattern to that of rabbit skeletal muscle enzyme upon SDS-PAGE (one main protein band of 61 kDa), although the Mr of both kinases were similar. In addition, the gizzard kinase was activated neither by cAMP-dependent protein kinase nor by autophosphorylation. Because of the complexity of the regulation of glycogenolysis by phosphorylase kinase it appeared of interest to investigate if the peculiar behaviour of gizzard phosphorylase kinase is a unique feature of primitive vertebrate tissues or it is a property shared also by mammalian smooth muscle cells. The aim of the present report is to describe recent data on the purification and characterization of bovine stomach smooth muscle phosphorylase kinase and to discuss the puzzling association of this kinase with an actin-related molecule.

Purification and characterization of phosphorylase kinase from bovine stomach smooth muscle

Phosphorylase kinase was purified 110-fold from bovine stomach smooth muscle by a procedure involving three steps: DEAE-cellulose chromatography at pH 7.5, ammonium sulfate fractionation and glycerol density ultra-centrifugation. The initial fractionation gives a single peak of phosphorylase kinase activity at about 0.2 M NaCl. It is essential to include protease inhibitors in the homogenization buffer, otherwise phosphorylase kinase is transformed to a new kinase form, eluted from the anion exchange column at 0.1 M NaCl, suggesting proteolytic degradation. Ammonium sulfate fractionation step takes advantage of the precipitation of the kinase at low salt concentrations (25% saturation), while ultracentrifugation is an efficient purification step because of the high molecular mass of smooth muscle kinase (1 MDa, as revealed by gel filtration analysis). The purified enzyme migrates on SDS-PAGE as a single protein band of 43 kDa and shows a close similarity with bovine aortic actin as revealed by amino acid analysis and sequencing of a tryptic deca-peptide fragment (His-Gln-Gly-Val-Met-Val-Gly-Met-Gly-Gln) (Vandekerchove and Weber, 1979). Moreover, the kinase preparation containing the 43 KDa protein band exhibits two other actin characteristics: (i) it shows a polydispersed profile upon glycerol density ultracentrifugation and non-denaturing PAGE and (ii) it is able to polymerize and hydrolyze ATP in presence of KCl and $MgCl_2$. In contrast, the 43 KDa protein is not immunoprecipitated by a polyclonal anti-actin antibody, it does not inhibit DNase I either before or after treatment with 1.5 M guanidine-HCl (Cooper and Pollard, 1982), while it shows a high sensitivity to proteolytic degradation. At this point it must be emphasized that although the proteolysis of our 43 KDa protein by trypsin is a very rapid process, producing a polydispersed mixture of low Mr peptides within a few minutes (Fig.1), F actin is totally resistant to trypsinolysis even after 19 h incubation (Kay et al, 1982), while G actin produces a main final product of 33 KDa (Movnet and Ue, 1984). Thus, the 43 kDa protein associated with bovine stomach phosphorylase kinase preparations differs widely from native actin in several important features, suggesting that in our case we deal

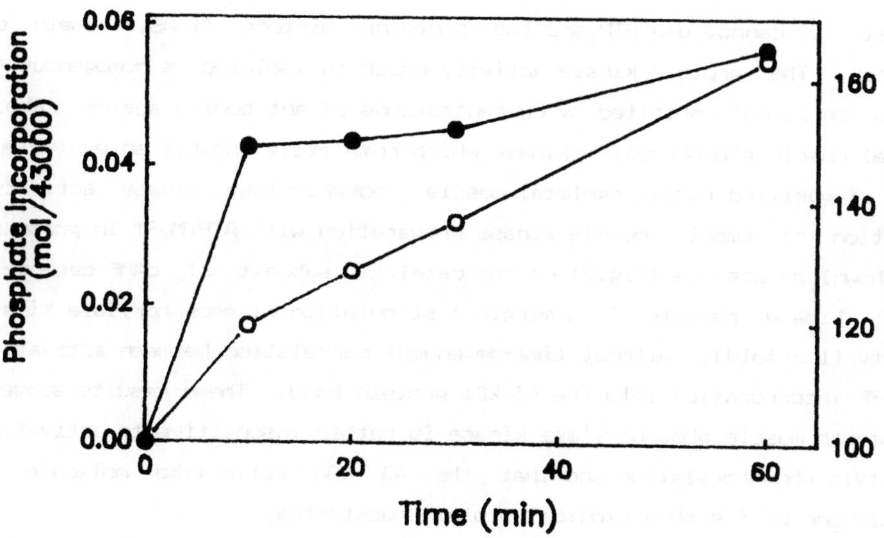

Fig. 1. Time course of proteolytic degradation of the 43 kDa bovine stomach smooth muscle actin-like protein. The 43 kDa protein purified up to ammonium sulfate fractionation step (0.25 mg/ml) was incubated with trypsin (2 µg/ml) in 25 mM glycerol-2-phosphate buffer pH 7.0, 0.1 mM EDTA, 1 mM 2-mercaptoethanol (22ºC). At different time intervals aliquots were taken and examined by SDS-PAGE (10% acrylamide). The time of trypsinolysis was: 1,0; 2, 0.5; 3,1; 4,2; 5,5; 6,10 and 7,30 min respectively.8, Mr markers.

with an actin-related protein.

In our effort to separate phosphorylase kinase activity from the 43 kDa protein we used a variety of chromatographic media, but both catalytic activity and protein either eluted superimposably (phosphorylase b-Sepharose, Blue-Sepharose) or they could not be eluted (phenyl-Sepharose, hydroxylapatite, protamine-Sepharose). Furthermore, protamine which is known to induce supramolecular actin-protamine structures (Grazi et al, 1982) totally precipitated the 43 kDa protein together with the kinase activity at a molar ratio of protamine to actin-like molecule of 0.5:1. In addition, in a similar experiment, our actin-like preparation was partially aggregated in presence of Mg^{2+} (6 mM) and Ca^{2+} (1 mM). After centrifugation (37000 xg, 60 min) about 10% of both protein and phoshorylase kinase activity were recovered in the pellet.

Bovine stomach phosphorylase kinase shows a pH 6.8/8.2 activity ratio of 0.23, it has an absolute requirement for Ca^{2+}, as demonstrated by almost complete inhibition in presence of EGTA, and it is activated 1.8-fold by Ca^{2+}/calmodulin and 3,0-fold by short-time trypsinolysis, at pH 6.8. It can utilize ATP as phosphoryl donor and rabbit phosphorylase b as substrate; by contrast, it cannot use GTP and the proteins histone II-AS, casein or phosvitin. The protein kinase activity which is isolated as homogenous 43 kDa protein is not inhibited by concentrations of antibodies against rabbit skeletal muscle phosphorylase kinase which completely inhibit an equivalent amount of purified rabbit skeletal muscle phosphorylase kinase activity. Incubation of smooth muscle kinase preparation with ATP/Mg^{2+} in presence (not shown) or absence (Fig.2) of the catalytic subunit of cAMP-dependent protein kinase results in a marginal stimulation of phosphorylase kinase activity (1.5-fold), without time-dependent correlation between activation and ^{32}P-incorporation into the 43 kDa protein band. These results suggest that smooth muscle phosphorylase kinase is rather insensitive to activation by protein phosphorylation and that the 43 kDa actin-like molecule is possibly one of its physiological protein substrates.

Bovine stomach smooth muscle phosphorylase kinase: A new isoform of the enzyme?

Our inability to separate phosphorylase kinase activity from the 43 kDa actin-related protein strongly suggests a functional association. On the other hand, the biochemical and immunological characterization of smooth

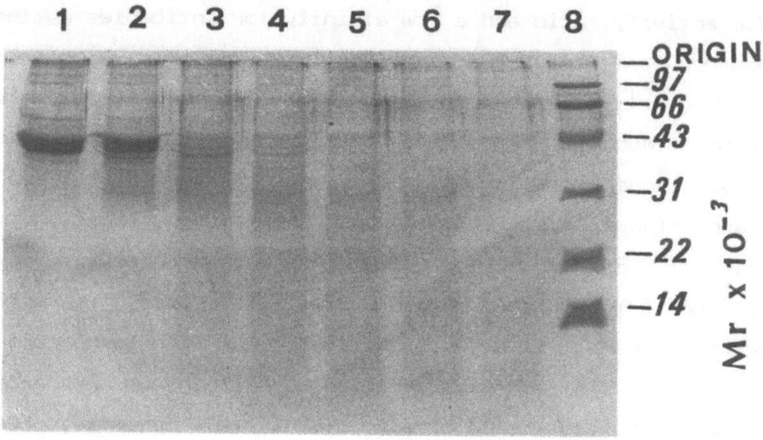

Fig.2. Time course of the autophosphorylation of bovine stomach phosphorylase kinase. Phosphorylase kinase purified up to glycerol density ultracentrifugation step was incubated at 30°C with 1 mM ATP (or[γ-^{32}P]ATP) and 5 mM Mg^{2+}, in presence of 60 mM Bes (pH 8.2), 20 mM NaF, 0.5 mM EGTA, 15 mM 2-mercaptoethanol and 0.55 mM Ca^{2+}. The reaction was started by the additon of ATP and at different time intervals aliquots were taken and assayed for phosphorylase kinase activity at pH 6.8 ( ) or for ^{32}P incorporation( ).

muscle kinase described above indicates a structural difference of phosphorylase kinase from bovine stomach and rabbit skeletal muscle. Thus, two main questions about smooth muscle phosphorylase kinase need to be addressed: (i) Does the enzyme from smooth muscle represent a new phosphorylase kinase entity and (ii) what would be the functional significance of the association of bovine stomach phosphorylase kinase with an actin-like protein?

There are several reports in the literature suggesting the existence of different phosphorylase kinase isoforms: (a) A protein kinase possessing phosphorylase b to a transforming activity was isolated from the trailing edge of a DEAE-cellulose chromatography fraction, referred to as Peak II, during purification of cAMP-dependent protein kinases. This kinase differed in several properties from both phosphorylase kinase and cAMP-dependent kinase (Reimann et al, 1971; Graves et al, 1975). (b) A Ca^{2+}-dependent protein kinase was detected in skeletal muscle membranes of I-strain and wild type mice, which converted phosphorylase b to a, and exhibited a high pH 6.8/8.6 activity ratio and a low affinity for antibodies against rabbit muscle phosphorylase kinase (Varsanyi et al, 1978). (c) A phosphorylase b to a converting activity has been shown associated with isolated homogenous calsequestrin which can be differentiated from skeletal muscle phosphorylase kinase (Varsanyi and Heilmeyer, 1979). (d) A trace residual phosphorylase kinase activity in mixed skeletal muscle of phosphorylase kinase deficient ICR/IAn mice had quite different properties from normal enzyme: It had a much higher pH 6.8/8.2 activity ratio, its activity was less dependent on Ca^{2+}, it was not activated by cAMP-dependent protein kinase or by trypsin and its elution behaviour on Sepharose 4B was remarkably polydispersed (Cohen et al, 1981). (e) Recently, Bender and Emerson (1987) demonstrated that skeletal muscle phosphorylase kinase catalytic subunit mRNAs are expressed in heart tissue but not in liver. This result suggests that the liver γ-subunit and skeletal muscle γ-subunit are encoded by separate genes. In light of the above observations and our experimental results it is reasonable to assume that bovine stomach phosphorylase kinase associated with an actin-like protein is a new isoform different from that associated with skeletal muscle glycogen complex.

A similar association of phosphorylase kinase activity with actin has been suggested by Fischer et al (1978), who showed a considerable homology between the γ-subunit of dogfish phosphorylase kinase and dogfish actin with respect to several structural and functional properties of this component of the contractile apparatus. The postulation by Paul (1989) that glycogenolysis in smooth muscle is linked to contractile filaments permits us to suggest that another form of phosphorylase kinase exists in this muscle type having a suitable structure which could allow the enzyme to position itself properly within the network of actin filaments. In this case phosphorylase kinase either utilizes a Ca^{2+}-binding actin-like molecule as a regulatory subunit and/or it serves as an actin-binding protein to regulate the arrangement and turnover of actin (Pollard and Cooper, 1986). An analogous hypothesis was also made recently by Putnam-Evans et al (1989) who indicated that a novel monomeric Ca^{2+}-dependent protein kinase is associated with actin filaments in plant cells. Of course one cannot exclude the possibility that the 43 kDa smooth muscle actin-like molecule itself possesses phosphorylase kinase activity. In this context, an actin like molecule, may (under certain conditions) present Ca^{2+}-dependent protein kinase catalytic activity. This appears reasonable, since it has nucleotide binding site and binding sites for divalent cations, although it has not the sequence motifs which are characteristic of other protein kinases (Pollard and Cooper, 1986). Nevertheless, it is known that there is at least one example of a protein kinase (isocitrate dehydrogenase kinase/phosphatase), which does not exhibit the typical characteristics of other protein kinases (LaPorte et al, 1989).

In summary, our data suggest that phosphorylase kinase from bovine stomach smooth muscle is associated with an actin-like molecule and it is possibly a different entity from the normal rabbit skeletal muscle enzyme. Further progress in our understanding of phosphorylase kinase function in smooth muscles will require the preparation of the corresponding catalytic subunit cDNA(s).

Acknowledgments

V.G.Zevgolis was supported by EMBO short-term fellowship 6120. We are grateful to Professor L.M.G. Heilmeyer Jr.(Ruhr-Universitat Bochum) for accepting V.G.Zevgolis to his Institute to perform the amino acid analysis and sequence experiments and for valuable discussions. We are indebted to Dr. H.E. Meyer for the assistance with the sequence experiments and to Dr. M.Varsanyi for helping in immunological experiments and for providing the antibodies to phosphorylase kinase. Material described herein is part of the Ph.D. Thesis of V.G.Zevgolis.

REFERENCES

Ballou LM, Fischer EH (1986) Phosphoprotein phosphatases. In: Boyer PD, Krebs EG (eds) The Enzymes vol 17, Academic Press, New York, p.43.

Bender PK, Emerson CP Jr (1987) Skeletal muscle phosphorylase kinase catalytic subunit mRNAs are expressed in heart tissue but not in liver. J Biol Chem, 262:8799-8805

Carlson GM, Bechtel PJ, Graves DJ (1979) Chemical and regulatory properties of phosphorylase kinase and cyclic AMP-dependent protein kinase. Adv Enzymol 50:41-115

Chan KFJ, Graves DJ (1984) Molecular properties of phosphorylase kinase. In: Cheung Y (ed) Calcium and cell function vol 5 Academic Press, New York, p1

Cohen P, Burshell A, Foulkes JG, Cohen PTW, Vanaman TC, Nairn AC (1978) Identification of the Ca^{2+}-depoendent modulator protein as the fourth subunit of rabbit skeletal muscle phosphorylase kinase. FEBS Lett 92:287-293

Cohen P (1989) The structure and regulation of protein phosphatases. Annu Rev Biochem 58:453-508

Cohen P, Holmes CFB, Tsukitani Y (1990) Okadaic acid: a new probe for study of cellular regulation. Tr Biochem Sci 15:98-102

Cohen PTW, Le Marchand Brustel Y, Cohen P (1981) Regulation of glycogen phosphorylase and glycogen synthase by adrenalin in soleus muscle of phosphorylase kinase-deficient mice. Eur J Biochem 115:619-625

Cooper JA, Pollard TD (1982) Methods to measure actin polymerization. Methods Enzymol 85:182-210

da Cruz EF, Cohen PTW (1987) Isolation and sequence analysis of a cDNA clone encoding the entire catalytic subunit of phosphorylase kinase. FEBS Lett, 220:36-42

Fischer eH, Alaba JO, Brautigan DL, Kerrick WGL, Malencik DA, Moeschler HJ, Picton C, Pocinwong S (1978) Evolutionary aspects of the structure and regulation of phosphorylase kinase. In: Li CH (ed) Versatility of proteins, Academic Press, New York, p133

Graves DJ, Carlson GM, Skuster JR, Parrish RF, Carty TJ, Tessmer GW (1975) Pyridoxal phosphate-dependent conformational states of glycogen phosphorylase as probed by interconverting enzymes. J Biol Chem 250: 2254-2258

Grazi E, Magri E, Pasquali-Ronchetti I (1982) Multiple supramolecular structures formed by interaction of actin with protamine. Biochem J 205:31-37

Hunter T, Cooper JA (1985) Protein-tyrosine kinases. Annu Rev Biochem 54:897-930

Hunter T (1989) Protein-tyrosine phosphatases:the other side of the coin. Cell 58:1013-1016

Kamm KE, Stull JT (1989) Regulation of smooth muscle contractile elements by second messengers. Ann Rev Physiol 51:299-313

Kay J, Siemankowski LM, Siemankowski RF, Greweling JA, Goll DE (1982) Degradation of myofibrillar proteins by trypsin-like serine proteinases. Biochem J 201:279-285

Krebs EG, Beavo JA (1979) Phosphorylation-dephosphorylation of enzymes. Annu Rev Biochem 48:923-959

Krebs EG (1986) The phosphorylation of proteins:a major mechanism for biological regulation. Biochem Soc Trans 13:813-820

Kilimann MW, Zander NF, Kuhn CC, Crabb JW, Meyer HE, Heilmeyer LMG Jr (1988) The α and β subunits of phosphorylase kinase are homologous: cDNA cloning and primary structure of the β subunit. Proc Natl Acad Sci USA, 85:9381-9385

LaPorte DC, Stueland CS, Ikeda TP (1989) Icocitrate dehydrogenase kinase/phosphatase. Biochimie 71:1051-1057

Mornet D, Ue K (1984) Proteolysis and structure of skeletal muscle actin. Proc Natl Acad Sci USA 81:3680-3684

Nikolaropoulos S, Sotiroudis TG (1985) Phosphorylase kinase from chicken gizzard. Partial purification and characterization. Eur J Biochem 151: 467-473

Paul RJ (1989) Smooth muscle energetics. Ann Rev Physiol 51:331-349

Pickett-Gies CA, Walsh DA (1986) Phosphorylase kinase. In: Boyer P.D, Krebs EG (eds) The Enzymes vol 17. Academic Press, New York, p395

Pollard TD, Cooper JA (1986) Actin and actin-binding proteins. A critical evaluation of mechanisms and functions. Ann Rev Biochem 55: 987-1035

Putnam-Evans C, Harmon AC, Palevitz BA, Fechheimer M, Cormier MJ (1989) Calcium-dependent protein kinase is localized with F-actin in plant cells. Cell Motil Cytosk 12:12-22

Rasmussen H (1986) The calcium messenger system. N Eng J Med 314:1094-1101 and 1164-1170

Reimann EM, Titani K, Ericsson LH, Wade RD, Fischer EH, Walsh KA (1984) Homology of the γ subunit of phosphorylase b kinase with cAMP-dependent protein kinase. Biochemistry 23:4185-4192

Reimann EH, Walsh DA, Krebs EG (1971) Purification and properties of rabbit skeletal muscle adenosine 3′,5′-monophosphate-dependent protein kinases. J Biol Chem 246:1986-1995

Schulman H (1988) The multifunctional Ca^{2+}/calmodulin-dependent protein kinae. In: Greengard P, Robinson GA (eds) Advances in second messenger and phosphoprotein research, vol 22. Raven Press, New York, p39

Sibley DR, Benovic JL, Garon MG, Lefkowitz RJ (1987) Regulation of transmembrane signaling by receptor phosphorylation. Cell 48:913-922

Skuster JR, Chan KFJ, Graves DJ (1980) Isolation and properties of the catalytically active γ subunit of phosphorylase b kinase. J Biol Chem 255:2203-2210

Sotiroudis TG, Nikolaropoulos S, Evangelopoulos AE (1986) Glycogen metabolism in smooth muscle. In: Heilmeyer LMG (ed) Signal transduction and protein phosphorylation, Plenum Press, New York, p243

Tonks NK, Charbonneau H (1989) Protein tyrosine dephosphorylation and signal transduction. Tr Biochem Sci 14:497-500

Vandekerchove J, Weber K (1979) The complete amino acid sequence of actins from bovine aorta, bovine heart, bovine fast skeletal muscle and rabbit slow skeletal muscle. Differentiation 14:123-133

Varsanyi M, Groschel-Stewart U, Heilmeyer LMG Jr (1978) Characterization of a Ca^{2+}-dependent protein kinase in skeletal muscle membranes of I-strain and wild-type mice. Eur J Biochem 87:331-340

Varsanyi M, Heilmeyer LMG Jr (1979) The protein kinase properties of calsequestrin. FEBS Lett 103:85-88

Yarden Y, Ullrich A (1988) Molecular analysis of signal transduction by growth factors. Biochemistry 27:3113-3119

Zander NF, Meyer H, Hoffmann-Posorske E, Crabb JW, Heilmeyer LMG Jr, Kilimann MW (1988) cDNA cloning and complete primary structure of skeletal muscle phosphorylase kinase (α subunit). Proc Natl Acad Sci USA 85:2929-2933

PROPERTIES OF PHOSPHOINOSITIDES IN MEMBRANES

Theodorus W.J. Gadella Jr., Jan Verbist[1] and Karel W.A. Wirtz

Centre for Biomembranes and Lipid Enzymology
State University of Utrecht
Transitorium III, Padualaan 8
NL-3584 CH Utrecht, The Netherlands

Introduction

The metabolism of phosphatidylinositol (PI), PI-4-phosphate (PIP) and PI-4,5-bisphosphate (PIP_2) is a key process in agonist-induced transmembrane signal transduction (Berridge and Irvine, 1989; Williamson and Hansen, 1987). Activation of receptors leads to the degradation of PIP_2 yielding inositol 1,4,5-trisphosphate and diglyceride as second messengers. Maintenance of the levels of hormone-sensitive PIP_2 in the plasma membrane requires the subsequent phosphorylation of PI and PIP. It remains to be established whether the PI- and PIP kinases involved in these reactions, are active at the level of the plasma membrane. Besides its role in receptor function, there is also good reason to accept that PIP_2 plays a role in the interaction of the cytoskeleton with the plasma membrane. So it was shown that PIP_2 forms a strong complex with glycophorin and that the association between this transmembrane protein and the membrane skeletal protein, protein 4.1, only occurs when PIP_2 is present (Anderson and Marchesi, 1985). Similarly, actin-binding proteins like gelsolin and profilin have a high affinity for PIP_2 (Janmey et al., 1987; Goldschmidt-Clermont et al., 1990). Because of this affinity, PIP_2 interferes with the interaction of profilin with actin monomers and, hence, may reverse the inhibition of actin polymerization by this protein (Lassing and Lundberg, 1985). These observations raise the possibility that membrane levels of PIP_2 have a direct

---

[1] Physiological Laboratory, Catholic University of Leuven, Campus Gasthuisberg, Herestraat 49, B-3000 Leuven, Belgium

NATO ASI Series, Vol. H 52
Biological Signal Transduction
Edited by E. M. Ross and K. W. A. Wirtz

effect on the stability of the cytoskeleton and consequently affect cell shape (Ferrell and Huestis, 1984). In this connection it is of interest that PI- and PIP kinases have been found to be associated with the cytoskeleton (Nahas et al., 1989). Despite their important role in the cell, relatively little is known about the organization and behavior of PI, PIP and PIP_2 in membranes. One of the main reasons for this gap in our knowledge is that, until very recently, no analogs of these phospholipids were available that carried a reporter group. Here we will briefly describe a method to synthesize fluorescent analogs and how these analogs may be used to obtain information on the dynamic behavior of phosphoinositides in membranes and on their affinity for intrinsic membrane proteins. However, we will first discuss some studies on the charge properties of these phosphoinositides.

Effect on membrane surface charge

One of the most striking features of PIP and PIP_2 is their high negative charge content making these phospholipids very acidic at physiological pH. By use of ^{31}P-NMR spectroscopy it could be estimated that PIP is fully ionized (three negative charges) above pH 6.5, and that PIP_2 undergoes a charge shift from -3 to -5 in the pH-range between 6 and 8 (Van Paridon et al., 1986). This implies that in contrast to PI and PIP, PIP_2 is not fully ionized at the pH of the cytosol of resting cells (pH 7.1-7.3). For more detailed discussion of this point, one is referred to Wirtz et al. (1987). Related to this high charge content, PIP_2 forms micelles upon dispersion in water. From the sedimentation velocity it could be derived that these micelles have an aggregation number of 82 (Sugiura, 1981).

In combination with other phospholipids, PIP_2 will normally be part of a bilayer structure. In a recent study, multilamellar vesicles consisting of phosphatidylcholine (PC) and either PIP_2 (5 mol%) or PI (5 mol%) were prepared in a solution of 0.1 M KCl (pH 7.0) and their electrophoretic mobility determined (Toner et al., 1988). From the difference in mobility it could be estimated that relative to PI (a charge of -1), PIP_2 in these vesicles has a net charge of -3. Since one proton is bound to PIP_2 at pH 7.0 (see above data), one potassium ion must have neutralized the other negative charge under these buffer conditions. In another set of experiments it was observed that one had to raise the calcium concentration above 10^{-4} M to measure an effect on the mobility of these vesicles. In fact, calcium was only about 10-fold more effective in reducing the surface charge on the vesicles that contained PIP_2 (6.4 mol%) than on the vesicles that contained PI (26 mol%). This shows that contrary to what was often assumed, PIP_2 has a low affinity for calcium ions (Toner et al., 1988). With an intracellular calcium concentration of about 10^{-6} M, it is clear that binding of calcium to PIP_2 does not play a significant role in calcium homeostasis. In fact, it is more likely that given their concentrations, magnesium and spermine act as the intracellular counter ions of PIP_2.

Synthesis of fluorescent analogs

Recently, methods have been developed for the synthesis of PI, PIP and PIP_2 that carry a pyrene-labeled fatty acyl chain in the sn-2-position (Somerharju et al., 1985; Gadella et al., 1990). Briefly, starting from yeast PI the hydroxyl groups of the ino-

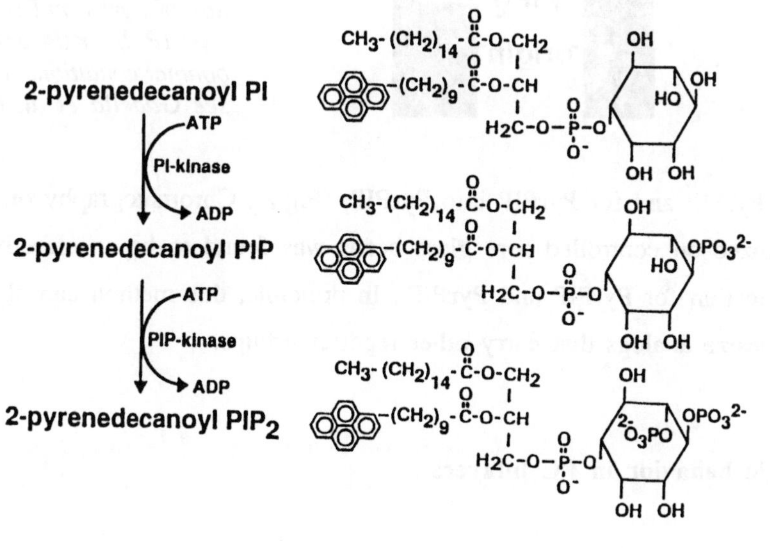

Fig. 1. Schematic representation of the enzymatic synthesis of PyrPIP and $PyrPIP_2$ by the action of PI- and PIP-kinase from bovine brain.

sitol moiety are protected by acetylation after which the sn-2-fatty acyl chain is cleaved off by degradation with phospholipase A_2. Then the lyso-compound is reacylated using the anhydride of the pyrene-fatty acid. After deacetylation by hydrazinolysis, the sn-2-pyrenylacyl-PI is obtained. This PI analog can be converted into the corresponding PIP and PIP_2 analogs using partially purified PI and PIP kinase preparations from bovine brain (Fig.1). Phosphorylation was most effective with PI that carries a pyrenyldecanoyl chain (PyrPI) (Gadella TWJ, unpublished observation). Under optimal conditions of incubation (2 mM ATP, 0.2% Triton X-100, 20 mM $MgCl_2$, 1 mM EDTA, 250 mM sucrose, 0.1% polyethylene glycol 20,000) enzymatic conversions of as high as 80% have been observed both for Pyr-

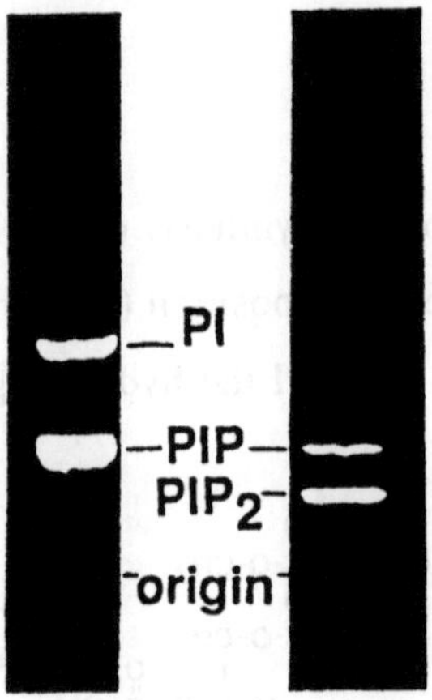

Fig. 2. Thin-layer chromatography of PyrPI, PyrPIP and $PyrPIP_2$ after phosphorylation of PyrPI by PI-kinase and PyrPIP by PIP-kinase under optimal conditions. For details, see Gadella et al. (1990).

PI into PyrPIP and for PyrPIP into $PyrPIP_2$ (Fig.2). Chromatography on neomycin immobilized on controlled-pore glass beads was found to be an effective method of purification for PyrPIP and $PyrPIP_2$. In principle, this method can also be used to synthesize analogs that carry other reporter groups.

Dynamic behavior in PC bilayers

By way of steady-state fluorescence spectroscopy we have investigated the dynamic behavior of PyrPI, PyrPIP and $PyrPIP_2$ in PC vesicle membranes (Gadella et al.,

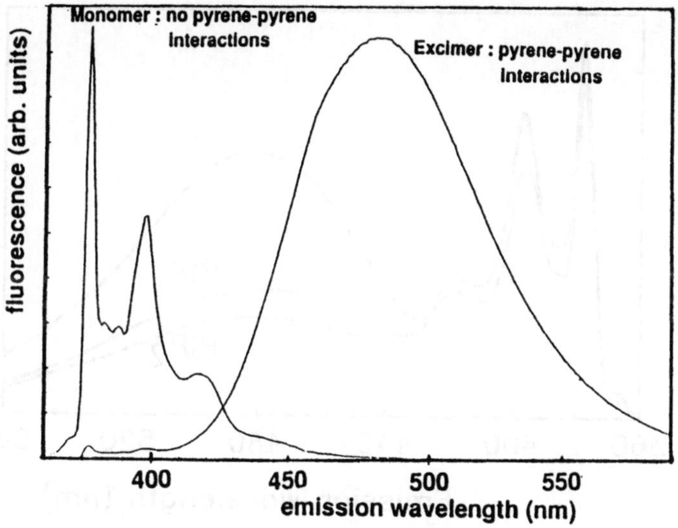

Fig. 3. *Monomer and excimer spectrum of PyrPI in PC bilayer. Wavelength of excitation is 345 nm.*

1990). The pyrenyl-chromophore is very well suited for this kind of measurements because upon excitation (for example, at 345 nm) pyrene may give rise to both a monomer and a excimer emission spectrum (Fig.3). The monomer spectrum is typical for the situation where pyrene in the excited state loses its energy by direct fluorescence emission before having collided with a ground-state pyrene molecule to form the excited-state dimer or excimer complex. When the excimer complex is formed within the fluorescence lifetime of the pyrene excited state, it may loose its energy by fluorescence emission at a longer wavelength resulting in the characteristic excimer spectrum. The ratio of excimer to monomer fluorescence (E/M) is proportional to the collision frequency of the pyrene moieties which is dependent on concentration and diffusion rate of the probe molecules (Förster, 1969). The emission spectra of PyrPI, PyrPIP and $PyrPIP_2$ (present in an amount of 9 mol% in PC vesicles) is shown in Fig.4. By measuring the fluorescence intensity at 475 nm and 377 nm, one finds that the E/M-value equals 0.69 for PyrPI, 0.37 for PyrPIP and 0.26 for $PyrPIP_2$. This shows that in PC bilayers the collision frequen-

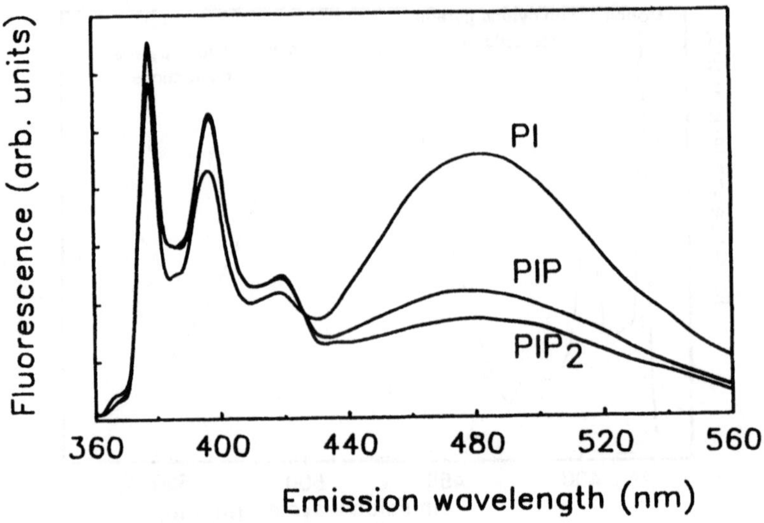

Fig. 4. Fluorescence emission spectra of PyrPI, PyrPIP and PyrPIP$_2$ present in egg PC vesicles. The vesicles contained 9 mole% pyrene phospholipid. For details, see Gadella et al. (1990).

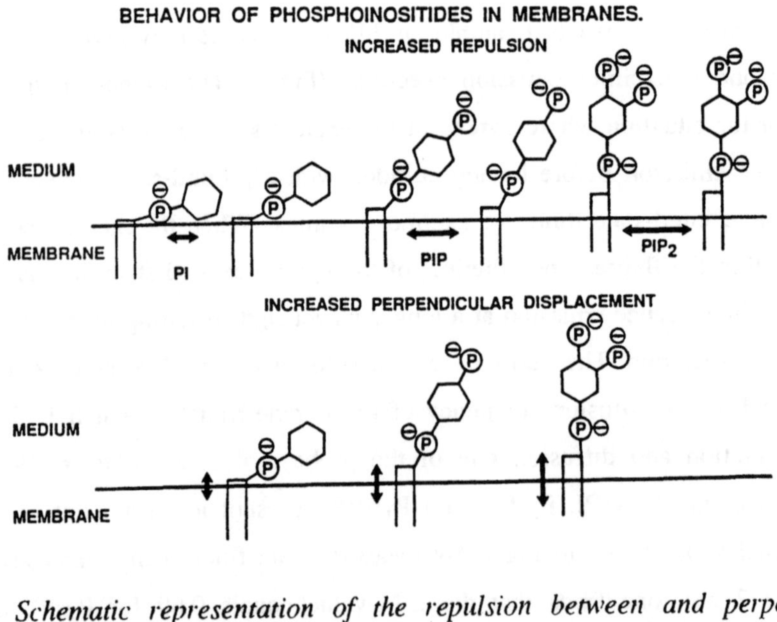

Fig. 5. Schematic representation of the repulsion between and perpendicular diplacement of PyrPI, PyrPIP and PyrPIP$_2$ present in a membrane interface.

cy decreased in the order PyrPI > PyrPIP > $PyrPIP_2$ as a result of increased charge repulsion between molecules. As shown in Fig.5, we have also evidence that the perpendicular displacement of these phosphoinositides at the membrane interface increases in the same order most likely due to the relative increase of polarity of the polar head group (Gadella, unpublished observation). This may have major implications for the efficacy with which, for example, the phosphoinositide-specific phospholipase C acts on PIP_2 at the interface upon agonist-induced receptor activation. The change in orientation of the polar headgroup from parallel to the interface for PI to perpendicular for PIP_2 is purely speculative and remains to be established. Ca^{2+}-levels had to be increased to above 0.1 mM to observe an increase in E/M for $PyrPIP_2$ (9 mole%) in PC vesicles (Fig.6). In agreement with the observations of Toner et al. (1988) relatively high concentrations of calcium are required to have an effective shielding of the negative charge on PIP_2. As shown in Fig.6, calcium barely has an effect on the collision frequency of PyrPIP and no effect on the diffusion behavior of PyrPI.

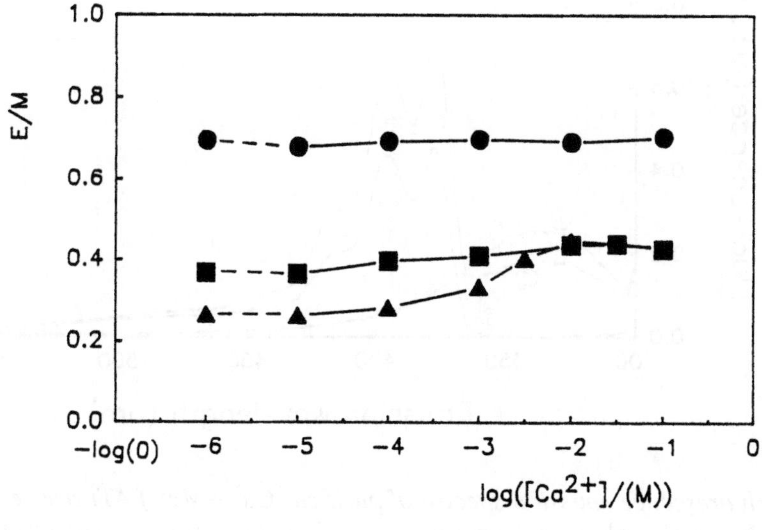

Fig. 6. Effect of Ca^{2+} on the E/M-ratio of PC-vesicles containing 9 mol % Pyr-PI (●), Pyr-PIP (■), or Pyr-PIP_2 (▲). For details see Gadella et al. (1989).

Affinity for (Ca^{2+}+Mg^{2+})-ATPase

In a previous study (Missiaen et al., 1989) it was shown that the activity of the plasma-membrane Ca^{2+}-transporting ATPase purified from smooth muscle and from erythrocytes, was stimulated by phosphoinositides in the order PIP_2 > PIP > PI. The interaction of these phosphoinositides with the ATPase was determined by reconstituting the enzyme in a mixture of PC and the pyrene-labeled analogs (1 mole%), followed by measuring fluorescence enery transfer between the tryptophanyl residues and the pyrene moieties present (Verbist et al., 1991). These measurements are possible as the tryptophanyl emission spectrum of ATPase (excitation at 290 nm) shows a good overlap with the pyrene-absorption spectrum. As shown in Fig.7, in a reconstituted system containing 1 mole% PyrPIP, the intensity of the tryptophan emission (spectrum 2) was decreased compared to the control (spectrum 1, PyrPIP is absent) whereas the intensity of the pyrene signal (spectrum

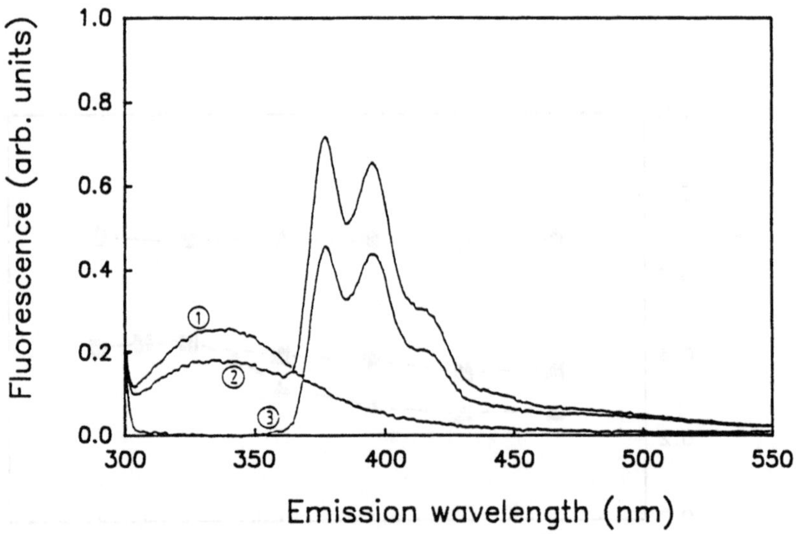

***Fig.** 7. Fluorescence emission spectra of purified (Ca^{2+}+Mg^{2+})-ATPase reconstituted in egg PC in the absence (spectrum 1) and in the presence (spectrum 2) of 1 mole% PyrPIP and of 1 mole% PyrPIP in egg PC in the absence of (Ca^{2+}+Mg^{2+})-ATPase (spectrum 3). Wavelength of excitation is 290 nm. For details, see Verbist et al. (1991).*

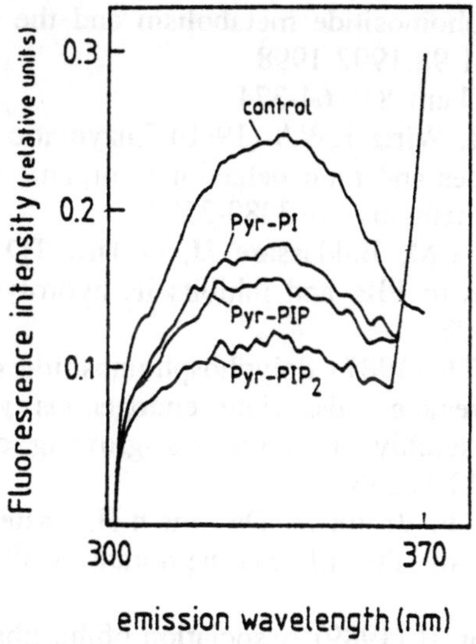

Fig. 8. Fluorescence emission spectra of purified (Ca^{2+}-Mg^{2+})-ATPase reconstituted in egg PC (control) and in egg PC that contains 1 mole% PyrPI, PyrPIP and $PyrPIP_2$. For details, see Verbist et al. (1991).

2) was increased compared to the control (spectrum 3, 1 mole% PyrPIP in the absence of ATPase). This indicates that tryptophanyl residues in the excited state loose energy by exciting pyrenyl residues which must be in close association with the ATPase. The effect of 1 mole% PyrPI, PyrPIP and $PyrPIP_2$ on the tryptophanyl fluorescence emission of the reconstituted ATPase is shown in Fig.8. Fluorescence quenching due to energy transfer increased in the order $PyrPIP_2$ > PyrPIP > PyrPI. This strongly suggests that (Ca^{2+}+Mg^{2+})-ATPase attracts PIP_2 more efficiently than PIP or PI which may explain the effective stimulation of ATPase activity by PIP_2 in the reconstituted system (Missiaen et al., 1989). This is the first direct evidence in support of the model that in the plasma membrane PIP_2 may not be homogeneously distributed in the inner aspect of the bilayer but rather may be concentrated around transmembrane proteins including ion pumps and receptors.

References

Anderson RA, Marchesi VT (1985) Regulation of membrane skeletal protein 4.1 with glycophorin by a polyphosphoinositide. Nature 318:295-298

Berridge MJ, Irvine RF (1989) Inositol phosphates and cell signalling. Nature 341:197-205

Ferrell JE, Huestis WH (1984) Phosphoinositide metabolism and the morphology of human erythrocytes. J Cell Biol 98:1992-1998

Förster T (1969) Excimere. Angew Chem 81:364-374

Gadella TWJ, Moritz A, Westerman J, Wirtz KWA (1990) Enzymatic synthesis of pyrene-labeled polyphosphoinositides and their behavior in organic solvents and phosphatidylcholine bilayers. Biochemistry 29:3389-3395

Goldschmidt-Clermont PJ, Machesky LM, Baldassare JJ, Pollard TD (1990) The actin-binding protein profilin binds to PIP_2 and inhibits its hydrolysis by phospholipase C. Science 247:1575-1578

Janmey PA, Iida K, Yin H, Stossel TP (1987) Polyphosphoinositide micelles and polyphosphoinositide-containing vesicles dissociate endogenous gelsolin-actin complexes and promote actin assembly from the fast-growing end of actin filaments. J Biol Chem 262: 12228-12236

Missiaen L, Raeymaekers L, Wuytack F, Vrolix M, De Smedt H, Casteels R (1989) Phospholipid-protein interactions of the plasma-membrane Ca^{2+}-transporting ATPase. Biochem J 263:687-694

Nahas N, Plantavid M, Mauco G, Chap H (1989) Association ofphosphatidylinositol kinase and phosphatidylinositol 4-phosphate kinase activities with the cytoskeleton in human platelets. FEBS Lett. 246:30-34

Somerharju PJ, Virtanen JA, Eklund KK, Vaino P, Kinnunen PKJ (1985) 1-Palmitoyl-2-pyrenedecanoyl glycerophospholipids as membrane probes: evidence for regular distribution in liquid-crystalline phosphatidylcholine bilayers Biochemistry 24:2773-2781

Sugiura Y (1981) Structure of molecular aggregates of PIP_2 in water. Biochim Biophys Acta 641:148-159

Toner M, Vaio G, McLaughlin A, McLaughlin S (1988) Adsorption of cations to PIP_2. Biochemistry 27:7435-7443

Van Paridon PA, de Kruijff B, Ouwerkerk R, Wirtz KWA (1986) Polyphosphoinositides undergo a charge neutralization in the physiological pH range; a ^{31}P-NMR study. Biochim Biophys Acta 877:216-219

Verbist J, Gadella TWJ, Raeymaekers L, Wuytack F, Wirtz KWA, Casteels R (1991) Phosphoinositide-protein interactions of the plasma-membrane Ca^{2+}-transport ATPase as revealed by fluorescence energy transfer. Biochim Biophys Acta In press

Williamson JR, Hansen CA (1987) Signalling systems in stimulus-response coupling. In: Litwack G (ed) Biochemical Action of Hormones, vol. 14. Academic Press, New York, pp. 29-80

Wirtz KWA, Van Paridon PA, Visser AJWG, De Kruijff B (1987) Structure and function of phosphoinositides in membranes and cells. In: Wirtz KWA (ed) Membrane Receptors, Dynamics and Energetics, vol. A133. Plenum Press, New York, pp. 341-351

GLYCOSYL-PHOSPHATIDYLINOSITOL MEMBRANE ANCHORS

Michael A.J. Ferguson
Department of Biochemistry
University of Dundee
Dundee DD1 4HN
Scotland

Introduction

The consensus opinion that cell surface proteins could be anchored in the plasma membrane via covalent linkage to phospholipid emerged in 1985 through an amalgamation of data from several laboratories. Since the mid-1970s data from the groups of Ikezawa and Low showed that the treatment of cells or membranes with bacterial phosphatidylinositol-specific phospholipase C (PI-PLC) results in the selective release of certain membrane proteins such as alkaline phosphatase, acetylcholinesterase (AChE) and 5'-nucleotidase (reviewed by Low, 1987). In an extensive study Low showed that PI-PLC released alkaline phosphatase could no longer re-bind to membranes and concluded that it was originally anchored through a tight, and probably covalent, linkage to phosphatidylinositol (PI) phospholipid (Low and Zilversmit, 1980). However in the absence of direct chemical evidence this proposal was largely ignored. The first chemical evidence for covalent linkage of phospholipid components to protein came from work on the rat Thy-1 antigen by Williams and colleagues (Campbell et al, 1981) who found stoichiometric amounts of ethanolamine and fatty acid attached to C-terminal proteolysis fragments. Subsequently Holder showed that ethanolamine is in amide linkage to the C-trerminal amino acidα-carboxyl group of Trypanosoma brucei variant surface glycoprotein (VSG) polypeptide and associated with a novel carbohydrate moiety (Holder and Cross, 1981; Holder, 1983). However the VSG as conventionally isolated was water soluble (sVSG) with no evidence of a hydrophobic membrane insertion site to account for its original stable association with the trypanosome plasma membrane. This enigma was was explained by the elegant study of Cardoso de Almeida and Turner (1983) who isolated the amphiphilic membrane binding form of VSG (mfVSG) for the first time. They showed that trypanosomes contain a potent

NATO ASI Series, Vol. H 52
Biological Signal Transduction
Edited by E.M. Ross and K.W.A. Wirtz

enzyme which rapidly converts mfVSG to sVSG unless specifically inhibited during VSG isolation. Together with Holder's data it was concluded from this work that the membrane anchor of VSG was lipid in nature. Subsequent chemical analysis supported this view (Ferguson & Cross 1984, Ferguson et al, 1985a) and a partial structure defining the term glycosyl-phosphatidylinositol (GPI) was elucidated (Ferguson et al, 1985b). At the same time compositional analyses of Torpedo AChE (Futerman et al, 1985), Thy-1 (Tse et al, 1985) and erythrocyte AChE (Roberts & Rosenberry, 1895) demonstrated the covalent association of GPI components with these proteins. Since then over forty examples of GPI anchored eukaryotic proteins have been described (reviewed recently by Ferguson & Williams, 1988; Low 1989; Cross, 1990).

The elucidation of GPI anchor structure

As an example of the analytical strategies employed to determine GPI structure the example of the GPI anchor of T.brucei VSG will be used. The GPI anchor of the group 1 VSG MITat1.4 was the first to be structurally solved. Based on the known partial structure shown at the top of Fig 1 a strategy was developed to solve the structure using proton nuclear magnetic resonance spectroscopy (^{1}H NMR), mass spectrometry and chemical and enzymatic modifications as principle tools (Ferguson et al, 1988a). Of these techniques, NMR, provides the most information but requires reasonable quantities (>50nmoles) of material in a suitable form, i.e. freely water soluble and fairly small (<10kD). Clearly mfVSG itself fails both of these criteria. It was necessary therefore to remove the diacyl-glycerol (DAG) portion, already known to be sn-1,2-dimyristyl glycerol (Ferguson et al, 1985a; Schmitz et al, 1986), to produce sVSG and subsequently remove all the protein except for the C-terminal amino acid by Pronase digestion. The resulting fragment known as the soluble form C-terminal glycopeptide (sCt-gp, see Fig 1) is ideal for NMR analysis.

High resolution Fourrier transform NMR using high-field (500MHz) instruments has become an invaluable tool in the determination of complex carbohydrate structure. Many of the modern techniques were introduced to solve asparagine-linked glycoprotein oligosaccharides. Where previously solved structures are available a one-dimensional NMR spectrum "fingerprint" can be compared with reference spectra to give rapid

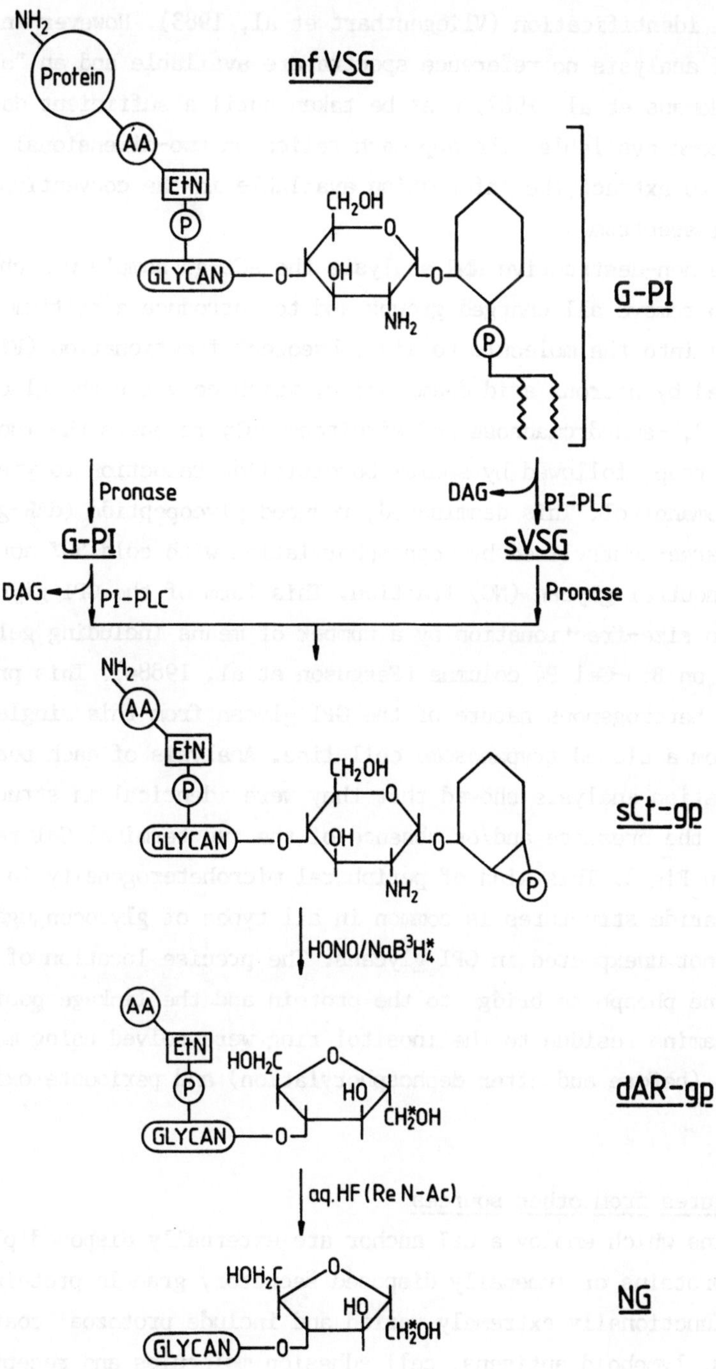

Figure 1. Strategy for generating various GPI fragments for characterisation by NMR and mass-spectrometry.

structural identification (Vliegenthart et al, 1983). However in cases such as GPI analysis no reference spectra are available and an "ab-initio" approach (Homans et al, 1987) must be taken until a sufficient data base of spectra become available. The approach relies on two-dimensional NMR techniques to extract the information available in the conventional one-dimensional spectrum.

Following non-destructive NMR analysis the sCt-gp sample was chemically modified to remove all charged groups and to introduce a tritium radiolabel selectively into the molecule to aid subsequent fractionation (Fig 1). This was acheived by nitrous acid deamination, which converts the glucosamine residue to 2,5-anhydromannose and simultaneously releases the inositol phosphate group, followed by sodium borotritiide reduction to yield labeled 2,5-anhydromannitol. This deaminated, reduced glycopeptide (dAR-gp) was further charge-neutralised by dephosphorylation with cold 50% aqueous HF to yield the neutral glycan (NG) fraction. This form of the GPI glycan is amenable to size-fractionation by a number of means including gel-filtration on Bio-Gel P4 columns (Ferguson et al, 1988a). This profile showed the heterogenous nature of the GPI glycan from this single VSG derived from a cloned trypanosome cell line. Analysis of each peak by NMR and methylation analysis showed that they were identical in structure except for the presence and/or absence of the two terminal Gal residues depicted in Fig 2. This kind of peripheral microheterogeneity in oligosaccharide structures is common in all types of glycoconjugates and therefore not unexpected in GPI glycans. The precise location of the ethanolamine phosphate bridge to the protein and the linkage position of the glucosamine residue to the inositol ring were solved using mannosidase digestions (before and after dephosphorylation) and periodate oxidation studies.

<u>GPI structures from other sources</u>

All proteins which employ a GPI anchor are externally disposed plasma membrane proteins or lumenally disposed secretory granule proteins. However they are functionally extremely varied and include protozoal coat proteins, hydrolases, lymphoid antigens, cell adhesion molecules and receptors. There is no clear correlation between protein function and the use of GPI anchors (Ferguson & Williams, 1988). One interesting correlation that does exist with the data available so far is the distribution of GPI usage at different stages of eukaryotic evolution. In mammals for example it seems that a significant minority of cell surface proteins employ GPI whereas in the protozoa GPI appears to be the predominant form of anchorage. Indeed, for the protozoa, in all cases where the mode of membrane anchorage has been studied directly GPI anchors have been found instead of single-pass transmembrane polypeptide domains. It may be assumed that multiple-pass transmembrane proteins exist in the protozoa such as nutrient and ion transporters.

This high frequency of GPI usage in the protozoa is also reflected in the expression of of structurally related glycolipids which have no clearly defined analogues in higher eukaryotes. These novel glycolipid species include the lipophosphoglycans (LPG) and glycoinositol phospholipids (GIPLs) of the Leishmania (Turco et al, 1989; McConville et al, 1990) and the lipopeptidophosphoglycan (LPPG) of Trypanosoma cruzi (Previato et al, 1990; De Lederkremer et al, 1990). All of these molecules have the sub-structure Manα1-4GlcNH$_2$$\alpha$1-6myo-Inositol-1-$PO_4$ in common and may therefore share some biosynthetic enzymes in their formation.

In general then it would seem that the parasitic protozoa rely rather heavily on GPI anchors and related molecules for their infectivity and survival (Ferguson & Homans, 1988). If this evolutionary skew in GPI usage is to be understood and possibly exploited then it is important to determine the full range of structures expressed by both parasite and host (mammalian) cells. So far only two other complete GPI structures have been reported, for rat brain Thy-1 antigen (Homans et al, 1988) and Leishmania major promastigote surface protease, PSP (Schneider et al, 1990), as well as a detailed partial structure for human erythrocyte AChE (Roberts et al, 1989b). The VSG and Thy-1 structures are shown together in Fig 2. The data for human erythrocyte AChE is consistent with the Thy-1 structure except that the AChE GPI does not contain the GalNAc or +/-Man residues. Similarly the PSP anchor glycan contains only the minimum Man$_3$-GlcN-Ino structure but without any extra ethanolamine phosphate groups (not shown).

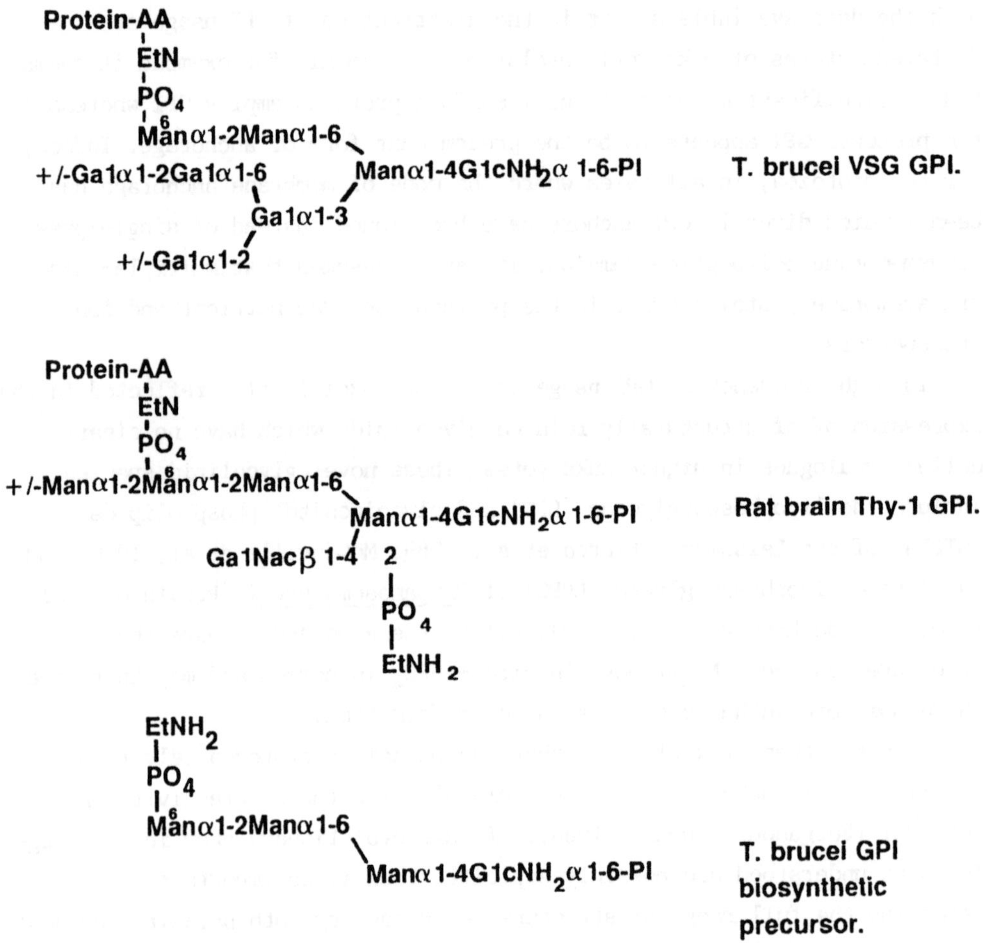

Figure 2. The GPI structures of T.brucei VSG, rat brain Thy-1 and the T.brucei GPI precursor.

On comparing these structures it appears that the core structure which actually links the protein to the lipid (i.e.-ethanolamine-PO_4-6Manα1-2Manα1-6Manα1-4GlcNH$_2$α1-6myo-Inositol-1-PO_4-) is conserved between the protozoal and mammalian GPIs. While more examples are needed to draw convincing conclusions the conservation of the core is striking and may suggest that the core region of GPI is conserved throughout the eukaryotes.

In contrast the side chains attached to this core region are quite different. The L.major PSP GPI anchor contains no side chains. The VSG GPI contains a large and unusual alpha-galactose branched sidechain linked α1-3 to the first Man residue whereas Thy-1 GPI contains a single N-acetyl galactosamine residue linked β1-4 to the the equivalent Man residue. In addition the Thy-1 structure can contain an extra Man residue at the end of the core region. Both Thy-1 and AChE bear a second ethanolamine phosphate group attached to the first Man residue. This extra ethanolamine phosphate, with a free amino group, appears to be common in mammalian GPIs (Fatemi et al, 1987, and references therein) but its function is unknown. Thus there appear to be conserved and species specific structures in GPI anchor carbohydrate.

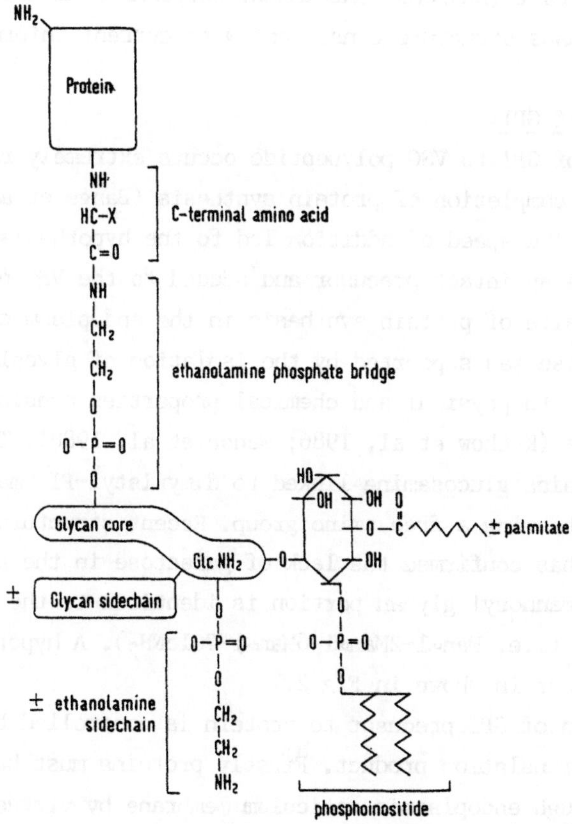

Figure 3. A concensus structure of GPI protein anchors based on known structural features of a variety of GPI anchors from protozoa and higher eukaryotes.

Another notable difference between VSG GPI and those from other sources is in the lipid moiety. The VSGs from T.brucei, T.congolense and T.equiperdum all contain exclusively myristic acid (Ferguson and Cross, 1984; Gurnett et al, 1986; Lamont et al, 1987; Ross et al, 1987) as sn-1,2,-dimyristyl glycerol (Ferguson et al, 1985a; Schmitz et al, 1986). In contrast GPIs of mammalian origin contain palmitic and stearic acids together with a variety of other saturated and unsaturated fatty acids (reviewed in Ferguson et al, 1988b), probably in the form of 1-alkyl-2-acyl glycerols, as found for human erythrocyte AChE (Roberts et al 1989a) and L.major PSP (Schneider et al, 1990). In addition ceramide lipid units have been described for GPI anchors in Dictyostelium dicoideum (Stadler et al, 1989) and in yeast (Andreas Conzelmann, personal communication). Figure 3 shows a concensus structure consistent with current information.

Biosynthesis of GPI

The addition of GPI to VSG polypeptide occurs extremely rapidly, within one minute of the completion of protein synthesis (Bangs et al, 1985; Ferguson et al, 1986). The speed of addition led to the hypothesis that GPI was synthesised as an intact precusor and addded to the VSG polypeptide en bloc close to the site of protein synthesis in the endoplasmic reticulum (Fig 4). This premise was supported by the isolation of glycolipid species from trypanosomes with physical and chemical properties consistent with being GPI precursors (Krakow et al, 1986; Menon et al, 1988). The putative precusor contains glucosamine linked to dimyristyl-PI, mannose and ethanolamine bearing a free amino group. Recent structural analysis (Mayor et al, 1990) has confirmed the lack of galactose in the precursor and shown that the tri-mannosyl glycan portion is identical to the core region of the mature anchor (i.e. Manα1-2Manα1-6Manα1-?GlcNH$_2$). A hypothetical structure for the precusor is shown in Fig 2.

The addition of GPI precusor to protein is controlled by two signals in the primary translation product. Firstly proteins must be sequestered across the rough endoplasmic reticulum membrane by virtue of a cleavable N-terminal signal sequence. Secondly an enzyme or enzyme complex recognises a C-terminal signal sequence which is cleaved and directly replaced by the GPI precusor to generate mfVSG. The nature of this transfer reaction is unknown although a transamidase activity has been suggested. The GPI addition C-terminal signal sequence shows considerable homology between VSG variants but there is no consensus in the signals when other eukaryotic sequences are compared (Ferguson & Williams, 1988). In general terms any C-terminal sequence which ends in a run of apolar amino acids, without any evidence of a polar cytoplasmic domain sequence following it, will serve as a GPI addition signal. The degeneracy in C-terminal GPI addition signals is similar to that found for N-terminal signal sequences.

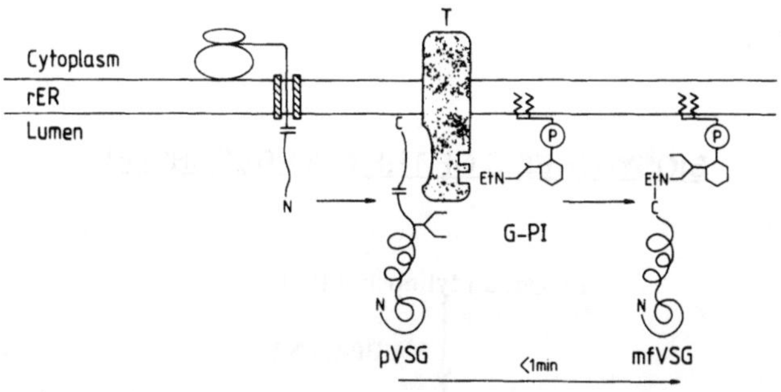

Figure 4. Model for the addition of a GPI precursor to a nascent protein in the endoplasmic reticulum.

Following the addition of GPI the VSG is transported through the Golgi apparatus and to the plasma membrane via the flagellar pocket (Duszenko et al, 1987). The transport from endoplasmic reticulum to the plasma membrane is relatively rapid with a $t_{1/2}$ of 15 minutes (Bangs et al, 1986). The addition of galactose to the GPI is believed to occur late in VSG processing in the Golgi stacks (Bangs et al, 1988).

Recently considerable progress has been made in the pathway of GPI precusor formation using cell-free trypanosome membrane preparations (reviewed in Doering et al, 1990 and Cross, 1990), see Figure 5. Essentially the pathway involves the sequential glycosylation of PI (Masterson et al. 1989, Menon et al 1990). Firstly GlcNAc is added from UDP-GlcNAc followed by de-N-acetylation to $GlcNH_2$-PI (Doering et al, 1989). Subsequently three mannose residues are added from GDP-Man via a Dol-P-Man donor (Menon et al, 1990). Next ethanolamine phosphate is added from an unknown donor and finally the fatty acids of the PI moiety are re-modelled to yield the dimyristyl-PI moiety of the mature trypanosome GPI precursor (Masterson et al, 1990). How similar the GPI biosynthetic pathway is in other eukaryotes is unknown. However the identical glycan core structure of the mammalian and protozoal examples suggests that the glycan core synthetic pathway is highly conserved. On the other hand the specific lipid remodelling described for T.brucei GPI is probably unique to the African trypanosome, since this is the only example of a dimyristyl-PI based anchor. However analogous lipid remodelling may occur in other eukaryotes since the alkyl/acyl content of mammalian GPI anchors is atypical compared with the cellular PI pool.

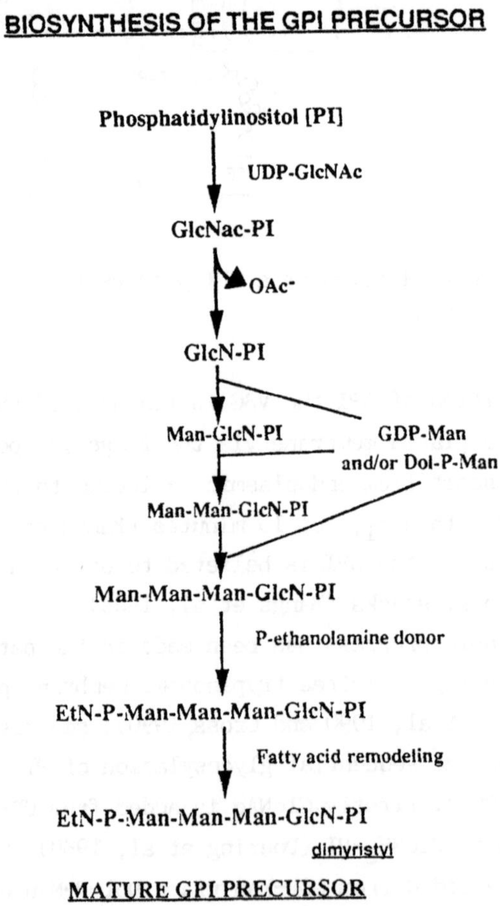

Figure 5. The biosynthetic pathway of GPI formation in trypanosomes.

An additional complication to the GPI biosynthetic pathway in trypanosomes is the presence of a mature GPI precursor species containing an extra acyl (palmitate) chain linked to the inositol ring (Krakow et al, 1989; Mayor et al, 1990). The significance of this GPI species is unclear but it is interesting to note that the mature anchor of human erythrocyte AChE (Roberts et al, 1989a) also contains this palmitate substitution. GPI anchors with this palmitate substitution are resistant to PI-PLCs and GPI-PLC.

Function of GPI anchors

Over the last five years there has been much speculation on the function of GPI anchors. Two of the most popular ideas are that GPI anchors impart a high degree of lateral mobility in the plane of the lipid bilayer and that GPI anchors can be selectively cleaved by GPI-specific phospholipases to release membrane proteins from living cells. It is probably fair to say that neither of these attractive hypotheses have been substantiated by experimental data.

In the case of lateral mobility it is clear that while some GPI anchored proteins display a high lateral mobility this is not a rule. For example the GPI anchored sperm glycoprotein PH-20 varies in its lateral mobility from extremely fast to average to immobile at different stages of sperm development (Phelps et al, 1988). In general the lateral mobility of a membrane protein (single pass transmembrane or GPI anchored) probably depends on the nature of its ectodomain and its interactions with other membrane components rather than any intrinsic properties of the membrane spanning or insertion domain.

The GPI-specific phospholipase mediated release hypothesis gained support from the observations that GPI-PLC and GPI-PLD enzymes do exist and that some GPI anchored proteins are sometimes found in an apparently soluble form in blood, tissue fluids or culture media. However in most cases which are investigated further the proteins are often found to be proteolytically cleaved close to the COOH-termminus or found to be still GPI anchored and present in small membrane vesicles. The idea that GPI-PLC activities are found in plasma membranes was first suggested when the *T.brucei* enzyme was purified and found to be membrane associated. However the enzyme has now been cloned and sequenced (Carrington et al, 1989) and localised by immuno-electron microscopy (Bulow et al, 1989). The results show that the enzyme has no identifiable signal sequence and that it localises to the cytoplasmic face of intracellular vesicles. The role of this GPI-PLC therefore remains obscure since it is not on the same side of the membrane as its putative VSG substrate. The more recently described serum GPI-PLD (Low and Prasad, 1988) is topologically in the correct place to act on GPI anchors but as yet no one has been able to demonstrate its activity except in detergent solutions of GPI anchored proteins. In summary then these GPI-specific phospholipases are intriguing enzymes with, at present, no known function.

The lack of evidence for phospholipase mediated release of GPI anchored proteins also calls into question the likelyhood of transmembrane signalling via the generation of diacyl glycerols (from GPI-PLC action) or phosphatidic acid (from GPI-PLD action) upon GPI cleavage. Nevertheless transmembrane signalling via GPI anchored proteins does occur and is well documented (for a recent review see Robinson, 1990). The best characterised transmembrane signalling involving GPI is the mitogenic response of T cells to cross-linking of GPI anchored proteins in the presence of phorbol esters. This effect has been studied by Robinson who made gene constructs based on the Qa-2 and H2-D T cell antigens. In normal cells cross-linking of GPI anchored Qa-2, but not transmembrane H2-D, caused mitogenesis. However constructs of Qa-2 bearing the H2-D transmembrane domain and H2-D bearing a GPI anchor produced the reverse effects (Robinson et al, 1989). Thus it would appear that the cross-linking of any protein ectodomain will cause T cell mitogenesis so long as it is anchored via a GPI moiety. The molecular mechanism and physiological significance of this type of transmembrane signalling remains to be investigated. One possible mechanism is shown in Figure 6. In this model the cross-linking of GPI anchored proteins by antibody is suggested to cause clustering of GPI anchors around a multi-valent, low affinity, receptor which recognises some feature of the GPI anchor. This putative transmembrane receptor might then transmit a signal to the interior of the cell when all GPI binding sites are occupied.

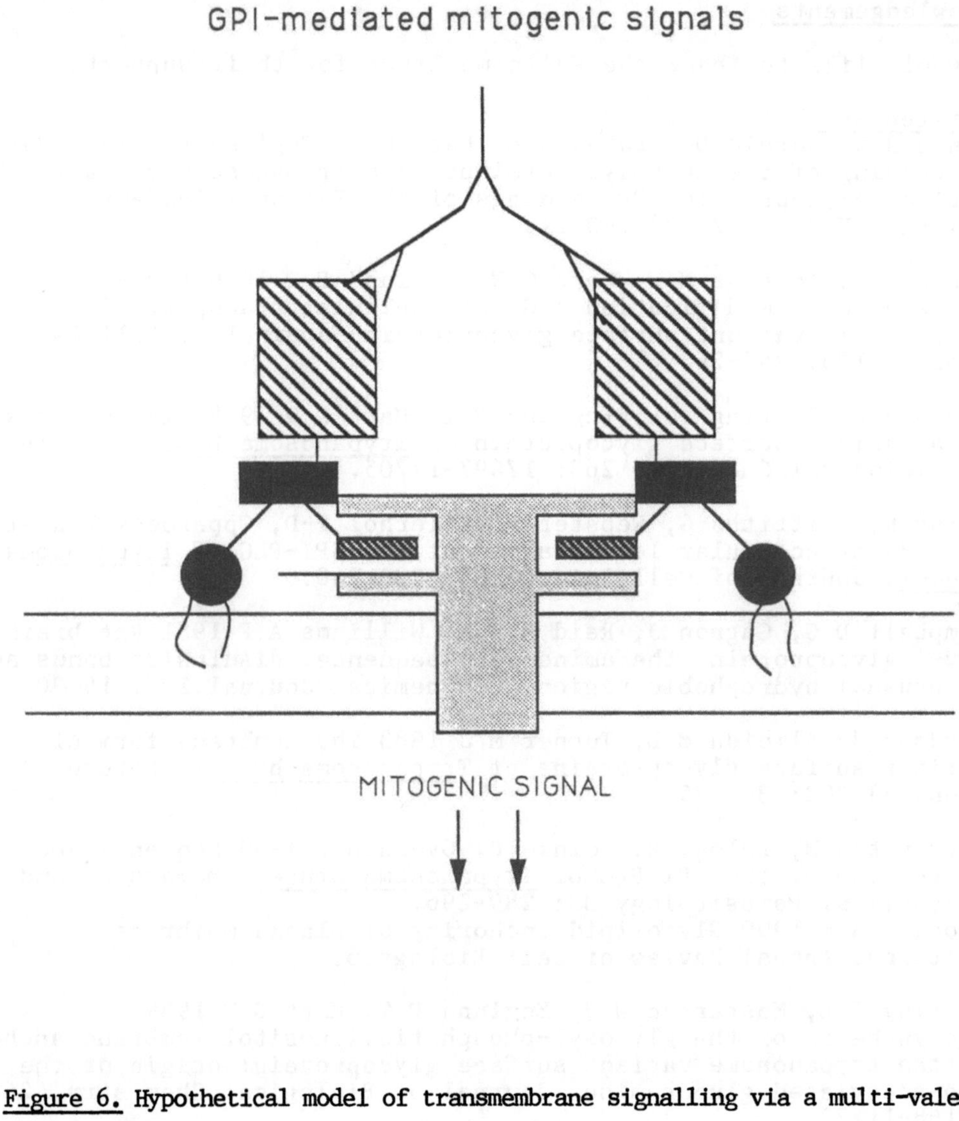

Figure 6. Hypothetical model of transmembrane signalling via a multi-valent GPI-receptor.

Acknowledgements

I would like to thank the Wellcome Trust for their support.

References.

Bangs J D, Hereld D, Krakow J L, Hart G W, Englund P T 1985 Rapid processing of the carboxyl terminus of a trypanosome variant surface glycoprotein. Proceedings of the National Academy of Science (U S A) 82: 3207-3211.

Bangs J D, Andrews N W, Hart G W, Englund P T 1986 Post-translational modification and intracellular transport of a trypanosome variant surface glycoprotein. Journal of Cellular Biology 103: 255-263.

Bangs J D, Doering T L, Englund P T, Hart G W 1988 Biosynthesis of a variant surface glycoprotein of Trypanosoma brucei. Journal of Biological Chemistry 263: 17697-17705.

Bulow R, Griffiths G, Webster P, Stierhof Y-D, Opperdoes F R et al 1989Intracellular localization of the GPI-PLC of Trypanosoma brucei. Journal of Cell Science 93: 233-240.

Campbell D G, Gagnon J, Reid K B M, Williams A F 1981 Rat brain Thy-1 glycoprotein: the amino acid sequence, disulphide bonds and an unusual hydrophobic region. Biochemical Journal 195: 15-30.

Cardoso de Almeida M L, Turner M J 1983 The membrane form of variant surface glycoproteins of Trypanosoma brucei. Nature (London) 302: 349-352.

Carrington M, Bulow, R, Reinke H, Overath P 1989 Sequence and expression of the GPI-PLC of Trypanosoma brucei. Molecular and Biochemical Parasitology 33: 289-296.
Cross G A M 1990 Glycolipid anchoring of plasma membrane proteins. Annual Review of Cell Biology 6.

Doering T L, Masterson W J, Englund P T, Hart G W 1989 Biosynthesis of the glycosyl-phosphatidylinositol membrane anchor of the trypanosome variant surface glycoprotein: origin of the non-acetylated glucosamine. Journal of Biological Chemistry, 264: 11168-11173.

Doering T L, Masterson W J, Hart G W, Englund P W 1990 Biosynthesis of glycosyl-phosphatidylinositol membrane anchors. Journal of Biological Chemistry 265: 611-614.

Duszenko M, Ivanov I E, Ferguson M A J, Plesken H, Cross G A M 1988 Intracellular transport of a variant surface glycoprotein in Trypanosoma brucei. Journal of Biological Chemistry 106: 77-86.

Ferguson M A J, Cross G A M 1984 Myristylation of the membrane form of Trypanosoma brucei variant surface glycoprotein. Journal of Biological Chemistry 259: 3011-3015.

Ferguson M A J, Haldar K, Cross, G A M 1985a Trypanosoma brucei variant surface glycoprotein has a sn-1,2-dimyristyl glycerol membrane anchor at its COOH-terminus. Journal of Biological Chemistry 260, 4963-4968.

Ferguson M A J, Low M G, Cross G A M 1985b Glycosyl-sn-1,2-dimyristyl phosphatidylinositol is covalently linked to Trypanosoma brucei variant surface glycoprotein. Journal of Biological Chemistry 260: 14547-14555.

Ferguson M A J, Duszenko M, Lamont G, Overath P, Cross G A M 1986 Biosynthesis of Trypanosoma brucei variant surface glycoproteins. Journal of Biological Chemistry 261: 356-362.

Ferguson M A J, Homans S W, Dwek R A, Rademacher T W 1988a Glycosyl-phosphatidylinositol moiety that anchors Trypanosoma brucei variant surface glycoprotein to the membrane. Science 239: 753-759.

Ferguson M A J, Homans S W, Dwek R A, Rademacher, T W 1988b The glycosyl-phosphatidylinositol membrane anchor of Trypanosoma brucei variant surface glycoprotein. Biochemical Society Transactions 16: 265-268.

Ferguson M A J, Homans S W 1988 Parasite glycoconjugates: towards the exploitation of their structure. Parasite Immunology 10: 465-479.

Ferguson M A J, Williams A F 1988 Cell surface anchoring of proteins via glycosyl-phosphatidylinositol structrures. Annual Reviews of Biochemistry 57: 285-320.

Futerman A H, Low M G, Ackerman K E, Sherman W R, Silman I 1985 Identification of covalently bound inositol in the hydrophobic membrane-anchoring domain of Torpedo acetylcholinesterase. Biochemical and Biophysical Research Communications 129: 312-317.

Gurnett A M, Raper J, Ward J, Turner M J 1986 Purification and characterisation of membrane form VSGs of Trypanosoma brucei. Molecular and Biochemical Parasitology 20: 1-13.

Holder A A 1983 Carbohydrate is linked through ethanolamine to the C-terminal amino acid of Trypanosoma brucei variant surface glycoprotein. Biochemical Journal 209: 261-262.

Holder A A, Cross G A M 1981 Glycopeptides from variant surface glycoproteins of Trypanosoma brucei. C-terminal location of antigenically cross reacting carbohydrate moieties. Molecular and Biochemical Parasitology 2: 135-150.

Homans S W, Dwek R A, Rademacher T W 1987 Solution conformations of N-linked oligosaccharides. Biochemistry 26: 6571-6578.

Homans S W, Ferguson M A J, Dwek R A, Rademacher T W, Anand R, Williams A F 1988 Complete structure of the glycosyl phosphatidylinositol membrane anchor of rat brain Thy-1 glycoprotein. Nature (London) 333: 269-272.

Krakow J L, Hereld D, Bangs J D, Hart G W, Englund P T 1986 Identification of a glycolipid precursor of the *Trypanosoma brucei* variant surface glycoprotein. Journal of Biological Chemistry 261: 12147-12153.

Krakow J L, Doering T L, Masterson W J, Hart G W, Englund P T 1989 A glycolipid from *Trypanosoma brucei* related to the variant surface glycoprotein membrane anchor. Molecular and Biochemical Parasitology 36, 263-270.

Lamont G S, Fox J A, Cross G A M 1987 Glycosyl-sn-1,2-dimyristylphosphatidylinositol is the membrane anchor for *Trypanosoma equiperdum* and *T.(Nannomonas) congolense* variant surface glycoproteins. Molecular and Biochemical Parasitology 24: 131-136.

De Lederkremer R M, Lima C, Ramirez M I, Thomas-Oates J E, Homans S W, Ferguson M A J 1990 Manuscript in preparation.

Low M G 1987 Biochemistry of the glycosyl-phosphatidylinositol membrane protein anchors. Biochemical Journal 244: 1-13.

Low M G 1987 Biochemistry of the glycosyl-phosphatidylinositol membrane protein anchors. Biochemical Journal 244: 1-13.

Low M G 1989 Glycosyl-phosphatidylinositol: a versatile anchor for cell surface proteins. Biochim Biophys Acta 988: 427-454.

Low M G, Zilversmit D B 1980 Role of phosphatidylinositol in attachment of alkaline phosphatase to membranes. Biochemistry 19: 3913-3918.

Low M G, Prasad A R S 1988 A phospholipase D specific for the phosphatidylinositol anchor of cell-surface proteins is abundant in plasma. Proc. Natl. Acad. Sci. USA 85: 980-984.

Masterson W J, Doering T L, Hart G W, Englund P T 1989 A novel pathway for glycan assembly: biosynthesis af the glycosyl-phosphatidylinositol anchor of the trypanosome variant surface glycoprotein. Cell 56: 793-800.

Masterson W J, Raper J, Doering T L, Hart G W, Englund P T 1990 Fatty acid remodeling: A novel reaction sequence in the biosynthesis of trypanosome GPI membrane anchors. Cell 62: in press.

Mayor S, Menon A K, Cross G A M, Ferguson M A J, Dwek R A, Rademacher T W, 1990a Glycolipid precursors for the membrane anchor of *Trypanosoma brucei* variant surface glycoproteins I.

Mayor S, Menon A K, Cross G A M, 1990b Glycolipid precursors for the membrane anchor of Trypanosoma brucei variant surface glycoproteins II.

McConville M J, Homans S W, Thomas-Oates J E, Dell A, Bacic A 1990 Structures of the Glycoinositolphospholipids from Leishmania major. Journal of Biological Chemistry 265: 7385-7394.

Menon A K, Mayor S, Ferguson M A J, Duszenko M, Cross G A M 1988 A candidate glycophospholipid precursor for the membrane anchor of Trypanosoma brucei VSG. Journal of Biological Chemistry 263: 1970-1977.

Menon A K, Schwartz R T, Mayor S, Cross G A M 1988, Cell-free synthesis of glycosyl-phosphatidylinositol precursors for the glycolipid membrane anchor of Trypanosoma brucei variant surface glycoprotein. Journal of Biological Chemistry 265, 9033-9042.

Phelps B M, Primakoff P, Koppel D E, Low M G, Mylews D G 1988 Restricted lateral diffusion of PH-20, a PI-anchored sperm membrane protein. Science 240: 1780-1782.

Previato J O, Gorin P A J, Mazurek M, Xavier M T, Fournet B, Wieruszesk J M, Mendonca-Previato L 1990, Primary structure of the oligosaccharide chain of lipopeptidophosphoglycan of epimastigote forms of Trypanosoma cruzi. Journal of Biological Chemistry 265, 2518-2526.

Roberts W L, Rosenberry T L 1985 Identification of covalently attached fatty acids in the hydrophobic membrane-binding domain of human erythrocyte acetylcholinesterase. Biochemical and Biophysical Research Communications 133: 621-627.

Roberts W L, Myher J J, Kuksis A, Low M G, Rosenberry T L 1989a Lipid analysis of the glycoinositol phospholipid membrane anchor of human erythrocyte acetylcholinesterase. Journal of Biological Chemistry 263: 18766-18775.

Roberts W L, Santikarn S, Reinhold V N, Rosenberry T L 1989 Structural characterization of the glycoinositol phospholipid membrane anchor of human erythrocyte acetylcholinesterase by fast atom bombardment mass spectrometry. Journal of Biological Chemistry 263: 18776-18784.

Robinson P J, Millrain M, Antoniou J, Simpson E, Mellor A L 1989 A glycophospholipid anchor is required for Qa-2 mediated T cell activation. Nature 342: 85-87.

Robinson P J 1990 Phosphatidylinositol membrane anchors and lymphocyte activation. Immunology Today, in press.

Ross C A, Cardoso de Almeida M L, Turner M J 1987 Variant surface glycoproteins of Trypanosoma congolense bloodstream and metacyclic forms are anchored by a glycolipid tail. Molecular and Biochemical Parasitology 22: 153-158.

Schmitz B, Klein R A, Egge H, Peter-Katalinic J 1986 A study of the membrane attachment site of the membrane form variant surface glycoprotein from Trypanosoma brucei brucei using lipid vesicles as a model of the plasma membrane. Molecular and Biochemical Parasitology 20: 191-197.

Schneider P, Ferguson M A J, McConville M J, Mehlert A, Homans S W, Bordier C 1990 Structure of the glycosyl-phosphatidylinositol membrane anchor of the Leishmania major promastigote surface protease. Journal of Biological Chemistry, in press.

Stadler J, Keenan T W, Bauer G, Gerisch G 1989 The contact site A glycoprotein of Dictyostelium discoideum carries a phospholipid anchor of a novel type. EMBO Journal 8: 371-377.

Tse A G D, Barclay N, Watts A, Williams A F 1985 A glycophospholipid tail at the carboxyl terminus of the Thy-1 glycoprotein of neurons and thymocytes. Science 230: 1003-1008.

Turco S J, Orlandi P A, Homans S W, Ferguson M A J, Dwek R A, Rademacher T W 1989 Structure of the phosphosaccharide-inositol core of the Leishmania donovani lipophosphoglycan. Journal of Biological Chemistry 264: 6711-6715.

Vliegenthart J F G, Dorland L, van Halbeek H 1983 High resolution ^1H NMR spectroscopy as a tool in the structural analysis of carbohydrates related to glycoproteins. Avances in Carbohydrate Chemistry and Biochemistry 41: 209-374.

CELL ADHESION IN MORPHOGENESIS

Jean Paul Thiery
Laboratoire de Physiopathologie du Développement CNRS
URA 1337 and Ecole Normale Supérieure 46, rue d'Ulm
75230 Paris cedex 05, France

All cells of the animal kingdom express adhesive molecules at their surfaces. Adhesive mechanisms are one of the primary processes involved in histogenesis; adhesion is also a key phenomenon in homeostasis, in wound healing, in regeneration and in numerous pathologies. Cells can interact with each other via intercellular ligands designated CAMs (Cell Adhesion Molecules). Cells also adhere to basal lamina and to extracellular matrices through the interaction of surface receptors and their ligands in the extracellular milieu; these interactive pairs are called SAMs (Substrate Adhesion Molecules). In this short article, I shall describe the cell adhesion molecules CAMs; SAMs have been reviewed recently (Hynes, 1990; Mosher, 1989; Edelman et al, 1990; Moller, 1990).

Although embryologists discovered early in this century that cells have the ability to recognize and adhere to each other to form defined tissues, it was not until the mid-seventies that the first CAM was discovered. An indirect immunologically based adhesion assay allowed the identification of cell surface polypeptides expressed by neural tissues in the chick embryo (Thiery et al, 1977). These molecules called N-CAM were found to be expressed in the nervous systems of all vertebrate embryos as well as transiently in many other tissues. N-CAM was found to be maintained in the adult mostly in nervous tissue and in muscle (Thiery et al, 1982; Crossin et al, 1985).

In the late seventies, other adhesive molecules were discovered using similar adhesion assays. The liver cell adhesion molecule (L-CAM) was isolated from embryonic chick hepatocytes (Gallin et al, 1983); related molecules were found in mammals (Takeichi, 1988). L-CAM appears very early during embryonic development on cells which also express N-CAM. Therefore, N-CAM and L-CAM were considered to be primary CAMs in contrast to other CAMs expressed only in a restricted number of cell types later in development (Edelman and Thiery, 1985). However, after gastrulation, many cells start to express either one of the two primary CAMs; N-CAM is more abundant in the neural primordium and in the axial mesoderm while L-CAM is found in developing epithelia derived from the three primary germ layers. It should be stressed, however, that cells of certain tissues will continue to express both CAMs transiently before reaching a stage at which only one of the two primary CAMs remains at the surface. More generally defined and often reiterated sequences of expression of

NATO ASI Series, Vol. H 52
Biological Signal Transduction
Edited by E. M. Ross and K. W. A. Wirtz

the two CAMs occurred during morphogenesis. The hypothesis was put forward that adhesion molecules are morphoregulators as their sequences of expression are correlated with morphogenetic processes rather than with the ligation of one defined cell type at a given stage of differentiation (Edelman, 1984).

Each embryonic cell carries several distinct adhesion systems which at least in part are utilized in different morphogenetic scenarios. During histogenesis and differentiation other adhesion molecules will complement the primary adhesion system. The hypothesis contrasts with those in which a vast repertoire of recognition molecules was considered necessary to achieve the correct assembly of cells into functional organs. This hypothesis is now largely validated by the data obtained on newly isolated adhesion molecules.

A. CALCIUM-INDEPENDENT ADHESION MOLECULES OF THE IMMUNOGLOBULIN SUPERFAMILY

1. N-CAM

N-CAM found in all vertebrates is the prototype member of adhesive molecule containing immunoglobulin (Ig)-like constant C2 domains. N-CAM includes a group of polypeptides derived from one single gene of at least 70 Kb (Cunningham et al, 1987). In their extracellular portions, all polypeptides contain five Ig-like domains and two motifs homologous to fibronectin type III repeats. However, alternative splicing of some of the 22 exons leads to different isoforms. The 140 kd polypeptides lack a 260 amino acid cytoplasmic segment present in the largest form of N-CAM (180 kd). The 120 kD polypeptide does not contain cytoplasmic and a transmembrane domains; it is anchored to the cell surface by phosphatidyl-inositol. Several exons in a cluster are specifically expressed in muscle tissues. The short sequences (37 amino acids in human) encoded by these exons may contribute to the function of a hinge region located just behind the fifth Ig-like domain. In this cluster another small exon was discovered (Gover et al, 1988) to code for a short 21 amino acid sequence followed by a step codon leading to a secreted form of N-CAM. A thirty nucleotide exon coding for 10 amino acids located in the variable region of the 4th constant Ig-like domain was also found to be alternatively spliced (Small et al, 1988). Altogether, as many as 18 different polypeptide chains are theoretically possible in the mouse (Santoni et al, 1989).

N-CAM on one cell binds directly to N-CAM on other cells via a homophilic mechanism (see Edelman 1988). Ligation between identical or different cell types occurs by homotypic and heterotypic mechanisms respectively. Although the three dimensional structure of N-CAM is not yet established, it is hypothesized that two N-CAMs interact through their last three amino-terminal Ig domains in cis or in trans (Becker et al, 1989). Studies of N-CAM inserted in lipid vesicles clearly demonstrate unusual kinetics of aggregation. The doubling of N-CAM concentration at the vesicle surface led to a 30-fold increase in the rate of aggregation suggesting that N-CAM polypeptides assembled into oligomers at the cells surface. Preliminary studies have shown that the affinity of the interaction of soluble N-CAM binding to insoluble N-CAM is of the order of 0.1 micromolar (Bock, personal communication). N-CAM may also contain binding sites for other macromolecules. A binding site for heparin has been localized within the 2nd Ig-like domain (Cole and Akeson, 1989). Such binding might occur in situations where N-CAM is released into the extracellular matrix. Heavily glycosylated forms of N-CAM have been characterized; as much as 30% of the molecular mass corresponds to carbohydrate in certain embryonic N-CAMs. Interestingly, most of the carbohydrates are polysialic acid attached to the fifth Ig-like domains; these unusual carbohydrates confer localized negative charges to N-CAM. In addition, these oligomers occupy an important partial specific volume possibly contributing to the orientation of N-CAM molecules. The oligomers are in fact located just above the hinge region previously mentioned. Highly sialylated embryonic forms of N-CAM are less adhesive than the much less glycosylated adult forms. It is notable that the sialylated forms progressively disappear from the developing nervous system possibly contributing to the strengthening of contacts between neuron cell bodies and processes (Chuong and Edelman, 1984). Interestingly, embryonic forms of N-CAM persist in the olfactory bulb which is known to generate continuously new neurons. These embryonic forms were also found in the mouse mutant staggerer which has a defective cerebellum (Edelman et al, 1983).

2. Other cell adhesion molecules of the Ig superfamily

   a) in the nervous system

Ng-CAM/L1. Several distinct transmembrane glycoproteins have been shown to participate in adhesive interactions between neurons and glia. The first one was discovered using an in vitro binding assay between a monolayer of glial cells and neurons in suspension (Grumet et al, 1983). The neuron cell adhesion molecule (Ng-CAM) in chicken is the equivalent of L1 in mammals (Moss et al, 1988); both proteins have 6 Ig-like domains and 5 fibronectin type III homologies. Mg-CAM/L1 glycoproteins have to be considered as secondary CAMs as they are expressed only on a restricted number of cell types. Ng-CAM appears during the early development of the neural primordium on post-mitotic neurons, particularly on neural processes (Thiery et al, 1985). Ng-CAM/L1 is also transiently expressed by immature Schwann cells. In vitro perturbation experiments have indicated that Ng-CAM/L1 is mostly involved in formation and maintenance of axon fascicules. In the cerebellum, the first internal granular layer cells form horizontal fibers on which newly formed neurons will adhere and migrate before descending along the Bergman glia. Antibodies to Ng-CAM or L1 applied to organotypic cerebellar slices perturb the migration of the post-mitotic external granular layer to the internal granular layer (Lindner et al, 1983; Chuong et al, 1987).

F11 glycoprotein anchored to the plasma membrane via a glycosylated phosphatidylinositol contains 6 Ig-like domains and 4 fibronectin type III repeats. This glycoprotein isolated from chick neural tissues is involved in neurite fasciculation and elongation. It is closely related to contactin which, however, has a transmembrane and a cytoplasmic domain (Brummendorf et al, 1989).

Contactin is a predominant cell surface protein of nervous tissue accounting for as much as 0.1% of total protein from adult chicken brain. It is associated with actin within a complex of 20 proteins suggesting a role in stable cell-cell contact. Perhaps contactin and F11 are produced by alternative splicing of the same primary transcript. F11 exhibits more extensive sequence identity to L1/Ng-CAM than to other members of the Ig superfamily. Interestingly there is some preliminary evidence that the FN type III homologies may promote contact between migratory growth cones and the extracellular matrix (Ranscht, 1988).

TAG-1 is a glycosylphosphatidylinositol-linked glycoprotein of the rat central and peripheral nervous systems. TAG-1 closely resembles F11 in that it has 6 Ig-like domains and four FN type III homologies. One of the type III homologies contains an RGD sequence motif known to be the primary recognition site of some integrins. TAG-1 has the interesting property of being expressed transiently on growing axons. Detailed studies of the commissural neurons of the developing spinal cord showed that TAG-1 is expressed on individual neurites elongating towards the floor plate. TAG-1 is replaced by L1 on the surface of these neurites shortly after they cross the floor plate and migrate anteriorly as nerve bundles in a perpendicular plane. TAG-1 may be involved in the initial extension of individual neurites interacting with a variety of neural epithelial cells. In vivo and in vitro TAG-1 can be released from neural cell surfaces; it can be used as a substrate for migration of sensory neurons in vitro and in vitro. It is hypothesized that TAG-1 may be released from the pioneer migratory axons; TAG-1 could then be used as a substrate for the axons that follow. The released TAG-1 molecules could also serve as competitors for TAG-1 to prevent homophilic interactions between axons thereby inhibiting fasciculation (Furley et al, 1990).

MAG (Myelin Associated Glycoprotein) is a minor component of central and peripheral myelin produced by oligodendrocytes and Schwann cells respectively. MAG is now known to be a cell adhesion molecule of the Ig superfamily with 5 Ig-like domains. MAG is one of the closest homologues to N-CAM although overall partial sequence identity does not exceed 25% (Johnson et al, 1989).

Po, one of the most abundant proteins of peripheral myelin, is a transmembrane glycoprotein containing one Ig-like domain. It may act as an homophilic adhesion molecule to maintain the multiple membrane layers in a compact form (Filbin et al, 1990).

Most interestingly, related adhesion molecules have been identified in Drosophila. This indicates that Ig domains appeared much prior to immunoglobulins which are present only in vertebrates. Sequence analyses have shown that the Ig-like domains of N-CAM are closer to constant domains in length but resemble V domains in terms of sequence. In addition, each domain is encoded by a single exon. It was postulated that an ancestral gene has given rise after duplication and divergence into C and V domains to a wide variety of molecules endowed with self-self and self-non self recognition functions (Edelman, 1987).

Fasciclin II, originally isolated from the grasshopper, is a transmembrane glycoprotein containing 5 Ig-like domains and two type III fibronectin repeats and thus closely resembles N-CAM. Fasciclin II plays a key role in the fasciculation of a subset of axons. In early development, fasciclin II is expressed in all ectodermal cells. Subsequently fasciclin II disappears from most parts of the neurectoderm but remains on the surface of some glial cells, and then is expressed in a very dynamic pattern on several newly developing neurons. Particularly interesting is the strong expression at the time axons that emerge from these neurons reach other axons on which they fasciculate. Antibodies to fasciclin II applied during this critical period prevent navigation in the correct direction as well as proper assembly with other subsets of axons (Harrelson and Goodman, 1988).

Amalgam is a protein that contains 3 Ig-like domains highly related to those of fasciclin II. However, Amalgam is expressed on all neurons of the Drosophila embryo just as for N-CAM in vertebrates (Seeger et al, 1988).

Neuroglian, discovered in Drosophila, is an integral membrane glycoprotein. The overall structure is closely related to Ng-CAM/L1 in vertebrates; neuroglian also has 6 Ig-like domains and five type III repeats of fibronectin (Bieber et al, 1989). The primary transcript of neuroglian can undergo alternative splicing leading to two membrane proteins differing in their cytoplasmic domains. The longer form is specifically expressed on the surface of neurons and some glial cells while the shorter version is more commonly expressed in the Drosophila embryo. Therefore the longer form of neuroglian which is restricted to neural tissues may be functionally related to Ng-CAM/L1 in the fasciculation mechanism, while the shorter form may carry a much wider adhesive function in different tissues of the Drosophila embryo. The additional 62 amino acids of the longer form may permit specific interaction with cytoskeletal elements as is the case for the longest form of N-CAM (Pollerberg et al, 1987). Some mutations in the neuroglian gene are lethal at the late embryonic stage, even though the pattern of fasciculation is almost normal, strongly suggesting that redundant mechanisms exist in axonal and glial adhesions.

b) in the hematopoietic system

CD4, CD8, CD2, LFA-3, I-CAM. Recently the mechanisms of T cell adhesion to target cells has been progressively elucidated. T cells adhere and recognize antigen presenting cells via their T cells receptor-associated molecules. In helper T cells, CD4 bearing 4 Ig-like domains binds to class II MHC while CD8, a homo or heterodimer containing 2 Ig-like domains, interacts with MHC class I antigens. In both cases, CD2 serves as an accessory adhesion molecule which binds to LFA-3 on the target cells; both molecules are transmembrane glycoproteins with 2 Ig-like domains. Another heterophilic interaction involves the integrin LFA-1, a heterodimeric glycoprotein which recognizes I-CAM-1 or I-CAM-2 on the target cells. This interactive pair is also involved in homotypic binding of lymphocytes. I-CAM-1 is a heavily glycosylated 90 kD transmembrane glycoprotein expressed by lymphocytes, endothelial cells and other cell types. It contains five Ig-like domains in the amino terminal region with a hinge region located after the second domain. I-CAM-2 contains only two Ig-like domains related to the two amino terminal domains of I-CAM-1 (Springer, 1990). I-CAM-1 expression can be induced by bacterial products and a variety of cytokines including IL-1, TNF, γInterferon, allowing binding of leukocytes to endothelial cells at sites of inflammation (Montavani and Dejana, 1989).

V-CAM-1, a transmembrane glycoprotein containing 6 Ig-like domains, is particularly expressed by vascular endothelial cells upon activation by a variety of cytokines. V-CAM-1 acts as a heterophilic adhesion molecule to the integrin VLA-4 ($\alpha 4\beta 1$) expressed by lymphocytes (Elices et al, 1990).

PECAM-1 (or Endo CAM; CD31) is a highly glycosylated transmembrane glycoprotein with 6 Ig-like domains. PECAM-1 is expressed on platelets, monocytes and on endothelial cells. Interestingly, PECAM-1 is localized at sites of endothelial cell-cell contacts and antibodies to PECAM-1 prevent formation of intercellular association (Newman et al, 1990; Albelda et al, 1990). There are some structural similarities between PECAM-1 and the PDGF receptor suggesting that this adhesion molecule could also transduce signals upon ligation with its receptor.

B cell antigen CD22 was recently cloned; it corresponds to a transmembrane glycoprotein with 5 Ig-like domains most closely related to MAG. Similar to the roles of CD2, CD4 and CD8 in T cell adhesion to target cells, CD22 may promote adhesion of B cells to their antigen presenting cells. CD22 may also transduce signals synergizing with the antigen receptor (Stamenkovic and Seed, 1990).

Cell CAM 105 is a transmembrane glycoprotein containing 4 Ig-like domains. It was originally discovered as an intercellular adhesion molecule in rat hepatocytes. It is a calcium-independent homophilic adhesion molecule expressed in a variety of tissues including several epithelia, vessel endothelia, granulocytes and platelets. Cell CAM 105 was recently shown to be identical to an ecto ATPase in rat liver (Aurivilliers et al, 1990). This enzyme is the homologue of human biliary glycoprotein 1 (BGP-1), which belongs to the carcinoembryonic antigen (CEA) family within the Ig-superfamily.

c) in malignant cells

I-CAM-1 is expressed in melanoma at stages that preceed the appearance of MUC18, another new member of the Ig-superfamily.

MUC18 is a transmembrane glycoprotein with 5 Ig-like domains which could possibly be considered as a marker of poor prognosis as it is expressed at stage 2 of Breslow (0.76-1.5 mm) (Lehmann et al, 1989).

The carcinoembryonic antigen (CEA) expressed on embryonic epithelia is found also in a variety of carcinomas particularly those of colon and breast. It is a highly glycosylated glycoprotein with 6 Ig-like domains. It has been shown recently to act as a calcium-independent adhesion molecule (Benchimol et al, 1989). This oncofetal antigen serves as a marker for tumor detection and recurrency. CEA belongs to a group of Ig-like domain-containing glycoproteins including the pregnancy specific b1 glycoproteins. Altogether 17 members have been found; it includes also the biliary glycoprotein (BGP) and the non-specific cross-reaction antigen (NCA). All these proteins share extensive partial sequence identity (e.g. 85% between BGP, CEA and NCA) (Khan, 1990).

d) as viral receptors

Human I-CAM-1 has been identified as a rhinovirus receptor; chimeric I-CAM-1 molecules (mouse and human) as well as site directed mutagenesis experiments have permitted the identification of the virus binding site in the most amino-terminal Ig-domain. Interestingly, I-CAM-1 which is a heterophilic adhesion molecule binds to the LFA-1 integrin to mediate cell-cell adhesion; the binding site for LFA-1 is also on the first Ig domain but is distinct for that for the virus (Staunton et al, 1990).

Another transmembrane glycoprotein containing 3 Ig-like domains was shown to bind to polio virus. CD4 is extensively studied as it is the primary receptor for HIV (human immunodeficiency virus) through its interaction with the viral envelope glycoprotein gp120 (White and Lithman, 1989).

A fibroblast growth factor receptor which has 3 Ig-like domains in the extracellular portion can serve as a receptor for the herpes simplex virus type 1 (Kaner et al, 1990). It is therefore important to consider that proteins carrying Ig-like domains can serve as growth factor receptors, intercellular adhesion molecules as well as viral receptors. Perhaps the single Ig domains of these molecules can fit best within the crevices present on the external surface of viral particles.

B. OTHER CALCIUM-INDEPENDENT ADHESION MOLECULES

Fasciclin I from Drosophila is a glycosylphosphatidylinositol glycoprotein. Fasciclin I is expressed on a subset of fasciculating central nervous system axons, all peripheral nervous system axons and some non-neural cells. Fasciclin I cDNA was transfected in the Drosophila cell line S2 which have low adhesive properties and are capable of growing in cell suspension. Fasciclin I acts as a homophilic calcium-independent CAM and sequence analysis has shown that it is unrelated to other CAMs. The fasciclin gene is located very close to the ultrabithorax complex locus. Mutations in the fasciclin I gene abolishing the expression of the protein do not alter the development of the nervous system providing further evidence for redundancy in the mechanism of selective adhesion. However, embryos doubly mutant for the fasciclin I gene and for the cytoplasmic tyrosine kinase encoded by the Abelson protooncogene homologue show extensive malformations of the central nervous system neuronal networks (Elkins et al, 1990).

Fasciclin III is a transmembrane glycoprotein of 80 kd expressed on a subset of axons as well as in several non-neural tissues in Drosophila embryo. Fasciclin III transfected into the S2 Drosophila cell line with a heat shock promotor induced S2 cell aggregation. The presence of fasciclin III expressing cells forming aggregates clearly proved a homophilic mechanism of adhesion. It was also shown that fasciclin III acts via a calcium-independent mechanism. The primary sequence of the molecule does not show any partial sequence identity with other CAMs; therefore both fasciclin I and fasciclin III define novel classes of homophilic CAMs. Fasciclin I expressing S2 cells can sort out from cells expressing fasciclin III (Elkins et al, 1990).

Neurotactin is a transmembrane glycoprotein of about 130 kD which is expressed already at the blastoderm stage; at gastrulation neurotactin is present in most cells. Subsequently neurotactin accumulates in the central nervous system but it is also found in mesodermal derivatives; the overall distribution of neurotactin is similar to that of Amalgam. Mutations on this gene do not provoke important defects in neurogenesis. The molecule is oriented opposite to most intrinsic proteins with its carboxy terminal regions in the extracellular space. It is intriguing that extracellular portion of neurotactin shares partial sequence identity with cholinesterase although it lacks the serine residue associated with the catalytic site. Neurotactin acts as an heterophilic adhesion molecule in a calcium-independent manner (De la Escalera et al, 1990); Barthalay et al, 1990 submitted).

C. CALCIUM-DEPENDENT ADHESION MOLECULES

1. Cadherins

A family of closely related transmembrane adhesive glycoproteins that function in a calcium-dependent manner has been characterized by several laboratories. These molecules named cadherins (calcium-dependent CAMs) by Takeichi (1988) have been found in several vertebrates. The extracellular portion of these molecules carries 3 repeats containing putative calcium binding sites; although the 130 kD glycoproteins share extensive partial sequence identity, they are all encoded by distinct genes. Each cadherin acts via a homophilic adhesion mechanism. The binding site has been localized on several cadherins by taking advantage of chimeric constructs and point mutations. The specificity of interaction lies within the first 113 amino-acids of the approximately 748 amino acid mature proteins (Nose et al, 1990). One of the critical recognition sites has been precisely located around amino acid 80; peptides derived from this region were able to inhibit homophilic interactions mediated by cadherins during early development (Blaschuck et al, 1990).

E-cadherin (also called uvomorulin, cell CAM120/80, Arc-1) has been isolated and characterized in mammals. It appears on fertilized eggs although it may play its first critical role during compaction at the 8-cell stage. Subsequently, E-cadherin distribution becomes restricted to epithelial structures. Particularly interesting is the fact that it is down regulated at sites of epithelial-mesenchymal cell conversion. A very similar and more detailed distribution has been established for L-CAM (chicken liver adhesion molecule) (Thiery et al, 1984). Perturbation experiments were carried out in organotypic cultures of embryonic chicken skin; the organization of feather placodes was greatly perturbed in the presence of anti L-CAM antibodies; scale-like rudiments were formed instead (Gallin et al, 1986). Work in progress in different laboratories indicates that an E-like cadherin is also present in Xenopus laevis; however its distribution may be more restricted than in mammals particularly during early development (Choi and Gumbiner, 1989; Levi, Gumbiner and Thiery, in preparation).

N-cadherin was first isolated from chicken embryo neural tissues. It appears early at the gastrulation stage particularly in newly formed mesodermal cells (Hatta et al, 1987). N-cadherin is up-regulated in cells that form transient epithelia such as somites. Indeed, antibodies to N-cadherin are particularly efficient in disrupting epithelial somites (Duband et al, 1987). The adherens junction CAM (A-CAM) originally isolated from chick cardiac intercalated disc (Volk and Geiger, 1984) is most likely identical to N-cadherin; its distribution and molecular weight are very similar to N-cadherin although the protein has not been sequenced yet (Duband et al, 1988). N-cadherin is becoming predominant in the neural plate; it is however down regulated at the surface of early emigrating neural crest cells (Thiery et al, 1990). N-cadherin may play an important role in neurite outgrowth; cells transfected with

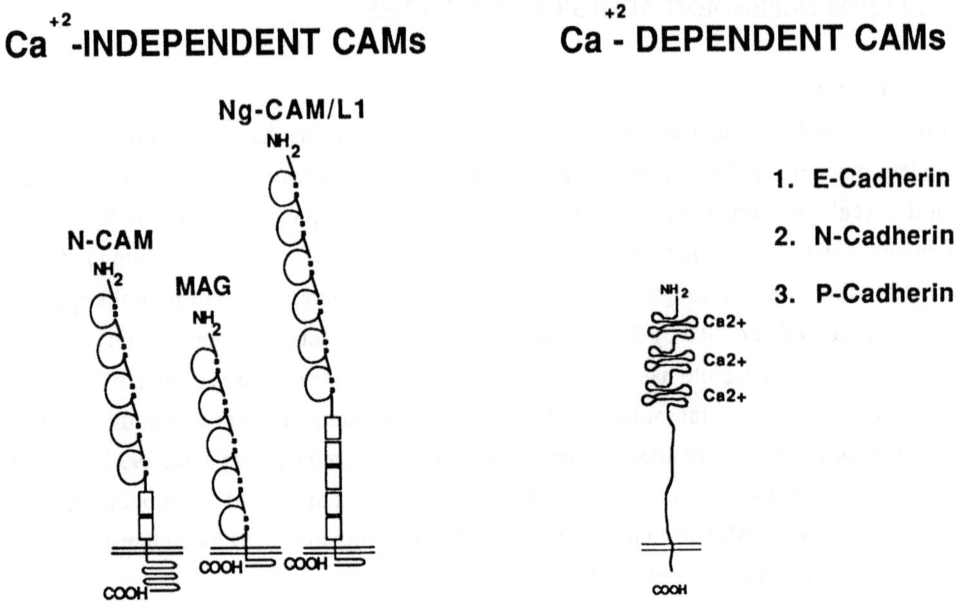
Ca+2-INDEPENDENT CAMs
Ca+2 - DEPENDENT CAMs
Ng-CAM/L1
N-CAM
MAG
NH2
COOH
1. E-Cadherin
2. N-Cadherin
3. P-Cadherin
Ca2+
Ca2+
Ca2+

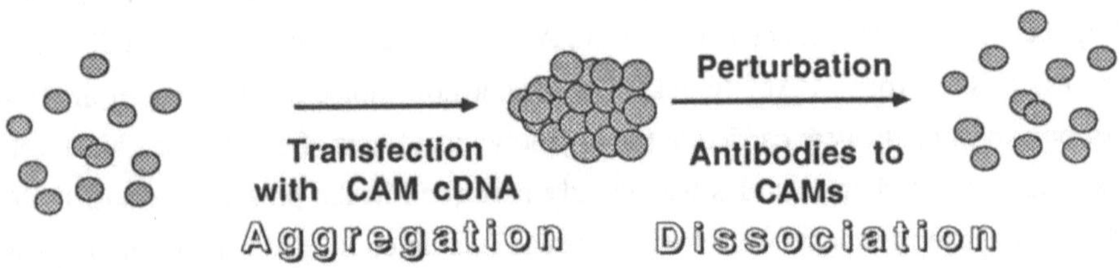
Transfection
with CAM cDNA
Aggregation
Perturbation
Antibodies to
CAMs
Dissociation

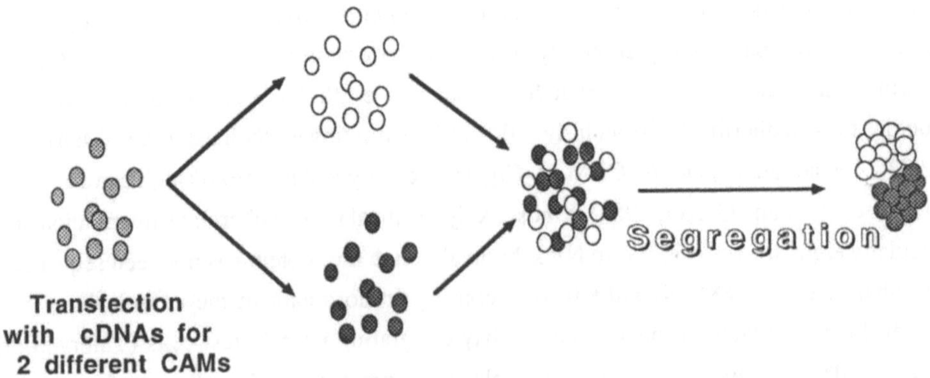
Transfection
with cDNAs for
2 different CAMs
Segregation

cDNA coding for N-cadherin can serve as a substrate for neurite elongation in vitro (Matsunuga et al, 1988; Bixby and Zhang, 1990). N-cadherin has been characterized recently in Xenopus (Detrick et al, 1990) and in the mouse (Miyatani et al, 1989); sequence comparison shows about 85% conservation in the mature proteins.

In Xenopus, N-cadherin appears 4 hours after induction in the Xenopus ectoderm from the late blastula stage. Injection on N-cadherin RNA in one blastomere at the 8 to 16 cell stage led to the expression of N-cadherin at ectopic sites particularly in ectoderm which partially detached from the underlying tissue; embryos which were able to gastrulate had severe defects in the neural tube.

P-cadherin was isolated from a partial yolk sac cell line in the mouse (Nose and Tacheichi, 1986). It is detected in extraembryonic layers, though later during development P-cadherin is expressed together with E-cadherin in the epidermis. The mesothelium which derives from the lateral plate mesoderm is also positive for P-cadherin. This cadherin was also isolated from human tissues, though it was not detected in the placenta (Shimoyama et al, 1989).

Preliminary experiments have shown that cells expressing different cadherins after transfection with appropriate cDNA will sort out from each other (Nose et al, 1988; Friedlander et al, 1989).

In addition and most importantly, sorting out was obtained when cells expressing cadherins were allowed to aggregate to cells expressing cadherin in the extracellular domain linked to N-CAM in the transmembrane and cytoplasmic domains. It was also shown that the cadherin cytoplasmic domain preferentially interacts with cytochalasin D sensitive microfilaments while the N-CAM cytoplasmic domain may be connected to microtubules. The sorting out mechanism (Jaffe et al, 1990) is therefore dependent upon the sequence of the binding site in the external domain but is also influenced by the type of connection in the cytoplasm. In addition the relative density of CAMs at the surface will modify the pattern of sorting out; high CAM expressors sort out from low expressors. Finally sorting can occur if doubly transfected cells were mixed with singly transfected cells (Friedlander et al, 1989).

More cadherins are going to be isolated in the coming years. One major immediate task is to obtain precise sequence comparisons to try to understand the degree of relatedness of these different cadherins and their precise distribution, function and regulation in the different species.

2. Other calcium-dependent adhesion molecules

Notch and Delta are transmembrane glycoproteins of Drosophila containing multiple EGF repeats involved in the mechanism controlling differentiation of ectodermal cells into neurons or epidermis. These two proteins encoded by the neurogenic loci were recently shown to interact with each other in a calcium-dependent fashion. These two proteins were also shown to interact on the same cell. Notch is expressed on neuronal processes during development and some mutations of this gene can produce serious defects in the pathfinding mechanisms in sensory axons (Fehon et al, 1990). These two molecules define a novel family of adhesion molecules characterized by EGF repeats.

D. MECHANISMS OF TRANSDUCTION OF ADHESIVE SIGNALS

One of the general features of transmembrane adhesion molecules is that their cytoplasmic domains are associated with specific proteins of the cytoskeleton. There is also accumulating evidence for modulation of second messengers during the adhesion process.

The largest form of N-CAM (180 kd) interacts specifically with fodrin; this type of interaction restricts its lateral mobility (Pollerberg et al, 1986). Cadherins also interact with cytoplasmic proteins. Three proteins of approximately 100 kd called catenins have been found to be tightly associated with cadherins (Osawa et al, 1990). It has also been postulated that spectrin can also interact with cadherins (Nelson, personal communication). Accumulation of vinculin in epithelial intermediate junctions is also accompanied by the clustering of A-CAM (Geiger 1990).

The rat pheochromocytoma cell line (PC12) expressing both N-CAM and L1 (Ng-CAM) can respond to monovalent and divalent antibodies to N-CAM and Ng-CAM by a reduction in intracellular levels of inositol phosphates, reduction of intracellular pH and an increase in intracellular calcium. It is likely that both N-CAM and Ng-CAM can be coupled to an inhibitory G protein which in turn will inhibit phospholipase C. Reduced levels of diacylglycerol will in turn decrease the activity of protein kinase C; a lower level of phosphorylation by PKC of calcium channels will induce calcium influx. The increase in calcium concentration can activate the ATP-dependent Ca^{2+}/proton antiporter leading to the acidification of the cytoplasm. Interestingly, direct PC12 cell-cell contact induced the same effects as anti N-CAM and anti Ng-CAM antibodies suggesting that adhesive processes trigger second messenger systems (Schuch et al, 1989)

E. CONCLUDING REMARKS

Research carried out by embryologists at the beginning of this century has laid the groundwork for the search for mechanisms of cell recognition and adhesion (Townes and Holtfreter, 1955). An overwhelming amount of data has been produced over the last decade showing that CAMs are morphoregulatory molecules (Edelman, Cunningham and Thiery, 1990). Most recent experiments in vertebrates as well as in insects show that CAMs directly contribute to recognition and morphogenetic events. The specificity of each CAM and its mode of action, particularly its signalling capacity to other cells, is under current investigation. Already with only a few CAMs cell sorting can occur and borders can form between tissue domains where a subset of CAMs are expressed. This situation is repeatedly found in embryogenesis.

More CAMs are certain to be discovered, and new families are already emerging, particularly in hemopoiesis. Homing receptors and addressins have been discovered to participate in the targeting of recirculating lymphocytes (Stoolman, 1989).

SAMs, the other major adhesion system also plays a key role in cell migration and more generally in morphogenetic events. In fact some integrins act as CAMs; overall perhaps several hundred of molecules may be involved in adhesive processes. One critical question is to understand the molecular basis of regulation of these different adhesion systems, and how a subset of those molecules expressed at a given time at defined sites in embryos will provide enough specificity to assemble cells progressively into tissues and functional organs. Perhaps the most difficult task of all will be to see how these molecules can contribute to the neuronal network.

ACKNOWLEDGEMENT: the author thanks Dr. Joseph Hill for reading the manuscript.

References

Albelda, S.M., Oliver, P.D., Romer, L.H. and Buck, C.A.(1990) Endo CAM: A novel endothelial cell-cell adhesion molecule. J. Cell. Biol, 110:1227-1237.

Aurivillius, M., Hansen, O.C., Lazrek, M.B.S., Bock, E. and Öbrink, B.(1990)The cell adhesion molecule Cell-CAM 105 is an ecto-ATPase and a member of the immunoglobulin superfamily. FEBS Letters.2:267-269.

Becker, J.W., Erickson, H.P., Hoffman, S., Cunningham, B.A. and Edelman, G.M.(1989). Topology of cell adhesion molecules. Proc.Natl. Acad. Sci. USA, 86 : 1088-1092.

Bieber, A.J., Snow, P.M., Hortsch, M., Patel, N.H., Jacobs, J.R., Traquina, Z.R., Schilling, J. and Goodman, C.S. (1989) Drosophila neuroglian: a member of the immunoglobulin superfamily with extensive homology to the vertebrate neural adhesion molecule L1. Cell. 59:447-460.

Benchimol, S., Fuks, A., Jothy, S., Beauchemin, N., Shirota, K. and Stanners, C.P.(1989). Carcinoembryonic antigen, a human tumor marker, functions as an intercellular adhesion molecule. Cell, 57 : 327-334.

Bixby, J.L. and Zhang, R.(1990) Purified N-cadherin is a potent substrate for the rapid induction of neurite outgrowth. J. Cell. Biol. 110: 1253-1260.

Blaschuck, O.W., Sullivan, R., David, S. and Pouliot, Y.(1990) Identification of a cadherin cell adhesion recognition sequence. Devl. Biol. 139: 227-229.

Brümmendrof, T., Wolff, J.M., Frank, R. and Rathjen, F.G.(1989). Neural cell recognition molecule F11 : homology with Fibronectin type III and Immunoglobulin type C domains. Neuron, 2 : 1351-1361.

Choi, Y.S. and Gumbiner, B.(1989) Expression of cell adhesion molecule E-cadherin in Xenopus embryos begins at gastrulation and predominates in the ectoderm. J. Cell Biol. 108: 2449-2458.

Chuong, C.M., Crossin, K.L., and Edelman, G.M.(1987). Sequential expression and differential function of multiple adhesion molecules during the formation of cerebellar cortical layers. J. Cell Biol., 104 : 331-342.

Chuong, C.-M., and Edelman, G.M.(1984). Alterations in neural cell adhesion molecules during development of different regions of the nervous system. J. Neurosci., 4 : 2354-2368.

Cole, G.J. and Akeson, R.(1989) Identification of a heparin binding domain of the neural cell adhesion molecule N-CAM using synthetic peptides. Neuron 2:1157-1165

Crossin, K.L., Chuong, C.-M., and Edelman, G.M.(1985). Expression sequences of cell adhesion molecules. Proc. Natl. Acad. Sci. USA. 82 : 6942-6946.

Cunningham,B.A., Hemperly, J.J., Murray, B.A., Prediger, E.A., Brackenbury, R., and Edelman, G.M.(1987). Neural cell adhesion molecule : structure, immunoglobulin-like domains, cell surface modulation, and alternative RNA splicing. Science, 236 : 753-884.

De la Escalera, S., Bockamp., E-O, Moya, F., Piovant, M. and Jimenez, F.(1990) Characterization and gene cloning of neurotactin, a Drosophila transmembrane protein homologous to cholinestrerases. submitted.

Detrick, R.J., Dickey, D; and Kintner, C.R.(1990) The effects of N-cadherin misexpression on morphogenesis in Xenopus embryos. Neuron. 4: 493-506.

Duband, J.L., Volberg, T., Thiery, J.P., and Geiger, B.(1988). Spatial and temporal distribution of the adherens-junction associated adhesion molecule A-CAM during avian embryogenesis. Development. 103,325-344.

Edelman, G.M., Hoffman, S., Cheng-Ming Chuong, Thiery, J.P., Brackenbury, R., Gallin, W.J., Grumet, M., Greenberg, M.E., Hemperly, J.J., Cohen, C. and Cunningham, B.A. (1983). Structure and modulation of neural cell adhesion molecules in early and late embryogenesis. Cold Spring Harbor Quantitative Symposium on Molecular Neurobiology, 48, 515-526.

Edelman, G.M.(1984). Cell adhesion and morphogenesis : The regulator hypothesis. Proc. Natl. Acad. Sci. USA, 81 : 1460-1464.

Edelman, G.M.(1987). CAMS and Igs: Cell adhesion and the evolutionary origins of immunity. Immunological Reviews, 100: 11-45.

Edelman, G.M.(1988). Topobiology an introduction to Molecular Embryology. Basic Books.

Edelman, G.M.(1988). Morphoregulatory Molecules. Biochemistry, 27, 3533-3543.

Edelman, G.M., Cunningham, B.A. and Thiery, J.P.(1990) Morphoregulatory Molecules. John Wiley and Sons.

Edelman, G.M., and Thiery, J.P.(1985) The Cell in Contact "Adhesions and Junctions as Morphogenetic Determinants. John Wiley and Sons, New York.

Elices, M.J., Osborn, L., Takada, Y., Crouse, C., Luhowskyj, S., Hemler, M.E. and Lobb, R.R. (1990) VCAM-1 on activated endothelium interacts with the leukocyte integrin VLA-4 at a site distinct from the VLA-4/ fibronectin binding site. Cell, 60: 577-584.

Elkins,T., Hortsch, M., Bieber. A.J., Snow, P.M., Goodman, C.S.(1990), Drosophila fasciclin I is a novel homophilic adhesion molecule that along with fasciclin III can mediate cell sorting. J. Cell Biol., 110 : 1825-1832.

Elkins, T., Zinn, K., McAllister,L., Hoffmann, F.M. and Goodman, C.S.(1990) Genetic analysis of a drosophila neural cell adhesion molecule:interaction of fasciclin I and Abelson tyrosine kinase mutations. Cell, 60: 565-575.

Fehon, R.G., Kooh, P.J., Rebay, I., Regan, C.L., Xu, T., Muskavitch, M.A.T., and Artavanis-Tsakonas, S.(1990). Molecular interactions between the protein products of the neurogenic loci notch and delta, two EGF-homologous genes in drosophila. Cell, 61 : 523-534.

Filbin, M.T., Walsh, F.S., Trapp, B.D, Pizzey, J.A. and Tennekoon, G.I.(1990) Role of myelin Po protein as a homophilic adhesion molecule. Nature, 344: 871-872.

Friedlander, D.R., Mege, R.M., Cunningham, B.A., and Edelman, G.M.(1989). Cell sorting-out is modulated by both the specificity and amount of different cell adhesion molecules (CAMs) expressed on cell surfaces. Proc. Natl. Acad. Sci. USA, 86: 7043-7047.

Furley, A.J., Morton, S.B., Manalo, D., Karagogeos, D., Dodd, J. and Jessel, T.M.(1990) The axonal glycoprotein TAG-1 is an immunoglobulin superfamily member with neurite outgrowth-promoting activity. Cell, 61: 157-170.

Gallin, W.J., Chuong, C.-M., Finkel, L.H., and Edelman, G.M.(1986). Antibodies to liver cell adhesion molecule perturb inductive interactions and alter feather pattern and structure. Proc. Natl. Acad. Sci. USA, 83 : 8325-8329.

Gallin, W.J., Edelman, G.M., and Cunningham, B.A.(1983). Characterization of L-CAM, a major cell adhesion molecule from embryonic liver cells. Proc. Natl. Acad. Sci. USA. 90: 1038-1042.

Geiger, B., Volberg, T., Sabalay, I. and Volk, T.(1990) A-CAM: an adherens junction-specific cell adhesion molecule. in Morphoregulatory Molecules.; Edelman, G.M. , Cunningham, B.A. and Thiery, J.P. Eds. Wiley and Sons. pp 57-79.

Gover, H.J., Barton, H., Elsom, V.L., Thompson, J., Moore, S.E., Dickson, G. and Walsh, F.S. (1988) Alternative splicing generates a secreted form of N-CAM in muscle and brain. Cell 55: 955-964.

Grumet, M., Rutishauser, U., and Edelman, G.M.(1983). Neuron-glia adhesion is inhibited by antibodies to neural determinants. Science, 222 : 60-62.

Harrelson, A.L. and Goodman, C.S.(1988). Growth cone guidance in insects : Fasciclin II is a member of the Immunoglobulin superfamily. Science, 242 : 700-708.

Hatta, K., Takagi, S., Fujisawa, H. and Takeichi, M.(1987) Spatial and temporal expression pattern of N-cadherin cell adhesion molecules correlated with morphogenetic processes of chicken embryos. Devl. Biol. 120: 215-227

Hemler, M.E.(1990) VLA proteins in the integrin family: structures, functions and their role in leukocytes. Ann. Rev. Immunol. 8: 365-400.

Hynes, R.O.(1990) Fibronectins. Springer Verlag New York

Jaffe, S.H., Friedlander, D.R., Matsuzaki, F., Crossin, K.L., Cunningham, B.A. and Edelman, G.M.(1990) Differential effects of the cytoplasmic domains of cell adhesion molecules on cell aggregation and sorting out. Proc. Natl. Acad. Sci. USA. 87: 3589-3593.

Johnson, P.W., Abramow-Newerly, W., Seilheimer, B., Sadoul, R., Tropak, M.B., Arquint, M., Dunn, R.J., Schachner, M. and Roder, J.C.(1989)Recombinant myelin-associated glycoprotein confers neural adhesion and neurite outgrowth function. Neuron, 3: 377-385.

Kaner, R.J., Baird, A., Mansukhani, A., Basilico, C;, Summers, B.D., Florkiewicz, R.Z. and Hajjar, D.P.(1990) Fibroblast growth factor receptor is a portal of cellular entry for Herpes simplex virus type 1. Science. 248: 1410-1413.

Kemler, R., Gossler, A., Mansouri, A. and Vestweber, D.(1990) The cell adhesion molecule Uvomorulin. in Morphoregulatory Molecules.; Edelman, G.M. , Cunningham, B.A. and Thiery, J.P. Eds. Wiley and Sons. pp 41-56.

Khan, W.N.(1990) The carcinoembryonic antigen gene family. Molecular analysis of the pregnancy-specific b1 glycoprotein subgroup. Thesis of the University of Umea: Department of Immunology.

Lehmann, J.M., Riethmüller, G. and Johnson, J.P.(1989) MUC18, a marker of tumor progression in human melanoma, shows sequence similarity to the neural cell adhesion molecules of the immunoglobulin superfamily. Proc. Natl.Acad. Sci. USA. 86:9891-9895.

Lindner, J., Rathjen, F.G., and Schachner, M.(1983). L1 mono- and polyclonal antibodies modify cell migration in early postnatal mouse cerebellum. Nature, 305 : 427-430.

Mantovani, A. and Dejana, E.(1989) Cytokines as a communication signals between leukocytes and endothelial cells. Immunology today. 10: 370-374.

Matsunaga, M., Hatta, K., Nagafuchi, A., and Takeichi, M.(1988). Guidance of optic nerve fibres by N-cadherin adhesion molecules. Nature, 334: 62-64.

Miyatani, S., Shimamura, K., Nagafuchi, A., Nose, A., Matsunaga, M., Hatta, K. and Takeichi, M.(1989) Neural cadherin: role in selective cell-cell adhesion. Science. 245: 631-635.

Moos, M., Tacke, R., Scherer, H., Teplow, D., Früh, K., and Schachner, M.(1988). Neural adhesion molecule L1 as a member of the immunoglobin superfamily with binding domains similar to fibronectin. Nature, 334: 701-703.

Möller, G.(1990) Adhesion molecules. Immunological Rev. 114. Munksgaard Copenhagen.

Mosher, D. F.(1990). Fibronectin. Academic press, Inc.

Newman, P. J., Berndt, M.C., Gorski, J., Whitell, G.C., Lyman, S., Paddock, C. and Muller, W.A.(1990) PECAM-1 (CD31) cloning and relation to adhesion molecules of the immunoglobulin gene superfamily. Science, 247: 1219-1222;

Nose, A., Nagafuchi, A., and Takeichi, M.(1988). Expressed recombinant cadherins mediate cell sorting in model systems. Cell, 54: 993-1001.

Nose, A., and Takeichi, M.(1986).A novel cadherin cell adhesion molecule :its expression patterns associated with implantation and organogenesis of mouse embryos. J. Cell Biol., 103 : 2649-2658.

Nose, A, Tsuji, K. and Takeichi, M.(1990). Localization of specificity determining sites in cadherin cell adhesion molecules. Cell, 61: 147-155.

Ozawa, M., Ringwald, M. and Kemler, R.(1990) Uvomorulin-catenin complex formation is regulated by a specific domain in the cytoplasmic region of cell adhesion molecule. Proc. Natl. Acad. Sci.USA. 87: 4246-4250.

Pollerberg, G.E., Schachner, M., and Davoust, J.(1986). Differentiation state-dependent surface mobilities of two forms of the neural cell adhesion molecule. Nature, 324 : 462-465.

Ranscht, B.(1988) Sequence of contactin, a 130kD-glycoprotein concentrated in areas of interneuronal contact, defines a new member of the immunoglobulin supergene family in the nervous system. J.Cell. Biol. 107: 1561-1573.

Seeger, M., Haffley, L. and Kaufman, T.C.(1988) Characterization of Amalgam: a member of the immunoglobulin superfamily from drosophila. Cell. 55: 589-600.

Santoni, M.J., Barthels, D., Vopper, G., Boned, A., Goridis, C. and Wille, W.(1989). Differential exon usage involving an unusual splicing mechanisms generates at least eight types of NCAM cDNA in mouse brain. The EMBO Journal, 8 : 385-392.

Schuch, U., Lohse, M.J. and Schachner, M.(1989) Neural cell adhesion molecules influence second messenger systems. Neuron, 3: 13-20.

Shimoyama, Y., Yoshida, T., Terada, M., Shimosato, Y., Abe, O., and Hirohashi, S.(1989). Molecular cloning of a human Ca^{2+} -dependent cell-cell adhesion molecule homologous to mouse placental cadherin : its low expression in human placental tissues. J. Cell Biol., 109: 1787-1794.

Small, S.J., Haines, S.L., Akeson, R.A.(1988). Polypeptide variation in an N-CAM extracellular immunoglobulin-like fold is developmentally regulated through alternative splicing. Neuron, 1 : 1007-1017.

Springer, T.A.(1990). Adhesion receptors of the immune system. Nature in press.

Stamenkovic, I. and Seed, B. (1990) The B-cell antigen CD22 mediates mocyte and erythrocyte adhesion. Nature, 345: 74-77.

Staunton, D.E., Dustin, M.L., Erickson, H.P. and Springer, T.A.(1990) The arrangement of the immunoglobulin-like domains of I-CAM-1 and the binding sites for LFA-1 and rhinovirus. Cell, 61: 243-254.

Stoolman, L.M.(1989). Adhesion molecules controlling lymphocyte migration. Cell, 56: 907-910.

Takeichi, M.(1988). The cadherins: cell-cell adhesion molecules controlling animal morphogenesis. Development, 102: 639-655.

Thiery, J.P., Brackenbury, R., Rutishauser, U. and Edelman, G.M.(1977). Adhesion among neural cells of the chick embryo. II. Purification and characterization of a cell adhesion molecule from neural retina. J. Biol. Chem. 252, 6841-6845.

Thiery, J.P., Duband, J.L., Rutishauser, U. and Edelman, G.M.(1982) Cell adhesion molecules in early chicken embryogenesis. Proc. Natl. Acad. Sci.,USA, 79: 6737-6741.

Thiery, J.P., Delouvée, A., Grumet, M. and Edelman, G.(1985). Initial appearance and regional distribution of the neuron-glia cell adhesion molecule (Ng-CAM) in the chick embryo. J. Cell Biol. 100, 442-456.

Thiery, J.P., Delouvée, A., Gallin, W., Cunningham, B.A. and Edelman, G.M.(1984). Ontogenetic expression of cell adhesion molecules: L-CAM is found in epithelia derived from the three primary germ layers. Dev. Biol. 102: 61-78.

Thiery, J.P., Duband, J.-L., Dufour, S.(1990). Adhesion systems in morphogenesis and cell migration during avian embryogenesis. In "Morphoregulatory Molecules" , Edelman, G.M., Cunningham, B.A.and Thiery, J.P. eds., John Wiley and Sons, New York, pp 597-625.

Townes, P.L. and Holtfreter, J.(1955) Directed movements and selective adhesion of embryonic amphibian cells. J. Exp. Zool. 128: 53-120.

White, J.M. and Littman, D.N.(1989) Viral receptors of the immunoglobulin superfamily. Cell. 56: 725-728.

INOSITOL PHOSPHATE METABOLISM AND FUNCTION

R.F. Irvine
Department of Biochemistry
AFRC Institute of Animal Physiology and Genetics Research
Babraham
Cambridge CB2 4AT
U.K.

Introduction

Ten years can be a long time in science, and it is certainly bewildering to compare the proposed inositol phosphate metabolism occurring as a result of stimulated inositide turnover in a review written in 1980 (Irvine et al. 1982), with that in a review from 1989 (Berridge & Irvine, 1989); even the latter is incomplete, as will be discussed below. This extraordinary proliferation of inositol phosphates - nearly one half of the 63 possible isomers of myoinositol phosphate have now been described as occurring naturally - illustrates to what extent evolution has played the variations on the inositol phosphate theme. The really taxing question is, just how many of these compounds have specific and precise intra (or extra) cellular functions? And how many are either anabolic or catabolic intermediates, or (as a sort of half-way house), perhaps are key intermediates on pathways and provide feedback inhibition or cross-talk between pathways?

This short review will address this question taking the simplistic approach used previously (Irvine et al. 1988), namely, that only a very few inositol phosphates actually have functions (i.e. have their presence registered by proteins which specifically bind them and which then do something once they have bound them). At the risk of over-simplifying, I will suggest that only Ins(1,4,5)$\underline{P}_3$ and Ins(1,3,4,5)$\underline{P}_4$ are acute second messengers, and only Ins$\underline{P}_6$ definitely possesses an intracellular function with possibly Ins(1,3,4,5,6)$\underline{P}_5$ doing something else in some cells. These will be considered in turn.

NATO ASI Series, Vol. H 52
Biological Signal Transduction
Edited by E. M. Ross and K. W. A. Wirtz

$Ins(1,4,5)P_3$

This is now so well-accepted as a second messenger with a well-defined action on a known receptor (Ferris et al. 1989; Furuichi et al. 1989; Mignery et al. 1989), that little space need be devoted to it in this context. Only two aspects will be discussed briefly. Firstly, recent evidence (summarized in Irvine, 1990b) has now begun to place $Ins(1,4,5)P_3$ as a second messenger in higher plants on a very much firmer basis than before. Red beet vacuoles appear to have an $InsP_3$ receptor similar to that in animals (Alexandre et al. 1990), and "caged" $Ins(1,4,5)P_3$ released in stomatal cells causes Ca^{2+} release and stomatal closure, but, as discussed (Irvine 1990b), the longer-term role of $Ins(1,4,5)P_3$ in plants is uncertain.

Secondly, there has been considerable interest taken recently in the strikingly unusual way in which $Ins(1,4,5)P_3$ releases Ca^{2+}] from intracellular stores. This has in fact been known (or at least with hindsight, the data have been staring us in the face) since the initial observations of $Ins(1,4,5)P_3$ as a second messenger; but it was Muallem et al. (1989) who must take the credit for observing that a sub-optimal dose of $Ins(1,4,5)P_3$ does not release all of the Ca^{2+} from the Ca^{2+} stores more slowly, but in fact releases a portion of it extremely quickly. The remarkable steepness of this release and the apparent "desensitization" was highlighted by recent data from Meyer & Stryer (1990). Taylor & Potter (1989) also documented this phenomenon, but in the absence of Ca^{2+} pumping (i.e. they removed the ATP from permeablised cells immediately before initiating the release of Ca^{2+}), and their data emphasise that the abrupt cessation of Ca^{2+} efflux once a certain amount of Ca^{2+} is released, is not dependent on the re-entry of Ca^{2+} into the pools.

Furthermore, as both (a) Muallem et al. (1989) showed by removing the $Ins(1,4,5)P_3$ after the Ca^{2+} had been released (in fact they let the cells' phosphatases remove it by omitting to protect it with the inhibitor 2,3bisphosphoglycerate) and (b) Meyer & Stryer (1990) showed by adding heparin to compete out the

Ins(1,4,5)$\underline{P}_3$, if the Ins(1,4,5)$\underline{P}_3$ receptor is truly "desensitized" by removing the Ins(1,4,5)$\underline{P}_3$, a re-uptake of Ca^{2+} into the stores occurs. This shows that as long as Ins(1,4,5)$\underline{P}_3$ is present, the receptor does not fully desensitize, but rather, it maintains a partially active state so that a new steady state is reached. However, even in this partially active state the half-life of Ca^{2+} in the Ins(1,4,5)$\underline{P}_3$-sensitive pool should be lower than if no Ins(1,4,5)$\underline{P}_3$ is present, and the data of Taylor & Potter (1989) do not reveal any such phenomenon, suggesting that if the Ca^{2+} in the pool continues to decrease (by leakage), then the Ins(11,4,5)$\underline{P}_3$-sensitive channels shut down altogether.

All these observations put together infer that it is the intraluminal Ca^{2+} which is limiting the ability of Ins(1,4,5)$\underline{P}_3$ to control Ca^{2+} efflux (i.e. as the intraluminal Ca^{2+} decreases it begins to "desensitize" the Ins(1,4,5)$\underline{P}_3$ receptor), and this in turn leads to the idea (Irvine 1990a) of a Ca^{2+}-binding site, with a low affinity, on the intraluminal surface of the Ins(1,4,5)$\underline{P}_3$ receptor, with a positive co-operativity between the binding site on the receptor and the binding site for Ins(1,4,5)$\underline{P}_3$. Computer-modelling of such a mechanism on a tetrameric receptor (R. Tregear & R.F.I. unpublished) has shown that phenomena similar to those of Muallem et al. (1989), Meyer & Styer (1990) and Taylor & Potter (1989) can be generated under particular defined conditions, though this does not, of course, prove that the model is exactly accurate.

Ins(1,3,4,5)P_4

If the model of Ins(1,4,5)$\underline{P}_3$ receptor action discussed above is correct, then it could help to shed light on the mysterious action of Ins(1,3,4,5)$\underline{P}_4$ (Irvine, 1990a). The complexities of the various reports of the actions of this compound have been discussed elsewhere (Irvine 1990 a,c; Berridge & Irvine 1989; Irvine et al. 1988; Cullen et al. 1990) and do not need re-iterating here. Suffice to say that our current picture of Ins(1,3,4,5)$\underline{P}_4$'s actions can be summarized by saying that we think all tissues possess aspects of their structure that are essentially similar to one part of skeletal muscle; in this tissue the dihydropyridine receptor in the plasma membrane and the ryanodine receptor in the sarcoplasmic reticulum interact to control intracellular Ca^{2+}, and I have suggested (Irvine 1990a) that the ins(1,3,4,5)$\underline{P}_4$ receptor and Ins(1,4,5)$\underline{P}_3$ receptor interact in a way analogous to these two muscle receptors respectively. A modulation of Ins(1,4,5)$\underline{P}_3$ receptor structure by Ca^{2+} levels inside the endoplasmic reticulum (see above) would explain the mysterious phenomenon whereby Ca^{2+} pools control Ca^{2+} entry (Putney 1986), and the idea of receptors for the two inositol phosphate interacting would explain their marked synergism in controlling both Ca^{2+} entry and mobilization (Irvine 1990a; Cullen et al. 1990).

The most obvious prediction of this set of ideas is that the receptor for Ins(1,3,4,5)$\underline{P}_4$ should (i) be localized in the plasma membrane (ii) resemble (be partially homologous with) the dihydropyridine receptor. We await further details on that particular protein with great anticipation, as if these predictions are borne out, it will help to clarify and strengthen the role of Ins(1,3,4,5)$\underline{P}_4$ as the second messenger controlling Ca^{2+} entry (Irvine et al. 1988; Irvine, 1990a,c).

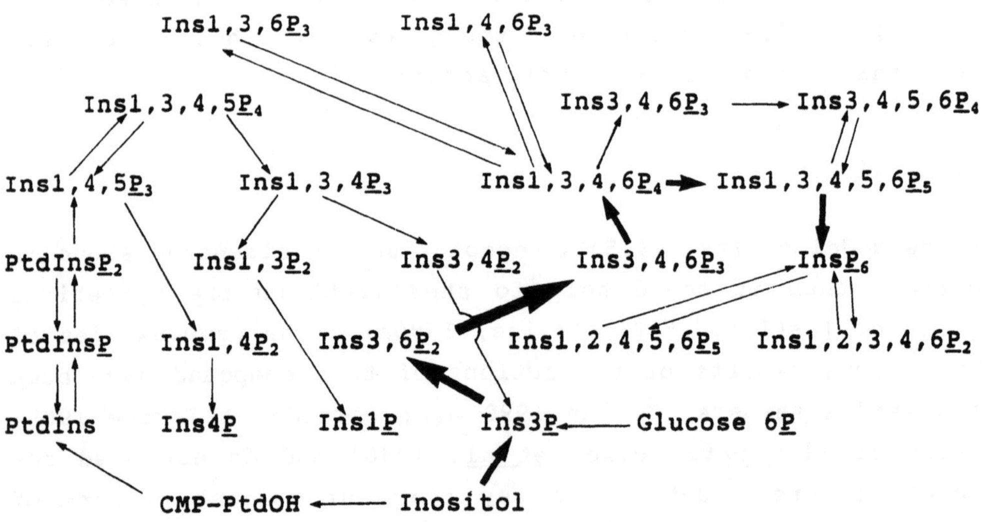

Figure 1 Inositol phosphate metabolism in eukaryotic cells

Ins(1,3,4,5,6)P_5 and InsP_6

If Ins(1,4,5)$\underline{P}_3$ and Ins(1,3,4,5)$\underline{P}_4$ are both second messengers, which other InsP's have functions? Probably only Ins(1,3,4,5,6)$\underline{P}_5$ and Ins$\underline{P}_6$. We know Ins(1,3,4,5,6)$\underline{P}_5$ has the discrete function of modulating O_2/haemoglobin interactions in avian and amphibian erythrocytes. Ins$\underline{P}_6$ has usually been assumed to be a store of phosphate (see Stephens & Irvine 1990 for references), but it could just as well be an antioxidant (Graff et al. 1987). There is also the enticing idea that both of them act as neurotransmitters (Vallejo et al. 1987), but even if true that does not tell us of their intracellular functions.

One problem up until recently is that we have not known how Ins$\underline{P}_6$ is synthesized - can it only be derived from Ins(1,4,5)$\underline{P}_3$? Can it only be synthesized from Ins(1,3,4,5,6)$\underline{P}_5$? Is there another route (avoiding Ins(1,4,5)$\underline{P}_3$) from inositol to these two higher inositol phosphates? At least for slime moulds we now have the answer to these questions, (Stephens & Irvine 1990), and Figure 1 summarizes not only this pathway, but also the complex interactions of inositol phosphates derived from Ins(1,4,5)$\underline{P}_3$ in avian erythrocytes (derived mostly from the studies of Stephens, Hawkins and co-workers - see Stephens & Downes, 1990 for most of the references).

Perhaps the most intriguing aspect of this complexity (apart from the scale of it!) is the apparent "futile" cycling which may be going on at various places. Note, in particular, that those cycles involving the 3-phosphates of Ins(1,3,4,6)$\underline{P}_4$ or Ins$\underline{P}_6$ are probably the most active; is there are specific function associated with the cycling of this particular grouping? And why, when there is a perfectly good kinase linking Ins(1,3,4,6)$\underline{P}_4$ to Ins(1,3,4,5,6)$\underline{P}_5$, do cells go round the houses via Ins(3,4,5,6)$\underline{P}_4$-kinase, the most active Ins$\underline{P}_4$ kinase (Stephens et al. 1988) in cells? Ins(3,4,5,6)$\underline{P}_4$ is an inhibitor of Ins(1,3,4)$\underline{P}_3$-6-kinase (Hughes et al. 1989), and it may be that this branch of metabolism serves to create cross-talk between receptor-stimulated and receptor-independent phosphate metabolism.

These areas are at present where the blackest boxes lie in inositol phosphate metabolism and function, and they highlight how desperately ignorant we still are. If Ins(1,3,4,5,6)$\underline{P}_5$ is the obligate precursor of Ins$\underline{P}_6$ in all organisms as it is in slime moulds (Stephens & Irvine 1990), then completely independent functions of these two might be tricky for the cell to manage (note the close interaction of functions of Ins(1,4,5)$\underline{P}_3$ and Ins(1,3,4,5)$\underline{P}_4$, one the obligate precursor of the other). Fig. 1 also highlights the careful nature of Ins(1,3,4,6)$\underline{P}_4$ as a metabolic intermediate - is this its only function?

It may be a long while before we get answers to these sort of questions, but it is certainly going to be fun trying.

References

Alexander J, Lassalles JP, Kado RT (1990) Opening of Ca^{2+} channels in isolated red beet root vacuole membrane by inositol 1,4,5-trisphosphate. Nature 343: 567-569.

Berridge MJ, Irvine RF (1989) Inositol phosphates and cell signalling. Nature 341: 197-205.

Cullen PJ, Irvine RF, Dawson AP (1990) Synergistic control of Ca^{2+} mobilisation in permeabilised mouse L-1210 lymphoma cells by inositol 2,4,5-trisphosphate and inositol 1,3,4,5-tetrakisphosphate. Biochem J (in press).

Ferris CD, Huganis R, Supattapone S, Snyder SH (1989 Purified inositol 1,4,5-trisphosphate receptor mediates calcium flux in reconstituted lipid vesicles. Nature 342: 87-89.

Furuichi T, Yoshikawa S, Miyawaki A, Wada K, Maeda N, Mikoshiba K (1989) Primary structure and functional expression of the inositol 1,4,5-trisphosphate-binding protein P_{400}. Nature 342: 32-38.

Graf E, Empson KL, Eaton JW (1987) Phytic acid. A natural antioxidant. J Biol Chem 262: 11647-11650.

Hughes PJ, Hughes AR, Putney JW Jnr, Shears SB (1989) The regulation of the phosphorylation of inositol 1,3,4-trisphosphate in cell-free preparations and its relevance to the formation of inositol 1,3,4,6-tetrakisphosphate in agonist-stimulated rat parotid acinar cells. J Biol Chem 264: 19871-19878.

Irvine RF (1990a) "Quantal" Ca^{2+} release and the control of Ca^{2+} entry by inositol phosphates - a possible mechanism. FEBS Letters 263: 5-9.

Irvine RF (1990b) Messenger gets the green light. Nature (in press).

Irvine RF (1990c) Inositol 1,4,5-trisphosphate and Inositol 1,3,4,5-tetrakisphosphate: Two second messengers in a duet of calcium regulation. In: Vanderhoek JY (ed). Biology of Cellular Transducing Signals. Plenum Press, New York pp. 213-225.

Irvine RF, Freinkel N, Dawson RMC (1982) Stimulated phosphatidylinositol turnover. A brief appraisal. In: Freinkel N (ed). Contemporary Metabolism Vol. 2. Plenum, New York and London, pp. 301-342.

Irvine RF, Moor RM, Pollock WK, Smith PM, Wreggett KA (1988) Inositol phosphates: proliferation, metabolism and function. Phil Trans Roy Soc Lond B 320: 281-298.

Meyer T, Stryer L (1990) Transient calcium release induced by successive increments of inositol 1,4,5-trisphosphate. Proc Nat Acad Sci USA 87: 3841-3845.

Mignery GA, Südhof TC, Takei K, De Camilli P (1989) Putative receptor for inositol 1,4,5-trisphosphate similar to ryanodine receptor. Nature 342: 192-195.

Muallem S, Pandol SJ, Beeker TG (1989) Hormone-evoked calcium release from intracellular stores is a quantal process. J Biol Chem 264: 205-212.

Putney JW Jnr (1986) A model for receptor-regulated Ca^{2+} entry. Cell Calcium 7: 1-12.

Stephens L, Downes CP (1990) Product-precursor relationships amongst inositol polyphosphates. Biochem J 265: 435-452.

Stephens LR, Hawkins PT, Morris AJ, Downes CP (1989) L-myoinositol 1,4,5,6-tetrakisphosphate-3-hydroxykinase. Biochem J 249: 283-292.

Taylor CW, Potter BVL (1990) The size of inositol 1,4,5-trisphosphate-sensitive Ca^{2+} stores depends on inositol 1,4,5-trisphosphate concentration. Biochem J: 189-194.

INOSITOL 1,4,5-TRISPHOSPHATE AND INOSITOL 1,3,4,5-TETRAKISPHOSPHATE: FUNCTIONS, METABOLISM AND INTERCONVERSION

M Teresa Estrada-Garcia, Andrew Craxton, Christopher J Kirk and Robert H Michell
School of Biochemistry
The University of Birmingham
P O Box 363
Birmingham B15 2TT
UK

Introduction

This paper discusses the metabolism of Ins(1,4,5)P_3 and Ins(1,3,4,5)P_3, with particular emphasis on the possible contribution of Ins(1,3,4,5)P_4 3-phosphatase to a cyclic interconversion of Ins(1,4,5)P_3 and Ins(1,3,4,5)P_4. The metabolism of the inositol phosphates produced in response to receptor activation, including the characteristics of Ins(1,4,5)P_3 3-kinase and of Ins(1,4,5)P_3/Ins(1,3,4,5)P_4 5-phosphatase, was recently reviewed in detail by Shears (1989). Current knowledge of the physiological role of these two inositol phosphates will also be outlined: for more extensive reviews, see Berridge (1987) and Berridge and Irvine (1989).

Second messenger roles of Ins(1,4,5)P_3 and Ins(1,3,4,5)P_4

Diverse receptors activate phosphoinositidase C, which catalyses the hydrolysis of phosphatidylinositol 4,5-bisphosphate, a quantitatively minor membrane phospholipid (Michell et al., 1981). This reaction leads to the formation of two second messengers; diacylglycerol, an activator of protein kinase C (Takai et al., 1979), and Ins(1,4,5)P_3, which mobilises calcium from a non-mitochondrial intracellular pool (Streb et al., 1983). In intensely stimulated cells, the limited size of the intracellular Ca^{2+} pool means that this is often a short-lived phenomenon (1 min or less). Only a proportion (30-50%) of the Ca^{2+} stored in non-mitochondrial pools is Ins(1,4,5)P_3-responsive (Streb et al., 1983; Taylor and Putney, 1985). In order to release Ca^{2+}, Ins(1,4,5)P_3

NATO ASI Series, Vol. H 52
Biological Signal Transduction
Edited by E. M. Ross and K. W. A. Wirtz

acts through a specific intracellular membrane receptor which is a transmembrane glycoprotein with the Ins(1,4,5)P_3 binding site in its cytoplasmic domain (Furuichi et al., 1989; Mignery et al., 1990).

Receptor-stimulated Ins(1,4,5)P_3 production has now been shown to mediate a variety of cellular processes. When Ins(1,4,5)P_3 is added to permeabilized cells or injected into intact cells it can trigger a number of the complex physiological processes normally evoked by receptor activation: for example, contraction, protein phosphorylation, modulation of the light response of photoreceptors and the cortical reaction of Xenopus eggs (Berridge, 1987).

In addition to mobilizing internal Ca^{2+} through Ins(1,4,5)P_3, many agonists can promote an influx of extracellular Ca^{2+}, and this Ca^{2+} entry may also be controlled by second messenger-operated channels (Pozzan and Meldolesi, 1987). Ins(1,4,5)P_3 has been implicated in controlling the influx of external calcium in some cells (Putney, 1986; Kuno and Gardner, 1987), whilst Ins(1,3,4,5)P_4, in the presence of Ins(1,4,5)P_3, has been implicated in others (Irvine et al., 1988; Irvine, 1990) discussed below.

Recently it has been reported that Ins(1,3,4,5)P_4 can stimulate Ca^{2+} release from a bovine adrenal microsome preparation by a mechanism independent of the Ins(1,4,5)P_3 receptor, although this response was quantitatively smaller than that to Ins(1,4,5)P_3 (Ely et al., 1990). It is possible that this Ca^{2+} release is mediated by a specific Ins(1,3,4,5)P_4 receptor/ion channel, since specific binding sites for Ins(1,3,4,5)P_4 have been identified in several systems (e.g. Bradford and Irvine, 1987; Donie and Reiser, 1989; Theibert et al., 1990). It therefore appears that Ins(1,3,4,5)P_4 may have a specific messenger function during hormone-induced Ca^{2+} signalling.

As mentioned above, Ins(1,3,4,5)P_4 in combination with Ins(1,4,5)P_3 has been implicated in intracellular Ca^{2+} regulation. For example, Ins(1,3,4,5)P_4 and Ins(1,4,5)P_3 have been shown to act synergistically to regulate Ca^{2+} levels within mouse lacrimal gland acinar cells (Morris et al., 1987) and pituitary membrane preparations (Spat et al., 1987). Possible mechanisms for this role of Ins(1,3,4,5)P_4 in controlling Ca^{2+} levels include: the control of Ca^{2+} transfer between Ins(1,4,5)P_3-sensitive and Ins(1,4,5)P_3-insensitive pools (Berridge and Irvine, 1989); or interaction of the Ins(1,4,5)P_3 receptor with a protein in the plasma membrane which controls Ca^{2+} entry, and which may be the Ins(1,3,4,5)P_4 receptor (Irvine, 1990). Ins(1,4,5)P_3 and Ins(1,3,4,5)P_4 may also interact in the regulation of the calcium oscillations observed in cells when the phosphoinositide pathway is stimulated (Berridge et al., 1988). These oscillations appear to constitute a frequency-encoded Ca^{2+} signalling system in which the conversion of Ins(1,4,5)P_3 to Ins(1,3,4,5)P_4 might act as a feed-forward signal to promote further calcium mobilization (Berridge and Irvine, 1989).

Ins(1,4,5)P_3 and Ins(1,3,4,5)P_3 therefore provide the cell with an exquisitely controlled mechanism regulating its internal Ca^{2+} levels, and determination of the possible metabolic pathways for interconversation of these compounds is of great importance.

Synthesis and metabolism of Ins(1,4,5)P_3 and Ins(1,3,4,5)P_4

Ins(1,4,5)P_3 is released into the cytosol following the hydrolysis of phosphatidylinositol 4,5-bisphosphate by the receptor-activated phosphoinositidase C. In most cells, some of this Ins(1,4,5)P_3 is then metabolised by a 3-kinase to yield Ins(1,3,4,5)P_4 (Batty et al., 1985).

Ins(1,4,5)P_3 and Ins(1,3,4,5)P_4 are both dephosphorylated by the same 5-phosphatase, producing Ins(1,4)P_2 (Downes et al., 1982) and Ins(1,3,4)P_3 (Batty et al., 1985). These 5-dephosphorylation reactions provide a mechanism for terminating the signalling functions of these compounds, since

$Ins(1,4)P_2$ does not mobilize Ca^{2+} (Berridge, 1987) and $Ins(1,3,4)P_3$ is either inactive or a very weak Ca^{2+} mobilizer (Irvine, 1989): neither has been convincingly implicated in calcium regulation (Berridge and Irvine, 1989).

However, the pathways for hydrolysis of $Ins(1,3,4,5)P_4$ are more complex that originally believed, since this compound can also be attacked by a recently described 3-phosphatase activity, regenerating $Ins(1,4,5)P_3$ (Cunha-Melo, 1988; Dean and Moyer, 1988; Doughney et al., 1988; Hoer et al., 1988). This $Ins(1,3,4,5)P_4$ 3-phosphatase activity could potentially provide an additional and unexpected source of $Ins(1,4,5)P_3$, so prolonging its signalling functions.

$Ins(1,4,5)P_3$ 3-kinase

$Ins(1,4,5)P_3$ 3-kinase activity, which catalyses the formation of $Ins(1,3,4,5)P_4$, was first described in rat cortical slices following carbacol stimulation of muscarinic receptors (Batty et al., 1985). 3-kinase activity has since been detected in a variety of cells. It is predominantly soluble, is highly specific for $Ins(1,4,5)P_3$ with a K_m in the range 0.2-1.5 μM, and is stimulated by physiological increases in intracellular Ca^{2+} concentration in a calmodulin-dependent manner (Shears, 1989).

In an elegant series of experiments it has recently been demonstrated that phosphorylation by cAMP-dependent protein kinase (PKA) enhanced the 3-kinase activity, whereas phosphorylation by protein kinase C (PKC) reduced the activity of a purified rat brain 3-kinase (Sim et al., 1990). PKC and PKA have also been implicated in the regulation of the 3-kinase in other cells (Biden et al., 1988a,b; Imboden and Pattison, 1987). The $Ins(1,4,5)P_3$ 3-kinase may also be activated by protein tyrosine kinases (Johnson et al., 1989).

Ins(1,4,5)P_3 and Ins(1,3,4,5)P_4 5-phosphatase

Specific Ins(1,4,5)P_3 5-phosphatase (Downes et al., 1982) and Ins(1,3,4,5)P_3 5-phosphatase (Batty et al., 1985) were first demonstrated in human erythrocytes. Since then, Mg^{2+}-dependent 5-phosphatase activity towards these two compounds has been described in a plethora of cell types (Shears, 1989) and it is believed that both Ins(1,4,5)P_3 and Ins(1,3,4,5)P_4 are hydrolysed by the same 5-phosphatase, for example in bovine brain (Erneux et al., 1989) and human erythrocyte membranes (A Craxton, unpublished). The 5-phosphatase is predominantly membrane-bound, but soluble forms also exist. It has a K_m for Ins(1,4,5)P_3 in the range 15-30 μM, and the K_m towards Ins(1,3,4,5)P_4 is in the range 0.8-1.2 μM, i.e. around 10-fold lower than for the Ins(1,4,5)P_4 (Shears, 1989).

The 5-phosphatase activity can be inhibited by: spermine (Seyfred et al., 1985); ATP (Hansen, 1987; Milani et al., 1988; Cunha-Melo et al., 1988; Shears, 1990); fructose 1,6-bisphosphate (Rana et al., 1986) and D-2,3-bisphosphoglycerate (Downes et al., 1982; Rana et al., 1986). In addition, 5-phosphatase activity is inhibited by increasing ionic strength, with all monovalent cations tested (K^+, Na^+ and Li^+) being equipotent (Hoer et al., 1988). The 5-phosphatase activity is probably unaffected by protein kinase C activation and by physiological concentrations of Ca^{2+} (Shears, 1989).

From the K_m values mentioned above, it seems that the affinity of the 5-phosphatase for Ins(1,3,4,5)P_4 is approximately 10-fold higher than for Ins(1,4,5)P_3, also that the affinity of 3-kinase for Ins(1,4,5)P_3 is approximately 10-fold greater than that of the 5-phosphatase. Therefore it seems that there is a potential for Ins(1,4,5)P_3 to be preferentially phosphorylated in the cell. Whether or not this happens in situ will, however, depend on the relative V_{max} values for phosphorylation and dephosphorylation, the mechanisms regulating the 3-kinase and 5-phosphatase activity, and possibly also on the intracellular localisations of the 3-kinase and 5-phosphatase.

Ins(1,3,4,5)P_4 3-phosphatase

Until recently, it was believed that the putative second messenger Ins(1,3,4,5)P_4 was only metabolized by an Ins(1,3,4,5)P_4 5-phosphatase, yielding Ins(1,3,4)P_3. However, it was then observed that Ins(1,3,4,5)P_4 was degraded rapidly to both Ins(1,3,4)P_3 and Ins(1,4,5)P_3 following cross-linking of the IgE receptor of rat basophilic leukemia (RBL-2H3) cells (Cunha-Melo et al., 1987). Subsequently, 3-phosphatase activity towards Ins(1,3,4,5)P_4 has been reported in several cells and tissues: human erythrocytes (Doughney et al., 1988), human platelets (Oberdisse et al., 1990), GH_3 rat pituitary tumour cells (Dean and Moyer, 1988), mouse lymphoma L1210 cells (Cullen et al., 1989), porcine brain (Hoer et al., 1988), ovarian follicles of Xenopus laevis (McIntosh and McIntosh, 1990) and rat liver (Shears, 1990). Such a 3-phosphatase activity could clearly have an important role in controlling Ins(1,4,5)P_3 and Ins(1,3,4,5)P_4 levels, and thus in regulating cell function.

The contribution of the 3-phosphatase to Ins(1,3,4,5)P_4 metabolism in vitro varies considerably with the incubation conditions chosen. For example, it was reported that the hydrolysis of Ins(1,3,4,5)P_4 by the 5-phosphatase activity in human erythrocyte membranes is ten times faster than by the 3-phosphatase under conditions of low ionic strength (Doughney et al., 1988). In contrast, when Ins(1,3,4,5)P_4 was incubated with human erythrocyte membranes in an isoionic medium (130 mM, KCl, 10 mM NaCl), the relative rates of the 3-phosphatase and 5-phosphatase were approximately 1:2 (Estrada-Garcia et al., in preparation). The 3-phosphatase activities of porcine brain cytosol (Hoer et al., 1988), permeabilized human platelets (Oberdisse et al., 1990) and human erythrocyte membranes (A Craxton, unpublished) are all dependent on ionic strength (K^+ > Na^+ > Li^+), being highly activated by physiological intracellular concentrations of KCl (100-130 mM) and very low in the absence of this cation. The 3-phosphatase activity appears to be unaffected by ATP (2-5 mM) (Cunha-Melo et al., 1988; Shears, 1990) or by physiological concentrations of Ca^{2+} (Oberdisse et al., 1990).

3-phosphatase activity in cell homogenates and partially purified cell fractions

Incubation of Ins(1,3,4,5)P_4 with various cell homogenates has revealed different relative rates of 3-phosphatase and 5-phosphatase activities. For example, ovarian follicle homogenates of Xenopus laevis preferentially hydrolyse Ins(1,3,4,5)P_4 to Ins(1,4,5)P_3 (McIntosh and McIntosh, 1990), whereas GH_3 cell homogenates incubated with Ins(1,3,4,5)P_4 produced approximately 97% Ins(1,3,4)P_3 and 3% Ins(1,4,5)P_3, albeit in a medium of low ionic strength (Dean and Moyer, 1988). RBL-2H3 cell homogenates (70 mM potassium glutamate, 30 mM KCl) also showed low 3-phosphatase activity. When Ins(1,3,4,5)P_4 was incubated with a membrane fraction from RBL-2H3 cells, approximately 30-55% Ins(1,4,5)P_4 was produced. In contrast, when the soluble fraction of the cells was studies, Ins(1,3,4)P_3 was almost the only product, suggesting that conversion of Ins(1,3,4,5)P_4 to Ins(1,4,5)P_3 was catalyzed by a membrane-bound enzyme (Cunha-Melo et al., 1988). However, Hoer et al., (1988) reported substantial 3-phosphatase activity in a partially purified cytosolic preparation of porcine brain after incubation of Ins(1,3,4,5)P_4 under physiological ionic strength: they did not, however, directly compare homogenate, membrane and cytosolic activities.

The 3-phosphatase activity in an isolated 100,000 x g particulate fraction was several-fold greater than that of parent homogenates in RBL-2H3 cells (Cunha-Melo et al., 1988), rat liver (Shears, 1990) and rat parotid glands (Hughes and Shears, 1990). This apparent activation may be due to the presence of an endogenous inhibitor of the 3-phosphatase in homogenates. Such an activity has been reported: for example, the 3-phosphatase activity of a rat parotid soluble fraction was increased upon dilution (Hughes and Shears, 1990) and that of rat liver cytosol was increased when the 3-phosphatase was partially purified (Hodgson and Shears, 1990).

It is possible that the observation of an active cytosolic 3-phosphatase activity in porcine brain (Hoer et al., 1988) was possible only because the partial purification of the 3-phosphatase had eliminated this endogenous inhibitor. The 3-phosphatase inhibitor is a heat-stable factor with a molecular mass above 14kDa (Hodgson and Shears, 1990).

In cell homogenates, intracellular compartmentation is at least partially destroyed and therefore the observation of Ins(1,3,4,5)P_4 3-phosphatase activity in an isolated fraction can only be taken as a partial indication of the potential pathways in which this enzyme might be implicated. Furthermore, reports of 3-phosphatase inhibitors in cell fractions may not necessarily be of physiological significance, since enzyme and inhibitor may not be present in the same cell compartments.

3-phosphatase activity in permeabilized and intact cells

Injection of Ins(1,3,4,5)P_4 into intact Xenopus laevis ovarian follicles is followed by very little formation of Ins(1,4,5)P_3. However, 3-phosphatase activity towards Ins(1,3,4,5)P_4 is observed either in the presence of PMA (60 nM) or following stimulation with 200 μM acetylcholine (McIntosh and McIntosh, 1990). Moreover, Ins(1,3,4,5)P_4 was predominantly dephosphorylated to Ins(1,4,5)P_3 when added to saponin-permeabilized platelets and this process was also activated by phorbol esters and thrombin, either of which increased the activity of the 5-phosphatase (Oberdisse et al., 1990). This suggests that Ins(1,3,4,5)P_4 3-phosphatase may be activated by PKC and its activity may therefore be increased as a consequence of receptor occupation. Moreover, the addition of Ins(1,3,4,5)P_4 at micromolar concentration to electroporated mouse L1210 lymphoma cells caused release of Ca^{2+}, but did not release Ca^{2+} from digitonin-permeabilized cells. The Ca^{2+} release was shown to be due to conversion of Ins(1,3,4,5)P_4 into Ins(1,4,5)P_3 by a 3-phosphatase. Interestingly, in cells permeabilized by electroporation, followed later by addition of digitonin, Ins(1,3,4,5)P_4 was still converted into Ins(1,4,5)P_3. Furthermore, Ins(1,3,4,5)P_4 also released Ca^{2+} if the digitonin-permeabilized cells were then subjected to electroporation (Cullen et al., 1989). It therefore seems likely that electroporation in some way leads to activation of a 3-phosphatase, but whether this is mediated by PKC activation is yet to be established.

In summary, therefore, Ins(1,3,4,5)P_4 3-phosphatase activity is present both in intact and permeabilized cells, and its activity appears to be subject to physiological control, in common with some other enzymes involved in Ins(1,4,5)P_3 metabolism.

Inhibition of Ins(1,3,4,5)P_4 3-phosphatase activity by inositol polyphosphates

The widely occurring cellular inositol polyphosphates, Ins(1,3,4,5,6)P_5 and $InsP_6$, potently inhibited 3-phosphatase activity of rat parotid gland cytosol with IC_{50} values of 1.0 μM and 0.5 μM respectively (Hughes and Shears, 1990). These compounds also inhibited the 3-phosphatase of human erythrocyte membranes isolated under isoionic conditions (Estrada-Garcia et al., in preparation). The potent inhibitory effects of Ins(1,3,4,5,6)P_5 and $InsP_6$ suggest that only a small fraction of the total cellular content of these relatively abundant polyphosphates would need to be soluble in order to inhibit Ins(1,3,4,5)P_4 3-phosphatase in situ, since physiological intracellular concentrations of these compounds are of the order of 5-50 μm (e.g. Pittet et al., 1989; French et al., 1988). It is therefore possible that these compounds are making a significant contribution to the inhibition of the 3-phosphatase activity observed in liver (Hodgson and Shears, 1990) and rat parotid glands (Hughes and Shears, 1990). It has also been suggested that either Ins(1,3,4,5,6)P_5 or $InsP_6$, or both, may be competing substrates for the 3-phosphatase: Hughes and Shears (1990) reported that Ins(1,3,4,5,6)P_5 was a competitive inhibitor of the 3-phosphatase, and dephosphorylation of Ins(1,3,4,5,6)P_5 by uncharacterized 3-

phosphatase species has been described (Balla et al., 1989; Stephens and Downes, 1990).

These observations have led to the suggestion that Ins(1,3,4,5)P_4 3-phosphatase may not be a quantitatively or physiologically important participant in Ins(1,3,4,5)P_4 metabolism in situ (Hughes and Shears, 1990). Nevertheless, the cellular localizations of $InsP_5$ and $InsP_6$ have yet to be establishedand it is therefore not clear whether they have access to the 3-phosphatase in cells. Moreover, it has recently been reported that the synthesis of $InsP_6$, in which Ins(1,3,4,5,6)P_5 is an intermediate, is independent of the metabolism of both phosphatidylinositol and Ins(1,4,5)P_3 in WRK1 cells (Kirk et al., 1990) and Dictyostelium (Stephens and Irvine, 1990).

Concluding remarks

In reviewing the known physiological roles of Ins(1,4,5)P_3 and Ins(1,3,4,5)P_4, we have emphasised the possible physiological relevance of an Ins(1,3,4,5)P_4 3-phosphatase, which may provide the cell with an additional pathway for the formation of the second messenger Ins(1,4,5)P_3, thus prolonging receptor-triggered effects and also providing a new regulatory mechanism for controlling Ins(1,3,4,5)P_4 levels. To date, Ins(1,3,4,5)P_4 3-phosphatase activity has been reported both in intact and in permeabilized cells, and it has been shown that the Ins(1,4,5)P_3 produced is capable of Ca^{2+} mobilization. Moreover, it seems that there is crosstalk with the other arm of the phosphoinositide signalling pathway, in that Ins(1,3,4,5)P_4 3-phosphatase activity is regulated by protein kinase C. However, it remains to be demonstrated that the 3-phosphatase and Ins(1,3,4,5)P_4 do indeed coexist in any cellular compartment. We have recently made progress in this direction by demonstrating that the active site of the Ins(1,3,4,5)P_4 3-phosphatase activity is exposed at the cytoplasmic surface of the human erythrocyte membrane (Estrada-Garcia et al., in preparation), where it could readily act on cytosolic inositol phosphates. At this site, it coexists with two other key enzymes of the inositol lipid signalling pathway, $Ins(1,4,5)P_3/Ins(1,3,4,5)P_4$ 5-phosphatase and phosphoinositidase C.

Acknowledgments

These studies were supported by the Medical Research Council and the Royal Society, also by a SERC Research Studentship to A Craxton.

REFERENCES

Balla T, Hunyady L, Baukal AJ and Catt KJ (1989) Structures and metabolism of inositol tetrakisphosphate and inositol pentakisphosphate in bovine adrenal glomerulosa cells. J. Biol. Chem. 264:9386-9390

Batty IR, Nahorski SR and Irvine RF (1985) Rapid formation of inositol 1,3,4,5-tetrakisphosphate following muscarinic receptor stimulation of rat cerebral cortical slices. Biochem. J. 232:211-215

Berridge MJ (1987) Inositol trisphosphate and diacylglycerol: two interacting second messengers. Ann. Rev. Biochem. 56:159:193

Berridge MJ and Irvine RF. (1989) Inositol phosphates and cell signalling. Nature 341:197-205

Berridge MJ and Michell RH (eds) (1988) Inositol Lipids and Transmembrane signalling. London, The Royal Society.

Berridge MJ, Cobbold PH and Cuthbertson KSR (1988) Spatial and temporal aspects of cell signalling. Phil. Trans. Roy. Soc. Lond. B 320 325-343

Biden TJ, Altin JG, Karjalainen A. and Bygrave FL (1988b) Stimulation of hepatic inositol 1,4,5-trisphosphate kinase activity by Ca^{2+}-dependent and independent mechanisms. Biochem. J. 256:697-701

Biden TJ, Vallar L and Wollheim CB. (1988a) Regulation of inositol 1,4,5-trisphosphate metabolism in insulin-secreting RINm5F cells. Biochem. J. 251:435-440

Bradford PG and Irvine RF (1987) Specific binding sites for [^{3}H]-1,3,4,5-tetrakisphosphate on membranes of HL60 cells. Biochem. Biophys. Res. Commun. 149:680-685

Cullen PJ, Irvine RF, Droback JK and Dawson AP (1989) Inositol 1,3,4,5-tetrakisphosphate causes release of Ca^{2+} from permeabilized mouse lymphoma L1210 cells by its conversion into inositol 1,4,5-trisphosphate. Biochem. J. 259:931-933

Cunha-Melo JR, Dean NM, Moyer JD, Maeyama K and Beaven MA (1987) The kinetics of phosphoinositide hydrolysis in rat basophilic leukemia (RBL-2H3) cells varies with the type of IgE receptor cross-linking agent used. J. Biol. Chem. 262:11455-11463

Cunha-Melo JR, Dean NM, Moyer JD, Ali H and Beaven MA (1988) Formation of inositol 1,4,5-trisphosphate and inositol 1,3,4-triphosphate from inositol 1,3,4,5-tetrakisphosphate and their pathways of degradation in RBL-2H3 cells. J. Biol. Chem. 263:14245-14250

Dean NM and Moyer JD (1988) Metabolism of inositol bis-, tris-, tetrakis- and pentakis-phosphates in GH_3 cells. Biochem. J. 250;493-500

Donie F and Reiser G (1989) A novel, specific binding protein assay for quantitation of intracellular inositol 1,3,4,5-tetrakisphosphate ($InsP_4$) using a high-affinity $InsP_4$ receptor from cerebellum. FEBS Lett. 454:155-158

Doughney C, McPherson MA and Dormer RL (1988) Metabolism of inositol 1,3,4,5-tetrakisphosphate by human erythrocyte membranes. Biochem. J. 251:927-929

Downes CP, Mussat MC and Michell RH (1982) The inositol trisphosphate phosphomonoesterase of the human erythrocyte membrane. Biochem. J. 203:169-177

Ely JA, Hunyady L, Baukal AJ and Catt KJ (1990) Inositol 1,3,4,5-tetrakisphosphate stimulates calcium release from bovine adrenal microsomes by a mechanism independent of the inositol 1,4,5-trisphosphate receptor. Biochem. J. 268:333-338

Erneux C, Lemos M, Verjans B, Vanderhaeghen P, Delvaux A and Dumont JE (1989) Soluble and particulate Ins(1,4,5)P_3-/Ins(1,3,4,5)P_4 5-phosphatase in bovine brain. Eur. J. Biochem. 181:317-322

French PJ, Bunce CM, Brown G, Creba JA & Michell RH (1988) Inositol phosphates in growing and differentiating HL60 cells. Biochem. Soc. Trans. 16:985-986

Furuichi T, Yoshikawa S, Miyawaki A, Wada K, Maeda N, Mikoshiba K. (1989) Primary structure and functional expression of the inositol 1,4,5,-trisphosphate-binding protein P_{400}. Nature 342:32-38

Hansen CA, Johanson RA, Williamson MT and Williamson JR (1987) Purification and characterization of two types of inositol phosphate 5-phosphomonoesterases from rat brain. J. Biol. Chem. 262:17319-17326

Hodgson ME and Shears SB (1990) Rat liver contains a potent endogenous inhibitor of inositol 1,3,4,5-tetrakisphosphate 3-phosphatase. Biochem. J. 267:831-834

Hoer D, Kwiatkowski A, Seib C, Rosenthal W, Schultz G and Oberdisse E (1988) Degradation of inositol 1,3,4,5-tetrakisphosphates by porcine brain cytosol yields inositol 1,3,4,-trisphosphate and inositol 1,4,5-trisphosphate. Biochem. Biophys. Res. Commun. 251:668-675

Hughes PJ and Shears SB (1990) Inositol 1,3,4,5,6-pentakisphosphate and inositol hexakisphosphate inhibit inositol 1,3,4,5-tetrakisphosphate 3-phosphatase in rat parotid glands. J. Biol. Chem. 265:9869-9875

Imboden JB and Pattison G (1987) Regulation of inositol 1,4,5-trisphosphate kinase activity after stimulation of human T cell antigen receptor. J. Clin. Invest. 79:1538-1541

Irvine RF (1989) Functions of inositol phosphates. In Inositol lipids in cell signalling. Michell RH, Drummond AH. and Downes CP (eds), pp. 135-161, Academic Press, London.

Irvine RF. (1990) "Quantal" Ca^{2+} release and control of Ca^{2+}

entry by inositol phosphates - a possible mechanism. FEBS Lett. 263:5-9

Johnson RM. Wasilenko WJ, Mattingly RR Weber MJ and Garrison JC (1989) Fibroblast transformed with v-*src* show enhanced formation of an inositol tetrakisphosphate. Science 246:121-124

Kirk CJ, Maccallum SH, Michell RH, and Barker CJ (1990) Inositol phosphates in receptor-mediated cell signalling: metabolic origins and interrelationships. Biotech and Applied Biochem. 12:489-495

Kuno N and Gardner P. (1987) Ion channels activated by inositol 1,4,5-trisphosphate in plasma membrane of human T-lymphocytes. Nature 334:499-504

Maeda N, Niinobe M and Mikoshiba K (1990) A cerebellar Purkinje cell P_{400} protein is an inositol 1,4,5-trisphosphate ($InsP_3$) receptor protein. Purification and characterization of $InsP_3$ receptor complex. EMBO J. 9: 61-67.

McIntosh RP and McIntosh JEA (1990) Metabolism of the biologically active inositol phosphates Ins(1,4,5)P_3 and Ins(1,3,4,5)P_4 by ovarian follicles of *Xenopus laevis*. Biochem J. 268:141-145

Michell RH, Kirk CJ, Jones LM, Downes CP and Creba JA (1981) The stimulation of phosphatidylinositol breakdown that accompanies calcium mobilization in stimulated cells: defined characteristics and unanswered questions. Phil. Trans. Roy. Soc. Lond. Ser. B 296:81-147

Mignery GA, Newton CL, Archer BT and Sudhof TC (1990) Structure and expression of the rat inositol 1,4,5-trisphosphate receptor. J. Biol. Chem. 265:12679-12685

Milani D, Volpe P and Pozzan T (1988) D-myo-inositol 1,4,5-trisphosphate phosphatase in skeletal muscle. Biochem. J. 254:525-529

Morris AP, Gallacher DV, Irvine RF and Petersen OH (1987) Synergism of inositol 1,3,4,5-tetrakisphosphate with inositol 1,4,5-trisphosphate in mimicking muscarinic receptor activation of Ca^{2+}-dependent K^+ channels. Nature 330:653-655

Oberdisse E, Nolan RD and Lapetina EG (1990) Thrombin and phorbol ester stimulate inositol 1,3,4,5-tetrakisphosphate 3-phosphomonoesterase in human platelets. J. Biol. Chem. 265:726-730

Penner R, Matthews G and Neher E (1988) Regulation of calcium influx by second messengers in rat mast cells. Nature 334:499-504

Pittet D, Schlegel W, Lew DP, Monod A and Mayr G (1989) Mass changes in inositol tetrakis- and pentakisphosphate isomers induced by chemotactic peptide stimulation in HL-60 cells J. Biol. Chem. 264:18489-18493

Putney JW Jr (1986) A model for receptor-regulated calcium entry. Cell Calcium 7:1-12

Rana RS, Sekar CM, Hokin LE and McDonald MJ (1986) A possible role for glucose metabolites in regulation of inositol-1,4,5-trisphosphate 5-phosphomonoesterase activity in pancreatic islets. J. Biol. Chem. 261:5237-5240

Seyfred MA, Farell LE and Wells WW (1984) Characterization of D-myo-inositol 1,4,5-trisphosphate phosphatase in rat liver plasma membranes. J. Biol. Chem. 259:13204-13208

Shears SB (1989) Metabolism of the inositol phosphates produced upon receptor activation. Biochem. J. 260:313-324

Shears SB (1990) Kinetic consequences of the inhibition by ATP of the metabolism of inositol 1,4,5-trisphosphate and inositol 1,3,4,5-tetrakisphosphate in liver. Different effects upon the 3- and 5-phosphatases. Cell Signalling 2:191-195

Sim SS, Kim JW and Rhee SG (1990) Regulation of D-myo-inositol 1,4,5-trisphosphate 3-kinase by cAMP-dependent protein kinase and protein kinase C. J. Biol. Chem. 265:10367-10372

Spat A, Lukacs GL, Eberhardt I, Kiessel L and Runnebaum B (1987) Binding of inositol phosphates and induction of Ca^{2+} release from pituitary microsomal fractions. Biochem. J. 244:493-496.

Stephens LR and Irvine RF (1990) Stepwise phosphorylation of myo-inositol leading to myo-inositol hexakisphosphate in *Dictyostelium*. Nature 346:580-583

Stephens LR, and Downes CP (1990) Product-precursor relationships amongst inositol polyphosphates. Biochem. J. 265:435-452

Streb HD, Irvine RF, Berridge MJ and Schulz I (1983) Release of Ca^{2+} from non-mitochondrial intracellular store of pancreatic acinar cells by inositol 1,4,5 trisphosphate Nature 306:67-68

Takai Y, Kishimoto A, Kikkawa U, Mori T and Nishizuka Y (1979) Unsaturated diacylglycerol as a possible messenger for the activation of calcium-activated phospholipid-dependent protein kinase system. Biochem. Biophys. Res. Commun. 91:1218-1224

Taylor CW and Putney JM Jr. (1985) Size of the inositol 1,4,5-trisphosphate-sensitive calcium pool in guinea pig hepatocytes. Biochem. J. 232:435-438

Theibert AB, Supattapone S, Ferris CD, Danoff SK, Evans RK and Snyder SH (1989) Solubilization and separation of inositol 1,3,4,5-tetrakisphosphate- and inositol 1,4,5-trisphosphate-binding proteins and metabolizing enzymes in rat brain. Biochem. J. 267:441-445.

HISTAMINE H_1-RECEPTOR-MEDIATED BREAKDOWN OF INOSITOL PHOSPHOLIPIDS IN PRIMARY NEURONAL CULTURES

L. Arbonés and A. García
Instituto de Biología Fundamental and Departamento de Bioquímica y Biología Molecular
Universidad Autónoma de Barcelona
08193 BELLATERRA
Spain

ABSTRACT

Histamine stimulation of phosphoinositide hydrolysis was measured in primary cultures of neurons obtained from cerebrum of fetal rats of 16 days gestation. In cells prelabeled with (^{3}H)inositol, histamine in the presence of lithium induces accumulation of (^{3}H)inositol monophosphate linearly for 30-40min. The effect is concentration-dependent (Emax=149±9% over basal, EC_{50}=5,8±0,9 µM, Hill coefficient=0,99±0,07, n=3) and shows pharmacological characteristics typical of an H_1-mediated response. The histamine effect increases with neuronal differentiation during the first week in culture and stabilizes thereafter. In contrast to astrocytes where the histamine response is abolished when Ca^{2+} is omitted from the incubation medium, in neurons it is necessary to quelate all the remnant ion to prevent the effect. Furthermore, the response to a depolarizing concentration of K^+ is additive to that of histamine and the stimulation by the amine is not affected by omission of Na^+, or by incubation with the voltage-dependent Na^+ channel blocker tetrodotoxin. All these observations indicate that the H_1-receptor-stimulated inositol phosphate formation in neurons is direct and not a consequence of a histamine-induced depolarization, increased Ca^{2+} permeability or release of other neurotransmitters.

L. Arbonés present address: Cell Genesys Inc. 344A Lakeside Drive. Foster City, CA, 94404. USA.

NATO ASI Series, Vol. H 52
Biological Signal Transduction
Edited by E. M. Ross and K. W. A. Wirtz

INTRODUCTION

Activation of histamine H_1-receptors in the central nervous system and in peripheral tissues stimulates the breakdown of membrane phosphoinositides and causes calcium mobilization and, in some cases, transmembrane calcium influx (Hill, 1990). Studies performed in brain tissue slices have demonstrated differences in the response mediated by H_1- and other receptors that stimulate inositol phospholipid hydrolysis (i.e. muscarinics) in the sensitivity of that response to extracellular calcium (Kendall and Nahorski, 1984; Alexander et al. 1990) and temperature (Carswell et al. 1987), in the effect of lithium in the pattern of inositol phosphates formed (Whitworth and Kendall, 1988) and in the modulation by other receptor systems (Hill and Kendall, 1989) suggesting that different metabolic pathways may be involved. However, responses measured in brain slices my be a combination of responses from the different cell types present in that heterogeneous preparation. In order to study the characteristics of the H_1-receptor response in a particular cell type,we have set up primary cultures from rat brain cerebrum enriched in astrocytes or neurons and have already described some characteristics of the H_1-mediated phosphoinositide hydrolysis in astrocytes (Arbonés et al. 1988). In these cells, as occurs in rat cortex slices (Kendall and Nahorski, 1984), the histamine-stimulated inositol phosphate formation shows a stronger extracellular calcium dependency than that reported for the responses to charbachol or noradrenaline (Pearce et al. 1986). In astrocytes, H_1-receptors also stimulate calcium entry (Arbonés et al. 1990) which may in turn be responsible for the stimulation of the phosphoinositide phospholipase C. This enzyme in brain can be activated by increased calcium or by a G-protein-linked receptor (Gonzales and Crews, 1988; Claro et al. 1989). In the present work, we have studied the histamine-stimulated inositol lipid breakdown in neuronal cultures.

MATERIALS AND METHODS

Neuronal primary cultures were obtained from 16-day-old rat embryos (Sprague-Dawley) as previously described (Agulló et al. 1990). Briefly, brain hemispheres were dissociated by successive passages through nylon cloths of 135 and 22 μm mesh openings and 1.2 x 10^6 viable cells in 2 ml Dulbecco's modified Eagle's medium -10% horse serum were seeded in 35 mm diameter plastic Petri dishes pretreated with poly-ornithine (0.01% wt/vol). After one day in culture, medium was replaced by serum-free glia-conditioned medium supplemented with insulin (5 μg/ml), transferrin (100 μg/ml), putrescine (100 μM), progesterone (20 nM) and Na_2SeO_3(30 nM).Two days later, cells were treated for 24 h with cytosine arabinoside (5 μM) and were used routinely after 7-12 days in culture.
The histamine stimulated phosphoinositide breakdown was monitored by measuring the formation of (^{3}H)inositol phosphates in the presence of 10 mM lithium in cells prelabeled with (^{3}H)inositol as described (Arbonés et al. 1988). The separation of the (^{3}H)inositol phosphates was performed after lipid extraction with acidified chloroform/methamol, by passing an aliquot of the aqueous phase through a column of 0.5 ml Dowex 1X-8 (formate form) resin and elution with 6 ml each of the following: water; 60 mM sodium formate/5 mM sodium tetraborate; 0.2 M ammonium formate/0.1 M formic acid; 0.6 M ammonium formate/0.1 M formic acid; 1 mM ammonium formate/0.1 M formic acid. When the effect of high K^+ and the absence of Na^+ were studied, the osmolarity of the buffers was maintained by modifying the concentration of NaCl or by substituting with choline chloride.

RESULTS

The incorporation of (^{3}H)inositol into the intracellular inositol and membrane lipid pools of neuronal cultures increased during the first 18 h reaching a steady state that was maintained for at least 10 more hours (not shown). After prelabeling the cultures for 24 h, time used in routine experiments, the amount of tritium incorporated into lipids ranged from 93 x 10^3 to 158 x 10^3 dpm/30 mm dish in the different preparations

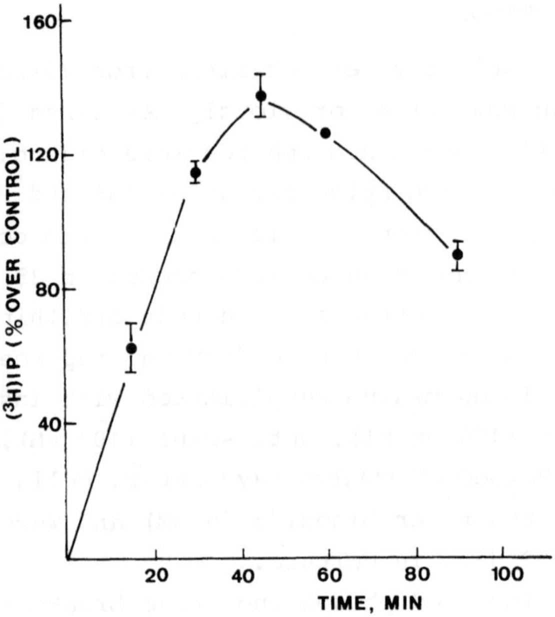

Fig. 1. Time course of histamine-stimulated (^{3}H)IP accumulation. Neuronal cultures prelabeled for 24 h with 1 μCi/dish of (^{3}H)inositol were incubated for the indicated times with 0.1 mM histamine in the presence of 10 mM LiCl. After lipid extraction, aqueous phases were chromatographed in columns of 0.5 ml Dowex 1-X8 and (^{3}H)IP eluted with 6 ml of ammonium formate/0.1 M formic acid. Results are presented as percent of controls (1.9±0.06 dpm/100 dpm in lipids) and are means±SE of triplicate determinations in a representative experiment that was repeated two more times.

used, which represents 55% of the total radioactivity incorporated into cells. Under these conditions and in the presence of 10 mM LiCl, the accumulation of (^{3}H)inositol monophosphate ((^{3}H)IP) induced by 0.1 mM histamine increased linearly for the first 40 min of incubation with the agonist and decreased at longer fimes (Fig. 1). The amount of (^{3}H)IP in control cultures remained constant for the time period studied. After 30 min of incubation, time chosen for the following experiments, no stimulation of the accumulation of higher (^{3}H)inositol phosphates was observed (not shown).

The histamine-induced (^{3}H)IP accumulation was concentration-dependent reaching maximal stimulation at a concentration of 0.1 mM. The parameters obtained from the linear Hill transfor-

TABLE I. Concentration-effect parameters for histamine stimulation of (^{3}H)IP formation in different preparations from rat brain.

Preparation	(^{3}H)IP ACCUMULATION E_{MAX} (% over basal)	EC_{50} (μM)	n_H
Neurons	149 ± 9	5,8 ± 0,9	0,99 ± 0,07
Astrocytes (1)	345 ± 12	1,7 ± 0,3	0,95 ± 0,10
Cortex (2)			
6 days (H_1comp)	65,8 ± 1,4	6,4 ± 0,2	1
Adult (3)	124,1 ± 14,2	94,7 ± 11,8	1,04 ± 0,05

Maximal effect (E_{max}), potency (EC_{50}) and Hill coefficient(n_H) for neurons are means±SEM from 3 concentration-effect curves and were estimated by fitting the data to a Hill equation using the BMDP-AR nonlinear iterative program implemented on a VAX 11/785 system. (1)Arbonés et al. 1988, (2)Claro et al.1987, (3)Claro et al. 1986.

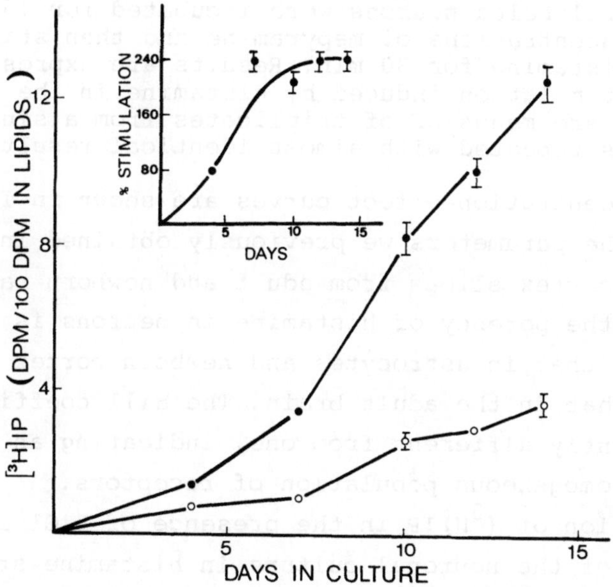

Fig. 2. Ontogenic development of histamine-induced (^{3}H)IP accumulation in cultured neurons. Prelabeled neurons maintained in culture for the indicated days were incubated for 30 min in the absence (O) or presence (●) of 0.1 mM histamine. Results are means±SE of triplicate determinations from an experiment that was repeated two more times with similar results. Inset: Data expressed as percent stimulation over controls.

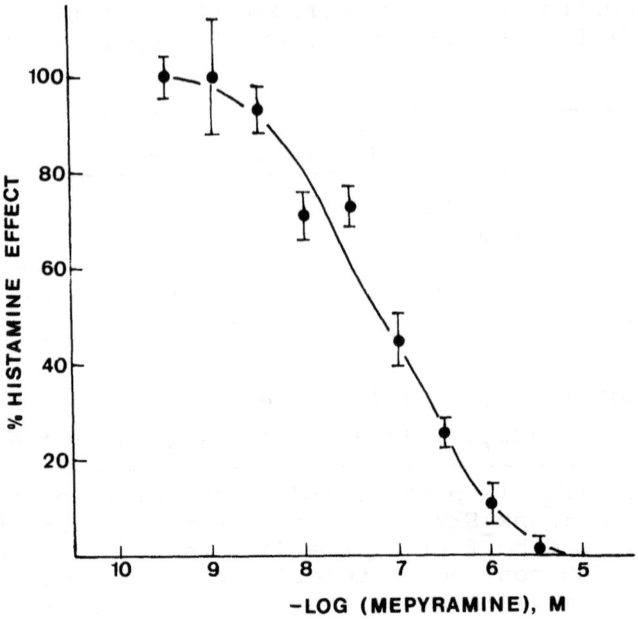

Fig. 3. Inhibition of histamine-induced (^{3}H)IP accumulation by mepyramine. Prelabeled neurons were incubated for 15 min with increasing concentrations of mepyramine and then stimulated with 0.1 mM histamine for 30 min. Results are expressed as per cent of the stimulation induced by histamine in the absence of inhibitor and are means±SE of triplicates from a single experiment that was repeated with almost identical results.

mation of concentration-effect curves are shown in Table I together with the parameters we previously obtained in astrocyte cultures and cortex slices from adult and newborn rats. As can be observed, the potency of histamine in neurons is of the same order of that in astrocytes and newborn cortex and much higher than that in the adult brain. The Hill coefficient was not significantly different from one, indicating an interaction with a homogeneous population of receptors.

The accumulation of (^{3}H)IP in the presence of LiCl increases with the age of the neuronal culture in histamine-stimulated as well as in control cells (Fig. 2). However, when the (^{3}H)IP formation induced by histamine is expressed as percent of that in controls, a maximum effect is reached by day 7 that is maintained al least until day 14 (Fig. 2 inset). Thus, maximal histamine stimulation is attained at an age of the culture

TABLE II. Effect of histaminergic agonists and antagonists on (^{3}H)IP accumulation.

Compound	% Histamine effect	N
Dimaprit (1 mM)	9 ± 9	2
2-Thiazolylethylamine (0.3 mM)	100 ± 0	2
HA (0.1 mM) + D-chlorpheniramine (10 μM)	4 ± 3	2
HA (0.1 mM) + L-chlorpheniramine (10 μM)	75 ± 3	2
HA (0.1 mM) + Cimetidine (100 μM)	125 ± 12	3
HA (0.1 mM) + Metiamide (100 μM)	96 ± 8	4

Prelabeled neurones were incubated for 30 min in the presence of the indicated compounds. Antagonists were added 15 min before histamine (HA). Results, expressed as percent of the stimulation by 0.1 mM histamine, are means±SEM of N experiments performed in triplicate.

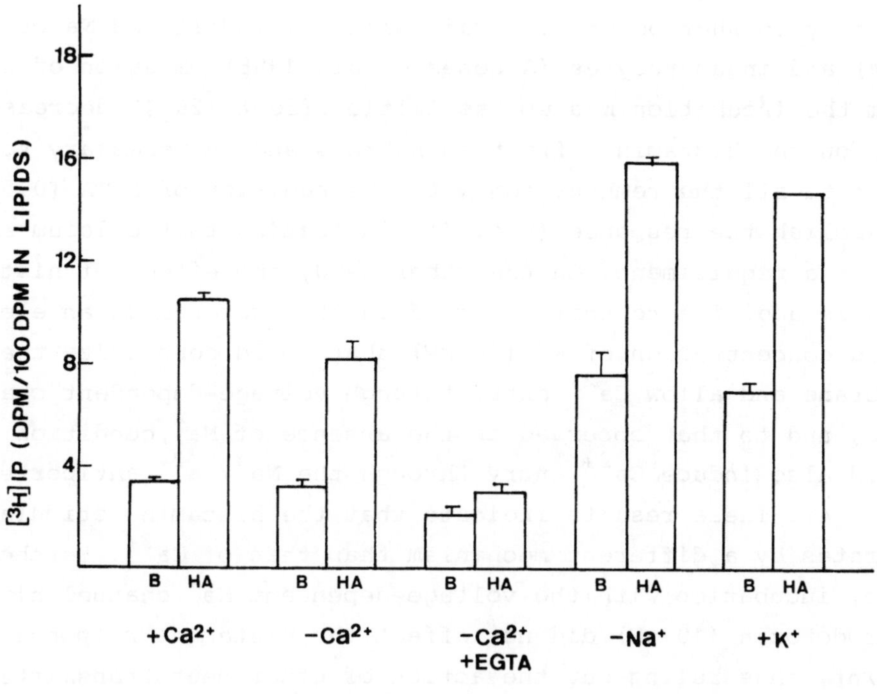

Fig. 4. Effect of extracellular Ca^{2+} on histamine-induced (^{3}H)IP accumulation. Prelabeled neurons were incubated for 30 min in the absence (B) or presence of 0.1 mM histamine in media containing: 2.5 mM Ca^{2+}; no added Ca^{2+}; no added Ca^{2+} and 0.5 mM EGTA; no NaCl and 116 mM choline-Cl; 18 mM K^+. Results are means±SE of triplicate determinations in one experiment that was repeated with very similar results.

when cells are still differenciating. Similarly, maximal stimulation in rat cortex slices is attained at postnatal day 15 when synaptogenesis is still in course (Claro et al. 1987). The effect of 0.1 mM histamine is mimicked by the H_1-agonist 2-thiazolylethylamine (0.3 mM) but not by the H_2-agonist dimaprit (1 mM) and is stereospecifically inhibited by the active isomer of the H_1-antagonist chlorpheniramine (10 μM) but not by the H_2-antagonists cimetidine and metiamide (100 μM)(Table II), as corresponds to a H_1-mediated effect. Furthermore, the H_1-antagonist mepyramine inhibits the histamine response (Fig. 3) with a Ki value, calculated from the IC_{50}s obtained in two experiments like the one shown in Fig. 3 using the equation $Ki=IC_{50}/(1+[histamine]/EC_{50})$, of 2.7±0.7 nM which is very close to that obtained in adult rat cortex slices (Claro et al.1986). Contrary to what occurs in brain slices (Kendall and Nahorski 1984) and in astrocytes (Arbonés et al. 1988) omission of Ca^{2+} from the incubation medium has little effect (26±4% decrease, n=2) on the histamine effect in neurons and is necessary to eliminate all the remnant ion with the addition of EGTA (0.5 mM) to abolish the response (Fig. 4), indicating that calcium entry is not a requirement. On the other hand, the effect of histamine is additive to that observed in the presence of an elevated concentration of K^+ (18 mM) that would depolarize the membrane and allow Ca^{2+} entry through voltage-dependent channels, and to that observed in the absence of Na^+,condition that would also induce Ca^{2+} entry through the Na^+/Ca^{2+} antiporter (Fig. 4). These results indicate that the histamine stimulation operates by a different mechanism than that of Ca^{2+}. Furthermore, incubation with the voltage-dependent Na^+ channel blocker tetrodotoxin (10 μM) did not affect the histamine response (not shown), thus ruling out the action of other neurotransmitters that could be potentially liberated.

DISCUSSION

The presence of histamine H_1 receptors in CNS neurons had been previously suggested in autoradiographic (Palacios et al.1981) and ontogenic and subcellular fractionation binding studies

(Tran et al. 1980) using (^{3}H)mepyramine. However, functional H_1-receptors mediating phosphoinositide hydrolysis have been only shown in rat cerebellar granule cells in culture (Xu and Chuang, 1987). In this work, we demonstrate the presence of H_1-receptors coupled to phosphoinositide phospholipase C in neuronal cultures from rat cerebrum. In cells prelabeled with (^{3}H)inositol and in the presence of 10 mM LiCl, exposure to histamine for 30 min induces a significant accumulation of (^{3}H)IP but not of the higher inositol phosphates, in agreement with similar studies in cortex slices (Claro et al. 1986) and astrocyte-rich cultures (Arbonés et al. 1988) from the same rat strain. The stimulation of (^{3}H)IP formation was linear with time for 30-40 min, decreasing at longer times most probably as a result of receptor desensitization, phenomenon that in cerebellar granule cells was shown to occur more rapidly for H_1- than for other receptors coupled to phosphoinositide hydrolysis (Dillon-Carter and Chuang, 1989).

The histamine-induced (^{3}H)IP accumulation in cerebral neurons was concentration-dependent and appeared to me mediated by a homogeneous population of H_1-receptors as indicated by the following observations: 1)the Hill coefficient of concentration-effect curves is not significantly different from one; 2)the effect of histamine is mimicked by an H_1- but not by an H_2-receptor agonist; 3)the effect is stereospecifically inhibited by the H_1-antagonist D-chlorpheniramine but not by H_2-antagonists; and 4)the calculated Ki for the H_1-antagonist mepyramine is very close to that obtained for the same effect in rat cortex slices (Claro et al. 1986) and for (^{3}H)mepyramine binding to membranes (Tran et al. 1978, Hill and Young, 1980).

The potency of histamine for stimulating (^{3}H)IP formation in neurons is of the same order of that found in astrocytes (Arbonés et al. 1988) and developing rat cortex (Claro et al.1987) but much higher than that determined in adult rat cortical slices (Claro et al. 1986). In this last preparation as well as in astrocyte cultures, the absence of spare receptors was demonstrated by using the irreversible antagonist phenoxybenzamine. Thus, although not demonstrated here, it is very unlike ly that the high potency of histamine in neurons is due to the existence of a receptor reserve. An explanation for the higher potency of histamine in both neurons and astrocytes compared to the adult cortex may arise from considering possible changes in receptor sensitivity as a function of the specific innervation of the tissue. We have previously observed that in rat cortical slices, both the potency and the efficacy of histamine for eliciting (^{3}H)IP accumulation decrease during ontogenic development (Claro et al. 1987) and the same phenomenon has been described for rat brain muscarinic receptors (Heacok et al. 1987). These observations suggest that a natural form of desensitization may take place as innervation develops. Plasticity of histamine H_1-receptors in rat brain was shown by the hypersensitivity that developed after denervation of cortical areas and that was manifested by the appearance of a higher potency component in the histamine-stimulated phosphoinositide hydrolysis (García-Ladona et al. 1987). In our cerebral cultures which are prepared from 16-day-old rat embryos, very few histaminergic neurons are expected to be present since at that age, histamine-immunoreactive neurons are located predominantly in the ventral mesencephalon and rhombencephalon (Auvinen and Panula, 1988). In fact, we have been unable to demonstrate (^{3}H)histamine formation in neurons prelabeled with (^{3}H)histidine (unpublished results).

The maximal histamine effect observed in neurons is less than half that in astrocytes. This is not due to an insufficient maturation of the system in the cultured neurons because maximal stimulation is reached by 7 days in culture, and is maintained at least, for the time period the cells are generally used. One possibility is that less H_1-receptors are present in neurons than in astrocytes. An observation that would support the hypothesis that astrocytes contribute more than neurons to the overall population of cerebral H_1-receptors, is the absolute dependency on extracellular Ca^{2+} that is common to the inositol phosphate response in cortical slices (Kendall and Nahorski, 1984) and astrocytes (Arbonés et al. 1988) but not in neurons. While in the two former preparations the histamine response is totally abolished by the omission of Ca^{2+} from the incubation medium, in neurons that condition only decreases the effect by 26%, being necessary to quelate all the remnant ion in order to eliminate it. This difference in Ca^{2+}-dependency may also be an indication of differences in the molecular mechanisms operating in the two cell types. It has been demonstrated in brain membranes that H_1-receptors are coupled to a GTP-binding protein-linked phospholipase C and can stimulate the enzyme at basal Ca^{2+} concentrations (approx. 0.1 μM) (Claro et al. 1989). On the other hand, the phospholipase C can also be stimulated by higher Ca^{2+} concentrations in an agonist-independent manner and there is evidence that this latter enzyme may be different from the G-protein-linked activity (Gonzales and Crews, 1988). In astrocytes, H_1-receptors stimulate Ca^{2+} entry (Arbonés et al. 1990) and this could trigger phosphoinositide hydrolysis, whereas in neurons the lesser Ca^{2+}-dependency suggests that a G-protein mediated mechanisms might predominate. In accordance with this hypothesis, conditions that increase intracellular Ca^{2+}, such as depolarizing concentrations of K^+ or the absence of Na^+, stimulate inositol phospholipid hydrolysis in neurons, as also occurs in other excitable tissue preparations (Eberhard and Holz, 1988), but their effects are additive to that of histamine. Furthermore, these results together with the lack of inhibition of the histamine response by the Na^+ channel-blocker tetrodotoxin argue against the effect of histamine being due to the liberation of other neurotransmitters, as appears to occur for carbachol and norepinephrine in striatal neurons (Weiss at el. 1988).

ACKNOWLEDGMENTS

This work was supported in part by a grant from DGICYT (PM 89-0079). We thank M.E. Palou for typing the manuscript.

REFERENCES

Agulló L, Picatoste F, García A (1990) Histamine stimulation of cyclic AMP accumulation in astrocyte-enriched and neuronal primary cultures from rat brain J Neurochem 55:1592-1598

Alexander SPH, Hill SJ, Kendall DA (1990) Differential effects of elevated calcium ion concentrations on inositol phospholipid responses in mouse and rat cerebral cortical slices Biochem Pharmacol 40:1793-1799

Arbonés L, Picatoste F, García A (1988) Histamine H_1-receptors mediate phosphoinositide hydrolysis in astrocyte-enriched primary cultures Brain Res 450:144-152

Arbonés L, Picatoste F, García A (1990) Histamine stimulates glycogen breakdown and increases $^{45}Ca^{2+}$ permeability in rat astrocytes in primary culture Mol Pharmacol 37:921-927

Auvinen S, Panula P (1988) Development of histamine-immunoreactive neurons in the rat brain J Comp Neurobiol 276:289-303

Carswell H, Galione AG, Young JM (1987) Differential effect of temperature on histamine- and carbachol-stimulated inositol phospholipid breakdown in slices of guinea-pig cerebral cortex Br J Pharmacol 90:175-182

Claro E, Arbonés L, García A, Picatoste F (1986) Phosphoinositide hydrolysis mediated by histamine H_1-receptors in rat brain cortex Eur J Pharmacol 123:187-196

Claro E, García A, Picatoste F (1987) Histamine-stimulated phosphoinositide hydrolysis in developing rat brain Mol Pharmacol 32:384-390

Claro E, García A, Picatoste F (1989) Carbachol and histamine stimulation of guanine-nucleotide-dependent phosphoinositide hydrolysis in rat brain cortical membranes Biochem J 261: 29-35

Dillon-Carter O, Chuang D (1989) Homologous desensitization of muscarinic cholinergic, histaminergic, adrenergic, and serotonergic receptors coupled to phospholipase C in cerebellar granule cells J Neurochem 52:598-603

Eberhard DA, Holz RW (1988) Intracellular Ca^{2+} avtivates phospholipase C Trends Neurosci 11:517-520

García-Ladona J, Claro E, García A, Picatoste F (1987) Denervation supersensitivity of histamine H_1-receptors in rat brain cortex Neurosci 284P

Gonzales RA, Crews FT (1988) Differential regulation of phosphoinositide phosphodiesterase activity in brain membranes by guanine nucleotides and calcium J Neurochem 50:1522-1528

Heacock AM, Fisher SK, Agranoff BW (1987) Enhanced coupling of neonatal muscarinic receptors in rat brain to phosphoinositide turnover J Neurochem 48:1904-1911

Hill SJ (1990) Distribution, properties and functional characteristics of three classes of histamine receptors Pharmacol Rev 42:45-83

Hill SJ, Kendall DA (1989) Cross-talk between different receptor-effector systems in the mammalian CNS Cell Signal 1:135 -141

Hill SJ, Young JM (1980) Histamine H_1-receptors in the brain of the guinea-pig and the rat: differences in ligand binding properties and regional distribution Br J Pharmacol 68: 687-696

Kendall DA, Nahorski SR (1984) Inositol phospholipid hydrolysis in rat cerebral cortical slices.II Calcium requirement J Neurochem 42:1388-1394

Palacios JM, Wamsely JK, Kuhar MJ (1981) The distribution of histamine H_1-receptors in the rat brain: an autoradiographic study Neuroscience 6:15-37
Pearce B, Morrow C, Murphy S (1986) Receptor-mediated inositol phospholipid hydrolysis in astrocytes Eur J Pharmacol 121: 231-243
Tran VT, Chang RSL, Snyder SH (1978) Histamine H_1-receptors identified in mammalian brain membranes with (^3H)mepyramine Proc Natl Acad Sci USA 75:6290-6294
Tran VT, Freeman AD, Chang RSL, Snyder SH (1980) Ontogenic development of histamine H_1-receptor binding in rat brain J Neurochem 34:1609-1613
Weiss S, Schmidt BH, Sebben M, Kemp DE, Bockaert J, Sladeczek F (1988) Neurotransmitter-induced inositol phosphate formation in neurons in primary culture J Neurochem 50:1425-1433
Whitworth P, Kendall DA (1988) Lithium selectively inhibits muscarinic receptor-stimulated inositol tetrakisphosphate accumulation in mouse cerebral cortex slices J Neurochem 51:258-265
Xu J, Chuang DM (1987) Serotonergic, adrenergic and histaminergic receptors coupled to phospholipase C in cultured cerebellar granule cells of rats Biochem Pharmacol 36:2353-2358

INTRACELLULAR CA^{2+} HOMEOSTASIS: RECEPTOR-ACTIVATED CA^{2+} CHANNELS AND INTRACELLULAR CA^{2+} POOLS

C. Fasolato[1*], M. Zottini[1], P. Chiozzi[2], S. Treves[2], A. Villa[3], E. Clementi[3], J. Meldolesi[3] and T. Pozzan[1]

1 Institute of General Pathology, University of Padova and Consiglio Nazionale delle Ricerche, Center of Biomembranes, Padova, Italy.
2 Institute of General Pathology, University of Ferrara, Italy.
3 Department of Pharmacology, University of Milano and Scientific Institute S. Raffaele, Milano, Italy.
* To whom correspondence should be addressed: Institute of General Pathology, Via Trieste 75, 35121 Padova.

INTRODUCTION

In eukaryotic cells, the intracellular free calcium concentration, $[Ca^{2+}]_i$, plays a critical role at almost every level. Key functions, such as excitability, contraction, metabolism and gene expression require changes in $[Ca^{2+}]_i$.

Over the last few years, our knowledge about the mechanisms of Ca^{2+} entry and discharge from intracellular stores has increased dramatically thanks to the introduction of new tecniques, such as patch clamping (Hamill et al. 1981) and intracellularly trappable Ca^{2+} indicators (Tsien et al. 1982, Grynkiewicz et al. 1985). There is now the opportunity and the need for integrating the informations obtained by these approaches with those obtained with molecular biology.

$[Ca^{2+}]_i$ is a very dynamic parameter in living cells. Following stimulation, it can increase 10-20 fold over basal

NATO ASI Series, Vol. H 52
Biological Signal Transduction
Edited by E. M. Ross and K. W. A. Wirtz

in very short periods of time (msec to sec). This initial rise is either followed by repetitive oscillations or is maintained elevated for several minutes, depending on the dose, the type of stimulus and the cell type investigated. Cells are able to cope for several hours with $[Ca^{2+}]_i$ increases 2-3 fold greater than resting values, while larger elevations (5-10 fold) are cytotoxic if maintained for these long periods of time. These transient or sustained increases in $[Ca^{2+}]_i$ are finely regulated by plasma membrane and intracellular Ca^{2+} store pumps and channels.

A number of agonists have been shown to increase plasma membrane Ca^{2+} conductance through pathways different from the classical voltage-operated Ca^{2+} channels (VOCs). The nature and the control of Ca^{2+} entry via receptor-regulated Ca^{2+} channels remain, however, largely obscure (Pietrobon et al. 1990). In the first part of this contribution we will discuss recent data on this type of Ca^{2+} entry.

In addition, both in excitable and non-excitable cells, different agonists rapidly increase the $[Ca^{2+}]_i$ by releasing Ca^{2+} from intracellular stores. In non-muscle cells the most popular candidate for triggering this release is Ins(1,4,5)P_3, which binds to and opens an intracellular Ca^{2+} channel localized in a membrane-enclosed Ca^{2+} pool. The mechanism leading to Ins(1,4,5)P_3 formation, is reasonably well characterized in its functional and molecular aspects (Majerus et al. 1988). The pharmacological, electrophysiological and molecular properties of the Ins(1,4,5)P_3 receptor are beginning to be clarified and recent data demonstrate that they are clearly different from the ryanodine receptors, the molecular complexes which mediate Ca^{2+} release from sarcoplasmic reticulum of striated muscles (Fukuiki et al. 1989; Ferris et al. 1989; Takeshima et al. 1989). Nevertheless, our knowledge on intracellular Ca^{2+} stores is still incomplete and numerous questions remain largely unanswered. In particular: a) Is $Ins(1,4,5)P_3$ the only mechanism for receptor-mediated release of internal Ca^{2+}? b) What is the nature of the $Ins(1,4,5)P_3$ sensitive Ca^{2+} store? Is it a separate organelle or a specialised element of the endoplasmic reticulum? c) Which is the mechanism of Ca^{2+} uptake? d) Is there a link between Ca^{2+} release and Ca^{2+} influx?

In this report we try to provide some answers to these questions.

RECEPTOR-MEDIATED CA^{2+} ENTRY

As opposed to Ca^{2+} VOCs which, under physiological conditions, are highly Ca^{2+} selective and open only in response to depolarization, receptor-mediated Ca^{2+} channels show a slight preference for Ca^{2+} over Na^+, are inhibited by depolarization and are insensitive to dihydropyridine agonists and antagonists (Pietrobon et al. 1990). Nevertheless, receptor-mediated Ca^{2+} channels can not be gathered together in an homogenous group and these channels may be tentatively classified on the basis of their gating mechanism (Meldolesi & Pozzan, 1987) :

- Receptor-operated Ca^{2+} channels, or ROCs. In these channels the receptor binding site is part of the molecular complex forming the channel itself. Ca^{2+} ROCs are exemplified by the NMDA receptor, in neurons (Novak et al. 1984), and the ATP receptor, in smooth muscle cells (Benham & Tsien 1987).

- Second messenger-operated Ca^{2+} channels or SMOCs. The cytosolic 2^ messenger has not been identified yet. Different activators have been postulated in different cell types, such as Ca^{2+} itself in neutrophils (Von Tscharner et al. 1987), Ins(1,4,5)P_3 in T lymphocytes (Kuno & Gardner 1987) and Ins(1,4,5)P_3 in concert with Ins(1,3,4,5)P_4, in lacrimal acinar cells (Morris et al. 1987).
- G-protein-activated Ca^{2+} channels or GOCs (Pietrobon et al. 1990). When observed in patch clamp studies (cell-attached configuration), opening of these channels is triggered by the agonist in the pipette but not in the bath. In addition, GOCs are also activated by GTP analogues, added to the cytoplasmic side of the excised patch (Matsumaga et al. 1988).

The data, here reported, briefly summarize the characteristics of two Ca^{2+} influx channels (activated by bradykinin and ATP respectively) in the cell line PC12, endowed with both VOCs and Ins(1,4,5)P_3 releasable Ca^{2+} stores. In this model the nonapeptide bradykinin, BK, strongly activates phosphatydil inositol-4,5-bisphosphate ($PtInsP_2$) hydrolysis, mobilizing more than 50% of the Ca^{2+} released by ionomycin. On the contrary, extracellular ATP, only sligthly increases the level of Ins(1,4,5)P_3 with a small release of Ca^{2+} from the Ins(1,4,5)P_3 sensitive stores (see Table I).

TABLE I. *Effect of ATP and BK on Ins(1,4,5)P_3 and Ca^{2+} redistribution.*

Cells, labelled with ^{3}H-inositol for 48h, were loaded with Fura-2 as described by Fasolato et al. (1988) and challenged with ATP (100 μM) or BK (100 nM) for 10 sec in Ca^{2+} -free, EGTA-containing, Krebs-Ringer medium buffered with Hepes (KRH). Parallel cell batches were used to measure ^{3}H-Ins(1,4,5)P_3 level and Ca^{2+} redistribution. Inositol phosphates were extracted and measured by HPLC with a Partisil-SAX column as described by Fasolato et al. (1988). Calibration of Ca^{2+} redistribution is expressed as % increase in Fura-2-Ca^{2+} complex (ΔF) compared to the total Ca^{2+} sensitive Fura-2 signal (Fmax). Data are mean values ± SD of 8 samples.

	^{3}H-Ins(1,4,5)P_3 (dpm/mg protein)	Ca^{2+} redistribution (% ΔF/Fmax)
Unstimulated	2890 ± 88	0
Bradykinin	8800 ± 106	39.4 ± 3
ATP	3650 ± 101	6.5 ± 4

Both agonists, however, share the ability of increasing plasma membrane Ca^{2+} permeability; in the case of BK this is characterized by a long-lasting $[Ca^{2+}]_i$ increase, while for ATP the Ca^{2+} influx is rapidly inactivated (see fig. 1 a,b).

Ca^{2+} influx, activated by both agonists, appears to be associated with an influx of Na^+ ions. In fact an inward current, revealed by membrane potential depolarization, is observable with both stimuli in the presence (not shown) and absence of Ca^{2+} in the medium (see fig 1 c,d), but not in sucrose-based medium devoid of permeant cations.

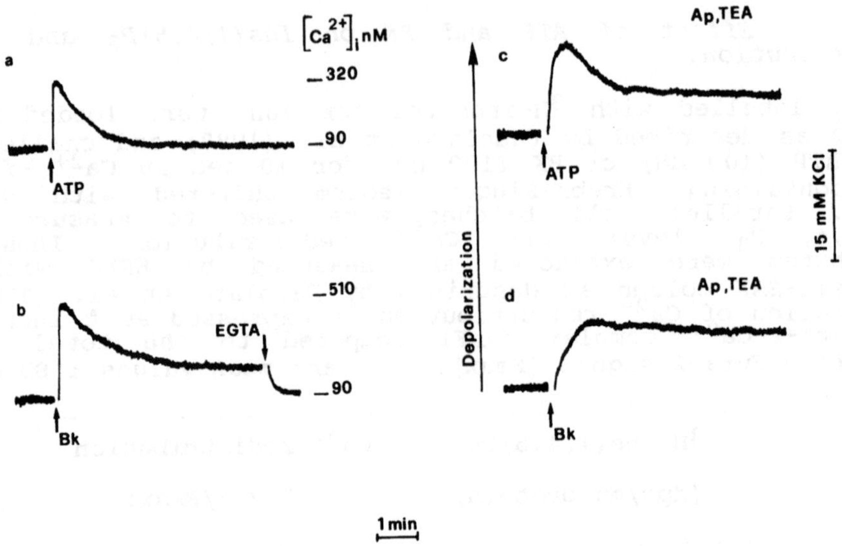

FIGURE 1. *Effect of ATP and BK on $[Ca^{2+}]_i$ (a,b) and membrane potential (c,d) in PC12 cells.*

a,b: KRH-medium containing 1 mM $CaCl_2$. Cells were loaded with Fura-2 as described in Table I and challenged with an optimal dose of ATP or BK. Where indicated ATP (100 µM), BK (100 nM) and EGTA (1 mM) were added. The calibrated $[Ca^{2+}]_i$ (nM) is reported on the right hand side.
c,d: KRH-medium, containing apamin (0.6 µM, Ap) and tetraethylammonium (10 mM, TEA). Membrane potential was measured with bis-oxonol as described by Di Virgilio et al. (1987). Where indicated ATP (100 µM) and BK (100 nM) were added.

Patch clamp data, in the whole cell configuration, confirm that BK activates an inward current. Activation of a low selective cation channel seems thus to be a common pathway utilized by both ATP and BK receptors (Fasolato et al. 1990).

Ca^{2+} redistribution and cation influx, triggered by BK, are due to the activation of the same receptor, a B_2 subtype, (Fasolato et al. 1988). In contrast, Ca^{2+} redistribution and influx, induced by ATP, are apparently mediated by two separate receptors. As shown in Fig 2, other nucleotide trisphosphates, such as UTP, inhibit the ATP-induced Ca^{2+} redistribution but leave the ATP-activated Ca^{2+} influx and depolarization almost completely unaffected.

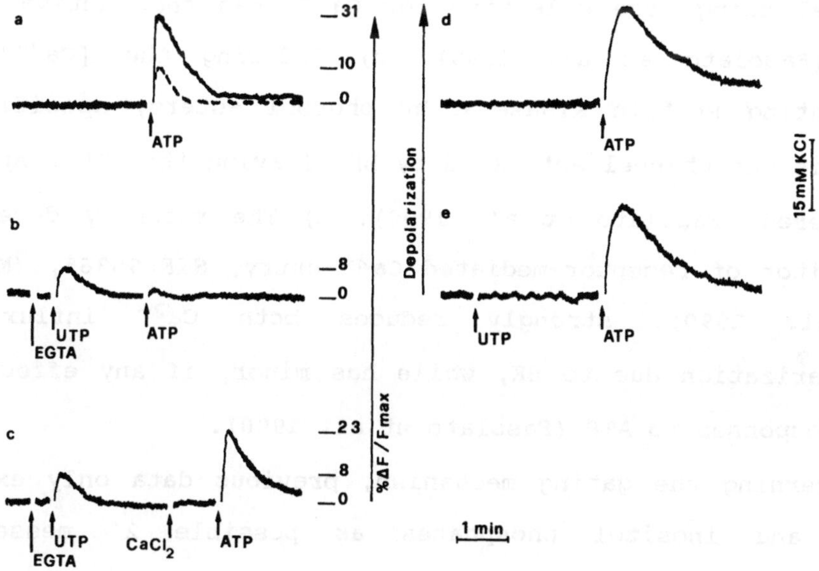

FIGURE 2. *Effect of UTP on cation influx induced by ATP*.
a-c: Effect of UTP on ATP-induced Ca^{2+} redistribution and Ca^{2+} influx. Conditions as in fig 1. *a:* KRH-medium containing 1 mM $CaCl_2$ (continuous trace) or 0.1 mM EGTA (dashed trace).
b,c: Ca^{2+} -free, KRH-medium. Where indicated EGTA (0.1 mM), UTP (10 µM), ATP (100 µM) and $CaCl_2$ (1 mM) were added.
d-e: Effect of UTP on ATP-induced depolarization; conditions as in Fig 1c,d. Ca^{2+} -free, KRH-medium. Where indicated UTP (10 µM) and ATP (100 µM) were added.

Ca^{2+} influx and depolarization, induced by ATP, show the same dose-dependence when plotted as function of the ATP^{4-} concentration ($EC_{50}=3x10^{-7}M$) (Fasolato et al. 1990). The minor effect on $Ins(1,4,5)P_3$ level, together with the inability of ADP or other nucleotide trisphosphates (except ATPγS) to mimick the ATP-induced increase in cation permeability suggest that this effect is not linked to a classical P_{2Y} receptor (Fasolato et al. 1990).

Several lines of evidence suggest that, although similar, ATP- and BK-activated cation channels are clearly distinct: a) Cation selectivity seems to be different, the BK-activated channel being more selective for Ca^{2+} than that activated by ATP (Fasolato et al. 1990). b) Reducing the $[Ca^{2+}]_i$ or activating Protein Kinase C by phorbol esters, specifically inhibit the channel activated by BK, leaving the ATP responses unaltered (Fasolato et al. 1990). c) The recently developed inhibitor of receptor-mediated Ca^{2+} entry, SKF 96365, (Merrit et al. 1990), strongly reduces both Ca^{2+} influx and depolarization due to BK, while has minor, if any effect, on the responses to ATP (Fasolato et al. 1990).

Concerning the gating mechanism, previous data only exclude Ca^{2+} and inositol phosphates as possible 2^ messengers mediating the BK responses (Fasolato et al. 1988). Direct modulation by G-protein or a capacitative like model (Putney, 1986; Takemura et al. 1989) appear thus the most plausible working hyphothesis to explain the mechanism of Ca^{2+} influx activated by BK. On the other hand, the rapid onset of ATP-activated channel, together with its pharmacological properties and ligand specificity, indicate that a strong similarity exists between this channel and the ROC type, ATP-activated channel, described by Benham & Tsien (1987), in smooth muscle cells.

$INS(1,4,5)P_3$ SENSITIVE STORES

Streb et al. (1983) first demonstrated that $Ins(1,4,5)P_3$ was capable of causing Ca^{2+} redistribution from a non-mitochondrial intracellular store in permeabilized pancreatic acinar cells. Binding studies, later, revealed that $Ins(1,4,5)P_3$ receptors copurify with microsomal fractions (Prentki et al. 1984). Hence the intracellular Ca^{2+} store was generally identified with the major component of microsomes, the endoplasmic reticulum (ER), despite the fact that numerous evidence suggest that this was probably an oversimplification (Meldolesi et al. 1990).

The functional characteristics of the $Ins(1,4,5)P_3$ sensitive stores closely resemble those of the sarcoplasmic reticulum (SR) of striated fibers. In fact both stores take up Ca^{2+} in an energy dependent way and release it rapidly upon stimulation. We have thus speculated that the molecular components of the $Ins(1,4,5)P_3$ sensitive stores should be, at least functionally, similar to those of SR. In particular we predicted (Volpe et al. 1988) that the $Ins(1,4,5)P_3$ sensitive stores should be endowed with: a) Ca^{2+} channels and Ca^{2+} accumulating mechanisms in their membrane b) Ca^{2+} binding proteins, functionally similar to muscle calsequestrin (CS), whithin their lumen.

Using two antibodies raised against the muscle Ca^{2+}-ATPase and CS, we found that in liver, pancreas and in a variety of cultured cell lines, the crossreactive antigens are not uniformly distributed throughout the ER, but rather are localized in small structures, often apposed to ER cisternae, but clearly distint from them (Volpe at al. 1988; Pozzan et al. 1988). These structures, named "calciosomes", were supposed to be the targets of Ins(1,4,5)P_3, although no formal proof for this hypothesis could be provided.

In the last few years, a large number of intracellular Ca^{2+} binding proteins have been purified from non-muscle cells. Ca^{2+} binding proteins, functionally similar to CS, now renamed "calreticulins", have been found in a large variety of eukaryotic tissues (Obendorf et al. 1988; Fliegel et al. 1989; Treves et al. 1990). Liver and skeletal muscle calreticulins have been cloned and their deduced aminoacid sequence reveals little homology with striated muscle CS. Interestingly, in spite of the presence in liver calreticulin of the C-terminal sequence KDEL, a property of resident ER proteins, calreticulin contains a N-linked oligosaccharide residue with a terminal galactose (Fliegel et al. 1989, Nguyen Van et al. 1989), a typical feature of proteins reaching the trans Golgi network. This observation suggests again that the calreticulin containing structures, or calciosomes, are distinct from typical ER cysternae.

Studies on subcellular fractions of HL60 cells, cerebellum, liver and parotid show that most of the Ins(1,4,5)P_3 binding is recovered in fractions enriched in Ca^{2+}-ATPase and calreticulin, while no or little correlation was found with typical ER markers (Volpe et al. 1988; Volpe et al. 1990; Chiozzi et al. unpublished). Purkinje neurons express extremely high level of Ins(1,4,5)P_3 receptors (> 50 fold higher than in other tissues); thus only in this cell type quantitative immmunolocalization has been possible to date.

The Ins(1,4,5)P_3 receptor was found to be localized primarily on stacked smooth cisternae, which, in the cell body, are often clearly continuous with rough ER (Satoh et al., 1989). One must keep in mind however, that Purkinje neurons exibit other peculiarities which make these cells unique. Indeed, calreticulin has been recently purified from rabbit and dog brain and cerebellum, but mammalian Purkinje neurons appear to express neither calreticulin nor calsequestrin to any significant extent (Satoh et al. 1990). On the other hand, chicken Purkinje neurons express high levels of a protein biochemically and antigenically identical to skeletal muscle CS (Volpe et al. 1990).

INS(1,4,5)P_3 INSENSITIVE STORES

In addition to the high level of Ins(1,4,5)P_3 receptors, Purkinje neurons also express high levels of ryanodine binding protein(s) (Ellisman et al. 1990). While the functional role of the ryanodine receptor in striated muscles has been widely investigated, its function and modulation in non-muscle cells is still rather obscure. The subcellular localization of the ryanodine receptors, in non-muscle tissues, has not been studied at the EM level yet. Using immunofluorescence, Ellisman et al. (1990) demonstrated intense labeling of Purkinje neurons at the level of the cell body and of the dendrites. Opposed to the anti-Ins(1,4,5)P_3 receptor however, no labeling of the dendritic spines was observed with anti-ryanodine receptor antibodies. In skeletal and cardiac muscles Ca^{2+} has been proposed as the natural activator of the ryanodine channel, via the so called "Ca^{2+} induced Ca^{2+} release", (CICR, Berridge 1990). The same mechanism has been proposed as the basis for the oscillatory behaviour of $[Ca^{2+}]_i$ induced by a large variety of agonists (Berridge 1990, Rink & Jacob 1989). Accordingly, the oscillations of $[Ca^{2+}]_i$ observed upon stimulation with agonists coupled to $Ins(1,4,5)P_3$ generation, are due to cycles of Ca^{2+} uptake and release by $Ins(1,4,5)P_3$ sensitive and CICR stores. Such a model predicts that the ryanodine receptors and the $Ins(1,4,5)P_3$ gated channels should be expressed in distinct Ca^{2+} pools. Indirect evidence in favour of two separate intracellular Ca^{2+} pools has been provided by various laboratories. For example Malgaroli et al. (1990) demonstrated that in rat chromaffin cells the spontaneous oscillations of $[Ca^{2+}]_i$ occur independently of a block of $Ins(1,4,5)P_3$ generation and, viceversa, that emptying of the CICR store does not block the $Ins(1,4,5)P_3$ induced Ca^{2+} mobilization. In bovine chromaffin cells Burgoyne et al. (1989) demonstrated that $[Ca^{2+}]_i$ increases induced by caffeine (acting on the CICR) and carbachol (producing $Ins(1,4,5)P_3$) cause spatially distinct rises in $[Ca^{2+}]_i$. Furthermore they demonstrated that two monoclonal antibodies, raised against the skeletal muscle Ca^{2+}-ATPase, labelled two distinct bands in Western blotting (MW 100 and 140 KDa respectively). The antibody directed against the 100 KDa band diffusely labelled the cytoplasm while that directed against the 140 KDa band decorated a specific region near the nucleus. Apart from the immunoreactivity, however, no proof was provided that the 140 KDa protein is a Ca^{2+}-ATPase.

In the last two years we have addressed the problem of different Ca^{2+} pools by two experimental approaches: on the one hand we have tried to determine whether one or more intracellular Ca^{2+}-ATPases are expressed in microsomal fractions; to this end we have utilized as a model the chicken cerebellum. On the other hand we have functionally characterized, in intact PC12 cells, the intracellular Ca^{2+} pools sensitive to Ins(1,4,5)P_3, caffeine or ryanodine.

In cerebellum microsomes only one protein band cross-reacts with a monoclonal anti cardiac Ca^{2+}-ATPase and is phosphorylated in the presence of Ca^{2+} and [^{32}P]-ATP. Although similar to the muscle isoforms, this Ca^{2+}-ATPase ($\simeq$100 KDa) is kinetically distinct from them and appears to be widely distributed in the cytoplasm, dendrites and axons of Purkinje neurons (Michelangeli et al. submitted).

Since Purkinje cells express high concentrations of both ryanodine and Ins(1,4,5)P_3 receptors, but apparently only one Ca^{2+}-ATPase is detectable by immunological, kinetic and biochemical criteria, it seems that CICR and Ins(1,4,5)P_3 sensitive stores are endowed with the same, or extremely similar, Ca^{2+}-ATPase(s).

Fig. 3 summarizes the functional characteristics of Ca^{2+} pools in intact PC12 cells. In these cells it seems that ryanodine and Ins(1,4,5)P_3 receptors colocalize in the same Ca^{2+} pool. In fact, treatments with doses of BK that completely empty the Ins(1,4,5)P_3 sensitive stores also cause depletion of the caffeine sensitive Ca^{2+} pools. Viceversa, treatments with caffeine plus ryanodine result in complete depletion of the Ins(1,4,5)P_3 sensitive stores.

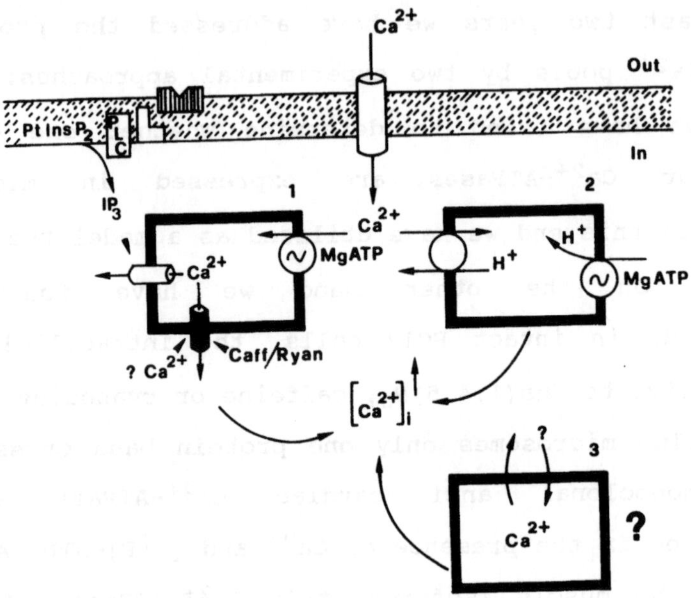

FIGURE 3. *Ca^{2+} pools in PC12 cells.*

Store 1: Ins(1,4,5)P_3 and caffeine sensitive store, (calciosome?) This store can be emptied by Ins(1,4,5)P_3, caffeine plus ryanodine or thapsigargin.
Store 2: This store is presumably represented by secretory granules. (Fasolato et al., in preparation) Release of Ca^{2+} from this store requires both ionomycin and collapse of pH gradients (monensin or NH_4Cl).
Store 3: This store is insensitive to Ins(1,4,5)P_3, caffeine, ryanodine, thapsigarin and to collapse of pH gradients but can be emptied by ionomycin. Its cytological nature is presently obscure.

In addition, PC12 cells contain two other slowly exchangeable Ca^{2+} pools, which are both Ins(1,4,5)P_3 and caffeine insensitive. One of these pools seems to coincide with secretory granules, the other is not identified yet and is mobilizable only by Ca^{2+} ionophores.

The Ins(1,4,5)P_3 and caffeine sensitive pools take up Ca^{2+} via a thapsigargin sensitive Ca^{2+}-ATPase (Tharstrup et al. 1990), while secretory granules utilize a pH driven Ca^{2+} accumulation mechanism. Ca^{2+} uptake by the ionophore sensitive pool has not been characterized yet.

Taken together these results indicate that Ca^{2+} handlying by intracellular stores is more complex than expected and that qualitative differences exist between apparently similar cell types. For example one can compare rat chromaffin cells and PC12 cells (a tumor cell line derived from the former): using the same experimental approach it was concluded that the caffeine sensitive and insensitive pools are partially distinct in the normal cells, but identical in the tumor cell line. It is temptying to speculate that this is the reason why in chromaffin cells the $[Ca^{2+}]_i$ oscillates both spontaneously and after stimulation (Malgaroli et al. 1990) while this does not occur, under any condition, in PC12 cells.

AKNOWLEDGEMENTS. We are in debt to G. Ronconi for skilfull technical assistance and K. Marchioro for typing. This work was supported by grants from the Italian Ministry of Education (40% and 60%), by the National Research Council "Target Project on Biotechnology and Bioinstrumentation" and by the Associazione Italiana per la Ricerca sul Cancro.

REFERENCES

Benham, C.D. and Tsien, R.W. (1987) Nature 328,275-278.

Berridge, M.J. (1990) J. Biol. Chem. 265,9583-9586.

Burgoyne, R.D., Cheek, T.R., Morgan, A., O'Sullivan, A. Moreton, R.B., Berridge, M.J., Mata, A., Colyer, J., Lee, A.G. & East, J.M. (1989) Nature 342,72-74.

Di Virgilio, F., Milani, D., Leon, A., Meldolesi, J. & Pozzan, T. (1987) J. Biol. Chem. 262,9189-9195.

Ellisman, M.H., Deerinck, T.J., Onyang, Y., Beck, C.F., Tanksley, S.J., Walton, P.D., Airey, J.A. and Sutko, J.L. (1990) Neuron 5,135-146.

Fasolato, C., Pandiella, A., Meldolesi, J. & Pozzan, T. (1988) J. Biol. Chem. 263,17350-17359.

Fasolato, C., Pizzo, P. and Pozzan, T. (1990) J. Biol. Chem. (in press).

Ferris, C.D., Hunganir, R.L., Supattapone, S., Snyder, S.H. (1989) Nature 342,87-89.

Fliegel, L., Burns, K., MacLennan, D.H., Reithmeier, R.A. F. & Michalak, M. (1989) J. Biol. Chem. 269,21522-21528.

Furuichi, T., Yoshikawa, S., Miyawaki, K., Wada. K., Maeda, M., Mikoshiba, K. (1989) Nature 342,32-38.

Grynkiewicz, G., Poenie, M. and Tsien, R.Y. 1985), J. Biol. Chem. 260,3440-3450.

Hamill, O.P., Marty, A., Neher, E., Sakmann, B., and Sigworth, F.J. (1981) Pflueger's Arch. 391,85-100.

Kuno, M. & Gardner, P. (1987) Nature 326,301-304.

Malgaroli, A., Fesce, R. & Meldolesi, J. (1990) J. Biol. Chem. 265,3005-3008.

Majerus, P.W., Conally, T.M., Bansall, V.S., Inhorn, R.C., Ross, T.S., Lips, D.L. (1988) J. Biol. Chem. 263,3051-3054

Matsumaga, H., Nishimoto, I., Kojima, I., Yamashita, N., Kurokawa, K. & Ogata, E. (1988) Am. J. Physiol. 255,C442-C446.

Meldolesi, J., Madeddu, L. & Pozzan, T. (1990) Biochim. Biophys. Acta 1055,130-140.

Meldolesi, J. and Pozzan, T. (1987) Exp. Cell Res. 171,271-283.

Merrit, J.E., Armstrong, W.P., Benham, C.D., Hallam, T.J., Jacob, R., Jaxa-Chamlec, A., Leigh, B.K., Mc Carthy, S.A., Moores, K.E. and Rink, T.J. (1990) Biochem. J. 271,515-522.

Morris, A.P., Gallacher, D.V., Irvine, R.F. & Peterson, O.H. (1987) Nature 330,653-655.

Nguyen Van, P., Peter, F. & Solig, H.D., (1989) J. Biol. Chem. 264,17494-17501.

Nowak, L., Bregetowsky, P., Asher, P., Herbert, A. and Prochiantz, A. (1984) Nature 307,462-465.

Obendorf, J.A., Lebeche, D., Head, J.F. & Kaminer, B. (1988) J. Biol. Chem. 263,6806-6809.

Pietrobon, D., Di Virgilio, F. & Pozzan, T. (1990) Eur. J. Biochem. 193,599-622.

Pozzan, T., Volpe, P., Zorzato, F., Bravin, M., Krause, K.H., Lew, D.P., Hashimoto, S., Bruno, B. & Meldolesi, J. (1988) J. Exp. Biol. 139,181-193.

Prentki, M., Biden, T.J., Janjic, D., Irvine, R.F., Berridge, R.J. & Wollheim, C.B. (1984) Nature 309,562-564

Putney, J.W. (1986) Cell Calcium 7,1-12.

Rink, T.J. & Jacob, R. (1989) Trends Neurosc. 12,43-46.

Satoh, T., Ross, C.A., Supattapone, S., Pozzan, T., Snyder, S.H. (1989) Nature 342,87-89.

Satoh, T., Ross, C.A., Villa, A., Supattapone, S. Pozzan, T., Snyder, S.H. & Meldolesi, J. (1990) J. Cell. Biol. 111,615-624.

Streb, H., Irvine, R.F., Berridge, M.J. and Shulz, I. (1983) Nature 306,67-69.

Takemura, H., Huges, A.R., Thastrup, O. & Putney, J.W. (1989) J. Biol. Chem. 264,12266-12271.

Takeshima, H., Nishimura, S., Matsumoto, T., Ishida, H., Kangawa, K., Matsuo, H., Veda, M., Hanaoka, M., Hirose, T. & Numa, S. (1989) Nature 339,439-445.

Thastrup, O., Cullen, P.J., Drhak, B.K., Hanley, M.R., Dawson, A.P. (1990) Proc. Natl. Acad. Sci. USA 87,2466-24770.

Treves, S., De Mattei, M., Lanfredi. M., Villa, A., Green, G., MacLennan, D.H., Meldolesi, J. & Pozzan, T. (1990) Biochem. J. 271,473-480.

Tsien, R.J., Pozzan, T. & Rink, T.J. (1982) Nature 295,68-71.

Volpe, P., Alderson, Lang, B.H., Madeddu, L., Damiani, E., Collins, J.H. & Margreth, A. (1990) Neuron (in press).

Volpe, P., Krause, K.H., Hashimoto, S., Zorzato, F., Pozzan, T., Meldolesi, J. & Lew, D.P. (1988) Proc. Natl. Acad. Sci. USA 85, 1091-1095.

Von Tscharner, V., Prod'hom, B., Baggiolini, M. & Reuter, H. (1986) Nature 324,369-372.

Multiple sources of *sn*-1,2-diacylglycerol in mitogen-stimulated Swiss 3T3 cells; evidence for activation of phosphoinositidase C and PtdCho-phospholipase D.

Simon J. Cook¶, Susan Palmer, Robin Plevin & Michael J.O. Wakelam§.
Molecular Pharmacology Group, Institute of Biochemistry, University of Glasgow, Glasgow, G128QQ, Scotland, U.K.

Introduction

A number of growth factors, exemplified by bombesin, bind to receptors and stimulate the hydrolysis of PtdIns(4,5)P_2 to generate Ins(1,4,5)P_3 and *sn*-1,2-diacylglycerol (DG) (reviewed by Berridge, 1987). These two molecules are potent second messengers: Ins(1,4,5)P_3 binds to a specific receptor, probably on the E.R., stimulating the release of sequestered calcium into the cytosol (reviewed by Berridge & Irvine, 1989) whilst DG remains within the inner leaflet of the plasma membrane where it binds to and activates protein kinase C (reviewed by Parker *et al.*, 1989). This pathway is implicated in the regulation of a number of the early events associated with entry into the cell cycle including increases in pH_i (Hesketh *et al.*, 1986), induction of the expression of c-*fos* and c-*myc* proto-oncogenes (Rozengurt & Sinnet-Smith, 1987) and activation of the ribosomal protein S6 kinase (Susa *et al.*, 1989). However PtdIns(4,5)P_2 hydrolysis is not a pre-requisite for stimulation of cell proliferation by all growth factors, e.g. EGF in Swiss 3T3 cells (Whitman & Cantley, 1988). In addition, it should be noted that the increase in Ins(1,4,5)P_3 is essentially transient, as is the associated increase in $[Ca^{2+}]_i$, declining towards control within 1-2 minutes whereas DNA synthesis requires the continued presence of the mitogen for up to 8-10 hours.

Recent evidence from a number of stimulus-response systems has suggested that increases in cellular DG content may be sustained after Ins(1,4,5)P_3 levels have declined

---

¶ Current Address: Dept of Molecular Biology, Cetus Corporation, Emeryville, CA, USA.
¶,§ Corresponding Authors.
Abbreviations: Cho, choline; ChoP, phosphocholine; DG, *sn*-1,2-diacylglycerol; GroPCho, glycerophosphocholine; PKC, protein kinase C; PLC, phospholipase C; PLD, phospholipase D; PMA, phorbol myristate acetate; PPH, PtdOH phosphohydrolase; PtdBut, phosphatidylbutanol; PtdOH, phosphatidic acid; PtdCho, phosphatidylcholine.

NATO ASI Series, Vol. H 52
Biological Signal Transduction
Edited by E. M. Ross and K. W. A. Wirtz

(Wright *et al.*, 1988; Matozaki *et al.*, 1990). Similarly, a number of DG/PKC-mediated signals are more sustained, e.g. phosphorylation of the 80kDa protein (Isacke *et al.*, 1986). These observations suggests that the DG/PKC arm of the signal may be more important in long term responses such as the G_0/G_1 to S-phase transition. This proposal is supported further by the finding that micro-injection of Balb/c 3T3 cells with DG, but not Ins(1,4,5)P_3, stimulates their proliferation (Suzuki-Sekimori *et al.*, 1989). Furthermore, considerable evidence exists to suggest that the source of sustained DG elevation is phosphatidylcholine (PtdCho) (Billah & Anthes, 1990).

We have used the Swiss 3T3 fibroblast cell line to examine the kinetics of bombesin-stimulated DG formation. Bombesin stimulates the hydrolysis of PtdIns(4,5)P_2 to generate Ins(1,4,5)P_3 and DG (Heslop *et al.*,1986); however, the increase in $[Ca^{2+}]_i$ induced by Ins(1,4,5)P_3 is transient and declines within 1-2 minutes. Therefore we sought to determine the contributions made by PtdIns(4,5)P_2 and PtdCho hydrolysis to the DG signal and to identify the pathway of PtdCho hydrolysis.

Methods

Swiss 3T3 cells were maintained as previously described (Cook & Wakelam, 1989).

Assay of Ins(1,4,5)P_3 and DG mass. Samples from stimulated Swiss 3T3 cells, prepared as described previously (Cook *et al.*, 1990), were assayed for Ins(1,4,5)P_3 mass or DG mass by the methods of Palmer *et al.*, (1989) and Preiss *et al.*, (1986), respectively.

Analysis of water soluble metabolites of [^{3}H]PtdCho hydrolysis. Assay of increases in [^{3}H]Cho, [^{3}H]ChoP and [^{3}H]GroPCho was as described by Cook & Wakelam (1989).

Assay of PLD transferase activity in whole cells. [^{3}H]palmitate labelled Swiss 3T3 cells were washed in DMEM/20mM Hepes (pH 7.4) containing 1% (w/v) bovine serum albumin (fraction V) (DMBH) at 37°C followed by DMBH containing 0.3% (v/v) butan-1-ol for 5 minutes. Cells were then incubated with DMBH containing 0.3% (v/v) butan-1-ol and bombesin (617nM), PMA (100nM) or vehicle. Incubations were terminated by the removal of the medium and addition of ice-cold methanol. The cell debris was harvested in 0.7ml of methanol and the lipids extracted for 15 minutes (in glass tubes) with 0.7ml of chloroform. Phases were resolved by addition of 0.585 ml of water. The lower chloroform layer was dried down *in vacuo*, re-dissolved in chloroform/methanol (19/1, v/v) and [^{3}H]PtdBut isolated by t.l.c. as described by Bonser *et al.*, (1989).

Analysis of [^{3}H]1,2-DG levels. [^{3}H]palmitate labelled Swiss 3T3 cells were stimulated with 100nM PMA and the lipids were extracted as described above for [^{3}H]PtdBut. [^{3}H]1,2-DG was isolated by t.l.c. with a solvent of hexane: diethyl ether: acetic acid (60:40:1, v/v).

Results

[1] Bombesin stimulates biphasic DG formation in Swiss 3T3 cells

Figure 1 shows the time course for bombesin (617nM)-stimulated increases in Ins(1,4,5)P_3 and DG mass in Swiss 3T3 cells. Ins(1,4,5)P_3 mass was maximal (typically 14-fold over control) after 5 sec stimulation and declined to pre-stimulated levels by 30 sec after which there was no further increase up to 15 min. The EC_{50} for bombesin-stimulated increases in Ins(1,4,5)P_3 mass at 5 sec was 5.88±3.66 nM (n=3 experiments, Cook *et al.*, 1990). In contrast, bombesin (617nM) stimulated a biphasic increase in DG mass;

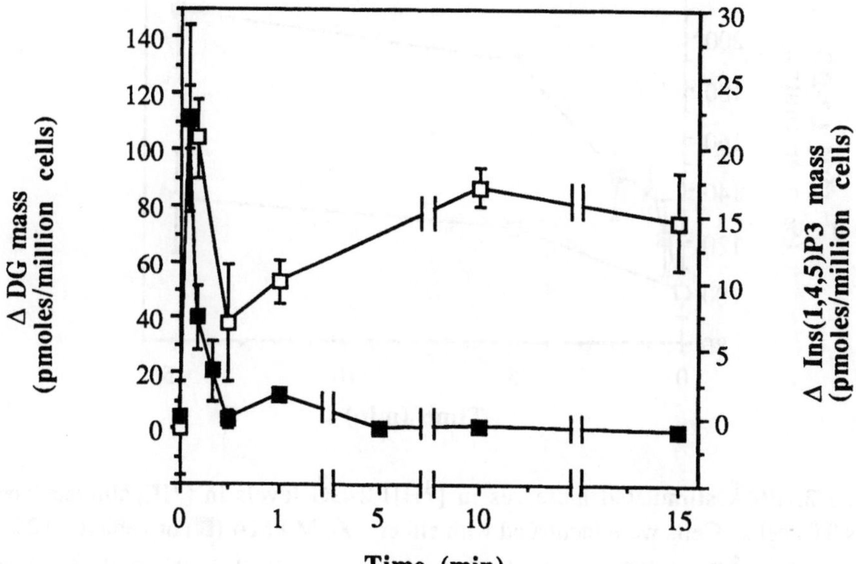

Figure 1. Bombesin-stimulated increases in mass of Ins(1,4,5)P_3(■) and DG (□) in Swiss 3T3 cells. After incubation for the times indicated with 617 nM bombesin, Ins(1,4,5)P_3 and DG were assayed as described. Results (mean±S.D., n=3) are expressed as the net increase in pmoles; note different axis scales.

the first phase was maximal after 5 sec and declined towards control levels between 30 sec and 1 min. From 1 min onwards, DG levels rose again in a pronounced second phase which was clearly divorced from any further increases in Ins(1,4,5)P_3 mass and was sustained for up to 30 min and in some cases 1 hour (data not shown). The EC_{50} values for the two phases of DG elevation were 1.66±1.15 nM at 5 sec and 0.8±0.28 nM at 5 min (Cook *et al.*, 1990). However, the net increase in DG mass at 5 seconds was approximately 4-fold greater than that for Ins(1,4,5)P_3 (note different axis scales).

Reports have suggested that addition of PKC-activating phorbol esters to cells results in increased DG levels in the absence of inositol lipid hydrolysis (Takuwa *et al.*, 1987). This was investigated in Swiss 3T3 cells labelled to a constant specific activity with [^{3}H]palmitic acid. Under these conditions the majority of label incorporated into phospholipids was found in PtdCho (53%) with less than 3% in inositol lipids (data not shown). Addition of 100nM PMA to [^{3}H]palmitate-labelled Swiss 3T3 cells resulted in the elevation of [^{3}H]1,2-DG which was significant at 2 min and sustained for up to 15 (Figure

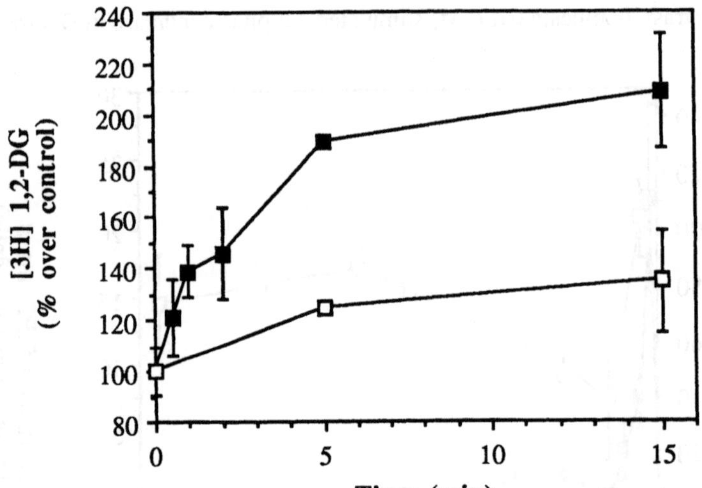

Figure 2. PMA stimulated increases in [^{3}H]1,2-DG levels in [^{3}H]palmitate labelled Swiss 3T3 cells. Cells were incubated with either 100nM PMA (■) or vehicle (0.2%, v/v DMSO) (□). [^{3}H]1,2-DG was resolved by t.l.c. as described in the methods. Results (means±S.D., n=3) are expressed as percentage of radioactivity in the zero-time controls.

2) and 30 min (data not shown). Furthermore, PMA-stimulated increases in DG mass were significant at 2 min and were maintained above control for at least one hour (data not

shown). Despite this, PMA alone had no effect on inositol phosphate formation and inhibited bombesin-stimulated increases in inositol phosphate accumulation (Brown *et al.*, 1987; Cook & Wakelam, submitted). Thus, PMA effectively dissociated DG formation from inositol lipid hydrolysis.

[2] Bombesin and PMA stimulate PtdCho hydrolysis

Recent reports have suggested that PtdCho hydrolysis of may be responsible for sustained increases in DG content in neutophils and IIC9 fibroblasts (Billah & Anthes 1990; Pessin & Raben, 1989). We examined the effects of bombesin and PMA on the formation of water-soluble [^{3}H]choline metabolites in Swiss 3T3 cells in which PtdCho had been labelled to a constant specific activity with [^{3}H]choline chloride. Both bombesin and PMA stimulated the formation of [^{3}H]Cho prior to any increases in [^{3}H]ChoP suggesting that PtdCho hydrolysis occurred by a phospholipase D-catalysed pathway (Fig 3). Bombesin-stimulated [^{3}H]Cho formation was significant at 30 sec, rose rapidly until 2

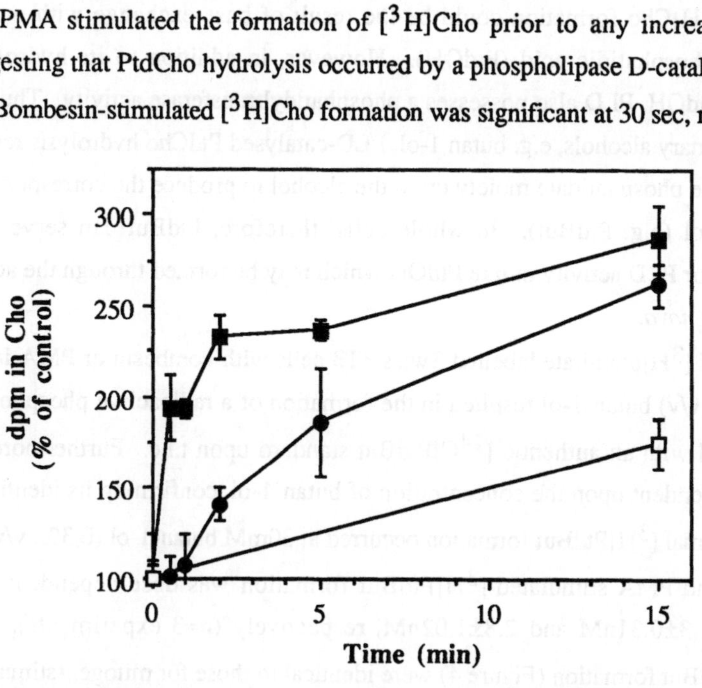

Figure 3. Bombesin- and PMA-stimulated release of [^{3}H]Cho from Swiss 3T3 cells labelled with [^{3}H]choline chloride. Following incubation with 617nM bombesin (■), 100nM PMA (●), or vehicle (□) [^{3}H]Cho was isolated as described by Cook & Wakelam (1989). Results are expressed as percentage of the radioactivity in the zero-time controls.

min and then increased more gradually, being maintained above control levels for 15 (Fig 3) and 30 min (data not shown). The EC_{50} value for bombesin-stimulated [^{3}H]Cho production was 1.44±1.02 nM (n=4 experiments, Cook & Wakelam, 1989). In contrast, PMA-stimulated the accumulation of [^{3}H]Cho after a lag of 1 min beyond which the response was sustained for up to 30 (Fig 3) and 60 min (data not shown). The EC_{50} for PMA-stimulated increases in [^{3}H]Cho was 4.95±3.63 nM (n=4 experiments, Cook & Wakelam, 1989). Thus, bombesin- and PMA-stimulated PtdCho hydrolysis appeared to occur initially by a PLD catalysed pathway at concentrations at which both exert their mitogenic effects.

[3] Bombesin and PMA stimulate the phosphatidyltransferase activity of PLD

Measuring release of Cho is not a definitive assay for PLD-catalysed PtdCho hydrolysis since [^{3}H]Cho formation could be the result of base-exchange with no net accumulation of phosphatidic acid (PtdOH). However, in addition to its hydrolytic activity, forming PtdOH, PLD also possesses a phosphatidyltransferase activity. Thus, in the presence of primary alcohols, e.g. butan 1-ol, PLD-catalysed PtdCho hydrolysis results in the transfer of the phosphatidate moiety on to the alcohol to produce the corresponding phosphatidylalcohol (e.g. PtdBut). In whole cells, therefore, PtdBut can serve as a definitive marker for PLD activity unlike PtdOH which may be formed through the action of DG kinase or *de novo*.

Stimulation of [^{3}H]palmitate labelled Swiss 3T3 cells with bombesin or PMA in the presence of 0.3% (v/v) butan-1-ol resulted in the formation of a radioactive phospholipid which co-migrated with an authentic [^{14}C]PtdBut standard upon t.l.c. Furthermore, its formation was dependent upon the concentration of butan-1-ol, confirming its identity as [^{3}H]PtdBut. Maximal [^{3}H]PtdBut formation occurred at 30mM butan-1-ol (0.3%, v/v).

Bombesin- and PMA-stimulated [^{3}H]PtdBut formation was dose-dependent with EC_{50} values of 1.3±0.31nM and 2.8±1.02nM, respectively (n=3 experiments). The kinetics of [^{3}H]PtdBut formation (Figure 4) were identical to those for mitogen-stimulated [^{3}H]Cho production. Bombesin-stimulated increases in [^{3}H]PtdBut were significant at 30sec, rose rapidly up to 2 min and continued at a reduced rate up to 15 min (Fig. 4). In contrast, PMA-stimulated increases in [^{3}H]PtdBut were not significant until after 1 min but increased in a sustained, near linear fashion for up to 15 (Fig. 4b) and 30 min (data not shown).

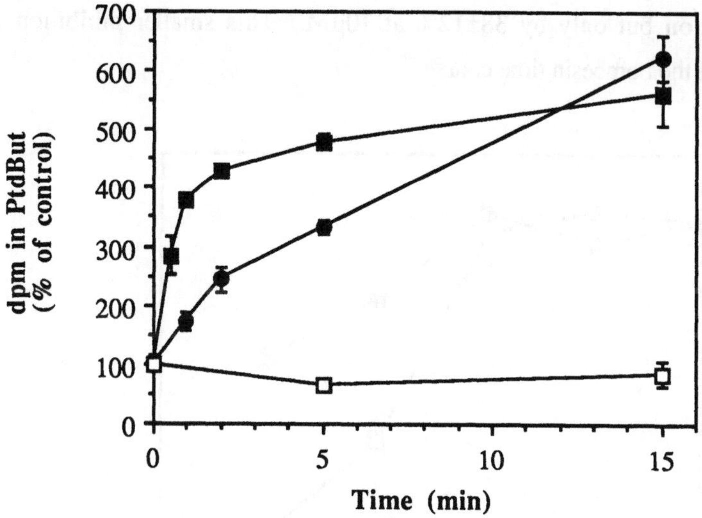

Figure 4. Bombesin- and PMA-stimulated [^{3}H]PtdBut formation in [^{3}H]palmitate labelled Swiss 3T3 cells. Cells were incubated with either 617nM bombesin (■), 100nM PMA (●), or vehicle (□) for the times indicated in the presence of 0.3% butan-1-ol and [^{3}H]PtdBut was resolved by t.l.c. as described in the methods section. Results (mean±S.D., n=3) are expressed as percentage of zero-time controls.

[4] A role for PKC in regulating PLD activity ?

Since both PMA and bombesin activate PKC in Swiss 3T3 cells (Erusalimsky *et al.*,1988), either directly (PMA) or via inositol lipid hydrolysis (bombesin), it seemed possible that PKC might play a role in activating PLD-catalysed PtdCho hydrolysis. Two strategies were adopted to investigate this possibilty.

Firstly, chronic pre-treatment of Swiss 3T3 cells with 400nM PMA for 48 hours results in the loss of cellular PKC activity (Rodriguez-Pena & Rozengurt, 1984), [^{3}H]PDBu binding (Collins & Rozengurt, 1984) and the protein itself by western blotting (Brown *et al.*, 1990). In these 'down-regulated' cells, the ability of either bombesin or PMA to stimulate increases in [^{3}H]Cho (Cook & Wakelam, 1989) or [^{3}H]PtdBut (data not shown) was completely abolished compared to the control cells.

Secondly, a five minute pre-treatment with the PKC inhibitor Ro-31-8220 (Davis *et al.*, 1989) resulted in a dose-dependent inhibition of [^{3}H]PtdBut formation in response to PMA (Fig 5). Inhibition of the PMA response was maximal at 10μM Ro-31-8220 with an IC_{50} of 0.7±0.3μM (n=3 experiments). Ro-31-8220 also inhibited bombesin-stimulated

$[^3H]$PtdBut formation but only by 38±12% at 10μM. This smaller inhibition was observed throughout the bombesin time course.

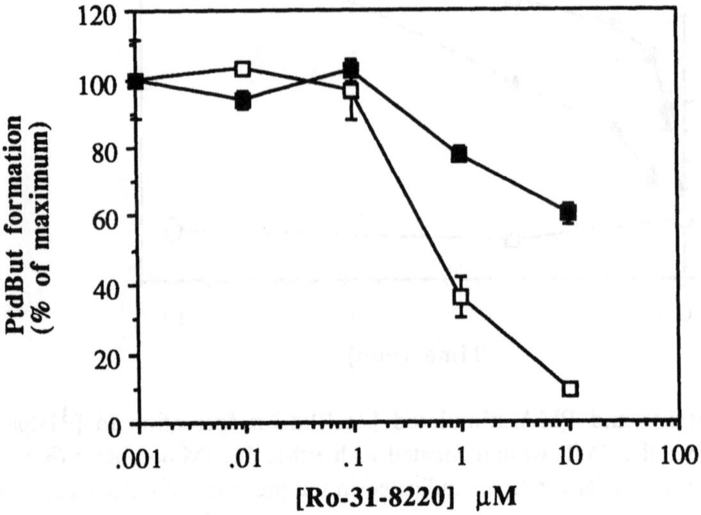

Figure 5. Effect of Ro-31-8220 upon bombesin- and PMA stimulated $[^3H]$PtdBut formation in $[^3H]$palmitate labelled Swiss 3T3 cells. After incubation for 5 min with the indicated dose of Ro-31-8220 + 0.3% butan-1-ol stimulations were started by addition of 100nM bombesin (■) or 100nM PMA (□). $[^3H]$PtdBut was resolved as previously and results are expressed as percentage of maximum response.

[5] PtdCho hydrolysis may be a common response to mitogens coupled to inositol lipid hydrolysis

Both $[Arg^8]$vasopressin ($[Arg^8]$Vp) and prostaglandin $F_{2\alpha}$ ($PGF_{2\alpha}$) are also potent mitogens for the Swiss 3T3 cell line (Corps *et al.*, 1985; MacPhee *et al.*, 1984) and stimulate PtdIns(4,5)P_2 hydrolysis by a similar mechanism to bombesin (Nanberg & Rozengurt, 1988; Black & Wakelam, 1990). We therefore assessed the abilty of these mitogens to stimulate inositol lipid and PtdCho hydrolysis. Both mitogens stimulated the accumulation of $InsP_t$, in the presence of 10mM LiCl, at concentrations at which they are mitogenic (EC_{50} values of 0.48±0.15nM for $[Arg^8]$Vp and 0.16±0.14μM for $PGF_{2\alpha}$ n=3 and 4 experiments respectively). In addition, both mitogens stimulated increases in $[^3H]$Cho in pre-labelled cells, suggesting activation of PLD. The EC_{50} values for $[^3H]$Cho release were 0.92±0.57nM for $[Arg^8]$Vp and 0.51±0.32μM for $PGF_{2\alpha}$.

Finally, when Swiss 3T3 cells were pre-treated with 400nM PMA for 48 hours the ability of both [Arg^8]Vp and PGF$_{2\alpha}$ to stimulate increases in [^{3}H]Cho was completely abolished (Cook & Wakelam, submitted).

Discussion

The results in Fig 1 show that bombesin stimulated a biphasic increase in DG mass, the first phase of which corresponded with the rapid increase in Ins(1,4,5)P$_3$ mass and presumably reflects PtdIns(4,5)P$_2$ hydrolysis. The rapid, but transient, increase in Ins(1,4,5)P$_3$ levels is fully consistent with its physiological role as a Ca^{2+} mobilizing second messenger. The decline in DG levels may reflect a high DG kinase activity, selective for *sn*-1-stearoyl-2-arachidonoyl glycerol as reported in Swiss 3T3 membranes (MacDonald *et al.*, 1988). However, DG levels rose again after 1 minute in the absence of a corresponding rise in Ins(1,4,5)P$_3$ and were sustained for 30-60 minutes. Indeed, a recent study has reported sustained increases in DG for up to 4 hours in bombesin-stimulated Swiss 3T3 cells (Takuwa *et al.*, 1989). These results clearly suggest that DG is generated from a source(s) other than PtdIns(4,5)P$_2$, the hydrolysis of which is rapidly desensitized within 30sec-1min of bombesin stimulation (Cook *et al.*, 1990; Palmer *et al*, submitted). In addition, the disparity in the increase in mass of DG and Ins(1,4,5)P$_3$ after 5 seconds of stimulation may indicate that some DG is derived from a source other than PtdIns(4,5)P$_2$ at these early time points. Alternatively a higher activity of Ins(1,4,5)P$_3$ 5' phosphatase and 3' kinase compared to DG kinase may result in an underestimation of mass increases of Ins(1,4,5)P$_3$ at these early time points.

An additional source of DG is supported further by the observation that the second sustained phase of the bombesin response is mimicked by PMA. In [^{3}H]palmitate labelled cells, PMA-stimulated increases in [^{3}H]DG were sustained for up to 30 min. In other experiments PMA-stimulated DG mass was elevated for at least one hour. Since phorbol esters do not stimulate, but actually inhibit agonist-stimulated hydrolysis of inositol lipids, by a PKC-mediated feedback pathway (Brown *et al.*, 1987), PMA effectively dissociated DG formation from PtdIns(4,5)P$_2$ hydrolysis.

In [^{3}H]palmitate-labelled Swiss 3T3 cells, the major labelled phospholipid is PtdCho and this, together with recent reports of agonist-stimulated PtdCho hydrolysis (Billah &

Anthes, 1990), suggested that PtdCho was a prime candidate for the alternative source of DG. In [^{3}H] choline labelled Swiss 3T3 cells, both bombesin and PMA stimulated the release of [^{3}H]Cho (Fig 3) prior to any changes in [^{3}H]ChoP, suggesting that PtdCho was being hydrolysed by a PLD catalysed pathway. Increases in [^{3}H]ChoP were slower in onset (Cook & Wakelam, 1989) and may represent PtdCho-PLC activity or the phosphorylation of liberated [^{3}H]Cho by choline kinase (Warden & Friedkin, 1985) as part of a PtdCho cycle to replenish the parent lipid. The kinetics of PMA-stimulated [^{3}H]Cho and DG formation are similar, whereas bombesin-stimulated [^{3}H]Cho release occurred after Ins(1,4,5)P_3 generation, but prior to the second phase of *sn*-1,2-DG elevation. The kinetics of PtdCho hydrolysis are therefore fully consistent with it contributing to the sustained elevation of DG by both mitogens.

To confirm that the release of [^{3}H]Cho did represent PLD activity, the ability of both bombesin and PMA to stimulate the transferase activity of PLD was investigated (Fig 4). The kinetics and dose-dependency of bombesin-stimulated [^{3}H]PtdBut formation were identical to those for [^{3}H]Cho formation confirming that bombesin did stimulate a PtdCho-PLD activity. Likewise, PMA stimulated the formation of [^{3}H]PtdBut from 1 minute onwards with the same EC_{50} as that for [^{3}H]Cho release. This EC_{50} was similar to that for activation of PKC and the IC_{50} for inhibition of inositol lipid hydrolysis (Brown *et al.*, 1987) suggesting that the three events might be causally related: activation of PKC by bombesin serving to inhibit PtdIns(4,5)P_2 hydrolysis and activate PLD-catalysed PtdCho hydrolysis.

A role for PKC in the activation of PtdCho-PLD was confirmed in two ways. Pre-treatment of Swiss 3T3 cells with 400nM PMA for 48 hours to 'down-regulate' PKC resulted in the abolition of agonist-stimulated [^{3}H]PtdBut formation and [^{3}H]Cho release (Cook & Wakelam, 1989). In the case of bombesin, this was not due to loss of receptors (Brown *et al*, 1987) or inositol lipid hydrolysis which was enhanced in 'down-regulated' cells (Cook & Wakelam, submitted). This, together with the inhibition of PMA- and bombesin-stimulated [^{3}H]PtdBut formation by the PKC inhibitor Ro-31-8220, clearly indicates a requirement for PKC in activation of PLD by these mitogens. The anomaly that PMA-stimulated [^{3}H]PtdBut formation was completely abolished by Ro-31-8220 whilst bombesin-stimulated [^{3}H]PtdBut was only inhibited by at most 50% is of interest since both responses were fully inhibited by 'down-regulation' of PKC. There may be a number of reasons for this result. If TPA is able to activate PLD directly as well as via PKC then PLD may also be 'down-regulated' during prolonged treatment with PMA. Bombesin may exert some effects upon PLD independently of PKC, for example by direct receptor coupling or by increases in intracellular [Ca^{2+}], which would not be sensitive to PKC inhibitors but would still be abolished if PLD were 'down-regulated'. Alternatively, some of the effects of bombesin upon PLD activity could be mediated by an isozyme of PKC which is differentially sensitive to either PKC inhibitors or 'down-regulation'. These possibilities are currently under investigation.

The fact that both [Arg^8]Vp and $PGF_{2\alpha}$ were able to activate PtdCho-PLD by a similar mechanism suggests that PtdCho hydrolysis may be a common response to mitogens which stimulate inositol lipid hydrolysis and PKC. Furthermore, with the agreement between EC_{50} values for increases in [^3H]$InsP_t$ accumulation, [^3H]Cho release and DNA synthesis it is tempting to speculate that PtdCho hydrolysis may be an integral part of the mitogenic signal for bombesin, [Arg^8]Vp and $PGF_{2\alpha}$.

The role of mitogen-stimulated PtdCho hydrolysis

One role for stimulated hydrolysis of PtdCho may be to provide an alternative source of DG and the results presented here suggest that this could be the case. If the sustained phase of DG is formed by a PLD/phosphatidate phosphohydrolase pathway then the PLD products should be apparent prior to the second phase of DG since a further reaction step, de-phosphorylation of PtdOH (Martin, 1988), is required. The results are consistent with this since the onset of bombesin-stimulated PtdCho-PLD activity (Fig. 3 & 4) is after the generation of Ins(1,4,5)P_3 and the first phase of DG but prior to the second phase of DG elevation (Fig. 1). This PLD/PPH pathway of DG formation has recently been shown to be required for f-Met-Leu-Phe-stimulated superoxide generation in neutrophils, indicating its importance in stimulus-response coupling (Bonser *et al.*, 1989). Activation of the PtdCho-PLD pathway in Swiss 3T3 cells appears to be dependent, in part, upon PKC activity which may represent the major link between PtdIns(4,5)P_2 and PtdCho hydrolysis and hence the two phases of DG formation. The sustained elevation of DG in the absence of increased [Ca^{2+}]$_i$ could support activation of isozymes of PKC such as PKC-ε (Schaap & Parker, 1990) and n-PKC (Ohno *et al.*, 1988). Alternatively, PtdCho derived DG species may activate different species of PKC to differing degrees. It may be that these isozymes phosphorylate distinct substrates which may be important for the mitogenic signal.

The turnover of a PtdCho cycle is energetically cheaper than the inositol lipid cycle and so a 'switch' from PtdIns(4,5)P_2 to PtdCho might be advantageous once the requirement for an increase in $[Ca^{2+}]_i$ and therefore Ins(1,4,5)P_3 has passed. However, the PLD/PPH pathway of DG formation requires the effective reversal of the DG kinase reaction; as such, the interplay between DGK and PPH must be tightly regulated to prevent futile cycling of DG/PtdOH and wastage of ATP. Since a much simpler pathway would be PLC-catalysed PtdCho hydrolysis generating DG directly, it is possible that activation of PtdCho-PLD may serve other cellular functions in addition to DG formation. This is of particular interest since Huang & Cabot (1990) have recently shown that whilst activation of PLD by phorbol ester is a common response in a variety of cell types, the PtdOH produced does not serve as a precursor for DG in all the cell types tested. Thus, it appears that, in some cells, the sustained elevation of DG may occur by PtdCho-PLC activity acting in parallel with PtdCho-PLD. In the light of such observations it seems pertinent to ask what other functions PtdOH may serve if not as a precursor of DG.

The lipid product of PLD, PtdOH, and its lyso-derivative both exhibit potent mitogenic effects when added exogenously to cultured cells (Moolenaar *et al.*, 1986; van Corven *et al.*, 1989). A recent report has also shown that PtdOH may replace PtdSer supporting PKC activity in the absence of Ca^{2+} (Epand & Stafford, 1990), whilst its fusogenic properties (reviewed by Wakelam, 1988) may modulate membrane fluidity and cytoskeletal arrangements. Furthermore, Tsai *et al.* (1989) recently reported that arachidonoyl-containing PtdOH is able to inhibit the ability of GTPase activating protein (GAP) to stimulate the GTPase activity of p21$^{\text{Ha-}ras}$. Since the result of inhibiting GAP activity is to increase the amount of active p21^{ras} (i.e., that which is GTP-bound) this may point to important messenger functions for PtdOH in controlling cell proliferation.

The observation that PtdCho-PLD is activated by diverse mitogenic stimuli in Swiss 3T3 cells including platelet-derived growth factor (Plevin *et al.*, submitted) and epidermal growth factor (Cook & Wakelam, in preparation) as well as those outlined above, clearly warrants consideration of PLD-catalysed PtdCho hydrolysis as an important and widespread mitogenic signal transduction pathway.

Acknowledgements

This work was supported by grants from the Medical Research Council (U.K.) and the Cancer Research Campaign. S.J.C. is in receipt of a Science & Engineering Research

Council C.A.S.E. award with the Wellcome Foundation p.l.c. We thank the Wellcome Laboratories, Beckenham, U.K. for provision of a [^{14}C]PtdBut standard. In addition we thank Roche U.K. for providing us with Ro-31-8220.

References

Berridge, M.J. (1987) Biochim. Biophys Acta. **907**, 33-45.

Berridge, M.J. & Irvine, R.F. (1989) Nature (Lond) **341**, 197-205.

Billah, M.M. & Anthes J.C. (1990) Biochem. J. **269**, 281-291.

Black, F.M. & Wakelam, M.J.O. (1990) Biochem. J. **266**, 661-667.

Bonser, R.W., Thompson, N.T., Randall, R.W. & Garland, L.G. (1989) Biochem. J. **264**, 617-620.

Brown, K.D., Blakeley, D.M., Hamon, M.H., Laurie, M.S. & Corp, A.N. (1987) Biochem. J. **245**, 631-639.

Brown, K.D., Littlewood, C.J. & Blakeley, D.M. (1990) Biochem. J. **270**, 557-560.

Castagna, M., Takai, Y., Sano, K., Kikkaw, U. & Nishizuka, Y. (1982) J. Biol. Chem. **257**, 7847-7851.

Collins, M.K.L. & Rozengurt, E. (1984) J. Cell. Physiol. **118**, 133-142.

Cook, S.J. & Wakelam, M.J.O. (1989) Biochem. J. **263**, 581-587.

Cook, S.J., Palmer, S. Plevin, R. & Wakelam, M.J.O. (1990) Biochem. J. **265**, 617-620.

Corps, A.N., Rees, L.H. Brown, K.D. (1985) Biochem. J. **231**, 781-784.

Davis, P.D., Hill, C.H., Keech, E., Lawton, G., Nixon, J.S., Sedgwick, A.D., Wadsworth, J., Westmacott, D. & Wilkinson, S.E. (1989) FEBS Lett. **259**, 61-63.

Epand, R.M., & Stafford A.R. (1990) Biochem. Biophys. Res. Commun. **171**, 487-490.

Erusalimsky, J.D., Friedberg, I. & Rozengurt, E. (1988) J. Biol Chem. **263**, 19188-19194.

Hesketh, T.R., Moore, J.P., Morris, J.D.H., Taylor, M.V., Rogers, J. Smith, G.A. & Metcalfe, J.C. (1986) Nature (Lond)**313**, 481-484.

Heslop, J.P., Blakeley, D.M., Brown, K.D. Irvine, R.F. & Berridge, M.J. (1986) Cell **47**, 703-709.

Huang, C. & Cabot, M.C. (1990) J. Biol. Chem. **265**, 14858-14863.

Isacke, C.M., Meisenhelder, J., Brown, K.D., Gould, K.L., Gould, S.J. & Hunter, T. (1986) EMBO. J. **5**, 2889-2898.

MacDonald, L.M., Mack, K.F., Williams, B.W., King, W.C. & Glomset, J.A. (1988) J.

Biol. Chem. **263**, 1584-1592.

MacPhee, C.H., Drummond, A.H., Otto, A.M. & Jimenez de Asua, L. (1984) J. Cell. Physiol. **119**, 35-40.

Martin, T.W. (1988) Biochim. Biophys. Acta. **962**, 282-296.

Matozaki, T., Göke, B, Tsunoda, T., Martinez, J., & Williams, J.A. (1990) J. Biol. Chem. **265**, 6247-6254.

Moolenaar, W.H., Kruijer, W., Tilley, B.C., Verlaan, I., Bierman, A.J., & de Laat, S.W. (1986) Nature (Lond) **323**, 171-173.

Nanberg, E. & Rozengurt, E. (1988) EMBO. J. **7**, 2741-2747.

Ohno, S. Akita, Y. Konno, Y. Imajoh, S. & Suzuki, K. (1988) Cell. **53**, 731-741.

Parker, P.J., Kour, G., Marais, R.M., Mitchell, F., Pears, C. Schaap, D., Stabel, S. & Webster, C. (1989) Mol. Cell. Endocrinol. **65**, 1-11.

Palmer, S, Hughes, K.T., Lee, D.Y. & Wakelam, M.J.O. (1989) Cell. Signal. **1**, 147-156.

Pessin, M.S. & Raben, D.M. (1989) J. Biol. Chem. **264**, 8729-8738.

Preiss, J., Loomis, C.R., Bishop, R.W., Stein, R., Niedel, J.E. & Bell, R.M. (1986) J. Biol. Chem. **261**, 8597-8600.

Rodriguez-Pena, A. & Rozengurt, E. (1984) Biochem. Biophys. Res. Commun. **120**, 1053-1059.

Rozengurt, E. & Sinnet-Smith, J.W. (1987) J. Cell. Physiol. **131**, 218-225.

Schaap, D. & Parker, P.J. (1990) J. Biol. Chem. **265**, 7301-7307.

Susa, M., Olivier, A.R., Fabbro, D. & Thomas, G. (1989) Cell. **57**, 817-824.

Suzuki-Sekimori, R., Matuoka, K. Nagai, Y. & Takenawa, T. (1989) J. Cell. Physiol. **140**, 432-438.

Takuwa, N. Takuwa, Y. & Rasmussen, H. (1987b) Biochem. J. **243**, 647-653.

Takuwa, N., Takuwa, Y., Yanagisawa, M.,Yamashita, K. & Masaki, T. (1989) J. Biol. Chem. **264**, 7856-7861.

Tsai, M.H., Yu, C.L., Wei, F.S. & Stacey, D.W. (1989) Science. **243**, 522-526.

Wakelam, M.J.O. (1988) Current Topics In Membranes & Transport. Academic Press **32**, 87-112.

Warden, C.H. & Friedkin, M. (1985) J. Biol. Chem. **260**, 6006-6011.

Whitman, M. & Cantley, L. (1988) Biochim. Biophys. Acta. **948**, 327-344.

Wright, T.M., Rangan, L.A., Shin, H.S. & Raben, D.M. (1988) J. Biol. Chem. **263**, 9374-9380.

van Corven, E.J., Groenink, A., Jalink, K., Eicholtz, T. & Moolenaar, W.H. (1989) Cell. **59**, 45-54.

PHORBOL ESTERS AND MUSCARINIC RECEPTOR AGONISTS ACTIVATE PHOSPHOLIPASE D IN HEART AND BRAIN

T. Holler, R. Lindmar and K. Löffelholz
Department of Pharmacology
University of Mainz
Obere Zahlbacher Str. 67
W 6500 Mainz
Germany

Abstract

Phospholipase D (PLD) hydrolyzes phosphatidylcholine and thereby seems to play a key role in a novel pathway of signal transduction. PLD activity in rat hippocampal slices and atria of rat, guinea pig and chicken hearts was determined by measuring the catalytic products choline (Ch), phosphatidic acid (PA) and, in the presence of a primary alcohol, phosphatidylpropanol or phosphatidylethanol. It was found that the PLD activity was high, even under resting conditions, in both tissues, especially in the hippocampus, and that the enzyme activity could be enhanced by activation of protein kinase C and by muscarinic receptor stimulation.

Introduction

For many decades the role of phospholipids in biological membranes was regarded as to maintain the membrane proteins in a proper conformation. This role was believed essential for allowing the membrane bound proteins to perform as enzymes, transporters, membrane pores or ion channels/pumps.

The present work was supported by a grant from the Deutsche Forschungsgemeinschaft.

NATO ASI Series, Vol. H 52
Biological Signal Transduction
Edited by E. M. Ross and K. W. A. Wirtz

This view, however, could not explain the multitude of individual phospholipid molecular species in their asymetrical position within the inner and outer membrane leaflets.

More than 30 years ago, Hokin and Hokin (reviewed by Hokin in 1985) showed that phosphatidic acid (PA) and phosphoinositides (PIs) underwent rapid turnover in biological sytems stimulated by a variety of agonists. This early observation has developed to an assignment of an essential role of PIs, the PI-specific phospholipase C (PI-PLC) and the catabolic product diacylglycerol (DAG), as the principle activator of protein kinase C (PKC) in signaling across biological membranes.

It has now become apparent that a role in signal transduction is not restricted to PIs, but also applies to the most common phospholipid, phosphatidylcholine (PC). PC hydrolysis is activated by various receptor agonists and also by PKC activation (reviewed by Billah and Anthes, 1990; Löffelholz, 1989, 1990). Moreover evidence is accumulating that phospholipase D (PLD) may play a crucial role in this phenomenon. - PLD catalyzes the reaction

Phosphatidyl-O-choline + R-OH → Phosphatidyl-O-R + Choline

When R-OH is water, PLD yields PA plus choline (Ch)(reaction 1). When R-OH is a primary alcohol, like ethanol or propanol, PLD yields phosphatidylethanol (PEth) or phosphatidylpropanol (PP) plus Ch (reaction 2).

In the present study we utilized both of the aforementioned reactions, the "hydrolytic" and the "transphosphatidylation" reactions, to study activation of PLD in intact tissues such as incubated heart atria and hippocampal slices.

Methods

Hippocampal slices from rats or atria from rat, guinea pig and chicken hearts were incubated at 37°C in Tyrode's solution which had the following composition (in mM): Na^+ 149.3, K^+ 2.7, Ca^{2+} 1.8, Mg^{2+} 1.05, Cl^- 145.5, HCO_3^- 11.9, $H_2PO_4^-$ 0.4; (+)-glucose 5.6. In some experiments, the Ca^2 concentration was reduced to 0.225 mM as indicated in the text. Some brain slices were superfused with Tyrode's solution at a rate of 0.75 ml/min and 34°C.

Choline was determined in the incubation or superfusion medium and in the tissue (after incubation or superfusion). In the experiments with brain slices, the acetylcholinesterase activity was blocked by incubation of the preparations in DFP-containing (0.1 mM) Tyrode's solution for 30 min that was followed by a 10 min washout. We determined Ch in the medium by a chemoluminescence assay (Israel and Lesbats, 1982) and Ch in the tissue by HPLC with a post column enzyme reactor and electrochemical detection after extraction from the tissue homogenate by acetone:1 M formic acid (85:15) (Damsma et al., 1985; Stadler and Nesselhut, 1986). Ch net formation was determined in hippocampal slices and in chicken atria by means of the following values obtained in the same organ:
(1.) total Ch efflux during the experimental period of 60 min, (2.) Ch tissue content at zero time and (3.) Ch tissue content after the 60 min under control conditions ("individual control value", Tab.1) and after exposure to a phorbol ester or a muscarinic receptor agonist.

Phospholipids, including PA, PEth and PP, in the tissues were determined after labeling the incubated slices and atria with 50 or 16 µCi, respectively, of [^{3}H]-myristic acid (40-60 Ci/mmol). The labeling period of 2 or 3 hours, respectively, was followed by a 15 min washout. Thereafter the phorbol ester PDB (see below) or muscarinic agonists were added throughout the experiment. The phospholipids were extracted as described by Folch et al. (1957). The tritiated phospholipids were separated by two-dimensional thin layer chromatography using chloroform/ methanol/25% ammonia (65:35:5) in the first dimension. In the second dimension, chloroform/acetone/ methanol/acetic acid /water (60:24:12:12:3) was used for the separation of [^{3}H]-PA and the upper phase of ethylacetate/ isooctane/acetic acid/ water (13:2:3:10) for [^{3}H]-PEth and [^{3}H]-PP. The spots were visualized with iodine vapor, and spots corresponding to PA, PEth and PP standards were scraped off and quantified by liquid scintillation counting. The amounts of these lipids were expressed as "percent of total labeled phospholipids", which was defined by means of TLC as sum of labeled sphingomyelin, phosphoinositides, phosphatidylserine, phosphatidylcholine, phosphatidylethanolamine, phosphatidylglycerol, phosphatidic acid and, in some experiments, phosphatidylethanol or phosphatidylpropanol.

<u>Drugs used:</u> 4ß-phorbol-12ß, 13α-dibutyrate (PDB) (PDB was solved in 0.01% dimethylsulfoxide (DMSO), which did not significantly alter the control values in any of the preparations used in this study); acetylcholine chloride; carbachol chloride (Sigma Chemie, Deisenhofen, Germany); arecaidine propargyl ester (APE) chloride (kindly donated from Prof.Dr.G. Lambrecht, Frankfurt); diisopropylfluoro-phosphate (DFP) (Fluka AG, Buchs, Switzerland).

Data are presented as means ± SEM of N experiments. Statistical significance ($P<0.05$) was evaluated by t-test or analysis of variance.

Results

<u>Resting conditions.</u> Under control conditions, rat hippocampal slices and heart atrial preparations of rats, guinea pigs and chickens released Ch into the incubation medium at constant rates of 1.46, 0.75, 1.08 and 1.88 nmol/g min, respectively. Hence, organ and species differences were small. However, in rat brain slices the total Ch efflux during the experimental period of 60 min was more than twice the initial tissue Ch content, whereas the efflux from isolated atria (chicken) was less than half of the initial Ch content (Tab.1). The Ch efflux from chicken atria was

Tab.1. Choline (Ch) efflux and net formation in hippocampal slices and in isolated atria under resting conditions and after PDB expressed as pmol/mg protein (hippocampus) and as nmol/g wet weight (atria).

Tissue	Ch content		Ch efflux	Ch net production
	Initial	Change	Total (60 min)	
Rat hippocampus/ 1.8 mM Ca				
- Control	478±44	+18±24	1287±38	1305±18 (102%)*
	(33)	(+1)	(88)	(89)**
- PDB (0.1 μM)	478±44	+53±120	1852±132 (S)	1906±251 (97%)
	(33)	(+4)	(126)	(130)**
Chicken atria/ 1.8 mM Ca				
- Control	293±27	-69±20	113±4	44±20 (39%)
- PDB (0.3 μM)	293±27	-57±19	183±14 (S)	125±17 (69%) (S)
Chicken atria/ 0.225 mM Ca				
- Control	286±34	-115±35	148±17	34±39 (23%)
- APE (1 μM)	284±29	-111±21	292±53 (S)	181±65 (62%) (SS)
- APE + atropine	295±100	-63±94	119±20	56±101 (47%)
- CCh (10 μM)	279±15	-17±33	223±9 (S)	206±27 (92%) (S)

* Percent values in parenthesis indicate Ch net formation in % of "total Ch efflux".
** Approximative expression of the data as nmol/g wet weight to allow comparison with the data obtained in the atria.
(S) Significantly different from general and individual control values.
(SS) Significantly different from individual control value (see Methods).
N was between 3 and 5, except for the control group "atria/0.225 mM Ca" (N=14).

preferentially due to translocation of free cellular Ch rather than to Ch net formation, whereas in the hippocampal slices the Ch efflux was totally due to net formation. It is concluded that the extremely rapid rate of Ch efflux in the brain slices under resting conditions was caused by a correspondingly high rate of phospholipid hydrolysis.

The tissues were labeled with [^{3}H]-myristic acid to study the formation of PA and, in the presence of propanol or ethanol, PP or PEth, respectively. The basal formation of [^{3}H]-PA and of [^{3}H]-PP or -PEth was at least 4-fold higher in hippocampal slices than in atria (Tab.2 and 3). The primary alcohols themselves caused a slightly and partially significant reduction of the formation of labeled PA in both, hippocampal slices and rat atria (Tab.2). At the same time, there was a significant formation of PP or PEth, respectively (Tab.3). Alcohol-induced decreases of PA and increases of PP or PEth were of the same order of magnitude.

The results show that Ch and PA were formed under resting conditions and that, in the presence of primary alcohols, PP and PEth were synthesized on the expense of PA. The data indicate a basal activity of PLD in both brain and heart preparations. However, basal phospholipid hydrolysis via PLD activity was considerably higher in the brain slices than in the atria.

Effects of PDB. The phorbol ester PDB (0.1-0.3 µM) added to the incubation medium for 60 min enhanced the total efflux of Ch from both brain slices and atria by about 50% (Tab.1). We confirmed the previous observation (Sandmann et al., 1990) that PDB caused a linear rise during the 60 min-period. The maximum PDB-evoked increases (in % of the individual resting rate) were 48% in rat (0.2 µM PDB) and 98% in chicken atria (0.3 µM PDB). The PDB-evoked acceleration of Ch efflux was totally due to an increase in Ch net production in both tissues, an increase that, in the chicken atria, even amounted to 300%.

An effect of PDB on the average formations of [^{3}H]-PA was observed in all investigated preparations, namely in rat hippocampus, and chicken and rat atria (Tab.2), but the effects did not reach statistical significance. Ethanol appeared to decrease the PDB-evoked rise of PA in chicken and rat atria (Tab.2), an effect that was balanced by a marked and significant formation of [^{3}H]-PEth in the atria of both

Tab. 2. Formation of [^{3}H]-PA in hippocampal slices and in isolated atria under resting conditions and after 0.2 µM PDB in the absence or presence of propanol (hippocampus) or ethanol (atria) expressed as % of total labeled phospholipids.

Tissue	Control (N)	Propanol/Ethanol (N)
Rat hippocampus		
- Control	11.3±0.4 (4)	10.6±0.2 (4)
- PDB	11.8±0.7 (4)	10.2±0.6 (4)
Rat atria		
- Control	1.58±0.37 (3)	1.26±0.39 (3)
- PDB	2.57±0.64 (3)	1.43±0.53 (3)
Chicken atria		
- Control	0.79±0.11 (4)	0.53±0.03 (5)
- PDB	1.01±0.12 (4)	0.67±0.09 (5)

Tab.3. Formation of [^{3}H]-PP in hippocampal slices and of [^{3}H]-PEth in isolated atria under resting conditions and after 0.2 µM PDB expressed as % of total labeled phospholipids.

Tissue	[^{3}H]-PP/-PEth	
	Control	PDB
Rat hippocampal slices	0.92±0.15 (4)	1.25±0.10 (4) (S)
Rat atria	0.28±0.04 (3)	0.85±0.07 (3) (S)
Chicken atria	0.26±0.04 (5)	0.89±0.15 (5) (S)

(S) Significantly different from Control.

species (Tab.3). In hippocampal slices, PDB significantly augmented the formation of labeled PP to almost the same value (1.25% of total phospholipids) as found for PEth in rat and chicken atria (0.85 and 0.89%, respectively). In comparison with the basal levels (Tab.3), the rise was only about one third in brain slices and about three-fold in atria. It appears, therefore, that the small effect of PDB in hippocampal slices was due to the very high resting activity of PLD (see above).

Nevertheless it can be concluded that the phorbol ester PDB enhanced the hydrolysis of Ch-containing phospholipids in both heart and brain via activation of PLD.

Effects of muscarinic receptor agonists. In chicken atria, the muscarinic receptor agonists arecaidine propargyl ester (APE) (1 μM) and carbachol (10 μM) added to the medium for 60 min significantly increased both total Ch efflux and net formation of Ch (Tab.1). In other words, the increase in efflux by APE or carbachol was due to an elevation of the net production of Ch. APE and carbachol, like PDB, caused a gradual rise of the Ch efflux during the 60 min-period. The maximum rises were 91% (APE) and 54% (carbachol). The source of this Ch must have been phospholipids, because the basal release of ACh, as the alternative Ch source, was by far too small (about 0.1 nmol/g min; see also Lindmar et al., 1986) to account for the muscarinic effect on Ch efflux.

In preliminary experiments, we studied the effect of muscarinic receptor stimulation on the formation of [^{3}H]-PA and -PEth using rat and guinea pig atria (Tab.4). ACh in the presence of the cholinesterase inhibitor DFP enhanced the formation of labeled PA, which, however, reached only borderline significance when compared with "ACh plus atropine". In contrast, ACh markedly increased the formation of [^{3}H]-PEth in both rat and guinea pig atria, an effect that was blocked in the presence of 1 μM atropine (Tab.4).

These preliminary data suggest that, in mammalian atria, muscarinic receptor stimulation enhances Ch phospholipid hydrolysis by activation of PLD.

Tab.4. Effects of 10 µM ACh on the formation of [³H]-PA and of [³H]-PEth in rat and guinea pig atria in the absence and presence of ethanol expressed as change in % of radioactivity per mg wet weight.

Tissue	Without ethanol	Ethanol	
	[³H]-PA	[³H]-PA	[³H]-PEth
Rat atria			
- ACh	+41±21% (4)	+50±20% (4) (s)	+129±29% (4) (S)
- ACh plus atropine	---	-13±12% (4)	+27±10% (4)
Guinea pig atria			
- ACh	+18±4% (3)	+13±26% (3)	+161±81% (3)

(S) Significantly different from the corresponding data obtained with "ACh plus atropine" and from the individual control values (100%) obtained in the absence of ACh and atropine. These 100% values ranged for PA from 124 to 394 dpm/mg (rat) and from 502 to 1013 dpm/mg (guinea pig); the corresponding ranges for PEth were 27-43 and 51-286 dpm/mg, respectively.

(s) Borderline significance (P=0.05), when compared with "ACh plus atropine".

Discussion

There is increasing evidence that many Ca^{2+}-mobilizing hormones, neurotransmitters and growth factors acting through the "PI-response" also stimulate the breakdown of phosphatidylcholine (reviewed by Billah and Anthes, 1990; Löffelholz, 1990). These effects seem to be mediated by various phospholipases yielding functionally important metabolites such as arachidonic acid, PA and DAG. The formation of arachidonic acid controls the generation of cyclooxygenase and lipoxigenase products (eicosanoids). PA, which itself has a number of marked potentially important intracellular effects, may be converted to other active metabolites such as lyso-PA (Jalink et al., 1990) and DAG. The latter product is well known as activator of PKC.

In 1986, receptor-mediated activation of PLD has been hypothesized by us on the basis of indirect evidence obtained in the perfused chicken heart (Lindmar et al., 1986, 1988). Moreover, also stimulation of PKC supposedly enhanced PLD activity in heart tissue (Sandmann et al., 1990). The present study provides unequivocal evidence in favor of our previous hypotheses. Moreover, these data fit to numerous literature data that appeared in the meantime and have been obtained in isolated cells or cell fractions (reviewed by Billah and Anthes, 1990; Löffelholz, 1990).

In the present study, PLD activity was determined by measuring net formation of Ch, formation of [^{3}H]-PA and, in the presence of a primary alcohol, of [^{3}H]-PP or -PEth (see Introduction) of labeling with tritiated myristic acid (reviewed by Billah and Anthes, 1990; Löffelholz, 1989, 1990). Labeled PP or PEth are conveniently used to determine PLD activity, because of their specificity, metabolic stability and low basal levels as compared to PA. Hence, in the present study, the PA levels of brain and heart tissues did not significantly respond to PDB and to muscarinic agonists, as the basal levels of PA were already high and additional PA might have been dephosphorylated.

The present results provide evidence for the following conclusions.

(1.) In intact tissues (heart and brain), the <u>basal</u> activity of PLD is extremely high, especially when compared with isolated cell preparations which essentially lack basal activity. The effects were observed without the presence of exogenous activating agents, such as oleic acid.

(2.) PLD activity was enhanced by the phorbol ester PDB in heart and brain tissues. These data in combination with previous results obtained by us in the perfused heart and in striatal slices (Sandmann et al., 1990) suggest that these effects were mediated by stimulation of PKC. Similar results have been obtained in NG 108-15 neuroblastoma X glioma hybrid cells (Liscovitch, 1989) and other cell preparations (reviewed by Billah and Anthes, 1990; Exton, 1990; Löffelholz, 1989, 1990).

3.) Muscarinic receptor stimulation activated PLD in heart preparations as hypothesized previously (see above). Muscarinic receptor-activated PLD has been demonstrated also in canine brain synaptosomes (Qian and Drewes, 1989) and in various isolated cells, such as rat pancreatic acini (Matozaki and Williams, 1989), 1321N1 astrocytoma cells (Martinson et al., 1989) and in human cholinergic neuroblastoma (LA-N-2) cells (Sandmann and Wurtman, 1990).

The possible physiological roles of PLD in signal transduction have been discussed in a number of recent review articles (Billah and Anthes, 1990; Exton, 1990; Löffelholz, 1989, 1990). Several metabolites, such as PA, lyso-PA, Ch and DAG, may be of critical importance for a variety of cellular functions. Much work remains to unravel the complex field of receptor-mediated phospholipid hydrolysis.

References

Billah MM, Anthes JC (1990) The regulation and cellular functions of phosphatidylcholine hydrolysis. Biochem J 269:281-291

Damsma G, Westerink BHC, Horn AS (1985) A simple, sensitive, and economic assay for choline and acetylcholine using HPLC, an enzyme reactor, and an electrochemical detector. J Neurochem 45:1649-1652

Exton JH (1990) Signaling through phosphatidylcholine breakdown. J Biol Chem 265:1-4

Folch J, Lees M, Stanley GHS (1957) A simple method for the isolation and purification of total lipids from animal tissues. J Biol Chem 226:497-500

Hokin LE (1985) Receptors and phosphoinositide-generated second messengers. Ann Rev Biochem 54:205-235

Israel M, Lesbats B (1982) Application to mammalian tissues of the chemoluminescent method for detecting acetylcholine. J Neurochem 39:248-250

Jalink K, Van Corven EJ, Moolenaar WH (1990) Lysophosphatidic acid, but not phosphatidic acid, is a potent Ca-mobilizing stimulus for fibroblasts. J Biol Chem 265:12232-12239

Lindmar R, Löffelholz K, Sandmann J (1986) Characterization of choline efflux from the perfused heart at rest and after muscarine receptor activation. Naunyn-Schmiedeberg's Arch Pharmac 332:224-229

Lindmar R, Löffelholz K, Sandmann J (1988) On the mechanism of muscarinic hydrolysis of choline phospholipids in the heart. Biochem Pharmacol 37:4689-4695

Liscovitch M (1989) Phosphatidylethanol biosynthesis in ethanol-exposed NG 108-15 neuroblastoma X glioma hybrid cells. J Biol Chem 264:1450-1456

Löffelholz K (1989) Receptor regulation of choline phospholipid hydrolysis. Biochem Pharmacol 38:1543-1549

Löffelholz K (1990) Receptors linked to hydrolysis of choline phospholipids: the role of phospholipase D in a putative mechanism of signal transduction. In: Current Aspects of the Neurosciences, vol 2 (Osborne NN, ed), pp 49-76, MacMillan Press London

Martinson EA, Goldstein D, Brown JH (1989) Muscarinic receptor activation of phosphatidylcholine hydrolysis. J Biol Chem 264:14748-14754

Matozaki T, Williams JA (1989) Multiple sources of 1,2-diacylglycerol in isolated rat pancreatic acini stimulated by cholecystokinin. J Biol Chem 264:14729-14734

Qian Z, Drewes LR (1989) Muscarinic acetylcholine receptor regulates phosphatidylcholine phospholipase D in canine brain. J Biol Chem 264:21720-21724

Sandmann J, Leissner J, Lindmar R, Löffelholz K (1990) The effects of phorbol esters on choline phospholipid hydrolysis in heart and brain. Eur J Pharmacol - Mol Pharmacol Sec 188:89-95

Sandmann J, Wurtman R (1990) Phospholipase D and phospholipase C in human cholinergic neuroblastoma (LA-N-2) cells: modulation by muscarinic agonists and protein kinase C. In: Advances in Second Messenger and Phosphoprotein Research, vol 24 (Nishizuka Y, ed), Raven Press New York

Stadler H, Nesselhut T (1986) Simple and rapid measurement of acetylcholine and choline by HPLC and enzymatic-electrochemical detec-tion. Neurochem Int 9:127-129

DEVELOPMENT OF A SINGLE CELL Ca^{2+} IMAGING SYSTEM TO STUDY THE ROLE OF PKC SUBSTRATE B-50 IN NEUROTRANSMITTER RELEASE AND NEURITE OUTGROWTH

J. Elands, W.H. Gispen, W.H. and P.N.E. De Graan[1]
Division of Molecular Neurobiology
Rudolf Magnus Institute
Padualaan 8
3584 CH Utrecht
The Netherlands

Introduction

One of the well-characterized substrates of protein kinase C (PKC) in neurons is the protein B-50. B-50 is a nervous tissue-specific substrate of PKC associated with the cytosolic face of the presynaptic membrane (see De Graan et al., 1991). Protein B-50 is identical to the growth-associated protein GAP-43, the calmodulin (CaM)-binding protein neuromodulin, and protein F1, which is implicated in long-term potentiation (for a review see Skene, 1989). B-50 is one of the abundant proteins in the neuronal growth cone (Skene, 1989) and has been implicated in signal transduction and the mechanism of neurite outgrowth during development and differentiation (Skene, 1989). In a series of studies we have shown that the degree of PKC-mediated phosphorylation of B-50 in hippocampal slices and synaptosomes is correlated with transmitter release (Dekker et al., 1991a; Heemskerk et al., 1989,1990). Based on these correlative studies and the fact that phorbol esters which stimulate PKC enhance neurotransmitter release (see Dekker et al., 1991a), we have suggested that PKC-mediated B-50 phosphorylation may be involved in the regulation of neurotransmitter release.

More direct evidence for a role of B-50 in noradrenaline (NA) release was obtained in studies using synaptosomes permeated with the toxin streptolysin-O (SL-O; Dekker et al., 1989;1991b; De Graan et al., 1991). NA release from permeabilized synaptosomes could be induced by elevation of the Ca^{2+} concentration in the permeation buffer. We have used these permeated synaptosomes to introduce antibodies to B-50, which are known to inhibit B-50 phosphorylation. The affinity purified anti-B-50 IgGs completely inhibited Ca^{2+}-induced NA release from the permeated synaptosomes (Dekker et al., 1989). Since the antibodies inhibit B-50 phosphorylation, these data indicate that B-50 phosphorylation is important for NA release. However, it is also possible that the antibodies inhibit other properties of the B-50

[1]To whom correspondence should be addressed.

NATO ASI Series, Vol. H 52
Biological Signal Transduction
Edited by E. M. Ross and K. W. A. Wirtz

molecule besides its PKC-mediated phosphorylation. In fact, we have evidence that PKC inhibitors do not inhibit Ca^{2+}-induced release from permeated synaptosomes (De Graan et al., 1991). An interesting possibility is the CaM binding property of B-50. It has been shown by Storm and co-workers that purified B-50 has a higher affinity for CaM in the absence of Ca^{2+} than in the presence (for a review see Liu and Storm, 1990) and that prephosphorylation of purified B-50 by PKC inhibits CaM binding. Based on these *in vitro* studies it has been suggested that *in vivo* B-50 may act as a local CaM store. CaM may be dissociated from B-50 after a depolarization-induced rise in intracellular Ca^{2+} and subsequently activate cellular processes involved in transmitter release, for instance CaM-dependent kinases and phosphorylation of synapsin I (Nichols et al., 1990). Once CaM and B-50 have been dissociated, B-50 may be phosphorylated by PKC, thus preventing reassociation of CaM and prolonging CaM action. Indeed, we could recently show (De Graan et al., 1990,1991) that using the chemical crosslinker disuccinimidyl suberate a complex can be detected between CaM and endogenous B-50 in native SPM and growth cone membranes. This complex formation is regulated by the free Ca^{2+} concentrations which occur under physiological conditions in neurons.

Interestingly, the concentrations of Ca^{2+} required to induce NA release and dissociation of the B-50/CaM complex (up to 10^{-5} M) in permeated synaptosomes are much higher than those measured in intact synaptosomes using the fluorescent Ca^{2+} indicator Fura-2. The high Ca^{2+} concentration necessary to induce release contrasts with the overall Ca^{2+} levels measured after K^+ depolarization of intact synaptosomes, which do not exceed the 0.5 x 10^{-6} M level (Verhage et al., 1988). There are two explanations for this apparent contradiction: (i) local high levels of Ca^{2+} are required to induce transmitter release, and (ii) high concentrations of Ca^{2+} are necessary in the permeated system to overcome SL-O-induced damage to the release machinery, or to overcome dilution of cytosolic components of the release system. Extremely high local Ca^{2+} concentrations would occur just below the plasma membrane immediately following depolarization when Ca^{2+} enters through voltage sensitive Ca^{2+} channels (VSCC). If B-50 is involved in the release mechanism the prediction would be that for instance changes in the degree of phosphorylation would parallel a Ca^{2+} influx rather than an overall increase in the Ca^{2+} concentration. A series of experiments using different depolarization techniques in intact synaptosomes indeed provide evidence for a close correlation between Ca^{2+} influx and B-50 phosphorylation (Heemskerk et al., 1991; De Graan et al., 1991).

The notion that B-50 function maybe directly regulated by local Ca^{2+} concentrations in synaptosomes and possibly also in growth cones has prompted us to develop a single cell Ca^{2+} imaging system to monitor local differences in Ca^{2+} concentrations in individual neurons and measure effects of modulation of the B-50/PKC/CaM system on Ca^{2+} parameters.

Single cell Ca^{2+} imaging system

Several techniques are available to study Ca^{2+} concentrations or Ca^{2+} flows; Ca^{2+} sensitive electrodes, $^{45}Ca^{2+}$ flow measurements, intracellular recording using patch clamp and Ca^{2+} sensitive fluorochromes. While patch clamp recording, Ca^{2+} electrodes and $^{45}Ca^{2+}$ flow measurements are promising techniques, the use of the second generation of fluorochromes in combination with video microscopy is the only technique enabling the measurement of Ca^{2+} concentrations in living cells with complex morphology under nearly physiological conditions.

The most successful compound in the first generation of fluorochromes was quin2, which has been widely used to measure Ca^{2+} level changes in dissociated and/or cultured cells, either using a spectrofluorimeter, fluorescence microscope equipped with a photomultiplier or a video camera. However, quin2 had several disadvantages over the second generation of Ca^{2+} fluorochromes (Grynkiewicz et al., 1985), of which fura-2 is the most commonly used. The new probes have a lower affinity for Ca^{2+}, thus enabling measurements at higher Ca^{2+} concentrations and also limiting buffering capacity, a higher total fluorescence yield, enabling reduction of the UV excitation light intensity and faster kinetics. Moreover, the selectivity for Ca^{2+} over other divalent metal cations (Mg^{2+}) has been substantially improved, while cations like Mn^{2+} can still be used to quench fura-2 completely (Grynkiewicz et al., 1985). Since cell membranes are impermeable for hydrophilic substances as fura-2 (and quin2), they have to be injected into the cell, or be presented in a more lipophilic form. The acetoxymethylester derivative, fura-2/AM, is able to pass the cell membrane to enter the cytoplasm where esterases convert it into free fura-2. This hydrolysis is not always complete (Schmidt et al., 1988), therefore Ca^{2+} insensitive but highly fluorescent metabolites of fura-2/AM may be present. The most important feature of fura-2 (and indo-1) over quin2 however is a shift in excitation (emission for indo-1) spectrum upon Ca^{2+} binding. Fura-2 can be more readily excited at 340 nm when it has bound Ca^{2+} and at 380 nm when it exists in the Ca^{2+} free state. The ratio of the fluorescence at these two wavelengths can be used to determine the exact Ca^{2+} concentration (Grynkiewicz et al., 1985; Tsien and Poenie, 1986). According to the method proposed by Tsien et al. (Grynkiewicz et al., 1985) the Ca^{2+} concentration follows from the following equation:

$$[Ca^{2+}] = K_d * \beta \left(\frac{R_t - R_{min}}{R_{max} - R_t} \right)$$

where the K_d represents the dissociation constant of fura-2 for Ca^{2+} (225 nm at 37°C), ß the ratio of the limiting minimal and maximal fluorescence levels at 380 nm (F_0/F_s), R_{min} and R_{max} the 340/380 nm fluorescence ratios at respectively minimal and maximal intracellular Ca^{2+} concentration and R_t the 340/380 nm fluorescence ratio at a given pixel and time in a (ratio) image. R_{max} can be measured after loading the cell with Ca^{2+} for instance by making the cell membrane permeable for Ca^{2+} by ionomycin. Then, using EGTA R_{min} can be determined.

When this approach is combined with computer assisted video microscopy the actual Ca^{2+} concentration can be visualized in a two-dimensional image (for review see Tsien 1988,1989). An alternative approach, also to be used as a control, is to prepare titrated EGTA/Ca^{2+} buffers and to measure fluorescence in vitro in flat glass capillaries or microscopic slides. The potential of the use of titrated buffers in ionomycin treated cells has also been put forward (Williams et al., 1985), but limitation of bulk volume Ca^{2+} transport through the ionomycin formed pores limits the use of this approach.

From the characteristics and dynamics of the above described B-50/PKC/CaM model several aspects of the ratio-imaging technique can be inferred. - Ca^{2+} concentrations in nerve cells are rapidly changing, the system should thus be able to capture data at a relatively high speed. - Since B-50 is a presynaptic protein, localized in axons, growth cones and in presynaptic terminals, measurements must be done in structures that are very small and will thus have a low fluorescence yield. A very sensitive video camera is necessary, in conjunction with a microscope generating virtually no background light. - Living cells are used that must be maintained for periods up to several hours under physiological circumstances. This can only be achieved in appropriate (specially designed) tissue culture chambers thermostatically controlled by a heating element and/or a perfusion system.

Technical aspects (Fig. 1)

Microscope

Essentially two types of microscopes exist for fura-2 measurements. The normal fluorescence and the epifluorescence microscope. The first type uses objectives with extra long working distance that are immersed into the cell culture medium. The excitation beam

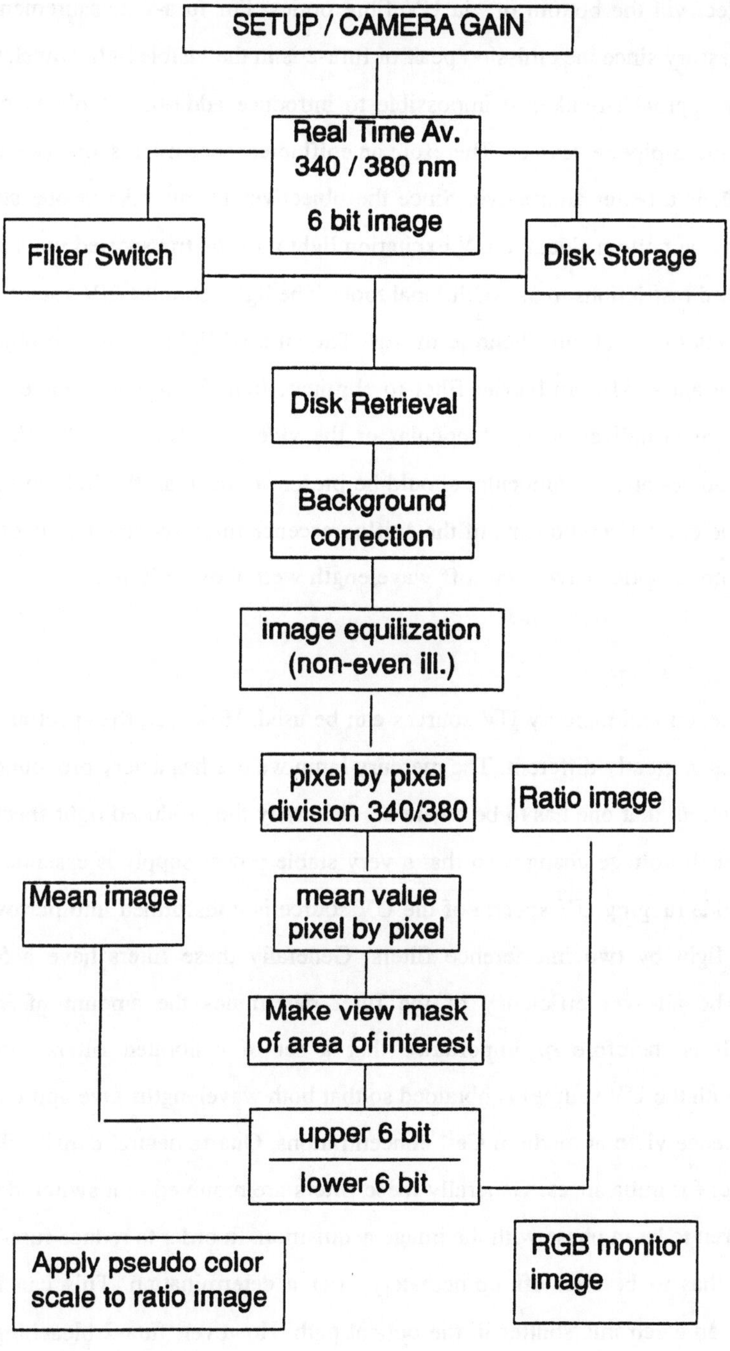

Fig. 1. A flow chart of the Ca^{2+} imaging equipment.

enters the object via the bottom or via UV fibre optics. For fura-2 measurements no fluor lenses are necessary since the emission peak of fura-2 is in the visible light wavelength range. However, this approach renders it impossible to introduce additional tools as patch-clamp electrodes or micro-pipette devices. Therefore an epifluorescence microscope (we use a Nikon Diaphot TMD) is a better alternative. Since the objectives in this microscope approach the specimen from the bottom side, the UV excitation light must be transported via the objectives in order to avoid restrictions to use additional tools. The light from the UV source is reflected to the objectives by a 400 nm dichroic mirror. The emitted light passes the objectives, the dichroic mirror and a 510 nm barrier filter to eliminate light from other sources than fura-2 and can than be visualized in the binocular or the video camera. A 100%-100% changer between the camera and the binocular should be present to have all the light available at the camera. All the optical components of the epifluorescence microscope must be of UV quartz glass since normal optics have a cut off wavelength well above 340 nm.

UV source

Both xenon and mercury UV sources can be used. However, the spectral distribution of both devices is greatly different. The mercury lamp we use has a very pronounced 360 nm line in its spectrum that one has to be aware of. Generally the produced light spectra are very sensitive to small voltage changes so that a very stable power supply is essential.

The wide ranging UV spectra of the UV source is transformed into narrow band 340 and 380 nm light by two interference filters. Generally these filters have a 5 to 20 nm bandwidth. The all-over efficiency of the filter determines the amount of light that is transmitted. It is therefore of importance that a set of calibrated filters (preferably in conjunction with the UV source) is obtained so that both wavelengths give approximately the same fluorescence yield at medium Ca^{2+} concentrations. Quartz neutral density filters can be used to correct for imbalances. Generally these filters are mounted in a switch device that is computer driven to be in phase with the image acquisition. In order to reduce fura-2 bleaching the UV light has to be shut off immediately after a determination. This can be done by incorporating an electronic shutter in the optical path. However, fura-2 bleaching may even then be a problem (Becker and Fay, 1987).

Video camera

To visualize fura-2 emission CCD intensified cameras are a good choice. In a relatively affordable way sensitivities of 10^{-7} lux can be obtained which are necessary for measurements in axons and growth cones. The class of cooled intensified CCD cameras is not only very sensitive but also have an extremely low dark-current (background light or noise) and a large dynamic range (> 1000). These cooled cameras are however rather slow (slow-scan camera), i.e. the generation of a single image needs a scan time in the order of the 100 msec range. The non-cooled intensified CCD cameras (we use a MX-II, HCS, NL), working at ambient temperature, have scan times of 10 to 30 msec, enabling to work at video rate (25 images per second in Europe, 30 in the USA). However, they have a limited dynamic range 60-200 and a higher dark-current. As will be discussed later the real dynamic range is also dependent on the video frame capture board (frame grabber) or the real time average device.

Intensified CCD cameras are easily damaged by exposure to extreme amounts of light. Therefore it is highly recommended to mount an electronic shutter between the camera and the microscope housing to avoid accidental overexposure. For a general review on video microscopy see Inoué (1987).

Video frame capture board or frame grabber

When cameras are used that function according to industrial video standards they generate an analogous output signal. To use such a signal for measurements and calculations it must first be converted into a digital form. A frame grabber converts analogous signals continuously into digital signals. The resulting digital images can be manipulated by a computer. However, at low light levels, a single digitized image is usually of a rather poor quality due to heavy noise influence. To reduce the noice averaging of several images in the time can be done. Frame grabber devices either provide a slow averaging mode (often in conventional computer memory) or a fast, real-time averaging mode. The latter mode has the advantage of speed and temporal resolution since no images are left unused for the averaging process.

The dynamic range of frame grabbers is defined as the number of bits available to code the incoming grey values. As the early systems sometimes used only 3 or 4 bits (dynamic range from 8 to 16, 2^n where n is the number of bits) most modern systems use 6 to 8 bits, with exceptions using 16 to 24 bits. We use a frame grabber (FG-100 board, Imaging Technology Inc., USA) that has an 8 bit dynamic range for normal purposes and a 6 bit range for real time averaging (total memory is 12 bits, for real time averaging 6 bits are used for the average buffer and 6 for the incoming signal). Although a 6 bit dynamic range seems limiting one must consider the dynamic range of the camera (± 100) in relation to the range of the frame grabber. One point worth mentioning in this respect is that the system needs careful tuning of the camera gain in order to adjust the dynamic range of the camera to that of the frame grabber.

Image storage, computers

The storage of images is dependent on the video frame grabber specifications (image size and number of bits dynamic range) and experimental conditions (acquirement speed or time resolution and (sub)image size). Standard images at this moment are 256^2 pixel images and 8 bit dynamic range, although the trend is towards 512^2 pixel images. 256^2 pixel images by 8 bit (which we use) give rise to 64 Kilobytes of information, while 512^2 pixel images result in 256 Kb data files. This relatively large image file size imposes restrictions on the speed of data storage on computers. When standard "state of the art" IBM-PC computer hardware is used only 1 to 4 image files (64 Kb) per second can be stored. This is sufficient to measure low speed Ca^{2+} changes. We use a 486 25 mHz PC with a fast 660 Mb harddisk. Generally data images are stored before processing. This saves time during the acquisition sequence when the original data images are to be stored. In case ratio images are directly determined, processing before storage can be advantageous in terms of speed but obviously access to the original "raw" data is lost. For high speed purposes recording onto high quality industrial video recorders (e.g. the Sony UMATIC with frame code generator we use, or industrial super VHS systems) and digitalization and analysis afterwards can provide a good alternative. Using these standard techniques (at video frame rate) maximal 25 images (30 in USA) per second can be recorded on tape. The disadvantages are obvious, no possibility of real time average mode (due to maximal speed storage) and loss of image quality.

Image processing software

The coordination of filter switch devices (we use the Lambda 10 switcher from Sutter Instrument Co., USA), electronic shutters and the acquirement and processing of image data is taken care of by a computer program. Such a program can either be an adapted version of existing image manipulation software (we used TIM digital image processing software, DIFA Measuring Systems, Breda, NL) or a highly specialized custom written package. The most essential functions needed are: image correction for background illumination and uneven illumination; ratio determination using the 340 and 380 nm images; pseudocolour display of images with a calibrated color scale; area, point and line measurements of Ca^{2+} concentrations. An enormous amount of nice and fancy tricks can be offered to the experimentator, i.e. 3 dimensional plots of images, combinations of plots and pseudocolour images etc. Although these options can be useful they are not essential for Ca^{2+} measurements using the ratio approach.

Biological aspects

Methodology

Ca^{2+} measurement by means of fluorescent probes is rather sensitive to temperature changes, while pH changes, when restricted to the range of pH 6.75 to 7.05 appear to have no significant influence on the fluorescence. However, it is advisable to strictly control for both pH and temperature. Buffer pH can be nicely set using a oxygenated Hepes Krebs-Ringer buffer, which, in combination with a good perfusion system, can also perfectly control the temperature. In addition the culture dish can be placed in a heating element with a thermostatic control. Such a controlled heating device can also be used to control temperature in non-perfused steady state conditions. The evaporation of medium can be overcome by a special coating of oil on top of the medium, that is permeable to gasses but not to water. The control of the temperature by the thermostatic device should not take place by switching on and off the current to the heating element since this can result in electronic noise rendering electrophysiological measurements impossible. Regulating the heating element voltage is a more appropriate method. Ideally the microscope objective should be at about the same temperature as the culture dish.

Fura-2/AM loading properties of cells are variable, and need to be assessed carefully since the way of loading may affect the Ca^{2+} measurements dramatically (Almers and Neher, 1985). Fura-2/AM also penetrates cell compartments, especially when the de-esterification is slow and incomplete. When micro-injection techniques or path-pipettes are used, fura-2 can be introduced directly into the cell.

Using our Ca^{2+}-imaging system, changes in intracellular Ca^{2+} concentrations have been monitored in different cell lines, e.g. in the rat pheochromocytoma cell line PC-12. Resting PC-12 cells, treated for 24 h with Nerve Growth Factor, have a basal intracellular Ca^{2+} concentration of approximately 100-200 nM (Fig 2a, c). Upon potassium (K^{+}, 20 mM)-induced depolarization intracellular Ca^{2+} levels quickly rise to 500-700 nM. The Ca^{2+}

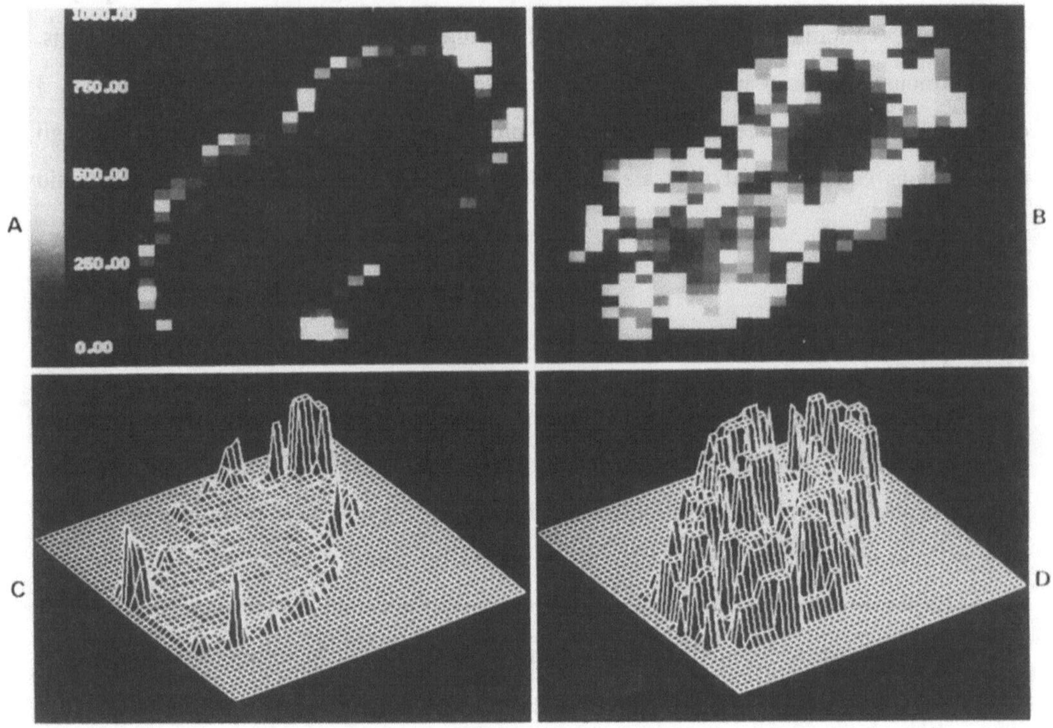

Fig. 2. PC-12 cells before (a, c) and after (b, d) a 20 mM K^+ stimulus. Upper panels, black and white representation of Ca^{2+} concentrations, black represents 0 nM while white represents concentrations above 1000 nM. Lower panels are 3D plots of the cells shown in the upper panels. The Ca^{2+} concentration is 100-200 nM under resting conditions (a,c). After the K^+ stimulus the Ca^{2+} concentration increases up to 700-1000 nM.

concentration begins to increase at the cell membrane region and thus anintracellular Ca^{2+} concentration gradient develops, eventually leading to a more uniform concentration. After several tens of seconds Ca^{2+} concentrations even ranged from 700 to 1000 nM (Fig. 2b, d). The Ca^{2+} concentration remained elevated for several minutes after K^+ addition and then slowly returned to basal levels. Not all cells respond in the same way, some cells were nearly not excitable, while some others reacted only faintly upon K^+ addition.

In conclusion the fura-2 ratio approach to measure Ca^{2+} concentrations in living cells is a powerful technique that can be combined with electrophysiological (patch-clamp) measurements. The addition of video microscopy enables to monitor changes in Ca^{2+} gradients with high spatial and temporal resolution. This technique greatly enhanced the understanding of the role of Ca^{2+} changes in strictly localized cell regions (Cheek et al., 1989; Connor et al., 1987; Connor, 1986; Williams et al., 1985). Although the method demands considerable investments a great variety of problems can be investigated, that are not only related to Ca^{2+} concentrations. In the same family of fluorochromes several other dyes, sensitive for pH , membrane potential, Na^+, Cl^- etc. are available (see Tsien, 1989 for review) and a new generation of dyes, possibly also for other second messengers, can be expected in the near future.

Acknowledgments

We thank DIFA Measuring Systems (Breda, NL) for their support and the development of the Ca^{2+} imaging software. J. Elands was supported by Tropon Werke GmbH (Cologne, Germany).

REFERENCES

Almers W, Neher E (1985) The calcium signal from fura-2 loaded mast cells depends strongly on the method of dye-loading. FEBS Lett 192: 13-18.

Becker PL, Fay FS (1987) Photobleaching of fura-2 and its effect on determination of calcium concentrations. Am J Physiol 253: C613-C618.

Cheek TR, O'Sullivan AJ, Moreton RB, Berridge MJ, Burgoyne RD (1989) Spatial localization of the stimulus-induced rise in cytosolic Ca^{2+} in bovine adrenal chromaffin cells. Distinct nicotinic and muscarinic patterns. FEBS Lett 247: 429-434.

Connor JA (1986) Digital imaging of free calcium changes and of spatial gradients in growing processes in single, mammalian central nervous system cells. Proc Natl Acad Sci USA 83: 6179-6183.

Connor JA, Cornwall MC, Williams GH (1987) Spatially resolved cytosolic calcium response to angiotensin II and potassium in rat glomerulosa cells measured by digital imaging techniques. J Biol Chem 262: 2919-2927.

De Graan PNE, Oestreicher AB, De Wit M, Kroef M, Schrama LH, Gispen WH (1990) Evidence for the binding of calmodulin to endogenous B-50 (GAP-43) in native synaptosomal plasma membranes. J Neurochem 55: in press.

De Graan PNE, Schrama LH, Oestreicher AB, Schotman P, Gispen WH (1991) Protein kinase C substrate B-50 (GAP-43) and neurotransmitter release. Progr Brain Res, in press.

Dekker LV, De Graan PNE, Oestreicher AB, Versteeg DHG, Gispen WH (1989) Inhibition of noradrenaline release by antibodies to B-50 (GAP-43). Nature 342: 74-76.

Dekker LV, De Graan PNE, Gispen WH (1991a) Transmitter release: target of regulation by protein kinase C? Progr Brain Res, in press.

Dekker LV, De Graan PNE, Pijnappel P, Oestreicher AB, Gispen WH (1991b) Nor-adrenaline release from streptolysin-O-permeated rat cortical synaptosomes. J Neurochem, in press.

Grynkiewicz G, Poenie M, Tsien RY (1985) A new generation of Ca^{2+} indicators with greatly improved fluorescence properties. J Biol Chem 260: 3440-3450.

Heemskerk FMJ, Schrama LH, Gianotti C, Spierenburg H, Versteeg DHG, De Graan PNE, Gispen WH (1989a) 4-Aminopyridine stimulates B-50/GAP-43 phosphorylation and [^{3}H]-noradrenaline release in rat hippocampal slices. J Neurochem 54: 863-869.

Heemskerk FMJ, Schrama LH, De Graan PNE, Gispen WH (1990) 4-Aminopyridine stimulates B-50 (GAP-43) phosphorylation in rat synaptosomes. J Mol Neurosci 2: 11-17.

Heemskerk FMJ, Schrama LH, Ghijsen WEHM, De Graan PNE, Lopes da Silva FH, Gis-pen WH (1991) Presynaptic mechanism of action of 4-aminopyridine: changes in $[Ca^{2+}]_i$ and its relationship to B-50 (GAP-43) phosphorylation. J Neurochem, in press.

Inoue S (1987) Video Microscopy. Plenum Press, NY.

Liu Y, Storm DR (1990) Regulation of free calmodulin levels by neuromodulin: neuron growth and regeneration. Trends Pharmacol Sci 11: 107-111.

Nichols RA, Sihra TS, Czernik AJ, Nairn A, Greengard P. (1990) Calcium/calmo-dulin-dependent protein kinase II increases glutamate and noradrenaline release from synaptosomes. Nature 343: 647-651.

Schmidt D, Papadopoulos MT, Crooke ST, Stassen FL (1988) Potential role of protein kinase C in the regulation of vasopressin (V_1) receptors of vascular smooth muscle cells (A10) Faseb J 2: A624.

Skene JHP (1989) Axonal growth-associated proteins. Ann Rev Neurosci 12: 127-156.

Tsien RY (1988) Fluoresence measurement and photochemical manipulation of cytosolic free calcium. TINS 11: 419-424.

Tsien RY (1989) Fluorescent indicators of ion concentrations. Meth Cell Biol 30: 127-156.

Tsien RY, Poenie M (1986) Fluoresence ratio imaging: a new window into intra-cellular ionic signalling. Trends Biochem Sci 11: 450-455.

Verhage M, Besselsen E, Lopes da Silva FH, Ghijsen WEJM (1988) Evaluation of the Ca^{2+} concentration in purified nerve terminals: relationship between Ca^{2+} homeostasis and synaptosomal preparation. J Neurochem 51: 1667-1674.

Williams DA, Fogarty KE, Tsien RY, Fay FS (1985) Calcium gradients in single smooth muscle cells revealed by the digital inmaging microscope using fura-2. Nature 318: 558-561.

REGULATED EXOCYTOSIS AS A PARADIGM OF BIOLOGICAL SIGNAL TRANSDUCTION.

B. D. Gomperts, Y. Churcher, A. Koffer, T. H. W. Lillie, P. E. R. Tatham and T. D. Whalley.
Department of Physiology
University College London
University Street
London WC1E 6JJ, UK.

Secretion is one of the basic processes of eukaryotic cells. It occurs by a mechanism called exocytosis by which the contents of secretory granules are released to the exterior following fusion of the perigranular and plasma membranes. Some cells secrete continuously at a rate which is determined by the rate of synthesis of the secreted materials (constitutive secretion). Others release previously packaged materials in response to external signals; this is called regulated exocytosis.

Although the final release stage of exocytosis, the fusion of perigranular membranes with the plasma membrane of cells is common to all secreting cell types, the regulatory pathway controlling this process varies widely. This allows secretory cells to serve a wide multiplicity of physiological tasks. Thus presynaptic membranes release neurotransmitters within milliseconds of depolarisation; secretion of parathyroid hormone [Nemeth and Scarpa 1986 Shoback et al. 1983], renin [Park et al. 1986 Fray et al. 1987] and human placental lactogen [Choy and Watkins 1976 Handwerger et al. 1981] is elicited in response to a decrement in cytosol Ca^{2+}; parotid cells release amylase in response to β-adrenergic activation and elevation of cyclicAMP [Dormer and Ashcroft 1974 Leslie et al. 1976 McMillian et al. 1988] and this allows them to release fluids separately, in response to muscarinic activation and elevation of cytosol Ca^{2+}. For most cells however, the regulated exocytotic process appears to be linked in some way to Ca^{2+} elevation. It also depends on intact metabolic pathways leading to the generation of ATP.

Early hints of these relationships came from experiments demonstrating constraints on secretion in metabolically inhibited cells and in cells deprived of calcium [Douglas 1968]. Following the advent of the Ca^{2+}-carrying ionophores [Reed and Lardy 1972], it was shown that nearly all secretory cells could be stimulated by the artificial introduction of

NATO ASI Series, Vol. H 52
Biological Signal Transduction
Edited by E. M. Ross and K. W. A. Wirtz

Ca^{2+} into the cytosol: the so-called Ca^{2+}-only stimulus. With the development of sensing dyes to measure the concentration (actually the activity) of cytosol Ca^{2+} it became apparent that the level of Ca^{2+} required to induce secretion in such a Ca^{2+}-only stimulus far exceeds the level achieved following induction of a commensurate degree of secretion following stimulation of cell surface receptors [Cockcroft and Gomperts 1988; Burgoyne 1984; Lew et al. 1986]. It was evident that another signalling pathway, initiated at the receptor level, is needed to produce a full response, and an obvious candidate for this was the generation of diglyceride, leading to the activation of protein kinase C with consequent phosphorylations [Haslam and Davidson 1984; Knight et al. 1984; Katakami et al. 1984].

Figure 1 illustrates the dependence on cytosol Ca^{2+} (measured using quin2) for secretion from rat basophilic leukemia cells stimulated by a) Ca^{2+}-ionophore A23187, b) A23187 plus phorbol ester, and c) a receptor directed (IgE) stimulus [Beaven et al. 1987].

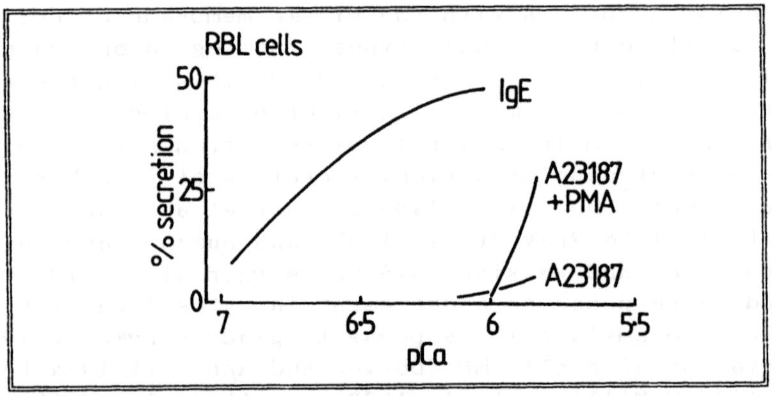

Figure 1. Dependence on intracellular Ca^{2+} for secretion from rat basophilic leukaemia cells following stimulation with calcium ionophore, calcium ionophore-plus-phorbol ester, or antigen.

Here it can be seen that while secretion commences as the concentration of intracellular Ca^{2+} rises above 10^{-7}M following receptor stimulation (in this case, the cross-linking by an antigen, of the receptors for IgE), very much higher concentrations of intracellular Ca^{2+} are required to induce a comparable extent of secretion following stimulation with a Ca^{2+}-ionophore (with or without the addition of PMA). From this it becomes evident that the combination of the Ca^{2+}-, and kinase C-pathways constitute only a partial stimulus to the activation mechanism. There must exist other pathways or mechanisms, not directly accessible when investigating intact cells, which cooperate with these more evident manifestations of cell activation to produce the full response. The possibility exists that these do not involve alterations in the concentrations of 2nd messengers, but rely on changes in the affinity of binding proteins for solutes (eg GTP) which remain essentially constant.

The above statements comprise about all that it is possible to say about the mechanism of exocytosis as gleaned from investigations of intact cells. To learn more it is necessary to work with cells in which the interacting membranes have been made accessible by techniques of permeabilisation [Gomperts and Fernandez 1985 Knight and Scrutton 1986]. There are many techniques which have been devised to achieve this and the effects these have on the cells to which they have been applied vary widely. In order to work with permeabilised cells it is vital to understand the nature of the membrane permeability lesions.

There are two main considerations. First it is necessary to know whether the lesions induced by any particular technique have dimensions which will allow the selective dialysis of ions and low molecular weight solutes (see table 1: HVD, α-toxin, and (for susceptible cells) ATP^{4-}), or whether the lesions will permit the leakage of cytosol proteins (high concentrations of ATP^{4-}, plant glycosides and SL-O). Among mechanical methods of accessing the interior of single cells, microinjection ensures that all cytosol components are retained, while the use of the patch-pipette in the whole cell mode allows extensive dilution of cytosol proteins. Second, it is important to be aware of the duration of the permeability lesions induced by any of these techniques. HVD treatment of adrenal chromaffin cells [Baker and Knight 1978], platelets [Knight and Scrutton 1980], pancreatic (exocrine [Knight and Koh 1984] and endocrine [Yaseen et al. 1982]) cells has the effect of generating semi-permanent (>1h) permeability lesions. Those formed in red blood cells [Kinoshita and Tsong 1977], macrophages and Hela cells reseal rather rapidly [Baker et al. 1985]. The lesions generated by ATP^{4-} reseal promptly following removal or conversion to the Mg^{2+} salt [Gomperts 1983; Tatham and Lindau 1989].

No one of these techniques has a monopoly of wisdom, and similar experiments carried out using different techniques (eg devised to induce dialysis or wash-out) can often give insights into mechanism which would not be evident by the use of one technique alone. An example of this is illustrated in figure 2 [Koffer and Gomperts 1989].

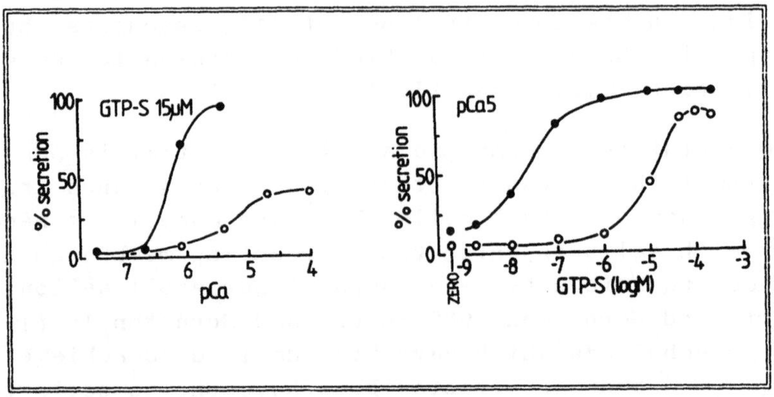

Figure 2. Dependence of Ca^{2+} (a: at 10μM GTP-γ-S) and GTP-γ-S (b: at pCa5) for secretion from mast cells permeabilised by SL-O (closed symbols) and ATP^{4-} (open symbols).

We have shown that in order to induce exocytosis from permeabilised mast cells it is necessary to provide both Ca^{2+} ions and a guanine nucleotide (such as GTP-γ-S): both are necessary, and together these are sufficient to ensure 100% release of histamine and lysosomal enzymes (we measure hexosaminidase) which are contained in the secretory granules of these cells) [Howell et al. 1987; Gomperts and Tatham 1988]. The dependence on the concentrations of the two activating effectors is very different depending on the technique used to achieve membrane permeabilisation [Koffer and Gomperts 1989]. For SL-O permeabilised cells from which cytosol proteins leak extensively and rapidly [Howell and Gomperts 1987; Gomperts et al. 1987], exocytosis occurs at concentrations of Ca^{2+} commensurate with the receptor stimulation of intact cells, and is sensitive to GTP-γ-S at concentrations as low as 10^{-8}M. Very much higher concentrations of both effectors are required following permeabilisation by ATP^{4-} (50μM, a concentration which permits slow and restricted leakage of cytosol proteins [Koffer and Gomperts 1989]). If, instead of measuring the secretory response at the time of permeabilisation, the stimulating effectors are applied later, then the effective affinity for the two effectors is found to increase with time, so that two hours after treatment with ATP^{4-} the sensitivity to both effectors is about the same as for SL-O permeabilised cells (figure 3).

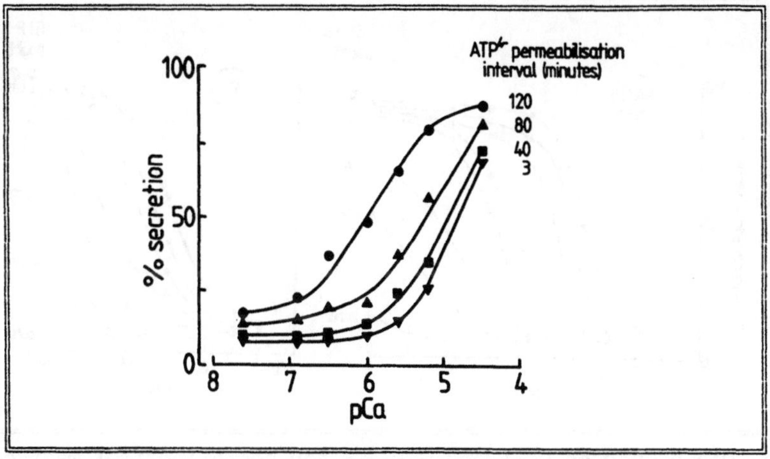

Figure 3. Dependence of Ca^{2+} for GTP-γ-S induced secretion from mast cells permeabilised with ATP^{4-} (50μM) stimulated after permeabilisation intervals ranging from 3 to 120 minutes.

During this time there is a gradual leakage of cytosol proteins (we measured the emergence of extracellular actin) and so it seems reasonable to conclude that among these is an inhibitor which when present in the cytosol, has the effect of suppressing the affinity for the two effectors. A high affinity situation is achieved rapidly (ie within seconds) in SL-O permeabilised cells, very slowly in ATP^{4-} permeabilised cells, and again rapidly following receptor stimulation of intact cells in which all cytosol proteins, by definition, are present in the cytosol. One of the pathways of cell activation by receptors must thus involve the attenuation of the suppressive constraint on effector affinity normally expressed by a soluble protein in the resting cells.

As demonstrated in the experiment illustrated in figure 2, both Ca^{2+} and a guanine nucleotide are required to initiate secretion from permeabilised mast cells. In this experiment, ATP (as the Mg^{2+} salt) was also provided in order to ensure the long term viability of the cells, and the permeabilising tetraanion ATP^{4-} was present as an equilbrium component. However, ATP is not required if the stimulus to exocytosis is applied immediately, or shortly after the time of permeabilisation by SL-O. Figure 4 illustrates the dependence on both Ca^{2+} and GTP-γ-S for exocytosis from mast cells permeabilised by SL-O in the absence and presence of MgATP.

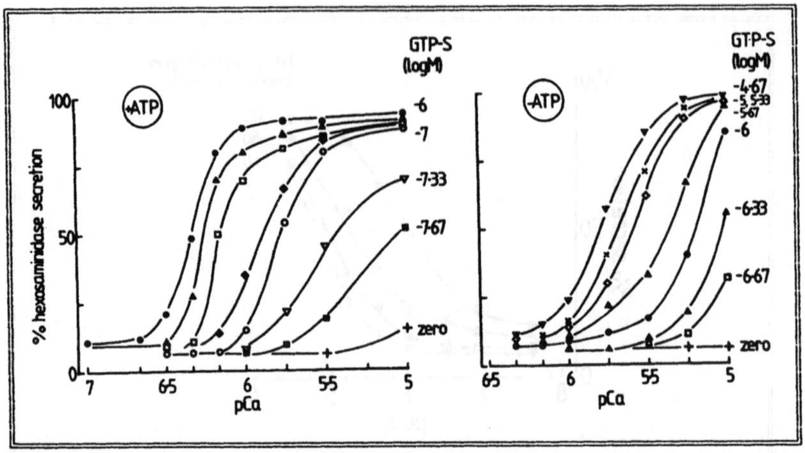

Figure 4. Dependence on Ca^{2+} and GTP-γ-S for secretion from mast cells permeabilised by SL-O in the presence and absence of ATP.

In these (and in all other) experiments the cells were treated, prior to permeabilisation, with metabolic inhibitors to the point where they become non-responsive to receptor-directed agonists, or to Ca^{2+} ionophores [Howell et al. 1987]. Even so, following permeabilisation full responsiveness can be induced by Ca^{2+}-plus-GTP-γ-S and the only evident effect of providing ATP is to enhance the apparent affinity of both of the effectors. A similar enhancement in effector affinity can be achieved even in the absence of ATP if the cells are pretreated (ie before applying metabolic inhibitors) with PMA [Howell et al. 1989]. For this reason we have suggested that modulation of affinity by ATP occurs as a result of phosphorylations mediated by protein kinase C. In addition to ATP, other nucleotides can support the affinity shift: these include ATP-γ-S, ITP, XTP, UTP and CTP; GTP and AppNHp are without effect in this respect [Lillie and Gomperts 1990].

Since exocytosis can occur in the total absence of ATP it follows that a phosphorylation reaction does not comprise a necessary step on the pathway to membrane fusion. It is worth questioning why the presence of ATP is an absolute requirement for exocytosis from intact cells? Apart from cellular housekeeping duties (eg Ca^{2+} and Na^{+} homeostasis), ATP is needed at a number of points in the complete stimulus-secretion sequence.

1) ATP is needed as a phosphoryl donor in the reactions catalysed by the phosphoinositide kinases to maintain the substrate of the phospholipase C reaction, PIP_2, and again as the donor in the protein kinase C mediated phosphorylations which enhance effector affinity. PIP_2 is the source of the initial 2nd messengers, IP_3 and diglyceride, which then allow the release of Ca^{2+} and the activation of protein kinase C. In the permeabilised cells, any IP_3 generated will leak rapidly from the cells and the Ca^{2+} is regulated by the use of a Ca.EGTA buffer system. Generation of diglyceride and the consequent activation of protein kinase C is, as we have already seen, best regarded as a modulator of the exocytotic process since it has the effect of enhancing affinities for Ca^{2+} and GTP. This may be a necessary event in the intact cells.

2) ATP is required to maintain the level of GTP through the nucleoside diphosphokinase reaction and GTP is required as a ligand for the G-protein G_P and the activation of phospholipase C. To induce exocytosis from the permeabilised cells it has been found necessary to provide GTP or one of its analogues. However, Ca^{2+}-plus-GTP-γ-S induced exocytosis persists under conditions in which hydrolysis of PIP_2 is fully inhibited [Cockcroft et al. 1987] (by neomycin, which can penetrate the intact cells, and which binds to the headgroups of the polyphosphoinositides). It follows that the guanine nucleotide is required at two sites in the full stimulus-secretion pathway, firstly as a ligand for G_P and then at a second site, G_E, which is closely linked to the terminal fusion event.

We have gained some insight into the possible function of the G_E protein by conditioning the cells so that they become responsive to stimulation by Ca^{2+} alone. If, the essential effectors are initially withheld following permeabilisation, and then delivered late, the secretory responsiveness declines systematically with time [Howell et al. 1989]. After 5 min of permeabilisation no secretion can be elicited by the combination of Ca^{2+}-plus-GTP-γ-S (both 10μM). If ATP is present throughout this time, or if the cells have been pretreated with PMA, then the period of responsiveness is considerably extended. Addition of ATP alongside the essential effectors can restore responsiveness to the refractory cells and such restoration can be prevented by inhibitors of protein kinase C. It is reasonable to think that responsiveness depends in some way on the phosphorylation state of the cells, its maintenance by protein kinase C and the tendency for dephosphorylation to occur spontaneously following permeabilisation in the absence of ATP.

The loss of responsiveness following permeabilisation, and its restoration following readdition of ATP can be better understood by examining the dependence on a) Ca^{2+}, and b) GTP-γ-S at each time point. The experiment in figure 5a illustrates the dependence on Ca^{2+} (applied together with a fixed concentration, 10μM, of GTP-γ-S) at various times after permeabilisation in the absence of ATP. It can be seen that the decline in responsiveness reflects a systematic shift in Ca^{2+} dependence so that higher and higher concentrations of Ca^{2+} are required to induce commensurate extents of secretion. Following the addition of ATP (at 5 min) the effective affinity for Ca^{2+} is restored to about its initial level. Recovery thus registers restoration of the ability of the exocytotic mechanism to sense the presence of Ca^{2+} at concentrations in the μmolar range. As we have seen, this is probably mediated through a protein kinase C mediated phosphorylation reaction.

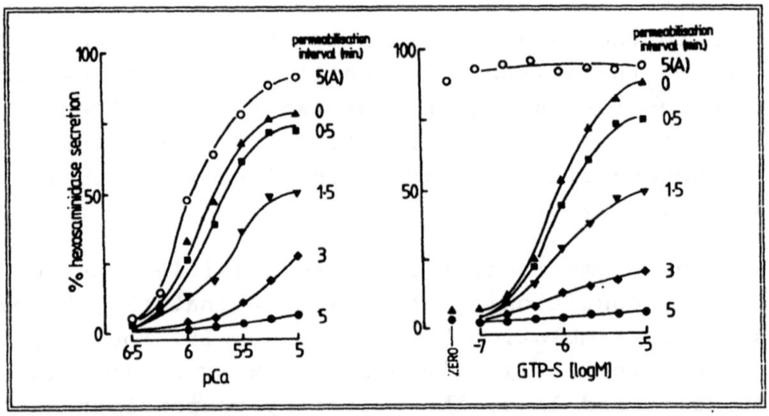

Figure 5. Dependence on a) Ca^{2+}, and b) GTP-γ-S for secretion from SL-O permeabilised mast cells stimulated after permeabilisation intervals. Open symbols indicate the effect of readdition of MgATP at 5 minutes.

A similar experiment, but one in which the concentration of Ca^{2+} was kept constant (pCa5.25: 5.6μM) and the dependence for secretion on the concentration of GTP-γ-S was measured at various times after permeabilisation, produced a very different result. Figure 6b shows the time dependent induction of the refractory condition but now, following the addition of ATP to restore responsiveness, exocytosis ensues in the absence of the guanine nucleotide. One possible explanation for this could be that the reaction which is normally dependent on the presence of the guanine nucleotide may have occurred spontaneously during the period between permeabilisation and stimulation by Ca^{2+} plus ATP. A likely candidate for such a reaction is a protein dephosphorylation. In cells permeabilised to the point of permitting rapid leakage of cytosol proteins, protein dephosphorylation might be expected to be enhanced to a rate exceeding that which occurs simply following metabolic deprivation, due to the loss of endogenous protein phosphatase inhibitors.

Supportive evidence for the idea that protein dephosphorylation might be an enabling step on the exocytotic pathway of mast cells came from a similar experiment, measuring the dependence on GTP-γ-S during the permeabilisation-induced decay and ATP-induced recovery, in the presence of okadaic acid [Churcher et al. 1990]. Okadaic acid is an inhibitor of protein phosphatases; it inhibits myosin phosphatase [Takai et al. 1987], and more generally those phosphatases (of the classes 1 and 2A) which catalyse the dephosphorylation of substrates phosphorylated by protein kinase C [Bialojan and Takai 1988; Haystead et al. 1989; Cohen et al. 1990]).

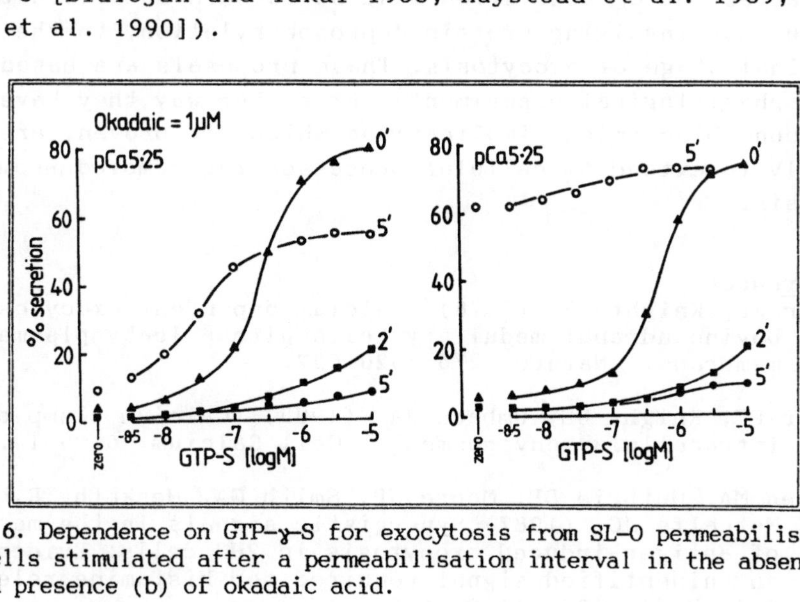

Figure 6. Dependence on GTP-γ-S for exocytosis from SL-O permeabilised mast cells stimulated after a permeabilisation interval in the absence (a) and presence (b) of okadaic acid.

In the experiment illustrated, it can be seen that okadaic acid has little effect on the characteristics of the decay in responsiveness following permeabilisation. However, following restoration with ATP, exocytosis from the treated cells remains strictly dependent on provision of a guanine nucleotide. The guanine nucleotide seems to ensure progression through steps which are inhibited by okadaic acid, and a reasonable conclusion to draw is that the target enzyme of the G-protein G_E might be a protein phosphatase [Churcher et al. 1990 Gomperts et al. 1990].

We and others have proposed a role for GTP, and hence a role for a GTP-binding protein (G_E) in the terminal pathway of exocytosis [Gomperts 1990]. Apart from the demonstration that the G-proteins mediating exocytosis and polyphosphoinositide hydrolysis (G_E and G_P) are distinct entities [Cockcroft et al. 1987; Stutchfield and Cockcroft 1988], the definition of G_E has relied entirely on measurements of secretion. The simplest conclusion to be drawn from the present results is that association of GTP with the hypothetical G_E results in protein dephosphorylation and that in permeabilised mast cells this late event is a necessary prelude to exocytosis. From this it follows that the effector unit of G_E is likely to be a phosphoprotein phosphatase, and that following activation by G_E the enzyme is no longer subject to inhibition by okadaic acid. Alternatively, if the inhibition of protein phosphatase by okadaic acid is irreversible then the effect of G_E activation must be to open an alternative route, not involving protein dephosphorylation, to the terminal stage of exocytosis. These proposals are based on cell-physiological experiments but either way they have profound biochemical implications which, if proven, are likely to extend to cellular processes far removed from exocytosis.

References

Baker PF, Knight DE (1978) Calcium dependent exocytosis in bovine adrenal medullary cells with a leaky plasma membrane. Nature 276: 620-622.

Baker PF, Knight DE, Umbach JA (1985) Calcium clamp of the intracellular environment. Cell Calcium 6: 5-14.

Beaven MA, Guthrie DF, Moore JP, Smith GA, Hesketh TR, Metcalfe JC (1987) Synergistic signals in the mechanism of antigen-induced exocytosis in 2H3 cells: Evidence for an unidentified signal required for histamine release. J Cell Biol 105: 1129-1136.

Bialojan C, Takai A (1988) Inhibitory effect of a marine-sponge toxin, okadaic acid, on protein phosphatases. Specificity and kinetics. Biochem J 256: 283-290.

Burgoyne RD (1984) The relationship between secretion and intracellular free calcium in bovine adrenal chromaffin cells. Bioscience Reports 4: 605-611.

Choy VJ, Watkins WB (1976) Effect of ionic environment on the release of human placental lactogen *in vitro*. J Endocrinonol 69: 349-358.

Churcher Y, Kramer IM, Gomperts BD (1990) Evidence for protein dephosphorylation as a permissive step in GTP-γ-S induced exocytosis from permeabilised mast cells. Cell Regul 1: 523-530.

Cockcroft S, Gomperts BD (1988) Some new questions concerning the role of calcium in exocytosis. in Calcium and Drug Action 305-338 (P F Baker, ed) Springer-Verlag, Heidelberg.

Cockcroft S, Howell TW, Gomperts BD (1987) Two g-proteins act in series to control stimulus-secretion coupling in mast cells: Use of neomycin to distinguish between g-proteins controlling polyphosphoinositide phosphodiesterase and exocytosis. J Cell Biol 105: 2745-2750.

Cohen P, Holmes CFB, Tsukitani Y (1990) Okadaic acid: A new probe for the study of cellular regulation. Trends in Biochem Sci 15: 98-102.

Dormer RL, Ashcroft SJH (1974) Studies on the role of calcium ions in the stimulation by adrenaline of amylase release from rat parotid. Biochem J 144: 543-550.

Douglas WW (1968) Stimulus-secretion coupling: The concept and clues from chromaffin and other cells. Brit J Pharmacol 34: 451-474.

Fray JCS, Park CS, Valentine AND (1987) Calcium and the control of renin secretion. Endocrine Reviews 8: 53-93.

Gomperts BD (1983) Involvement of guanine nucleotide-binding proteins in the gating of Ca^{2+} by receptors. Nature 306: 64-66.

Gomperts BD (1990) GTP-binding proteins and exocytotic secretion. in G Proteins 601-637 (R Iyengar and L Birnbaumer, eds) Academic Press.

Gomperts BD, Churcher Y, Koffer A, Kramer IM, Lillie THW, Tatham PER (1990) The role and mechanism of the GTP-binding protein, G_E in the control of regulated exocytosis. Biochem Soc Symp 56: 85-101.

Burgoyne RD (1984) The relationship between secretion and intracellular free calcium in bovine adrenal chromaffin cells. Bioscience Reports 4: 605-611.

Choy VJ, Watkins WB (1976) Effect of ionic environment on the release of human placental lactogen *in vitro*. J Endocrinonol 69: 349-358.

Churcher Y, Kramer IM, Gomperts BD (1990) Evidence for protein dephosphorylation as a permissive step in GTP-γ-S induced exocytosis from permeabilised mast cells. Cell Regul 1: 523-530.

Cockcroft S, Gomperts BD (1988) Some new questions concerning the role of calcium in exocytosis. in Calcium and Drug Action 305-338 (P F Baker, ed) Springer-Verlag, Heidelberg.

Cockcroft S, Howell TW, Gomperts BD (1987) Two G-proteins act in series to control stimulus-secretion coupling in mast cells: Use of neomycin to distinguish between G-proteins controlling polyphosphoinositide phosphodiesterase and exocytosis. J Cell Biol 105: 2745-2750.

Cohen P, Holmes CFB, Tsukitani Y (1990) Okadaic acid: A new probe for the study of cellular regulation. Trends in Biochem Sci 15: 98-102.

Dormer RL, Ashcroft SJH (1974) Studies on the role of calcium ions in the stimulation by adrenaline of amylase release from rat parotid. Biochem J 144: 543-550.

Douglas WW (1968) Stimulus-secretion coupling: The concept and clues from chromaffin and other cells. Brit J Pharmacol 34: 451-474.

Fray JCS, Park CS, Valentine AND (1987) Calcium and the control of renin secretion. Endocrine Reviews 8: 53-93.

Gomperts BD (1983) Involvement of guanine nucleotide-binding proteins in the gating of Ca^{2+} by receptors. Nature 306: 64-66.

Gomperts BD (1990) GTP-binding proteins and exocytotic secretion. in G Proteins 601-637 (R Iyengar and L Birnbaumer, eds) Academic Press.

Gomperts BD, Churcher Y, Koffer A, Kramer IM, Lillie THW, Tatham PER (1990) The role and mechanism of the GTP-binding protein, G_E in the control of regulated exocytosis. Biochem Soc Symp 56: 85-101.

Gomperts BD, Cockcroft S, Howell TW, Nüsse O, Tatham PER (1987) The dual effector system for exocytosis in mast cells: Obligatory requirement for both Ca^{2+} and GTP. Bioscience Repts 7: 369-381.

Gomperts BD, Fernandez JM (1985) Techniques for membrane permeabilisation. Trends in Biochem Sci 10: 414-417.

Gomperts BD, Tatham PER (1988) GTP-binding proteins in the control of exocytosis. Cold Spring Harbor Symp Quant Biol 53: 983-992.

Handwerger S, Conn PM, Barrett J, Barry S, Golander A (1981) Human placental lactogen release *in vitro*: Paradoxical effects of calcium. Am J Physiol 240: E550-E555.

Haslam RJ, Davidson MML (1984) Potentiation by thrombin of the secretion of serotonin from permeabilised platelets equilibrated with Ca^{2+} buffers. Biochem J 222:351-361.

Haystead TAJ, Sim ATR, Carling D, Honnor RC, Tsukitani Y, Cohen P, Hardie DG (1989) Effects of the tumour promoter okadaic acid on intracellular protein phosphorylation and metabolism. Nature 337: 78-81.

Howell TW, Cockcroft S, Gomperts BD (1987) Essential synergy between Ca^{2+} and guanine nucleotides in exocytotic secretion from permeabilised mast cells. J Cell Biol 105: 191 - 197.

Howell TW, Gomperts BD (1987) Rat mast cells permeabilised with streptolysin-O secrete histamine in response to Ca^{2+} at concentrations buffered in the micromolar range. Biochim Biophys Acta 927: 177-183.

Howell TW, Kramer I, Gomperts BD (1989) Protein phosphorylation and the dependence on Ca^{2+} for GTP-γ-S stimulated exocytosis from permeabilised mast cells. Cellular Signalling 1: 157-163.

Katakami Y, Kaibuchi K, Sawamura M, Takai Y, Nishizuka (1984) Synergistic action of protein kinase C and calcium for histamine release from rat peritoneal mast cells. Biochem Biophys Res Commun 121: 573-578.

Kinoshita K, Tsong TY (1977) Formation and resealing of pores of controlled sizes in human erythrocyte membrane. Nature 268: 438-441.

Knight DE, Koh E (1984) Ca^{2+} and cyclic nucleotide dependence of amylase release from isolated rat pancreatic acinar cells rendered permeable by intense electric fields. Cell Calcium 5: 401-418.

Knight DE, Niggli V, Scrutton MC (1984) Thrombin and activators of protein kinase C modulate secretory responses of permeabilised human platelets induced by Ca^{2+}. Eur J Biochem 143: 437-446.

Knight DE, Scrutton MC (1980) Direct evidence for a role for Ca^{2+} in amine storage granule secretion by human platelets. Thrombosis Res 20:437-446.

Knight DE, Scrutton MC (1986) Gaining access to the cytosol: The technique and some applications of electropermeabilisation. Biochem J 234: 497-506.

Koffer A, Gomperts BD (1989) Soluble proteins as modulators of the exocytotic reaction of permeabilised rat mast cells. J Cell Sci 94: 585-591.

Leslie BA, Putney JW, Sherman JM (1976) α-Adrenergic, β-adrenergic and cholinergic mechanisms for amylase secretion by rat parotid gland in vitro. J Physiol (Lond) 260: 351-370.

Lew PD, Monod A, Waldwogel FA, Dewald B, Baggiolini M, Pozzan T (1986) Quantitative analysis of the cytosolic free calcium dependency of exocytosis from three subcellular compartments in intact human neutrophils. J Cell Biol 102: 2197-2204.

Lillie THW, Gomperts BD (1990) Modulation of the exocytotic reaction of permeabilised rat mast cells by ATP and other nucleotides. (submitted for publication)

McMillian MK, Soltoff SP, Talamo BR (1988) Mediation of norepinephrine effects on free cytosolic calcium in rat parotid acinar cells by α_1-adrenergic receptors. Biochem Pharmacol 37: 3790-3793.

Nemeth EF, Scarpa A (1986) Cytosolic Ca^{2+} and the regulation of secretion in parathyroid cells. FEBS Lett 203: 15-19.

Park CS, Honeyman TW, Chung ES, Lee JS, Sigmon DH, Fray JCS (1986) Involvement of calmodulin in mediating inhibitory action of Ca^{2+} on renin secretion. Am J Physiol 251: F1055-F1062.

Reed PW, Lardy HA (1972) A23187: A divalent cation ionophore. J Biol Chem 247: 6970-6977.

Shoback DM, Thatcher JG, Leombruno R, Brown EM (1983) Effects of extracellular Ca^{++} and Mg^{++} on cytosolic Ca^{++} and PTH release in dispersed bovine parathyroid cells. Endocrinology 113: 424-426.

Stutchfield J, Cockcroft S (1988) Guanine nucleotides stimulate polyphosphoinositide phosphodiesterase and exocytotic secretion from HL-60 cells permeabilised with streptolysin O. Biochem J 250: 375-382.

Takai A, Bialojan C, Troschka M, Ruegg JC (1987) Smooth muscle myosin phosphatase inhibition and force enhancement by black sponge toxin. FEBS Lett 217: 81-84.

Tatham PER, Lindau M (1989) ATP-induced pore formation in the plasma membrane of rat peritoneal mast cells. J Gen Physiol 95: 459-476.

Yaseen MA, Pedley KC, Howell SL (1982) Regulation of insulin secretion from islets of Langerhans rendered permeable by electric discharge. Biochem J 206: 81-87.

Mannose 6-phosphate Receptors and their Role in Protein Sorting along the Pathway to Lysosomes

S. R. Pfeffer
Department of Biochemistry B400
Beckman Center for Molecular & Genetic Medicine
Stanford University School of Medicine
Stanford, California 94305-5307
USA

Mannose 6-phosphate receptors (MPRs) carry newly synthesized lysosomal enzymes from the trans Golgi network (TGN) to pre-lysosomes, and then return to the TGN to carry out another round of lysosomal enzyme delivery. While clathrin-coated vesicles mediate the export of MPRs from the TGN, nothing is known about the transport vesicles used to carry these receptors back to the TGN. Using two different *in vitro* assays, we have shown that an antibody that interferes with clathrin assembly blocks receptor-mediated endocytosis of transferrin, but has no effect on the recycling of the 300kD MPR from pre-lysosomes to the TGN. These results suggest that the transport of MPRs from pre-lysosomes to the TGN does not involve clathrin. In addition to cycling between the TGN and endosomes, MPRs can also undergo conventional receptor-mediated endocytosis. We constructed chimeric receptors to test whether the MPR cytoplasmic domain contained sufficient information to direct a cell surface receptor into both of these transport pathways. Our experiments demonstrate that the MPR cytoplasmic domain is not sufficient to alter the distribution of the EGF receptor, and demonstrate a role for extracellular and transmembrane domains in MPR routing.

Newly synthesized, soluble lysosomal enzymes undergo a unique posttranslational modification that facilitates their delivery to lysosomes (see Kornfeld & Mellman, 1989 for review). As lysosomal enzymes traverse the secretory pathway, their N-linked oligosaccharides acquire one or two mannose 6-phosphate (man6P) residues. The presence of man6P enables lysosomal enzymes to bind to MPRs in the TGN; the receptor-ligand complexes are then thought to be carried in clathrin-coated vesicles to pre-lysosomes. The acidic interior of pre-lysosomes (also termed "late endosomes") triggers the release of the enzymes from MPRs, and lysosomal enzymes later appear in lysosomes.

NATO ASI Series, Vol. H 52
Biological Signal Transduction
Edited by E. M. Ross and K. W. A. Wirtz

Meanwhile, MPRs are carried back to the TGN to complete the transport cycle. A small fraction of MPRs are also present at the cell surface, where they can bind extracellular lysosomal enzymes and deliver them to lysosomes by conventional receptor-mediated endocytosis (Figure 1).

We have recently reconstituted the recycling of the 300kD MPR from late endosomes to the TGN in a cell-free system (Goda & Pfeffer, 1988). The assay

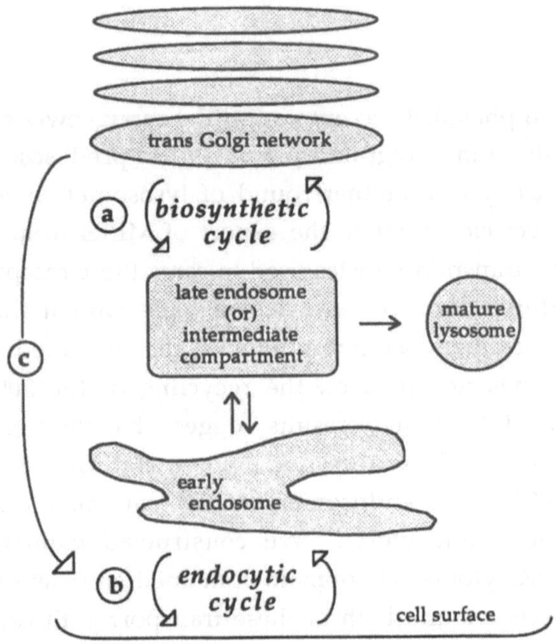

Figure 1. Intracellular transport pathways taken by cell surface and intracellular MPRs. (a) MPRs carry lysosomal enzymes from the trans Golgi network to late endosomes, and then return to the TGN to com-plete a "biosynthetic" cycle. Most cell surface receptors do not utilize this pathway. (b) Cell surface MPRs bind extracellular ligands, deliver them to early endosomes, and may then re-cycle to the surface (endocytic cycle). Upon ligand binding, native EGF receptors fail to recycle and are instead transferred to lysosomes via late endosomes. (c) Nascent cell surface receptors are transported directly to the surface after passing through the TGN. It is not known whether MPRs use this pathway.

takes advantage of the localization of sialyltransferase to the trans Golgi and TGN, and utilizes a mutant cell line in which glycoproteins are not sialylated. Radiolabelled MPRs, present in late endosomes in a mutant cell extract, acquire sialic acid residues when they are transported to the TGN of wild type Golgi complexes which are present in reaction mixtures. The acquisition of sialic acid by MPRs in this system reflects a vesicular transport process, since it is time, temperature, ATP, and cytosol dependent, and also requires GTP hydrolysis (Goda & Pfeffer, 1988). Furthermore, MPRs and sialyltransferase remain in sealed membrane compartments throughout the reaction, and non-specific membrane fusion is ruled out by several criteria (Goda & Pfeffer, 1988). We present here data which indicate that clathrin is not involved in MPR recycling. In addition, we have found that the MPR extracellular domain plays a key role in the intracellular transport of this protein, and finally, that the TGN is insensitive to the action of BFA.

Nothing is known about the transport vesicles that carry MPRs from late endosomes to the TGN. We used a monoclonal antibody (X19) that recognizes an epitope on the clathrin heavy chain to investigate the possible role of clathrin and clathrin-coated vesicles in MPR recycling (Draper et al., 1990). X19 antibody has been shown to inhibit both fluid phase endocytosis and receptor-mediated endocytosis when introduced into the cytoplasm of African green monkey kidney cells (Doxsey et al., 1987).

To verify that antibody X19 could inhibit a clathrin-dependent transport process *in vitro*, we tested the effect of X19 on transferrin endocytosis using an assay described by Smythe et al. (1989). In this assay, A431 human epidermoid carcinoma cells are incubated with radiolabelled transferrin at 4°C to label cell surface transferrin receptors. Cells are then gently broken by scraping, and the "semi-intact" cells are incubated at 37°C in the presence of cytosol and ATP to permit clathrin-coated vesicle formation and endocytosis. Transferrin endocytosis is then assessed using an immunoassay that measures the acquired inaccessibility of transferrin to transferrin antibodies (Smythe et al., 1989). *In vitro* endocytosis is ATP and cytosol dependent. Slight modifications were incorporated to make the *in vitro* endocytosis conditions as similar as possible to those of the MPR recycling assay.

The results of three independent endocytosis experiments are summarized in Fig. 2 (closed symbols), and demonstrate that receptor-mediated endocytosis was inhibited by clathrin-specific antibodies in a dose-dependent manner, as would be expected from previously published experiments. In contrast, the same concentrations (and preparations) of X19 IgG had no effect on the recycling of the 300kD MPR from late endosomes to the TGN (Fig. 2, open triangles), relative to the amount of transport measured in the presence of an equal concentration of control IgG.

X19 IgG was raised against human clathrin, and has been shown to block clathrin function in African green monkey kidney cells. It was important to rule out the possibility that X19 IgG failed to inhibit MPR recycling because it did not recognize CHO cell clathrin. Control experiments showed that X19 IgG binds to human and CHO clathrin with similar affinities. In addition, A431 and CHO cells contain roughly equivalent amounts of clathrin. To demonstrate unequivocally that X19 IgG could inhibit clathrin function in semi-intact CHO cells, we adapted the endocytosis assay (Smythe et al., 1989) described above for

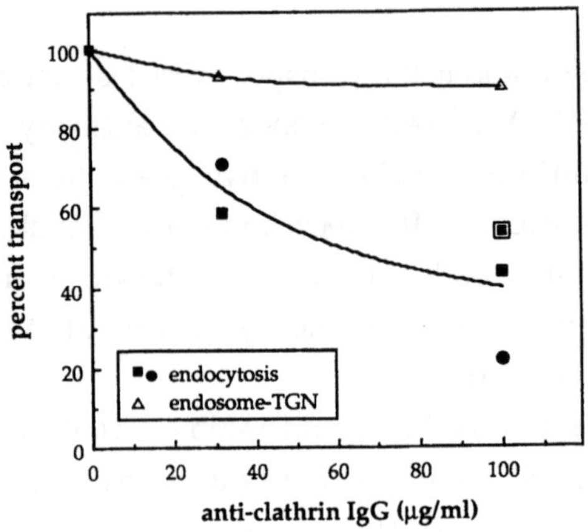

Figure 2. Anti-clathrin IgG inhibits endocytosis but not MPR recycling to the TGN *in vitro*. Endocytosis (closed symbols) was measured as described in Smythe et al., 1989; the different symbols refer to 3 inde-pendent endocytosis experiments. Transport of 300kD MPRs from endosomes to the TGN (open triangles) was assessed in 3 inde-pendent experiments by transport-coupled sialylation of MPRs in reaction volumes of 200μl; a typical result is shown. Data are presented as the percent transport observed relative to transport measured in the presence of a control IgG.

use with CHO cells expressing human transferrin receptors. X19 IgG inhibited the ATP-dependent endocytosis of transferrin by semi-intact CHO cells by as much as 65%, using cells scraped from polylysine-coated dishes. Somewhat less inhibition (36%) was observed if cells were scraped from collagen-coated dishes; this is likely to reflect a difference in antibody accessibility to the cytoplasmic components in each type of broken cell. In summary, these data show that X19 IgG inhibits receptor-mediated endocytosis in broken CHO cells to a similar extent as that observed using broken human A431 cells or intact monkey kidney cells.

Clathrin-coated vesicles mediate fluid phase and adsorptive endocytosis. Clathrin is also thought to function at the TGN, in the diversion of lysosomal enzymes and secretory storage granule content away from constitutively secreted and membrane proteins bound for the cell surface. Our results represent the first direct indication that in a transport cycle in which clathrin mediates the forward reaction (TGN export), another type of transport vesicle may be utilized for the reverse transport step (TGN import). In addition, since the transport of proteins from late endosomes to the TGN is a selective process, these data rule out a model in which the protein, clathrin, coats all transport vesicles involved in selective intracellular transport steps. Our findings spotlight an interesting problem, namely, how clathrin coated-vesicle components recognize and capture MPRs at the TGN and not in pre-lysosomes. Identification of the proteins which recognize MPRs in prelysosomes and direct them to the TGN represents an important challenge for the future.

MPR domains determine its intracellular routing

To investigate the role of the 300kD MPR cytoplasmic domain in determining the intracellular transport and distribution of that receptor, a hybrid protein was constructed by joining the extracellular and transmembrane domains of the human EGF receptor to the intracellular domain of the bovine 300kD MPR (Dintzis & Pfeffer, 1990; Figure 3). The expression construct was both transiently and stably transfected into human 293 kidney fibroblast cells which contain endogenous EGF receptors. The construct was also stably introduced into mouse NIH 3T3/2.2 fibroblasts which lack endogenous EGF receptors. Most of the experiments described here were carried out using stably transfected, "3T3c" cells, which represent a pool of cells selected first for neomycin resistance and subsequently, for viability in the presence of 1μM methotrexate.

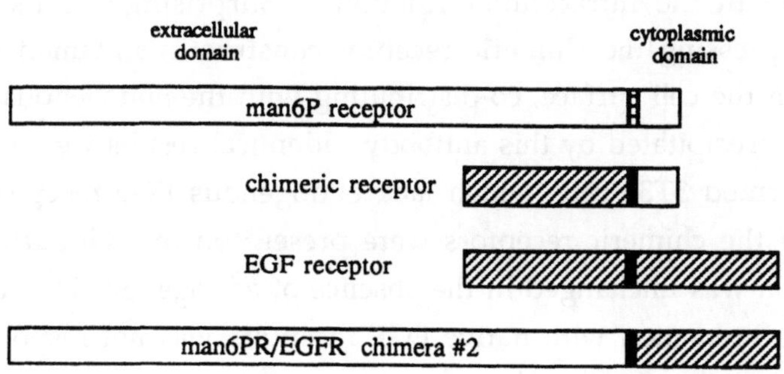

Figure 3. Chimeric mannose-6-phosphate/EGF receptors constructed

Chimera-expressing cells synthesized a protein of 124 kD that was specifically immunoprecipitated with an EGF receptor antibody, and was not detected in the parental 3T3/2.2 cell line. If present on the cell surface, a properly folded, functional chimera should have the capacity to both bind and internalize EGF. 3T3c cells bound EGF in a saturable manner; as would be expected, parental 3T3/2.2 cells displayed only non-specific levels of EGF binding. Scatchard analysis revealed a single dissociation constant of $\sim 0.9 \times 10^{-9}$ M and the presence of approximately 135,000 receptors on the surface of each 3T3c cell. Thus, at least a portion of the chimeric receptors were targetted to the cell surface, where they formed high affinity EGF binding sites. In addition to the ability to bind EGF with high affinity, cells expressing chimeric receptors could be shown to internalize EGF. Since cytoplasmic domain sequences residing 24-75 amino acid residues from the membrane are essential for MPR endocytosis (Lobel et al., 1989), the efficiency with which the chimera internalized EGF demonstrated that at least this portion of the cytoplasmic domain was correctly folded.

Cellular Distribution and Trafficking of the Chimeric Receptor

Unlike the EGF receptor, the 300kD MPR is predominantly an intracellular receptor, located in the Golgi complex and endosomes of most cell types. At steady state, less than 5-10% of MPRs are present on cell surfaces. If cytoplasmic domain sequences alone determine the cellular localization of this receptor, one would predict that the chimeric receptor would also be localized primarily to endosomes and the TGN; a small percent would be present on the cell surface, where the receptors could mediate EGF binding and uptake.

We employed a quantitative antibody assay to determine more precisely the cellular distribution of the chimeric receptor. As would be expected, the EGF receptor was present almost exclusively on the surfaces of cells transiently expressing a cDNA encoding that protein. The endogenous MPR was located predominantly in the intracellular fraction. Surprisingly, cells which were transiently expressing the chimeric receptor construct maintained almost all of this protein on the cell surface, co-distributing with the endogenous 170 kD EGF receptor, also precipitated by this antibody. Identical results were obtained with stably transformed 3T3c cells which lack endogenous EGF receptors. In these cells, >85% of the chimeric receptors were present on the cell surface. Because the distribution was unchanged in the absence of endogenous EGF receptors, we conclude that interaction with native EGF receptors was not responsible for the chimeric receptor distribution observed.

To test directly for ligand-independent internalization and recycling, we monitored the uptake of monovalent, monoclonal Fab fragments directed against the extracellular domain of the EGF receptor (Fig. 4). At 4°C, ^{125}I-Fab fragments bound to the surfaces of 3T3c cells, and were released by a low pH wash, consistent with their expected cell surface localization. However, cells warmed to 37°C internalized a significant fraction of the Fab fragments, as determined by their resistance to a low pH wash. Only insignificant binding was observed for the parental, NIH 3T3/2.2 cell line at either temperature. These results demonstrate that chimeric receptors are efficiently endocytosed, even in the absence of ligand.

We constructed a chimeric EGF/MPR to test whether the MPR cytoplasmic domain, upon transplantation, could alter the routing of a typical cell surface receptor (Dintzis & Pfeffer, 1990). Since cytoplasmic domains play a key role in

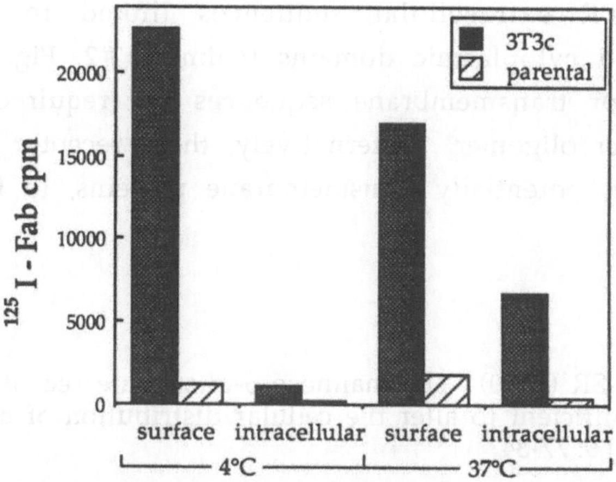

Figure 4. Chimeric receptors undergo endocytosis in the absence of ligand. 3T3c cells (solid bars) or parental NIH 3T3/2.2 cells (striped bars), plated in 6cm dishes, were incubated with ^{125}I-Fab fragments at 4°C for 90 minutes. Unbound antibodies were washed away; surface and intracellular receptors were then determined as described in Materials and Methods. Left, cells were incubated at 4°C; right, cells were warmed to 37°C for 15 minutes after prebinding Fab fragments for 90 minutes at 4°C.

the ability of proteins to enter clathrin-coated pits at the cell surface, we expected that the MPR cytoplasmic domain might carry sufficient information to direct the chimeric receptor into both endocytic and biosynthetic pathways. To our surprise, when expressed in fibroblasts, the hybrid receptor behaved precisely like a cell surface receptor, recycling continuously within the endocytic pathway for most, if not all, of its lifetime.

While the MPR cytoplasmic domain permitted efficient cycling of the chimera through the endocytic pathway, its presence was not sufficient to direct this protein for transport between endosomes and the Golgi complex at any appreciable rate. Moreover, since the receptor was rapidly internalized and recycled to the cell surface, it appeared that the receptor may not have even reached late endosomes. We favor a model in which the chimera had lost sorting information normally present in the MPR extracellular and/or transmembrane domains, leading to its predominance at the cell surface and within the endocytic pathway. We propose that native MPR extracellular and transmembrane domain sequences contain information that in some way facilitates the delivery of MPRs from early endosomes to late endosomes, a prerequisite event for their subsequent delivery to the TGN (Dintzis & Pfeffer, 1990). We have now demonstrated this directly by creating a chimeric receptor comprised of MPR extracellular sequences linked to EGF-receptor transmembrane and cytoplasmic domains (chimera #2, Fig. 3). Perhaps extracellular and/or transmembrane sequences are required to form an appropriate receptor oligomer. Alternatively, these receptor domains may interact with other, potentially transmembrane proteins, to facilitate MPR sorting.

References

Dintzis SM, Pfeffer SR (1990) The mannose 6-phosphate receptor cytoplasmic domain is not sufficient to alter the cellular distribution of a chimeric EGF receptor. EMBO J 9: 77-84

Doxsey SJ, Brodsky FM, Blank GS, Helenius A (1987) Inhibition of endocytosis by anti-clathrin antibodies. Cell 50: 453-63

Draper RK, Goda Y, Brodsky FM, Pfeffer SR (1990) Anti-clathrin antibodies inhibit endocytosis but not receptor recycling to the trans Golgi network in vitro. Science 248: 1539-1541

Goda Y, Pfeffer SR (1988) Selective recycling of the mannose 6-phosphate/IGF-II receptor to the trans Golgi network in vitro. Cell 55: 309-20

Kornfeld S, Mellman I (1989) The biogenesis of lysosomes. Ann Rev Cell Biol 5: 483-525

Lobel P, Fujimoto K, Ye RE, Griffiths G, Kornfeld S (1989) Mutations in the cytoplasmic domain of the 275 Kd mannose-6-phosphate receptor differentially alter lysosomal enzyme sorting and endocytosis. Cell 57: 787-96

Smythe E, Pypaert M, Lucocq J, Warren G (1989) Formation of coated vesicles from coated pits in broken A431 cells. J Cell Biol 108: 843-53

PROTEIN SORTING IN ENDOSOMES

Willem Stoorvogel

Laboratory for Cell Biology, University of Utrecht, AZU, rm H02.314, Heidelberglaan 100, 3584 CX, Utrecht, The Netherlands

Receptor-mediated endocytosis is a process in which macromolecules (ligands) first bind to specific receptors at the plasma membrane. These ligand-receptor complexes are subsequently transported into the cell within vesicles which pinch off from the plasma membrane (for reviews see Goldstein et al., 1985; VanDeurs et al., 1989). Except for the mature red blood cell, all eucaryotic cells are involved in endocytosis. Among the many different types of molecules that enter the cell by means of receptor-mediated endocytosis are growth factors (such as epidermal growth factor and insulin like growth factor), hormones (such as insulin and luteinizing hormone), transport proteins (such as transferrin for iron transport, low-density lipoprotein for cholesterol transport, and transcobalamin II for vitamin B_{12} transport), proteins which are destined for degradation (such as a2 macroglycoprotein and asialoglycoproteins), and proteins which have to be transported across a polarized cell (such as polymeric immunoglobulin A). In addition to the uptake of these physiological molecules, toxins and viruses also enter cells by receptor-mediated endocytosis. Each ligand-receptor complex follows a distinct intracellular transport route, which is dependent on the final destination of the ligand. However, the initial events for endocytosis of most receptor-ligand complexes appears to be identical.

The first step in receptor-mediated endocytosis is binding of the ligands to their receptor. Many ligands, such as asialoglycoproteins and transferrin bind their receptors along the entire cell surface (see below). The receptors move laterally across the plasma membrane, until they are trapped within clathrin-coated pits. Ligands are after complexation with their receptor internalized via these clathrin coated pits. Many different ligand types have been visualized to enter cells via the same coated pits, suggesting a common route of entry for receptor-bound ligands (Pastan and Willingham 1981; Willingham et al., 1981; Carpentier et al., 1982; Geuze et al., 1984a; Courtoy et al., 1985). Coated pits form coated vesicles during internalization, which are generally assumed to be involved

NATO ASI Series, Vol. H 52
Biological Signal Transduction
Edited by E. M. Ross and K. W. A. Wirtz

in the uptake of ligand-receptor complexes (Bretscher and Pearse, 1984; Wileman et al., 1985). Soon after uptake, coated vesicles uncoat, and fuse directly with endosomes or with one an other to form endosomes (Marsh and Helenius, 1980) which have also been termed receptosomes (Willingham and Pastan, 1980), or compartment of uncoupling receptor and ligand (CURL) (Geuze et al., 1983a). Within this compartment the pathways of different receptor and/or ligand types diverge. A major intrinsic characteristic of endosomes is their acidic interior (Tycko and Maxfield, 1982; Mellman et al., 1986; Rudnick, 1986; Yamashiro and Maxfield, 1984; Anderson and Orci, 1988). This low pH is essential for dissociation of some receptor-ligand complexes (such as asialoglycoproteins, lysosomal hydrolases, and LDL), for the release of iron from transferrin, and for entry of toxins and viruses into the cytosol. This low pH is maintained by an ATP-dependent proton pump (Yamashiro et al., 1983). CURL (endosomes) consist of vesicles with attached tubular membranous extensions (Geuze et al., 1983a). Receptors once released from ligand recycle to the plasma membrane, probably within vesicles pinching of from these tubular extensions (Geuze et al., 1983a; 1987). After recycling, receptors may then bind and take up a new ligand molecule, and thus may be reutilized many times. Dissociated ligands such as asialoglycoproteins (ASGP) (see below) and low density lipoproteins (Goldstein et al., 1985) remain within the vesicular compartment of CURL. Ultimately, they are transported to the lysosomes where they are degraded.

Some ligands however do not dissociate from their receptors within CURL. Transferrin (Tf) remains associated with its receptor, and recycles receptor bound to the cell surface. The recycling pathway of the transferrin receptor (TfR) is most intensively studied, because it remains assosiated with its ligand during the entire intracellular pathway (Klausner et al., 1983; Dautry-Varsat et al., 1983), and can therefore be followed by means of labeled ligand. Iron saturated Tf binds to the TfR with a high affinity (Kd 10^{-8}M) at neutral pH. After uptake via coated pits and coated vesicles, Tf loses its affinity for Fe^{3+} due to the acidic endosomal environment whereas the apotransferrin remains receptor bound because of its high affinity for the receptor at low pH. Apotransferrin remains TfR bound during recycling but loses its affinity when exposed to the neutral pH of the extracellular medium and dissociates. Most TfR's probably recycle directly from endosomes to the plasma membrane. Some TfR's however seem to pass the Golgi complex /*trans*-Golgi reticulum during their recycling (Stoorvogel et al., 1988). In contrast to the TfR, many other receptor types, such as the ASGPR, dissociate from their ligand before they recycle to the plasma membrane. ASGP's are after dissociation transported to lysosomes and are therefore excellent markers for the endocytic pathway to lysosomes.

The ASGPR mediates the endocytosis of plasma glycoproteins from which the terminal sialic acid of the carbohydrate has been removed, exposing galactose or N-acetylgalactosamine residues (ASGP) (Ashwell and Morell, 1974). ASGPR is found only in hepatic parenchymal cells. ASGPR's bind asialoglycoproteins at the plasma membrane, after which the complex is internalized by coated vesicles (Geuze et al., 1983a, 1983b) with kinetics that are very similar to those of the TfR (Tolleshaug et al., 1980; Bridges et al., 1982; Schwartz et al., 1982; Ciechanover et al., 1983b). Dissociation of the complex and sorting of receptor and ligand occurs intracellularly (Bridges et al., 1982; Simmons and Schwartz, 1984) in CURL (Geuze et al., 1983a). The ASGPR recycles to the plasma membrane, presumably via the tubular CURL elements whereas the ligand is transported to the lysosomes by means of the vesicular component of CURL (Geuze et al., 1983a; 1987). The recycling pathway of ASGPR to the plasma membrane is not precisely known, but may similarly to the TfR involve the *trans*-Golgi reticulum (Geuze et al., 1984b, 1987; van den Bosch et al., 1986; Zijderhand-Bleekemolen et al.,1987). After delivery to lysosomes ASGP's are degraded by acid hydrolases. These enzymes are delivered to lysosomes by mannose 6-phosphate receptors via a pathway that diverts from the biosynthetic secretory route and merges with the endocytic pathway.

Two mannose 6-phosphate receptor types are now known. Each functions in the transport of newly synthesized acid hydrolases to lysosomes. The first one to be isolated is now termed the cation-independent mannose 6-phosphate receptor and has a molecular weight of 275 kDa (Sahagian et al., 1982). It is present in all cell types. A second type of mannose 6-phosphate receptor, is 46 kDa, cation-dependent, and has not been found in all cells examined (Hoflack and Kornfeld, 1985). Here we will only discuss the cation-independent mannose 6-phosphate receptor (MPR). MPR has a dual function in that it is also the receptor for insulin-like growth factor II (Kiess et al., 1988). After synthesis in the endoplasmic reticulum and transport to the Golgi complex, most lysosomal enzymes are phosphorylated in the 6-hydroxy position of one or more mannose residues. They exit the secretory route complexed to either MPR, or the cation-dependent mannose 6-phosphate receptor. Lysosomal enzymes leave the biosynthetic route by a distinctive class of coated vesicles developing from the TGR (Geuze et al., 1985). These vesicles probably mediate transport of the enzyme-receptor complexes to the endocytic pathway. After dissociation of the complex in the acidic endosomal environment, MPR escapes the endocytic pathway, and recycles to the biosynthetic pathway, whereas the ligand-enzymes are delivered to lysosomes. The MPR itself is not found in the lysosomal membrane and thus must be efficiently sorted before reaching lysosomes (Griffiths et al., 1988; Geuze et al., 1988). Since 10% of MPR is located on the plasma membrane (Pfeffer, 1987; Braulke et al., 1987), lysosomal enzymes can also be transported to lysosomes by receptor-mediated endocytosis. This population of receptors appears to be in equilibrium with the total cellular MPR population (von Figura et al., 1984). Therefore, the endocytic and the biosynthetic pathways of lysosomal enzymes appear to be connected within a prelysosomal compartment.

As outlined above, much is already known about the intracellular pathways of many proteins that are involved in or related to endocytic pathways. Much less is known about the relationship between the pathways of these proteins. Here we will address these relations within the endocytic pathways.

The subcellular localization of proteins can be detected either morphologically or biochemically. By using biochemical methods, the distribution of a protein, with respect to its intra-or extracellular localization can be resolved. In addition, its subcellular distribution can be compared to that of known cellular organelle specific marker proteins. In such studies, radio-labeled or to enzyme-conjugated proteins are used as probes of intracellular transport as a function of time. However, the intracellular localization of specific intracellular marker proteins is rarely restricted to a single subcellular site. None the less, by means of morphological determinations, the subcelluar localization of individual proteins can be determined. Drawbacks of this approach include the intensive effort necessary for quantitative studies. In addition, pulse-chase kinetic studies are more difficult at a morphological than at a biochemical level. Thus, biochemical and morphological methods are often complementary in the elucidation of intracellular transport pathways.

The classical way to determine whether two proteins are localized within the same subcellular compartments is to disrupt the cells and fractionate their membrane contents by making use of the intrinsic properties of each organelle type. Organelles have been sorted by size, equilibrium density, and the electric charge of their surface. However even with these methods one can never be sure whether two different proteins are present within the same organelle, or are present within different ones which share similar characteristics. Several methods have recently been developed, in which the separation characteristics of cellular organelles are determined by the presence of the molecules at issue. Here, a concomittant change in the seperation characteristics of a second protein is strong evidence for its colocalization with the first. These methods rely on either immunoaffinity, loading of a magnetic charge, or density perturbation. Several methods have been described to increase the density of organelles. Karnovsky-Roots reagents deposits iron-copper in acetylcholine-esterase containing coated vesicles, thus allowing seperation on density gradients of enzyme-containing secretory vesicles from endocytic coated vesicles (Helmy et al., 1986; Fishman and Fine, 1987). Colloidal gold bound to anti-transferrin receptor antibodies has been used to isolate endosomes (Beardmore et al., 1986). A third method exploits the enzymatic activity of horseradish peroxidase, to specifically load peroxidase-containing endosomes with DAB polymer (Courtoy et al., 1984).

Horseradish peroxidase (HRP) can catalyze the hydroxylation of a variety of aromatic compounds, including phenolic substrates such as 3,3'-diaminobenzidine (DAB), as well as tyrosine, phenylalanine, and sialic acid. These latter compounds may partly substitute for DAB as reductant. DAB cytochemistry has been well established for the localization of HRP in fixed tissue (Graham and Karnovsky, 1966). This same principle has been used by Courtoy et al. (1984), to distinquish non-fixed HRP-containing endosomes from cellular organelles of a similar size and equilibrium density. They made use of a conjugate of asialoglycoprotein and HRP to specifically label rat liver endosomes, which are involved in the receptor mediated uptake of asialoglycoproteins, with peroxidase activity. Later, we applied this technique after labeling tissue culture cells with asialoglycoprotein-HRP and transferrin-HRP conjugates, as well as fluid phase-endocytosed HRP (Stoorvogel et al 1987; 1988; 1989; Geuze et al 1988). Two major effects of DAB cytochemistry were observed. 1. Intravesicular DAB polymer is formed, and trapped within the vesicle. Due to the high density of the DAB-polymer, HRP-DAB containing vesicles are recovered at a much higher equilibrium density in a density gradient following centrifugation than nonHRP-containing microsomes (Fig. 1). At increasing DAB concentrations this density shift becomes more pronounced. After DAB cytochemistry, proteins present within the HRP-containing compartment can no longer be extracted in a soluble form after lysis of the microsome. Encapsulation of proteins by DAB polymer as well as chemical crosslinking may play a role in this effect. The latter possibility is probably the major factor, since optimal protein cross-linking occurs at a low DAB concentration at which little DAB polymer is formed. The nature of this bond has not been defined as yet, however aromatic residues of (glyco)proteins appear to substitute for DAB as reducing agents.

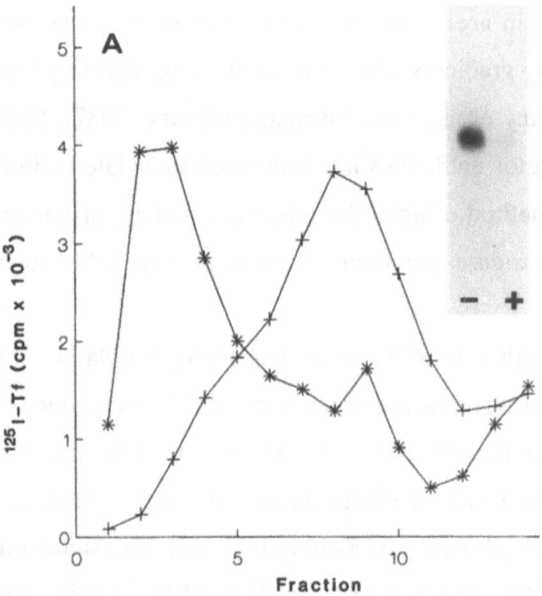

Figure 1. The effect of DAB cytochemistry on the density distribution and detergent extractebility of endosomal 125I-Tf. HepG2 cells were incubated for 30 min at 37C in medium containing both ^{125}I-Tf and a conjugate of Tf and HRP. Excess ligand was washed away. The cells were collected, and equal portions of the postnuclear supernatant of the homogenate were incubated with (*) or without (+) DAB and H_2O_2 prior to fractionation on Percoll gradients. Inset: Samples containing equal amounts of [125I]Tf from density shifted- (fraction 2+3) (+) and non DAB treated (fraction 8+9) (-) microsomes were immunoprecipitated using rabbit anti-human Tf and were analyzed by 10% PAGE and detected by fluorography.

We have used human hepatoma (HepG2) cells, and applied the DAB-induced density shift and cross-link principles to address questions related to the interactions between endocyic pathways. A model developed to summarize these interactions (Fig. 2) is described in the following paragraphs.

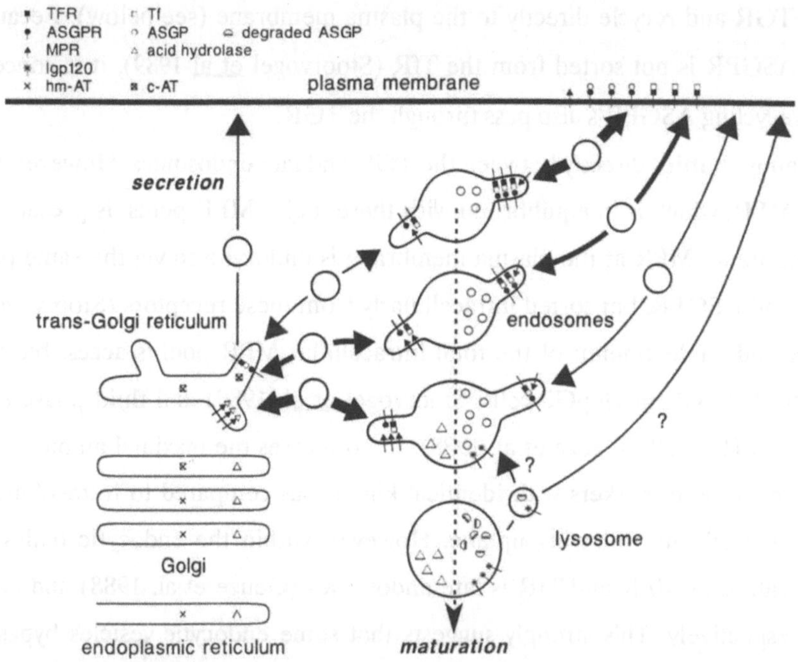

Figure 2. Receptor transport routes. Receptor containing vesicles budding from the plasma membrane fuse with all, but preferably with early endosomes. Vesicles budding from the tubular extensions of early endosomes preferably fuse with the plasma membrane. Endosomes mature during this fusion and fission process. During this process they lose their competence for interaction with the plasma membrane, while interaction with the TGR increases. TfR and ASGPR preferentially exit directly after arrival from endosomes and consequently shuttle mainly between the plasma membrane and endosomes. In contrast, MPR's preferentially exit late endosomes and multivesicular bodies, and are thus mainly shuttling between these organelles and TGR.

Simultaneously endocytosed Tf and ASGP are sorted intracellularly within minutes after uptake (Stoorvogel et al. 1987). This sorting process can be inhibited with primaquine, consistent with the notion that an acidic endosomal environment is a prerequisite for sorting. Endocytosed ASGPR reacycles to the plasma membrane within the same vesicles as TfR (Stoorvogel et al. 1989).

During intracellular trafficking, at least some of the TfR is transported to the secretory pathway (Stoorvogel et al. 1988). Among other secretory glycoproteins, alpha1-antitrypsin (AT) is converted from a high mannose (hm-AT) to a complex glycosylated form (c-AT) within the Golgi stack. Thereafter it is transported to the *trans*-Golgi reticulum (TGR) where it contacts recycling TfR. In the present model, however, most TfR's bypass the TGR and recycle directly to the plasma membrane (see below). Because endocytosed ASGPR is not sorted from the TfR (Stoorvogel et al 1989), it is conceivable that some recycling ASGPR's also pass through the TGR.

MPR mainly shuttles directly between the TGR and late endosomes. However, a small amount of MPR, which is in equilibrium with these major MPR pools, is present on the plasma membrane. MPR at the plasma membrane is endocytosed via the same pathway as the TfR and ASGPR, but sorted intracellularly from these receptors (Stoorvogel et al. 1989). A considerable amount of the total intracellular MPR pool is accessible to both endocytosed Tf/HRP (in HepG2 cells, Stoorvogel et al. 1989) and fluid phase endocytosed HRP (in H4S cells, Geuze et al. 1988). In both cases the maximal amount of MPR was sensitive to these markers with identical kinetics as compared to intracellular TfR, between 5 and 10 min following uptake. However, within the endocytic pathway, the main localization of MPR and TfR is late endosomes (Geuze et al. 1988) and early endosomes, respectively. This strongly suggests that some endocytic vesicles bypass early endosomes and fuse directly with late endosomes. In such a model, plasma membrane-derived receptor-containing vesicles fuse with all, but preferably with early, endosomes. Similarly, early endosome-derived vesicles fuse preferably with the plasma membrane. Endosomes mature during this fusion and fission process, during which they gradually decrease their interaction with the plasma membrane. Simultaneously, the intensity of membrane exchange with the TGR is increased (Geuze et al. 1988). TfR and ASGPR preferentially exit from endosomes immediately after arrival therein (Stoorvogel et al. 1987), and consequently shuttle mainly between the plasma membrane and early endosomes. MPR's however preferentially exit from late endosomes. Thus, endocytosed MPR may remain in contact with recycling TfR by means of a "slipping coupling" mechanism: endocytosed MPR is sorted from TfR after as long as 20-45 min of uptake (Stoorvogel et al. 1989) by sorting from the TfR recycling pathway rather than from simultaneously endocytosed TfR.

The lysosomal glycoprotein lgp120 is mainly localized within late endosomes and lysosomes. After synthesis, lgp120 is probably transported via late endosomes to lysosomes. About half of the total lgp120 is located within endosomes (Geuze et al. 1988). Since lgp120 has a considerable halflife, this large pool can not be composed exclusively of newly synthesized lgp120 which is transported to a final destination in the lysosome. Therefore, it seems likely that lgp120 transport to lysosomes is not unidirectional. Lgp120 transport from lysosomes to endosomes, either directly or via the plasma membrane, may induce the high lgp120 concentration in late endosomes. MPR and lgp120 are sorted within late endosomes, the former is transported to the TGR and the latter remains in the degradative pathway until late endosomes mature into lysosomes.

In summary, these studies revealed at least four distinct sites for sorting between integral membrane proteins: 1) specific receptor uptake at the plasma membrane, 2) sorting of ASGPR and TfR from MPR during the maturation of early endosomes into late endosomes, 3) sorting between MPR and lgp120 in late endosomes, and 4) sorting of MPR from the secretory pathway in TGR. In addition, a connection between the endocytic and the secretory pathway in the TGR was shown.

REFERENCES

Anderson, R.G.W., and l. Orci. 1988. A view of acidic intracellular compartments. J. Cell Biol. 106: 539-543.

Ashwell G, and A.G. Morell. 1974. The role of surface carbohydrates in the hepatic recognition and transport of circulating glycoproteins. Adv Enzymol 41: 99-128.

Beardmore, J., K.E. Howell, K. Miller, and C.R. Hopkins. 1986. Isolation of an endocytic compartment from A431 cells using a density modification procedure employing a receptor-specific monoclonal antibody complexed with colloidal gold. J. Cell Sci. 87: 495-506.

Braulke, T., C. Gartung, A. Hasilik, and K. von Figura. 1987. Is movement of mannose 6-phosphate-specific receptor triggered by binding of lysosomal enzymes? J. Cell Biol. 104: 1935-1742.

Bretscher, M.S., and B.M.F. Pearse. 1984. Coated pits in action. Cell 38: 3-4.

Bridges, K., J. Harford, G. Ashwell, and R.D. Klausner. 1982. Fate of receptor and ligand during endocytosis of asialoglycoproteins by isolated hepatocytes. Proc. Natl. Acad. Sci. USA. 79: 350-354.

Carpentier J.L., P. Gordon, R.G.W. Anderson, J.L. Goldstein, M.S. Brown, S. Cohen, and L. Orci. 1982. Colocalization of ^{125}I-epidermal growth factor and ferritin-low density lipoprotein in coated pits: A quantitative electron microscopic study in normal and mutant human fibroblasts. J. Cell Biol. 95: 73-77.

Ciechanover, A., A.L. Schwartz, and H.F. Lodish. 1983b. The asialoglycoprotein receptor internalizes and recycles independently of the transferrin and insulin receptors. Cell 32: 267-275.

Courtoy, P.J., J. Quintart and P. Baudhuin. 1984. Shift of equilibrium density induced by 3,3'-diaminobenzidine cytochemistry: A new procedure for the analysis and purification of peroxidase-containing organelles. J. Cell Biol. 98: 870-876.

Courtoy, P. J., J. Quintart, J. N. Limet, C. de Roe, and P. Baudhuin. 1985. Polymeric IgA and galactose-specific pathways in rat hepatocytes: evidence for intracellular ligand sorting. in Endocytosis, 163-194. I, Pastan and M. C. Willingham, editors. Plenum press, New York.

Dautry-Versat, A., A. Ciechanover, and H.F. Lodish. 1983. pH and the recycling of transferrin receptor during receptor-mediated endocytosis. Proc. Natl. Acad. Sci. USA. 80: 2258-2262.

Fishman, J.B., and R.E. Fine. 1987. A trans Golgi-derived exocytic coated vesicle can contain both newly synthesized cholinesterase and internalized transferrin. Cell 48: 157-164.

Geuze, H.J., J. W. Slot, G. J. Strous, H. F. Lodish, and A. L. Schwartz. 1983a. Intracellular site of asialoglycoprotein receptor-ligand uncoupling: double-label immunoelectron microscopy during receptor-mediated endocytosis. Cell 32: 277-287.

Geuze H.J., J. W. Slot, G, J. Strous, and A. L. Schwartz. 1983b. The pathway of the asialoglycoprotein-ligand during receptor-mediated endocytosis: a mophological study with colloidal gold/ligand in the human hepatoma cell line, HepG2. Eur. J. Cell Biol. 32: 38-44.

Geuze, H. J., J. W. Slot, G. J. Strous, J. Peppard, K. von Figura, A. Hasilik, and A. L. Schwartz. 1984a. Intracellular receptor sorting during endocytosis: comparative immunoelectron microscopy of multiple receptors in rat liver. Cell. 37: 195-204.

Geuze, H.J., J.W. Slot, G.J. Strous, J.P. Luzio, and A.L. Schwartz. 1984b. A Cyclohexamide-resistant pool of receptors for asialoglycoproteins and mannose 6-phosphate residues in the Golgi complex of rat hepatocytes. EMBO J. 3: 2677-2685.

Geuze, H.J., J.W. Slot, G.J. A.M. Strous, A. Hasilik, and K. von Figura. 1985. Possible pathways for lysosomal enzyme delivery. J. Cell Biol. 101: 2253-2262.

Geuze, H.J., J.W. Slot, and A.L. Schwartz. 1987. Membranes of sorting organelles display lateral heterogeneity in receptor distribution. J. Cell Biol. 104: 1715-1723.

Geuze, H.J., W. Stoorvogel, G.J. Strous, J.W. Slot, J.E. Bleekemolen, and I. Mellman. 1988. Sorting of mannose 6-phosphate receptors and lysosomal membrane proteins in endocytic vesicles. J. Cell Biol. 107: 2491-2501.

Goldstein J.L., M.S. Brown, R.G.W. Anderson, D.W. Russell, and W.J. Schneider. 1985. Receptor-mediated endocytosis: concepts emerging from the LDL receptor system. Annu. Rev. Cell Biol. 1: 1-39.

Graham RC Jr., and M.J. Karnovsky. 1966. The early stages of absorbtion of injected horseradish peroxidase in the proximal tubules of mouse kidney: ultrastructural cytochemistry by a new technique. J. Hist. Chem. 14: 291-302.

Griffiths, G., B. Hoflack, K. Simons, I. Mellman and S. Kornfeld. 1988. The mannose 6-phosphate receptor and the biogenesis of lysosomes. Cell 52: 329-341.

Helmy, S., K. Porter-Jordan, E.A. Dawidowicz, P. Pilch, A.L. Schwartz, and R.E. Fine. 1986. Separation of endocytic from exocytic coated vesicles using a novel cholinesterase mediated density shift technique. Cell 44: 497-506.

Hoflack, B., and S. Kornfeld. 1985. Purification and characterization of a cation-dependent mannose 6-phosphate receptor from murine P3 88D1 macrophages and bovine liver. J. Biol. Chem. 260: 12008-12014.

Hopkins, C.R., and I.S. Trowbridge. 1983. Internalization and processing of transferrin and the transferrin receptor in human carcinoma cells A431. J. Cell Biol. 97: 508-521.

Kiess, W., G.D. Blickenstaff, M.M. Sklar, C.L. Thomas, S.P. Nissley, and G.G. Sahagian. 1988. Biochemical evidence that the type II insulin-like growth factor receptor is identical to the cation-independent mannose 6-phosphate receptor. J. Biol. Chem. 263: 9339-9344.

Klausner R.D., G. Ashwell, J. van Renswoude, J.B. Harford, and K.R. Bridges. 1983. Binding of apotransferrin to K562 cells: explanation of the transferrin cycle. Proc. Natl. Acad. Sci. USA 80: 2263-2266.

Marsh, M., and A. Helenius. 1980. Adsorbtive endocytosis of Semliki Forest Virus. J. Mol. Biol. 142: 439-454.

Mellman, I., R. Fuchs, and A. Helenius. 1986. Acidification of the endocytic and exocytic pathways. Annu. Rev. Biochem. 55: 663-700.

Pastan I. H., and M.C. Willingham. 1981. Receptor-mediated endocytosis of hormones in cultured cells. Ann. Rev. Physiol. 43: 239-250.

Pfeffer, S.R. 1987. The endosomal concentration of a mannose 6-phosphate receptor is unchanged in the absence of ligand synthesis. J. Cell Biol. 105: 229-234.

Rudnick, G. 1986. ATP-driven H^+ pumping into intracellular organelles. Annu. Rev. Physiol. 48: 403-415.

Sahagian, G.G., J.J. Distler, and G.W. Jourdian. 1982. Characterization of a membrane-associated receptor from bovine liver that binds phosphomannosyl residues of bovine testicular galactosidase. Proc. Natl. Acad. Sci. USA. 78: 4289-4293.

Schwartz, A.L., S.E. Fridovich, and H.F. Lodish. 1982. Kinetics of internalization and recycling of the asialoglycoprotein receptor in a hepatoma cell line. J. Biol. Chem. 257: 4230-4237.

Simmons, C.F. and A.L. Schwartz. 1984. Cellular pathways of galactose-terminal ligand movement in a cloned human hepatoma cell line. Mol. Pharmacol. 26: 509-519.

Stoorvogel, W., H.J. Geuze, and G.J. Strous. 1987. Sorting of endocytosed transferrin and asialoglycoprotein occurs immediately after internalization in HepG2 cells. J. Cell Biol. 104: 1261-1268.

Stoorvogel, W., H.J. Geuze, J.M. Griffith, and G.J. Strous. 1988. The pathways of endocytosed transferrin and secretory protein are connected in the *trans*-Golgi reticulum. J. Cell Biol. 106: 1821-1829.

Stoorvogel, W., H.J. Geuze, J.M. Griffith, A.L. Schwartz, and G.J. Strous. 1989. Relations between the intracellular pathways of the receptors for transferrin, asialoglycoprotein, and mannose 6-phosphate in human hepatoma cells. J. Cell Biol. 108: 2137-2148.

Tolleshaug, H., T. Berg, and K. Holte. 1980. Kinetics of internalization and degradation of asialoglycoproteins in isolated rat hepatocytes. Eur. J. Cell Biol. 23: 104-109.

Tycko, B., and F.R. Maxfield. 1982. Rapid acidification of endocytotic vesicles containing 2-macroglobulin. Cell 28: 634-651.

Van den Bosch, R.A., H.J. Geuze, and G.J. Strous. 1986. Presence of the asialoglycoprotein receptor in the Golgi complex in the absence of protein synthesis. Exptl. Cell Res. 162: 231-242.

VanDeurs B., O.W. Petersen, S. Olsnes, and K. Sandvig. 1989. The ways of endocytosis. Int. Rev. Cytol. 117, 131-177.

Von Figura, K., V. Gieselmann, and A. Hasilik. 1984. Antibody to mannose 6- phosphate specific receptor induces receptor deficiency in human fibroblasts. EMBO J. 3: 1281-1286.

Wileman, T., C. Harding, and P. Stahl. 1985. Receptor-mediated endocytosis. Biochem. J. 232: 1-14.

Willingham M.C., and I. Pastan. 1980. The receptosome: An internal intermediate organelle of receptor-mediated endocytosis in cultured fibroblasts. Cell 21: 67-77.

Willingham, M.C., I.H. Pastan, G.G. Sahagian, G.W. Jourdian, and E.F. Neufeld. 1981. Morphologic study of the internalization of a lysosomal enzyme by the mannose 6-phosphate receptor in cultured Chines hamster ovary cells. Proc. Natl. Acad. Sci. USA. 78: 6967-6971.

Yamashiro, D.J., S.R. Fluss, and F.R. Maxfield. 1983. Acidification of endocytic vesicles by an ATP-dependent proton pump. J. Cell Biol. 97: 929-934.

Yamashiro, D.J., and F.R. Maxfield. 1984. Acidification of endocytic compartments and the intracellular pathways of ligands and receptors. J. Cell. Biochem. 26: 231-246.

Zijderhand-Bleekemolen, J.E., A.L. Schwartz, J.W. Slot, G.J.Strous, and H.J. Geuze. 1987. Ligand and weak base-induced redistribution of asialoglycoprotein receptors in hepatoma cells. J. Cell Biol. 104: 1647-1654.

NOTE TO THE ADDENDUM

The following two contributions were received by the Editor after the complete manuscript had been sent to the Publisher.

GUANYLATE CYCLASE IN SIGNAL TRANSDUCTION OF DICTYOSTELIUM DISCOIDEUM

Conchita C.G.M. Schulkes, Cor D. Schoen[1], Jos C. Arents, and Roel van Driel
E.C. Slater Institute for Biochemical Research
University of Amsterdam
Plantage Muidergracht 12
1018 TV Amsterdam
The Netherlands

Cell signalling through cGMP

External signals that regulate cell functions are often detected by cell surface receptors and transmitted via intracellular second messengers. Such messenger molecules control a wide variety of cellular processes, such as differentiation, proliferation, intermediary metabolism, secretion, changes in cell shape and cell motility. Two of these second messengers are cyclic 3',5'-adenosine monophosphate (cAMP) and cyclic 3',5'-guanosine monophosphate (cGMP). While the signalling system using cAMP has been studied extensively, relatively little is known about signal transduction pathways involving cGMP.

Components of the cGMP signalling system include guanylate cyclase, intracellular cGMP receptors and cGMP phosphodiesterase. Guanylate cyclase is responsible for the synthesis of cGMP, whereas cGMP receptors detect and process the signal that is transduced by cGMP. Known intracellular cGMP receptors are cGMP-dependent protein kinase, cGMP-specific and nonspecific phosphodiesterases, and certain ion-channels.

The physiological processes that are regulated by cGMP are equally diverse as the cell types where the cyclic nucleotide is found. The presence of cGMP has for instance been reported for bovine lung (Chrisman et al., 1975), cerebellum (Schlichter et al., 1978), smooth muscle tissue (Fiscus et al., 1985), intestinal mucosa (Garbers et al., 1974; De Jonge, 1976), blood platelets (Böhme et al., 1974), pheromone antennae of silkmoths (Ziegelberger et al., 1990), and rod outer segments of the retina (Pannbacker et al., 1972). Physiological processes that are controlled by cGMP include natriuresis, diuresis, and vasorelaxation induced by ANP (atrial natriuretic peptide) (Fiscus et al., 1985), transmission of signals between neurons (Sugden and Klein, 1987), visual excitation (Cook and Kaupp,

[1] Present address: Research Inst. for Plant Protection, Binnenhaven 12, 6709 PD Wageningen, The Netherlands

NATO ASI Series, Vol. H 52
Biological Signal Transduction
Edited by E. M. Ross and K. W. A. Wirtz

1986), and the induction of chemotaxis of spermatozoa by the peptides resact and speract (Garbers, 1989). Additionally in the lower eukaryote Dictyostelium discoideum, a cellular slime mould, cGMP is an important second messenger. It is required for the cAMP-induced chemotactic response, which has a key function during the developmental cycle of D. discoideum.

In this chapter we describe some aspects of the cGMP signal transduction system of Dictyostelium. Part of this overview will concern our studies on the regulation of guanylate cyclase.

Components of the cGMP pathway

Ubiquitous components of the cGMP signal transduction pathway are guanylate cyclase, as well as intracellular cGMP binding proteins, like cGMP-dependent protein kinase and cGMP phosphodiesterase.

Guanylate cyclase

The synthesis of cGMP from guanosine triphosphate (GTP) is catalyzed by the enzyme guanylate cyclase (GTP pyrophosphate lyase (cycling) EC 4.6.1.2). Two types of guanylate cyclase activities can be distinguished: soluble and membrane-bound enzyme. Two of the most thoroughly investigated membrane-bound guanylate cyclases are those activated by the peptide hormone ANP (atrial natriuretic peptide) and the sperm-activating peptides resact and speract. These are transmembrane proteins, localized in the plasmamembrane and consisting of three functional domains. One domain is exposed to the extracellular space and acts as a receptor (Kuno et al., 1986). One of the cytoplasmic domains carries the guanylate cyclase activity, whereas the other has some sequence homology with protein kinases (Sing et al., 1988). Another type of guanylate cyclase is found in rat pinealocytes (Klein et al., 1983; Ho et al., 1987) and is regulated via ß-adrenergic receptors. A G-protein-like transducing factor seems involved in signal transduction, since cholera toxin mimics the effect of ß-adrenergic receptor agonists on guanylate cyclase (Sugden and Klein, 1987).

Calcium is also a potent regulator of guanylate cyclase activity. Its effect can be either stimulatory or inhibitory. In Paramecium and Tetrahymena the membrane associated guanylate cyclase is stimulated by Ca^{2+} ions (Preston and Saimi, 1990). In Tetrahymena Ca^{2+} probably acts via calmodulin (Nagano et al., 1990). In rod outer segments Ca^{2+} inhibits the enzyme. Here too a separate Ca^{2+}-binding protein is thought to be required for the inhibitory effect (Koch and Stryer, 1988). The observation that the phorbol ester PMA (4ß-phorbol 12-myristate 13-acetate) activates guanylate cyclase suggests that protein kinase C may also be involved in the regulation of cGMP synthesis (Ho et al., 1987; Duda and Sharma, 1990).

Receptors for cGMP and cGMP degrading enzymes

The most thoroughly investigated cGMP-receptor is cGMP-dependent protein kinase. This cGMP-stimulated ser/thr kinase activity is high in cerebellum, lung, and smooth muscle (Walter et al., 1981). The substrate specificity in vitro is similar, though not identical to that of cAMP-dependent protein kinase. One of the substrates in vivo is the enzyme itself (Lincoln and Corbin, 1983). Auto-phosphorylation has been reported for bovine lung cGMP-dependent protein kinase (De Jonge and Rosen, 1977).

Various other intracellular types of cGMP-binding activities have been described. In rod outer segments of retina cells a cation-specific channel, that is directly controlled by cGMP binding, (Cook and Kaupp, 1986) and a cGMP-specific phosphodiesterase (Yamazaki et al., 1984) have been found. Visual excitation requires closure of the cGMP-controlled channel. This is achieved by a decrease in the cGMP concentration, due to the activation of a cGMP-specific phosphodiesterase by the light receptor rhodopsin (Yamazaki et al., 1984).

With the possible exception of the retina, the cGMP signal transduction pathway downstream of the intracellular cGMP receptor is still far from clear in most cell types.

Dictyostelium discoideum as a model for the study of signal transduction

Signal transduction in the cellular slime mould Dictyostelium discoideum shows many similarities with that of higher eukaryotes. These similarities include the functioning of cell surface receptors, mechanisms of adaptation and desensitization, and the functioning of signal transduction pathways involving cAMP, cGMP and phosphoinositides. Dictyostelium has been the subject of many cell biological studies in signal transduction, because agonist induced chemotaxis, differentiation and pattern formation are key phenomena during growth and development of this organism. The life cycle of D. discoideum is depicted in figure 1.

The Dictyostelium discoideum life cycle

Dictyostelium discoideum cells live in the soil as solitary amoeba and feed by phagocytosis of bacteria. When food is exhausted a specific developmental program is initiated (Loomis, 1982; Janssens and Van Haastert, 1987). After about six hours cells start to synthesize and secrete pulses of cAMP, which serves as a chemoattractant for surrounding cells. Excited cells respond to cAMP in two distinct ways. Firstly, they chemotactically move towards the cAMP source. Secondly, they synthesize and secrete cAMP. In this way a population of cells is able to relay cAMP signals over large distances. This process results in the formation of multicellular aggregates, involving up to 10^5 cells. In the aggregate, called pseudoplasmodium or 'slug', tight cell-cell contacts are formed. Slugs are true multicellular organisms. They are able to migrate over large distances and show chemotactic, phototactic, and thermotactic behaviour. At the slug stage cells differentiate into two major cell types, prestalk cells and prespore cells. The ratio

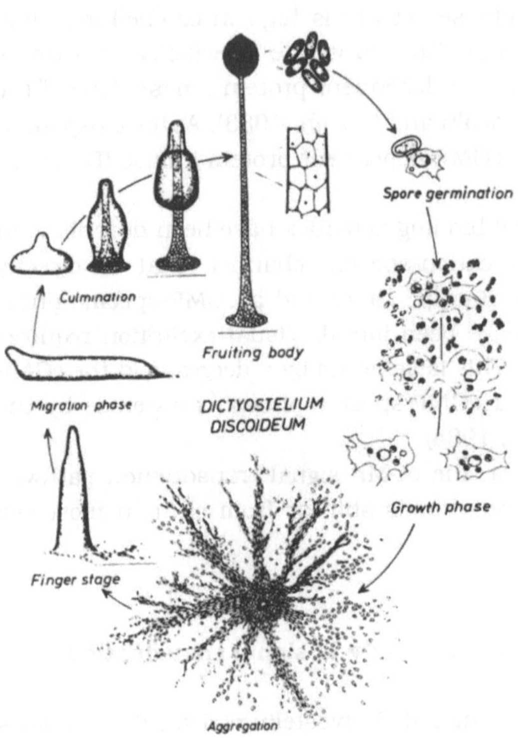

Figure 1 Life cycle of D. discoideum

between the number of prestalk and prespore cells in an aggregate is genetically determined: about 1 to 4, respectively. Interestingly, this ratio is homeostatically regulated. If it is disturbed by cutting off part of the slug, the remaining cells redifferentiate and re-establish the correct ratio. A specific cell pattern is formed and maintained in the slug: the prestalk cells are located in the anterior tip, the prespore cells in the posterior part. The cell aggregate undergoes several morphogenetic changes and finally forms a fruiting body, consisting of a spherical head of spores, held aloft by a slender stalk. This stalk is formed by highly vacuolized, rigid stalk cells. If sufficient food is available, spores can germinate and give rise to new amoeboid cells, which grow and proliferate as long as nutrients are available, and start a new developmental cycle upon starvation.

Cyclic AMP has a key role in D. discoideum development (Janssens and Van Haastert, 1987). Not only is it a potent chemoattractant, it is also a hormone-like substance that is required for cell differentiation and pattern formation. Extracellular cAMP is detected by highly specific high affinity receptors on the cell surface. The expression of these receptors is developmentally regulated, like most other components of the cAMP relay system (Sun et al., 1990). Occupation of the receptor by cAMP results in the immediate activation of adenylate cyclase, guanylate cyclase and a phospholipase C-like activity. Evidently, at least three different signal transduction pathways are activated via cAMP receptors. For this reason Dictyostelium discoideum is an attractive model system to study signal transduction resulting in, among others, chemotactic events and cell differentiation.

Cyclic GMP-mediated signal transduction in Dictyostelium discoideum

THE cGMP RESPONSE IN DICTYOSTELIUM

The cGMP response in cAMP-stimulated cells is extremely fast and transient (Wurster et al., 1977; Mato et al., 1977). Within seconds after receptor occupation, guanylate cyclase is activated and the intracellular cGMP concentration increases.

In stimulated cells a rapid adaptation mechanism is triggered, which inactivates guanylate cyclase with a $t_{1/2}$ of 2 to 3 s. Due to a cGMP-specific phosphodiesterase activity, the cGMP concentration rapidly decreases to the prestimulation level after guanylate cyclase inactivation. The entire process takes only 30 seconds, showing a cGMP concentration peak at about 10 seconds.
After receptor stimulation actin and myosin are redistributed in the cell, probably reflecting the chemotactic response. Actin polymerization is triggered, resulting in the formation of pseudopodia that contain bundles of parallel actin fibres. Changes in the aggregation state of myosin in Dictyostelium have been shown to be correlated with myosin heavy chain phosphorylation-de-phosphorylation events, triggered by occupation of the cAMP receptor (Hall et al., 1989).
Mutants have been isolated that directly implicate cGMP in Dictyostelium chemotaxis. These mutants, called streamer F (stm F) mutants, lack the cGMP-specific phosphodiesterase (Van Haastert et al., 1982). They show a cGMP response that lasts longer and reaches a higher peak level than wild type cells. At the same time their chemotactic response lasts longer. In stm F mutants, both myosin phosphorylation and myosin polymerization are delayed compared to wild type cells, suggesting that these events are controlled by cGMP (Liu and Newell, 1988). No significant difference in the actin response is seen in stm F mutants when compared to wild type cells. These observations strongly indicate that cGMP is a second messenger that transduces the chemotactic signal.

RECEPTORS FOR CGMP IN DICTYOSTELIUM

For Dictyostelium it is still not clear what the direct intracellular target for cGMP is. Besides the afore-mentioned cGMP-specific phosphodiesterase, at least one other cGMP binding protein is present with high specificity and high affinity (Mato et al., 1979). This putative target for cGMP is a soluble, cytoplasmic protein with still unknown function. It is conceivable that in Dictyostelium the receptor for cGMP is a protein kinase, as is found in many higher eukaryotic cells. Up to now, however, attempts to show cGMP-stimulated protein kinase activity have been unconclusive (however, see Wanner and Wurster, 1990). In contrast, cAMP-dependent protein kinase activity has been identified in D. discoideum cells (Arents and Van Driel, 1982).

Regulation of guanylate cyclase in Dictyostelium discoideum

Key questions in signal transduction via cGMP are how cAMP receptor occupation is coupled to activation of guanylate cyclase and how adaptation results in inactivation of the enzyme. We have studied the regulation of guanylate cyclase that uses the physiological substrate Mg-GTP. Experiments were performed in cell lysates and in electro-permeabilized cells. In the latter system, cells are permeabilized by brief exposure to a strong electric field (Schoen et al., 1989). Under carefully selected conditions only low molecular weight components can pass the permeabilized cell membrane, whereas all macromolecules remain inside. Moreover, membranes of organelles remain intact. Schoen et al. (1989) demonstrated that equilibration of the cytoplasmic compartment with the extracellular medium is rapid for small metabolites. Importantly, pores are stable for at least five minutes. Since the cells remain intact and regulatory macromolecules remain inside, the permeabilized cell system can be considered to represent a near-in vivo situation.

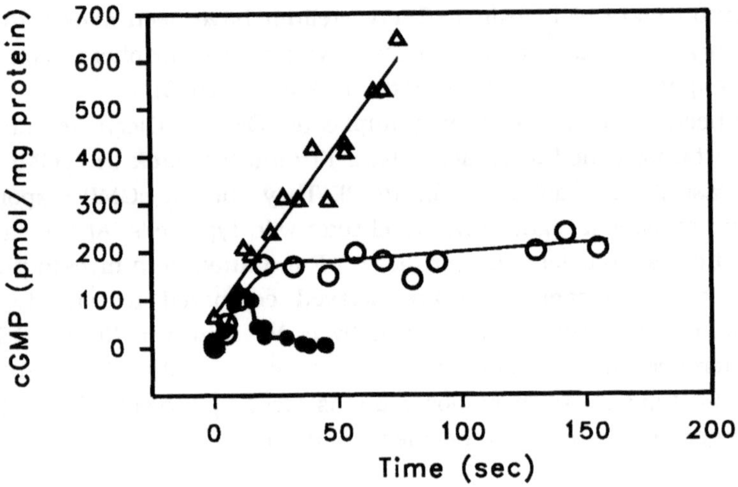

Figure 2 Kinetics of cGMP synthesis under in vivo, near-in vivo (i.e. electro-permeabilized cells), and in vitro conditions.
Cells in suspension (10^8 cells/ml) were permeabilized or homogenized (as described by Schoen et al., 1989) in the presence of 0.2 mM GTPγS. Cyclic GMP synthesis was started at t=0. Guanylate cyclase activity of permeabilized cells (O) is compared to that of an homogenate (Δ), and of intact cells (●). In (O) and (Δ) the phosphodiesterase activity was completely suppressed by 5 mM dithiothreitol (Green and Newell, 1975). In intact cells cGMP synthesis was started by addition of the cAMP agonist deoxy-cAMP (10 μM), in both other preparations by addition of GTP substrate.

GUANYLATE CYCLASE ACTIVITY UNDER IN VITRO AND NEAR-IN VIVO CONDITIONS

Previously, it has been shown that in broken cell preparations guanylate cyclase activity is stimulated by the non-hydrolyzable GTP-analogue GTPγS (guanosine 5'-O-(3-thio)-triphosphate; Janssens et al., 1988). Schoen et al. (manuscript in preparation) presented evidence that in permeabilized Dictyostelium cells guanylate cyclase activity is completely dependent on the presence GTPγS. In intact cells the chemotactically induced cGMP response is rapid and transient. Under these conditions the intracellular concentration of cGMP rapidly rises and decreases to the prestimulation level within 30 seconds after addition of cAMP. The resulting decrease is due to an adaptation mechanism where guanylate cyclase is switched off, combined with the degradative activity of cGMP phosphodiesterase. Essentially all cyclic nucleotide phosphodiesterase activity in D. discoideum can be inhibited by millimolar concentrations of dithiothreitol (Green and Newell, 1975). Under these conditions, after cell stimulation, the intracellular cGMP level remains constant after 30 seconds, the time point after which the guanylate cyclase is completely inactivated due to adaptation.

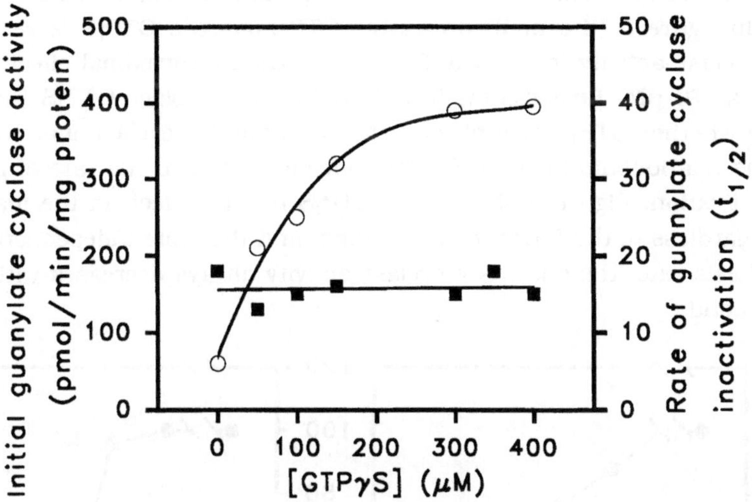

Figure 3 Effect of GTPγS concentration on the activation and inactivation of guanylate cyclase in permeabilized cells.
Cells in suspension (10^8 cells/ml) were permeabilized in the presence of different GTPγS concentrations. The guanylate cyclase reaction was started by the addition of 5 mM GTP substrate. The initial reaction rate (O) and the rate of inactivation (■) of guanylate cyclase are shown as a function of the GTPγS concentration. The rate of guanylate cyclase inactivation is given as $t_{1/2}$ (in seconds), i.e. the time required to reduce the reaction rate to 50% of the initial value.

The kinetics of cGMP production and subsequent adaptation of the cGMP response in permeabilized cells closely resembles that seen in vivo, as shown in figure 2. The results strongly suggests that in electro-permeabilized cells an adaptation-like mechanism is still operative. In cell-homogenates this adaptation-like inactivation of guanylate cyclase is not observed. in this case cGMP production continues for at least several minutes (fig. 2).
Several conceivable, simple reasons for inactivation of the enzyme in the permeabilized cell system have been checked. None of these possibilities were the cause for a decrease in guanylate cyclase activity that was seen at 10 to 30 seconds after starting cGMP synthesis. For instance, we found that the enzyme is stable for at least several minutes and the pores in the cell membrane remain open for a comparable time. Also, the product of the guanylate cyclase reaction, cGMP, did not inhibit the enzyme. From these experiments it can be concluded that the inactivation of guanylate cyclase is not an artefact induced by the experimental conditions. This implies that the adaptation-like inactivation process observed in permeabilized cells is an intrinsic property of the enzyme in the cell. Evidently, the

inactivation mechanism is lost in cell homogenates (fig. 2). These observations allow a more detailed analysis of the molecular mechanisms that underlay adaptation of the cGMP response.

GTPγS IS REQUIRED FOR GUANYLATE CYCLASE ACTIVITY, BUT NOT FOR ITS INACTIVATION

As shown in figure 3, the nonhydrolyzable GTP analogue GTPγS is required for guanylate cyclase activity in permeabilized cells. The half-maximal effect of GTPγS is observed at 70 μM. Remarkably, 5 mM GTP cannot replace GTPγS (see below). To examine whether GTPγS also plays a role in guanylate cyclase inactivation, we analyzed the relationship between GTPγS concentration and the rate of guanylate cyclase inactivation. Figure 3 shows that GTPγS has no effect on the inactivation process. Regardless of the GTPγS concentration, and therefore independently of the cGMP synthesis rate, the guanylate cyclase activity always decreases with a $t_{0.5}$ of about 15 seconds.

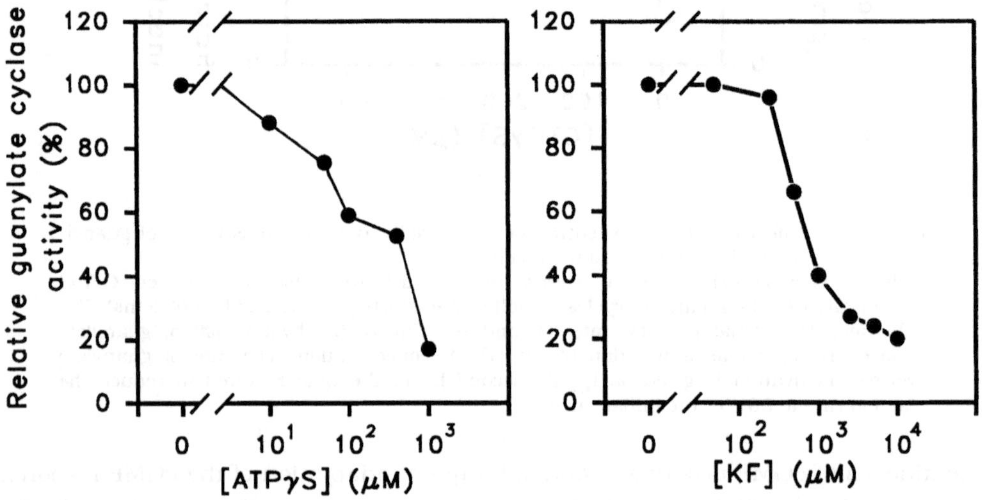

Figure 4 Effect of conditions that favour protein phosphorylation on the guanylate cyclase reaction.
A cell homogenate was preincubated with either ATPγS or potassium fluoride for 90 s. Subsequently the guanylate cyclase reaction was started by the addition of GTP substrate. The initial reaction rates are shown as a function of effector concentration.

INACTIVATION OF GUANYLATE CYCLASE MAY BE DUE TO PROTEIN PHOSPHORYLATION

Enzyme activity is often regulated via changes in phosphorylation state. This has concurrently been described for several types of guanylate cyclase (Garbers, 1989). We have analyzed whether the activity of guanylate cyclase in D. discoideum is also modulated by phosphorylation-dephosphorylation events. We made use of the ATP analogue ATPγS. Protein kinases are able to transfer the thio-phosphate group to ser or thr residues of proteins. Such thio-phosphate esters are resistant to phosphatases. Therefore, the thio-phosphorylated state of proteins is much more stable than the normal phosphoprotein. Concurrently, potassium fluoride was added, since fluoride ions at millimolar concentrations are potent inhibitors of protein phosphatases. Figure 4 shows that potassium fluoride and ATPγS strongly inhibit guanylate cyclase activity. Half-maximum effects are obtained at 0.8 mM and 0.1 mM, respectively. Incubation with the non-hydrolysable ATP analogue AppNp (adenosine 5'-[ß,γ-imido]-triphosphate), which is a competitive inhibitor of the protein kinase reaction, results in an increase in guanylate cyclase activity (not shown). These results show that conditions which favour protein phosphorylation inhibit guanylate cyclase activity, suggesting that protein phosphorylation is involved in inactivation (i.e., adaptation) of guanylate cyclase in Dictyostelium.

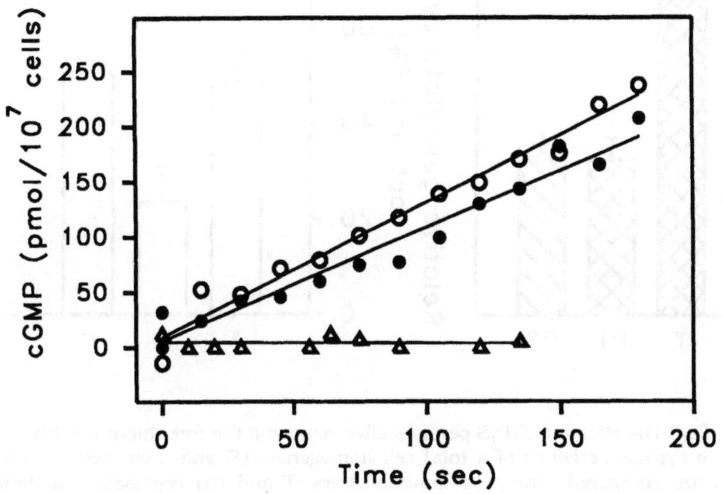

Figure 5 Cellular localization of guanylate cyclase.
A cell homogenate was separated in a particulate fraction and supernatant (2 min, 10,000 x *g*). The rate of cGMP synthesis was measured in the total homogenate (○), the pellet (●) and the supernatant (Δ).

ACTIVATION OF GUANYLATE CYCLASE IN DICTYOSTELIUM

We have analysed in more detail the role of GTPγS in guanylate cyclase activation in cell in homogenates. Figure 5 shows that the enzyme activity is localized exclusively in the particulate cell fraction if Mg-GTP is used as a substrate (Janssens et al., 1989). If Mn-GTP is used, activity is also found in the soluble cell fraction (Janssens et al., 1989). Figure 3 shows that even high concentrations of GTP (5 mM) cannot substitute for the stimulatory effect of submillimolar levels of GTPγS. Moreover, GTPγS has to be added *before* the GTP substrate, otherwise GTPγS is ineffective (data not shown). These observations suggest that guanylate cyclase activity is controlled by a G-protein-like entity, which hydrolyzes GTP to tightly bound GDP. Once the guanine nucleotide binding site is occupied by non-exchanging GDP, the nucleotides GTPγS or GTP can no longer be bound and the G-protein cannot be activated.

The stimulatory effect of GTPγS persisted even after washing the particulate fraction to remove all free thionucleotide (fig. 6). Washing itself did reduce the total guanylate cyclase activity by about 50%. Evidently, part of the guanylate cyclase, or some factor required for cGMP synthesis, is losely bound. When a particulate cell fraction was washed first, and subsequently assayed for GTPγS-stimulated guanylate cyclase activity, only a relatively low activity was observed (fig. 7).

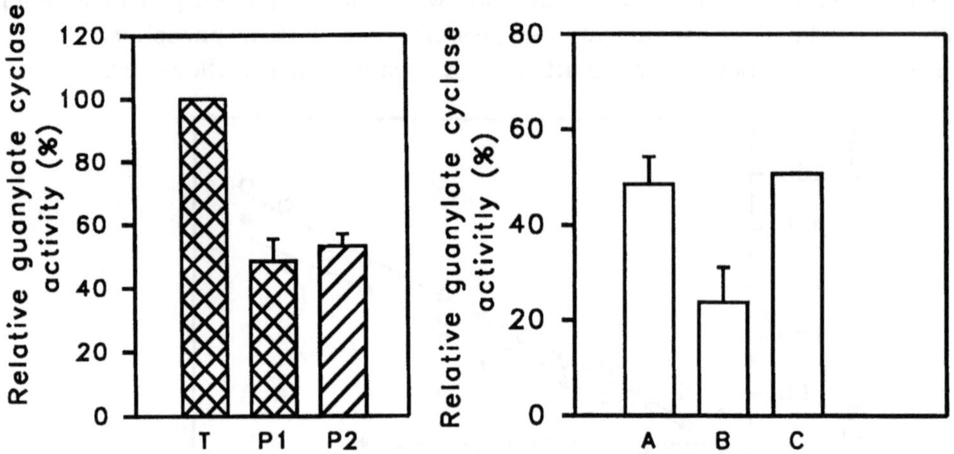

Figure 6 *(left)* The effect of GTPγS persists after removing the free thionucleotide. The guanylate cyclase activities of a total cell homogenate (T) and a washed particulate fraction (P1) are compared. The cross-hatched bars (T and P1) represent reactions in which GTPγS was present during homogenization and the guanylate cyclase assay. The hatched bar (P2) represents the activity in a particulate fraction after removing free GTPγS by washing.

Figure 7 *(right)* Guanylate cyclase activity requires a soluble factor. Bars represent the guanylate cyclase activity in the particulate cell fraction, isolated in different ways. The activity of a total homogenate prepared in the presence of GTPγS is set to 100% (see fig. 6 bar T). (A) Activity in a particulate fraction isolated in the presence of GTPγS. (B) Particulate fraction isolated in the absence of GTPγS, however, the thionucleotide was added before the reaction was started by addition of GTP substrate. (C) Same as B, however, the supernatant fraction was added together with GTPγS.

This indicates that a soluble factor that is required for GTPγS stimulation of the enzyme was lost upon washing. From these experiments we conclude that a soluble component is involved in the activation of guanylate cyclase. Significantly, GTPγS sensitivity in a washed particulate fraction could be restored by addition of the soluble cell fraction (fig. 7). Treatment of the soluble cell fraction with protease prevents reconstitution, suggesting that the active component in the reconstituted system is a protein (not shown). It is still unclear whether the soluble factor is the GTPγS binding protein or an additional component required for guanylate cyclase activity.

Conclusion

These studies show that the cAMP receptor-controlled regulation of guanylate cyclase in Dictyostelium discoideum is complex. At least two mechanisms are involved. Firstly, a soluble GTPγS binding component, possibly a G-protein, is required for the activation of the enzyme. On the other hand, inactivation is probably regulated via protein phosphorylation. Both mechanisms may not be unique for Dictyostelium.

Literature references

Arents JC, Van Driel R (1982) Soluble, cytoplasmic cyclic AMP-binding proteins of Dictyostelium discoideum. FEBS Letters 137:201-204

Böhme E, Jung R, Mechler I (1974) Guanylate cyclase in human platelets. Meth. in Enzymol. 28:199-202

Chrisman TD, Garbers DL, Parks MA, Hardman JG (1975) Characterization of particulate and soluble guanylate cyclase from rat lung. J. Biol. Chem. 250:374-381

Cook NJ, Kaupp UB (1986) The cGMP-dependent cation channel from vertebrate photoreceptors: purification and function reconstitution. Photobiochem. Photobiophys. 13:331-343

De Jonge HR (1976) Cyclic nucleotide-dependent phosphorylation of intestinal epithelium proteins. Nature (London) 262:590-591

De Jonge HR, Rosen OM (1977) Self-phosphorylation of cGMP-dependent protein kinase from bovine lung. J. Biol. Chem. 252:2780-2783

Duda T, Sharma RK (1990) Regulation of guanylate cyclase activity by atrial natriuretic factor and protein kinase C. Molec. Cell. Biochem. 93:179-184

Fiscus RR, Rapoport RM, Waldman SA, Murad F (1985) Atriopeptin II elevates cyclic GMP, activates cyclic GMP-dependent protein kinase and causes relaxation in rat thoracic aorta. Biochim. Biophys. Acta 846:179-184

Garbers DL (1989) Guanylate cyclase, a cell surface receptor. J. Biol. Chem. 264:9103-9106

Garbers DL, Dyer E, Hardman JG (1974) Determination of guanosine 3',5'-monophosphate in tissues and of guanylyl cyclase in rat intestine. J. Biol. Chem. 244:6371-6376

Green AA, Newell PC (1975) Evidence for the existence of two types of cAMP binding sites in aggregating cells of Dictyostelium discoideum. Cell 6:129-136

Hall AL, Warren V, Condeelis J (1989) Transduction of the chemotactic signal to the actin cytoskeleton of Dictyostelium discoideum. Developm. Biol. 136:517-525

Ho AK, Chik CL, Klein DC (1987) Protein kinase C is involved in adrenergic stimulation of pineal cGMP accumulation. J. Biol. Chem. 262:10059-10064

Janssens PMW, Van Haastert PJM (1987) Molecular basis of transmembrane signal transduction in Dictyostelium discoideum. Microbiol. Rev. 51:396-418

Janssens PMW, De Jong CCC (1988) A magnesium-dependent guanylate cyclase in cell-free preparations of Dictyostelium discoideum. Biochem. Biophys. Res. Commun. 150:405-411

Janssens PMW, De Jong CCC, Vink AA, Van Haastert PJM (1989) Regulatory properties of magnesium-dependent guanylate cyclase in Dictyostelium discoideum membranes. J. Biol. Chem. 264:4329-4335

Klein DC, Sugden D, Weller JL (1983) Postsynaptic α-adrenergic receptors potentiate the ß-adrenergic stimulation of pineal serotonin N-acetyltransferase. Proc. Natl. Acad. Sci. USA 80:599-603

Koch KW, Stryer L (1988) Highly cooperative feedback control of retinal rod guanylate cyclase by calcium ions. Nature (London) 334:65-66

Kuno T, Andresen JW, Kamisaki Y, Waldman SA, Chang LY, Saheki S, Leitman DC, Nakane M, Murad F (1986) Copurification of an atrial natriuretic factor receptor and particulate guanylate cyclase from rat lung. J. Biol. Chem. 261:5817-5823

Lincoln TM, Corbin JD (1983) Characterization and biological role of the cGMP dependent protein kinase. Adv. Cyclic Nucl. Res. 15:139-192

Liu G, Newell PC (1988) Evidence that cyclic GMP regulates myosin intraction with the cytoskeleton during chemotaxis of Dictyostelium. J. Cell Sci. 90:123-129

Loomis WF (1982) The development of Dictyostelium. Acad. Press, New York.

Mato JM, Krens FA, Van Haastert PJM, Konijn TM (1977) 3',5'-cyclic AMP-dependent 3',5'-cyclic GMP accumulation in Dictyostelium discoideum. Proc. Natl. Acad. Sci. USA 74:2348-2351

Mato JM, Woelders H, Konijn TM (1979) Intracellular cyclic GMP-binding proteins in cellular slime molds. J. Bacteriol. 137:169-172

Nagao S, Matsuki S, Kanoh H, Ozawa T, Yamada K, Nozawa Y (1990) Site-directed mutagenisis of glutamine residue of calmodulin. J. Biol. Chem. 265:5926-5929

Pannbacker RG, Fleischman DE, Reed RW (1972) Cyclic nucleotides phosphodiesterase: high activity in mammalian photoreceptors. Science 175:757-758

Preston RR, Saimi Y (1990) Calcium ions and the regulation of motility in Paramecium. Ciliary and Flagellar Membranes:173-200

Schoen CD, Arents JC, Bruin T, Van Driel R (1989) Intracellular localization of secretable cAMP in relaying Dictyostelium discoideum cells. Exp. Cell Res. 181:51-62

Schlicher DJ, Casnellie JE, Greengard P (1978) An endogenous substrate for cGMP-dependent protein kinase in mammalian cerebellum. Nature (London) 273:61-62

Sing S, Lowe DG, Thorpe DS, Rodriguez H, Kuang WJ, Dangott LJ, Chinkers M. (1988) Membrane guanylate cyclase is a cell surface receptor with homology to protein kinases. Nature (London) 334:708-710

Sugden D, Klein D (1987) A cholera toxin substrate regulates cyclic GMP content of rat pinealocytes. J. Biol. Chem. 262:7447-7450

Sun TJ, Van Haastert PJM, Devreotes PM (1990) Surface cAMP receptors mediate multiple responses during development in Dictyostelium discoideum: evidence by antisense mutagenesis. J. Cell Biol. 110:1549-1554

Van Haastert PJM, Van Lookeren Campagne MM, Ross FM (1982) Altered cGMP phosphodiesterase activity in chemotactic mutants of Dictyostelium discoideum. FEBS letters 147:149-152

Walter U (1981) Distribution of cGMP-dependent protein kinase in various rat tissues and cell lines determined by a sensitive and specific radioimmune assay. Europ. J. Biochem. 118:339-346

Wanner R, Wurster B (1990) Cyclic GMP-Activated Protein Kinase from Dictyostelium discoideum. Biochim. Biophys. Acta 1053:179-184

Wurster B, Schubiger K, Wick U, Gerisch G (1977) Cyclic GMP in Dictyostelium discoideum. FEBS Lett. 76:141-144

Yamazaki A, Uchida S, Stein PJ, Wheeler GL, Bitensky MW (1984) Enzyme regulation and GTP-binding protein: An algorithm of control that includes physical displacement of an inhibitory protein. Adv. Cyclic Nucleotide and Protein Phosphorylation Res.:16,381-392

Ziegelberger G, Vandenberg MJ, Kaissling KE, Klumpp S, Schultz JE (1990) Cyclic GMP levels and guanylate cyclase activity in pheromone-sensitive antennae of the silkmoths Antheraea polyphemus and Bombyx mori. J. of Neurosci. 10:1217-1225

EPIDERMAL GROWTH FACTOR MEDIATED SIGNAL TRANSDUCTION IN RAT HEPATOCYTES

Lijun Yang, György Baffy, Sue Goo Rhee[†] and John R. Williamson

Dept. of Biochemistry and Biophysics
University of Pennsylvania
Philadelphia, PA 19104 U.S.A.

[†]Laboratory of Biochemistry, NHLBI Institute
NIH, Bethesda, MD 20892, U.S.A.

SUMMARY

Treatment of rat hepatocytes with epidermal growth factor (EGF) produced an enhanced tyrosine phosphorylation of the EGF receptor and phospholipase C-γ (PLC-γ) in conjunction with the mobilization of Ca^{2+}. Approximately 30% of the total PLC-γ was tyrosine phosphorylated with a maximum being reached after 30s of incubation with EGF. Pretreatment of the rats with pertussis toxin prior to isolation of the hepatocytes blocked EGF-induced tyrosine phosphorylation of PLC-γ and Ca^{2+} mobilization but had no effect on autophosphorylation of the EGF receptor or Ca^{2+} responses elicited by angiotensin II or phenylephrine. Under these conditions Gi-protein α subunits were fully ADP-ribosylated. Antibodies specific to PLC-γ coimmunoprecipitated phosphorylated proteins with molecular masses of 190kDa, 96kDa, 47kDa and 25kDa but not the EGF receptor. A 41kDa Gi-protein α subunit was found to be present in the anti-PLC-γ immune complex as shown by *in vitro* ADP-ribosylation using [^{32}P]NAD$^+$ and activated pertussis toxin. Antibodies specific for the EGF receptor also coimmunoprecipitated a Gi-protein α subunit. Treatment of hepatocytes with EGF caused a decrease in the amount of Gi-protein α subunit associated with the EGF receptor and an increase in its association with PLC-γ. These results indicate that a pertussis toxin-sensitive Gi-protein is intimately involved in the signal transduction pathway mediating EGF-induced activation of PLC-γ and Ca^{2+} mobilization in hepatocytes.

INTRODUCTION

Epidermal growth factor (EGF) causes an enhanced breakdown of phosphatidylinositol 4,5-bisphosphate (PIP_2) in a variety of cell types with the production of Ins 1,4,5-P_3 and 1,2-diacylglycerol. These compounds are generated by activation of phospholipase C and serve as second messengers for Ca^{2+} mobilization and activation of protein kinase C, respectively (2,35,38). A number of recent studies have demonstrated that a likely mechanism for this effect is by phosphorylation of phospholipase C-γ (PLC-γ), mediated by the tyrosine kinase activity of the EGF receptor (20,22,36,37). PLC-γ is also selectively phosphorylated by the purified EGF

NATO ASI Series, Vol. H 52
Biological Signal Transduction
Edited by E. M. Ross and K. W. A. Wirtz

receptor (25,26,36) but studies with purified PLC-γ *in vitro* have failed to show an enhanced catalytic activity of the enzyme upon tyrosine phosphorylation (15). These results have prompted the suggestion that activation of PLC-γ by growth factors is modulated by other proteins (29). Potential candidates are proteins that have been shown to co-immunoprecipitate with PLC-γ by anti-PLC-γ antibodies in studies with 3T3 fibroblasts and A-431 cells (20,22,37).

In liver, a G-protein appears to be involved in coupling the EGF receptor to activation of phospholipase C since EGF-mediated increases of inositol phosphates and Ca^{2+} mobilization are inhibited by pertussis toxin (12,13). A pertussis toxin-sensitive G-protein involvement has also been reported for EGF-mediated growth effects in a human breast cancer cell line (6) and EGF-induced stimulation of phosphoinositide metabolism in renal epithelial cells (32). In murine epithelial A-431 cells, smooth muscle cells, liver epithelial cells and fibroblasts, however, effects of growth factors on phosphoinositide breakdown and inositol phosphate production were not attenuated by pertussis toxin (5,10,11,18,27,31), although one study with vascular smooth muscle cells has provided evidence in favor of a pertussis toxin-insensitive G-protein coupling to PDGF receptors (11). It thus appears that different cell types may utilize different mechanisms for coupling the EGF receptor to activation of phospholipase C.

The purpose of the present experiments was to investigate the involvement of a pertussis toxin-insensitive G protein in EGF-induced phosphorylation of PLC-γ. We demonstrate a rapid and transient EGF-stimulated increase in PLC-γ tyrosine phosphorylation, which together with Ca^{2+} mobilization is blocked by pertussis toxin pretreatment. The formation of an EGF-dependent complex between PLC-γ and a G-protein α subunit belonging to the Gi family was revealed by the ability of activated pertussis toxin to promote ADP-ribosylation of a G-protein in the immunoprecipitate formed with anti-PLC-γ antibodies. In addition, a complex between the EGF receptor and a Gi-protein is demonstrated that was dissociated by receptor occupancy.

EXPERIMENTAL PROCEDURES

Materials - Anti-phosphotyrosine IG_2 hybridoma cell line was obtained from American Type Culture Collection. Monoclonal anti-phosphotyrosine antibodies were produced from BALB/c mouse ascites, purified using an affinity column of phosphotyrosyl-Sepharose 4B and were subsequently covalently coupled to 15 mg/ml CNBr-activated Sepharose 4B. Anti-rat EGF receptor antibodies were produced from the hybridoma cell line 151-8AE obtained from Dr. A Hubbard, Johns Hopkins Univ., and were linked to Sepharose 4B. Monoclonal antibodies to PLC-γ were prepared as previously described (30). [^{32}P]-orthophosphate, ^{125}I-goat anti-mouse or rabbit IgG was purchased from ICN Biomedical. Mouse epidermal growth factor (EGF) was from Upstate Biotechnology, Inc. Protein A-Sepharose Cl-4B was from Sigma. Pertussis toxin (PT) was either purchased from List Biological Research,Inc. or was kindly provided by Dr. Eric L. Hewlett, University of Virginia. Giα-general polyclonal anti-serum (8730) was a gift from Dr. David Manning, University of Pennsylvania. Triton X-100 was from Sigma and goat anti-mouse-alkaline phosphatase was from Promega.

Isolated hepatocytes and pertussis toxin (PT) treatment - Hepatocytes were prepared from the livers of normal or PT-treated, 150-250g, fed, male Sprague-Dawley rats basically according to previously published procedures (21). To obtain PT-treated hepatocytes, intact rats were given an intraperitoneal injection of 50 μg of toxin per 100g of body weight, 60-72h prior to isolation of hepatocytes (28). The hepatocytes were washed twice with Hank's buffer and resuspended in L-15 medium. Cell viability was determined by Trypan blue exclusion and only preparations with greater than 90% viability were used.

Cell labelling, treatment, lysis, immunoprecipitation and Western blotting - For labelling proteins with ^{32}P, isolated hepatocytes were washed twice with phosphate-free Dulbecco's modified Eagle medium (DMEM) then resuspended at a concentration of 10mg dry weight/ml and incubated with 0.5mCi/ml of [^{32}P]-orthophosphate-free DMEM plus 0.5% dialysed calf serum for 90min at 37°C, in an atmosphere of 95% O_2/5% CO_2.

Hepatocytes were treated with 200g/ml EGF at 37°C for the indicated times, immediately washed twice with a 10-fold volume of ice-cold Ca^{2+}-Mg^{2+}-free phosphate buffered saline and lysed with cold solubilization buffer containing 1% Triton X-100, 10% glycerol, 50mM Hepes (pH 7.5), 150 mM NaCl, 1 mM EGTA, 1 mM Na orthovanadate, 100mM NaF and 10 mM Na pyrophosphate with 10 μg/ml aprotinin, 10μg/ml leupeptin and 1mM PMSF added just before use (20). Lysates were incubated on ice for 5 min and were clarified by centrifugation at 27,000 x g for 30min at 4°C. The proteins of interest were immunoprecipitated from clarified lysates by addition of antibody and 100 μl of 20% protein A-sepharose followed by overnight incubation at 4°C. Immune complexes were washed three times with solubilization buffer when anti-phosphotyrosine, anti-EGF receptor and anti-Giα antibodies were used. Anti-PLC-γ antibody immune complexes were washed five times with 1% Triton X-100, 1% deoxycholate, 0.1% SDS, 150mM NaCl, and 50mM Tris, pH 8.5 (17). Immunoprecipitated proteins were released by the addition of 50μl 2x SDS sample buffer followed by boiling for 5min and separated by electrophoresis in either 7.5% or 10% SDS-polyacrylamide gel (SDS-PAGE). The proteins were transferred to nitrocellulose paper and visualized by autoradiography or Western blotting.

For Western blotting, the nitrocellulose paper was blocked with 3% BSA or 3% goat serum in Tris buffered saline, pH 7.2 for 1h at room temperature, then incubated with the appropriate primary antibody. Blots are developed with either alkaline phosphatase-linked secondary antibodies and visualized with 5-bromo-4 indolyl phosphate/nitroblue tetrazolium as chomogenic substrates or with [^{125}I] goat anti-mouse IgG.

ADP-ribosylation of Giα and PLC-γ immunoprecipitates - Untreated or EGF-treated hepatocytes were lysed with 1% Triton X-100 solubilization buffer and immunoprecipitated using Giα polyclonal and PLC-γ monoclonal antibodies as above described. The immune complexes were washed twice with 1ml solubilization buffer. The pellet was incubated with [^{32}P]NAD (20,000 cpm/pmol) and preactivated pertussis toxin (final concentration 20μg/ml) for 90min at 30°C (4). The reaction was stopped by the addition of 20μl 4x SDS sample buffer and boiled for 5min. The immune complexes were electrophoresed on 10% SDS-PAGE, transferred to nitrocellulose paper and visualized by radioautography. The bands of [α-^{32}P] ADP-ribosylated protein were excised and counted in a liquid scintillation spectrometer.

Ca^{2+} measurement - Hepatocytes were resuspended at 2 x 10^5 cells/ml in L-15 medium supplemented with 20mM Hepes and 5.5mM glucose, plated on glass coverslips coated with poly-L-lysine (5μg/cm^2) and incubated for 1h at 37°C with 5μM fura-2/AM (24). The cytosolic Ca^{2+} concentration was determined by measuring the fluorescence of fura-2-loaded cells using a 100 X Nikon Fluor objective and a Nikon Diaphot epifluorescent microscope illuminated with alternating (100 Hz) 340nm and 380nm excitation light with fluorescence emission collected at 520nm (50nm half band width) by a photomultiplier tube. Output from the fluorometer was digitized, ratioed and analyzed by an AT-based analysis system (Indec Systems). Calibration of the fluorescent signal and calculation of cytosolic free Ca^{2+} was performed by the method of Grynkiewicz *et al* (9).

RESULTS

Effect of EGF on tyrosine phosphorylation of PLC-γ

Isolated rat hepatocytes (10 mg/ml) were labelled at 37°C for 90 min with [^{32}P] inorganic phosphate (0.5 mCi/ml) in phosphate-free DMEM, and incubated with EGF (200 ng/ml) for 30 sec. Fig. 1 shows the results of an experiment in which cellular proteins were denatured by extraction with solubilization buffer plus 1% SDS and boiled for 5 min. The extracts were diluted to a final concentration of 0.2% SDS and tyrosine phosphorylated proteins were immunoprecipitated with anti-phosphotyrosine antibody matrix, separated by SDS-PAGE, transferred to nitrocellulose paper and visualized by radioautography. The results in control cells illustrate that hepatocytes normally contain a large number of tyrosine phosphorylated proteins. EGF increased the tyrosine phosphorylation of bands with molecular weights of 170 kDa, 148 kDa and several proteins of lower molecular weight. The 148 Kda protein corresponded to PLC-γ, as shown by Western blotting with anti-PLC-γ antibodies. The 170 Kda tyrosine phosphoprotein is presumed to be EGF-receptor since its molecular weight is in the appropriate range and its tyrosine phosphorylation occurred specifically in response to EGF.

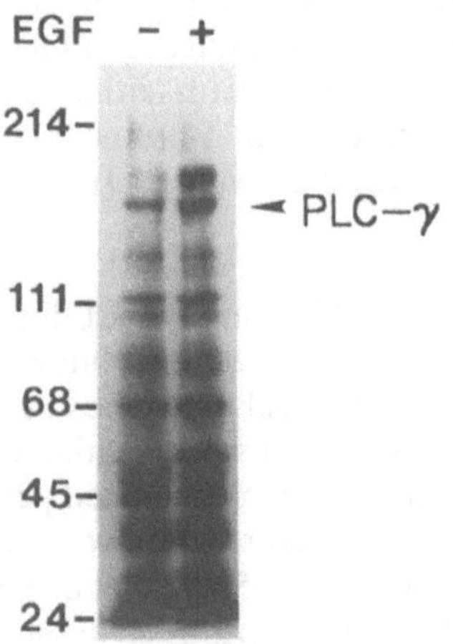

Fig. 1 EGF-induced tyrosine phosphorylation of PLC-γ in rat hepatocytes.
Isolated hepatocytes (10 mg/ml) were labelled at 37°C with [^{32}P]-orthophosphate for 90 min in phosphate-free DMEM. Cells were treated without (-) or with (+) EGF 200 ng/ml for 30s at 37°C, lysed and boiled with solubilization buffer plus 1% SDS for 5min. The lysates were diluted to a final concentration of 0.2% SDS and immunoprecipitated with anti-phosphotyrosine matrix. The immunoprecipitated tyrosine phosphoproteins were separated by 7.5% SDS-PAGE, transferred to nitrocellulose, and exposed at -70°C for 24 hr for radioautography. Each lane represents material from 5 mg dry weight of hepatocytes.

The kinetics of tyrosine phosphorylation of PLC-γ in hepatocytes were determined by measuring the amount of ^{32}P incorporated into PLC-γ different times after addition of EGF (Fig. 2). After extracting the proteins with solubilization buffer, phosphotyrosine proteins were immunoprecipitated with anti-phosphotyrosine antibodies and analyzed by SDS-PAGE and radioautography as described in Experimental Procedures. The bands corresponding to PLC-γ as identified by Western blotting (see inset) were excised and counted. Fig. 2 shows that PLC-γ was rapidly and transiently tyrosine phosphorylated with the peak of the phosphorylation occurring within 30s of EGF addition.

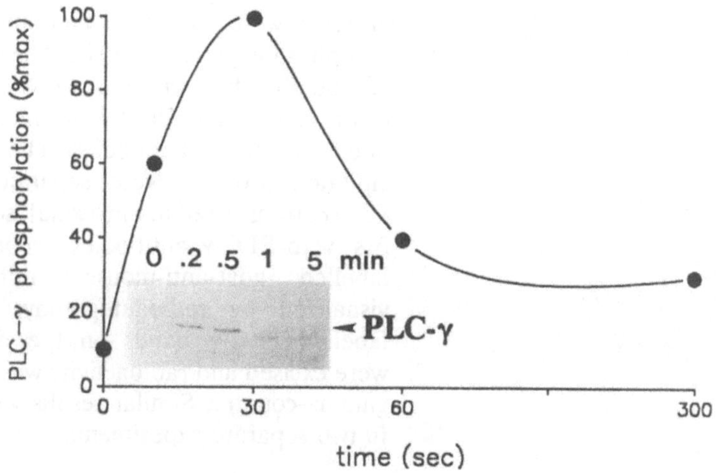

Fig. 2 Kinetics of tyrosine phosphorylation of PLC-γ in rat hepatocytes. [^{32}P]-labelled hepatocytes were treated for different times with EGF (200ng/ml), quenched with ice-cold PBS, extracted with solubilization buffer and the phosphoproteins were immunoprecipitated with anti-phosphotyrosine matrix. The bands on the nitrocellulose corresponding to PLC-γ were excised and quantitated by Cerenkov counting. Values are expressed as a percentage of the value at 30s of EGF treatment which was taken as the maximal response. The inset shows the bands of PLC-γ recognized by PLC-γ Western blotting after anti-phosphotyrosine immunoprecipitation. Similar results were obtained in three experiments.

Studies with fibroblasts (20,22) and the carcinoma-derived cell line HSC-1 (36) have shown that approximately 70% of the total PLC-γ becomes tyrosine phosphorylated within 2min of EGF addition. In order to estimate the proportion of PLC-γ that is tyrosine phosphorylated by EGF in hepatocytes, a two-step immunoprecipitation method was employed. Hepatocytes were treated with 200ng/ml of EGF for 30s at 37°C. Following extraction of the proteins with solubilization buffer, anti-phosphotyrosine antibodies were added to immunoprecipitate tyrosine phosphorylated PLC-γ and anti-PLC-γ antibodies were then added to the supernatants to immunoprecipitate the remaining PLC-γ. Proteins in the PLC-γ immunoprecipitate were separated by SDS-PAGE, transferred to nitrocellulose and Western blotted with anti-PLC-γ antibody. The amount of PLC-γ was quantitated by using [^{125}I] goat anti-mouse secondary antibody followed by radioautography, excision of the band and counting in a gamma spectrometer. Fig. 3 shows the results of the radioautograph, which illustrates the EGF-induced loss of PLC-γ from the supernatant. The counts in the band decreased from 7,187 cpm to 5,047 cpm, indicating that 30% of the total PLC-γ was tyrosine phosphorylated 30s after EGF addition.

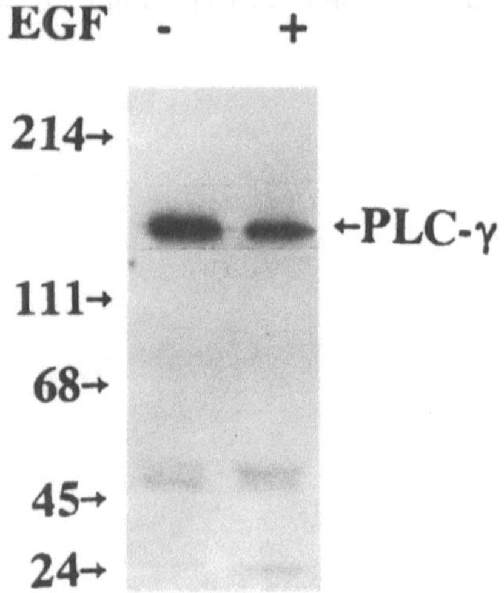

Fig. 3 Proportion of total PLC-γ that is tyrosine phosphorylated by EGF in rat hepatocytes
Hepatocytes were treated without (-) or with (+) EGF (200ng/ml) for 30s at 37°C. Cellular proteins were extracted and absorbed onto anti-phosphotyrosine matrix. The supernatants of non-absorbed proteins were separated by centrifugation and PLC-γ was immunoprecipitated with PLC-γ antibodies. The immunoprecipitated proteins were separated by SDS-PAGE, transferred to nitrocellulose and blotted first with PLC-γ antibodies, second with ^{125}I-labelled goat-anti-mouse antibodies and visualized by radioautography. The ^{125}I-labelled PLC-γ bands on the nitrocellulose were excised and radioactivity was assayed in a gamma-counter. Similar results were obtained in two separate experiments.

Complex of PLC-γ with other phosphoproteins

In order to investigate whether PLC-γ forms complexes with the EGF receptor or other cellular phosphoproteins as observed in A-431 cells (20,37) and WB cells (10), hepatocytes were labelled with [^{32}P]orthophosphate, treated with EGF (200 ng/ml) for different times up to 30min and PLC-γ was immunoprecipitated with anti-PLC-γ antibodies. Fig. 4 shows that a number of proteins exhibiting enhanced phosphorylation after EGF addition were co-immunoprecipitated with PLC-γ. The molecular weights of these proteins were 190kDa, 148kDa, 96kDa, 47kDa and 25kDa. The band at 148kDa was identified as PLC-γ by Western blotting with anti-PLC-γ antibodies. The 190kDa band was not phosphorylated in control cells and showed a transiently increased phosphorylation with EGF, reaching a peak after 30s. The intensities of the phosphoprotein bands at 96kDa, 47kDa and 25kDa also increased after EGF addition, with that at 96kDa being the most prominent. Radioactivity in the band below the 24,000 molecular weight marker did not transfer to nitrocellulose paper, suggesting that it may be a phospholipid. The phosphorylation of PLC-γ was rapid, but unlike the transient kinetics observed for its phosphorylation on tyrosine residues (Fig. 2), Fig. 4 shows that the ^{32}P content of PLC-γ was increased for up to 30min. This suggests that PLC-γ is also phosphorylated on serine residues as observed in A-431 cells (37). Under the conditions used for immunoprecipitation with anti-PLC-γ antibodies, an association of the EGF receptor with PLC-γ was not observed.

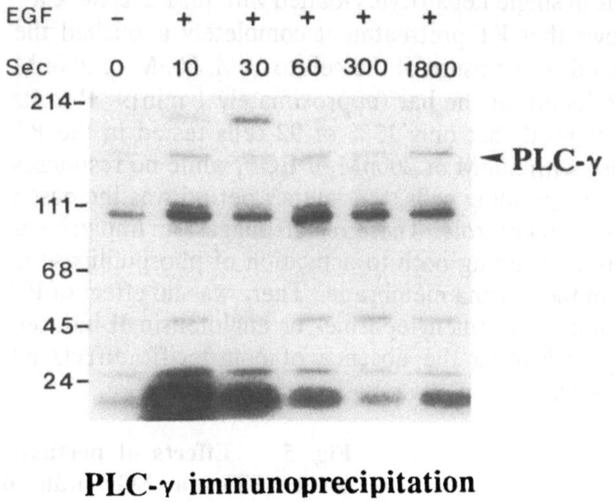

Fig. 4 Association between PLC-γ and phosphoproteins after immunoprecipitation with anti-PLC-γ antibodies. [^{32}P]-labelled hepatocytes were treated with EGF (200ng/ml) for different times at 37°C. The cells were lysed and immunoprecipitated with PLC-γ antibodies. Proteins were separated by SDS-PAGE and visualized by radioautography.

Effects of pertussis toxin on Ca^{2+} mobilization and EGF-mediated tyrosine phosphorylation of PLC-γ

Measurement of the EGF concentration-dependency of Ca^{2+} changes in control hepatocytes (Fig. 5A) showed that at low EGF concentrations there was a long latency period after the initiation of EGF perfusion before an increase in Ca^{2+} was observed. With 0.5nM EGF this extended beyond the termination of the EGF pulse. With increasing EGF concentrations, the latency period decreased and the rise of Ca^{2+} became greater. The rate of increase of Ca^{2+}, however, was relatively independent of the EGF concentration. An evaluation of responses obtained from 51 cells in the control group showed that 25% of the cells were unresponsive to EGF. With 1nM EGF, the mean latency period was 42±5s (n=15), which decreased to 20±3s (n=10) with 50μM EGF and to 18±2s (n=16) with 200nM EGF. Peak Ca^{2+} increased from a mean value of 110±8nM to 151±16nM, 310±30nM and 336±15nM with 1nM, 50nM and 200nM of EGF, respectively. This initial rapid increase of Ca^{2+} was also observed when the extracellular Ca^{2+} concentration was diminished to about 1μM by addition of 3mM EGTA and can be attributed to the release of Ca^{2+} from Ins 1,4,5-P_3-sensitive Ca^{2+} stores. On the other hand, the steady state level of increased Ca^{2+}, or plateau phase, was abolished by removal of extracellular Ca^{2+}, indicating that it originated from augmented Ca^{2+} influx.

Previous studies by Johnson *et al* (12) and Johnson and Garrison (13) with rat hepatocytes have shown that EGF-mediated increases of inositol phosphate formation and Ca^{2+} mobilization were inhibited when G-proteins of the Gi family were ADP-ribosylated by pretreatment of the rat with pertussis toxin. A similar protocol was followed in the present experiments, since it is difficult to accomplish a complete ADP-ribosylation of Gi-proteins by addition of pertussis toxin to isolated hepatocytes. Pretreatment of the rats for 72h with 50μg of toxin/100g of body weight by intraperitoneal injection caused a 100% modification of the hepatic Gi-proteins, as shown by the lack of any further ADP-ribosylation of membrane proteins upon incubation of crude membranes with [^{32}P]NAD^+ and activated pertussis toxin (data not shown). The effects of PT on EGF-mediated changes of cytosolic free Ca^{2+} were investigated by making observations in single hepatocytes loaded with fura-2 as the Ca^{2+} indicator. Fig. 5 B,C and D shows that PT pretreatment completely abolished the increase of cytosolic free Ca^{2+} elicited by exposure of the cell to 1nM, 50nM or 200nM EGF for the time indicated by the length of the bar (approximately 1 min). Results from three different experiments showed that only 13% of 92 cells tested in the PT pretreated group showed a response with 50nM or 200nM of EGF, while no responses were obtained with 1nM EGF. In responding cells, the latency period was increased and the peak Ca^{2+} diminished relative to controls. These results suggest an impairment by PT pretreatment of EGF receptor coupling both to activation of phospholipase C and to activation of Ca^{2+} channels in the plasma membrane. There was no effect of PT on the basal free Ca^{2+} concentration or on phenylephrine- or angiotensin II-induced Ca^{2+} responses (data not shown), indicating the absence of nonspecific effects on cellular Ca^{2+} homeostatic mechanisms.

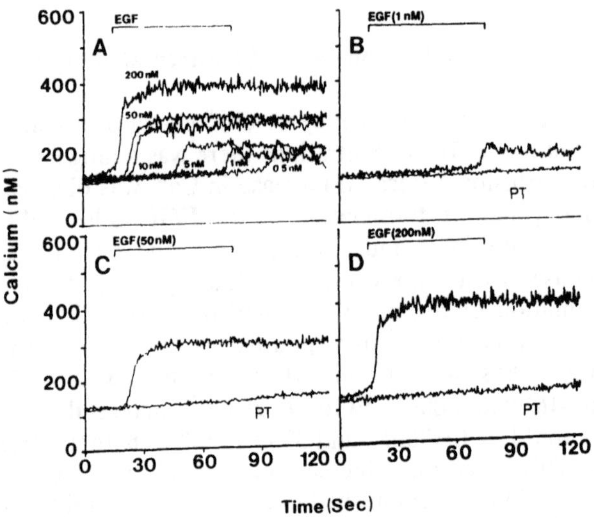

Fig. 5 Effects of pertussis toxin (PT) on EGF-induced cytosolic free Ca^{2+} changes in single rat hepatocytes
Hepatocytes were loaded with 5μM fura-2/AM for 45min. The time of EGF microperfusion was 60s (started at 10s). Ca^{2+} transients induced by EGF in hepatocytes from rats pretreated with PT for 72h (50μg/100g b.w.) are compared with hepatocytes from normal rats. EGF concentration, 1nm (B), 50nM (C) and 200nM (D). The dose response of Ca^{2+} transients in control hepatocytes is shown in (A). Each trace shown in Fig. 5 is representative of results from 15-20 individual cells.

In order to determine whether PT pretreatment affected EGF-mediated phosphorylation of PLC-γ, extracts from [^{32}P]-labelled normal or PT pretreated hepatocytes were used for immunoprecipitation with anti-phosphotyrosine or anti-PLC-γ antibodies, and the proteins were separated by SDS-PAGE as described in Experimental Procedures. The tyrosine phosphoproteins and phosphorylated PLC-γ were visualized by radioautography after transfer to nitrocellulose paper (Fig. 6). Fig. 6A and B show results using anti-phosphotyrosine antibodies. The band corresponding to PLC-γ was identified by Western blotting with anti-PLC-γ antibodies. In the absence of PT, there were two proteins of higher molecular weight, namely, 170kDa and 190kDa, that showed enhanced tyrosine phosphorylation with EGF. Tyrosine phosphorylation of the protein at 190kDa was not observed in hepatocytes pretreated with PT (Fig. 6B) or when the samples were denatured prior to addition of the antibody (Fig. 1). Pertussis toxin did not block tyrosine phosphorylation of the 170 kDa protein presumed to be the EGF receptor, but greatly inhibited tyrosine phosphorylation of PLC-γ (Fig. 6B). This finding was confirmed using anti-PLC-γ antibodies for immunoprecipitation, which showed that the EGF-induced enhanced phosphorylation of PLC-γ was blocked by PT pretreatment (Figs. 6C and 6D).

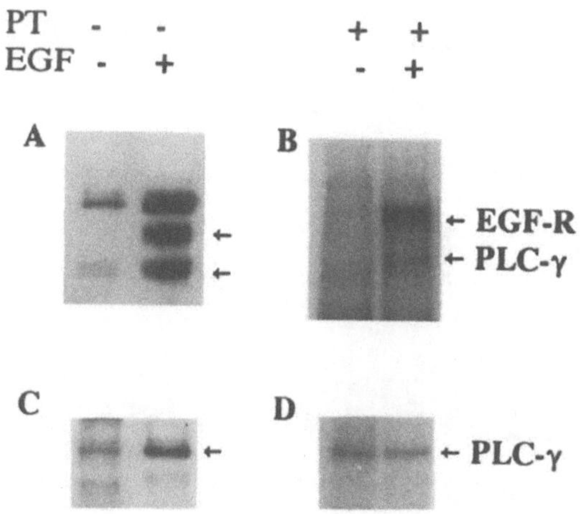

Fig. 6 Effects of pertussis toxin (PT) on EGF-induced phosphorylation of PLC-γ and EGF-receptor autophosphorylation in rat hepatocytes
[^{32}P]-labelled hepatocytes from normal (Fig. 6 A and C) or PT pretreated rats (Fig. 6 B and D) were incubated without (-) or with (+) EGF (200 ng/ml) for 30s at 37°C. Cell extracts were immunoprecipitated with anti-phosphotyrosine antibodies (Fig. 6 A and C), or anti-PLC-γ monoclonal antibodies (Fig. 6 C and D). The immunoprecipitates were analyzed by SDS-PAGE and transferred to nitrocellulose for radioautography. Similar results were obtained in three separate experiments.

Association of a Gi-protein α subunit with PLC-γ and the EGF receptor

The above data suggest that a PT-sensitive G-protein may mediate signal transduction from the EGF receptor to PLC-γ in hepatocytes. If PLC-γ serves as the effector enzyme for the G-protein, it is to be expected that a complex would be formed between PLC-γ and the PT-sensitive G-protein α subunit. In order to determine whether such a complex could be demonstrated, hepatocytes were incubated for 3min with or without EGF (200ng/ml), extracted with solubilization buffer and the proteins of interest were immunoprecipitated using anti-PLC-γ antibodies or an anti-peptide antibody (8730) to a carboxy-terminal sequence common to the Gi-protein family (KNNLKDCGLF). The immunoprecipitated proteins were incubated in buffer containing [^{32}P]NAD$^+$ and activated PT for *in vitro* ADP-ribosylation, followed by SDS-PAGE and visualization by radioautography after transfer to nitrocellulose. Fig. 7, lanes 1 and 2 shows total amounts of Gi-protein α subunits immunoprecipitated by Giα antibody in the absence and presence, respectively, of EGF. Fig. 7, lanes 3 and 4 shows the presence of Gi-protein α subunits in the PLC-γ immunoprecipitate. The greater intensity of the band in lane 4 compared with lane 3 illustrates that the complex between PLC-γ and the αi subunits was promoted by EGF. Quantitation of the radioactivity by excising and counting the 41 kDa bands showed that the association of Gi α subunits with PLC-γ was increased 56% by EGF (lane 4 versus lane 3). In the presence of EGF, approximately 30% of the Giα subunits were complexed with PLC-γ (lane 4 versus lane 2). Further experiments showed that the association of the Giα-subunit with PLC-γ was maximal 1min after EGF addition. PLC-γ was also detected in the Giα immunoprecipitate as revealed by Western blot analysis with anti-PLC-γ antibodies (data not shown). The fact that anti-PLC-γ antibodies are able to co-immunoprecipitate a significant part of the total Gi α subunit in hepatocytes strongly suggests that the two proteins become physically and functionally associated in an EGF-dependent manner.

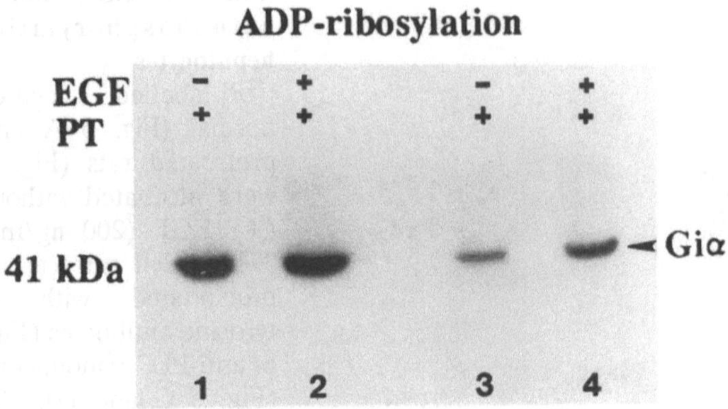

Fig. 7 Coimmunoprecipitation of a PT-sensitive Gi-protein α-subunit with PLC-γ in rat hepatocytes
Anti-Giα (lanes 1 and 2) and anti-PLC-γ (lanes 3 and 4) immunoprecipitates were prepared from hepatocytes, 5mg dry weight per lane without (+) or with (-) EGF (200 ng/ml). The immune complexes were gently washed twice with extraction buffer, then were reacted for 90min with [^{32}P]-NAD$^+$ and activated pertussis toxin for *in vitro* ADP-ribosylation. The exposure time was 24 hr at -70°C using XAR film with an intensifying screen. [^{32}P]-labelled PT-sensitive Giα protein bands are indicated with an arrow. These bands were excised and radioactivity was counted in a scintillation counter. The results shown are representative of three experiments.

The possibility of an association between the Gi-protein and the EGF receptor was investigated by use of anti-EGF receptor antibodies coupled to Sepharose 4B. The immunoprecipitate was subjected to *in vitro* ADP-ribosylation, as above, and the proteins were separated by SDS-PAGE, transferred to nitrocellulose and radioactive bands were visualized by radioautography. Fig. 8 shows that a band at 41 kDa was present in control cells, which decreased within 1 min after addition of EGF. This contrasted with several other radioactive proteins that were apparently non-specifically ADP-ribosylated and showed no change with EGF. Parallel experiments using Giα antibodies for immunoprecipitation followed by *in vitro* ADP-ribosylation identified the 41 kDA band as a Giα subunit. These results show that in control cells part of the Gi-protein is associated with the EGF receptor and that receptor occupancy promotes its dissociation.

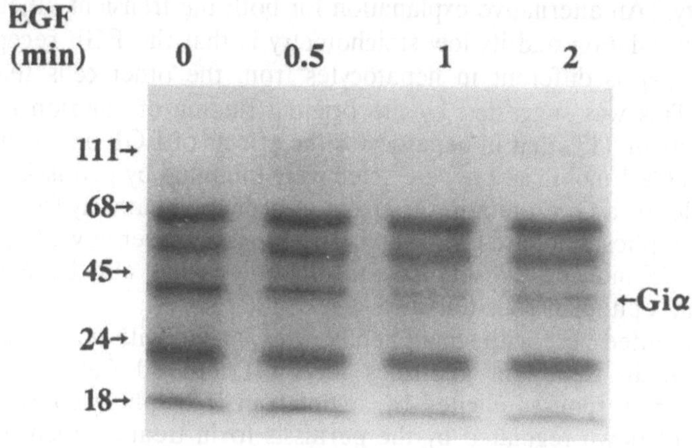

Fig. 8 The presence of a Giα subunit in the EGF receptor immunoprecipitate in rat hepatocytes
Anti-EGF-receptor immunoprecipitates were prepared from hepatocytes (5mg dry weight per lane) treated with EGF (200ng/ml) for different times. Immune complexes were treated as described in Fig. 7. The 41 kDa [^{32}P]-labelled PT-sensitive Giα-protein is indicated by an arrow.

DISCUSSION

The results presented here demonstrate that in isolated rat hepatocytes EGF induces a rapid tyrosine phosphorylation of PLC-γ. This observation is in accordance with a number of studies that have demonstrated an enhanced EGF-mediated tyrosine phosphorylation of PLC-γ in 3T3 fibroblasts (19,22), A-431 cells (20,37), human squamous carcinoma-derived HSC-1 cells (36) and liver epithelial WB cells (10). The results with hepatocytes, however, differ in several respects from EGF-mediated responses reported in other cell types. Notably, PLC-γ tyrosine phosphorylation is transient in hepatocytes, reaching a peak within 30s and declining thereafter to values only slightly greater than prestimulated levels. It is unlikely that this finding is caused by an inadequate inhibition of tyrosine phosphate phosphatases in the solubilization/extraction buffer since a similar biphasic increase of PLC-γ tyrosine phosphorylation was also observed when the cell proteins were rapidly denatured by boiling in 1% SDS prior to immunoprecipitation. An enhanced EGF-mediated incorporation of ^{32}P into PLC-γ was sustained for 30min, suggesting that serine residues on PLC-γ also become phosphorylated after EGF stimulation of hepatocytes, as observed after EGF treatment of A-431 cells (22,23,25,37). Purified EGF receptor, however, is only able to phosphorylate PLC-γ on tyrosine residues (16,25), indicating that a secondary activation of a serine/threonine kinase is responsible for the prolonged phosphorylation of PLC-γ.

A second difference relates to the relatively small amount of total PLC-γ that is tyrosine phosphorylated in hepatocytes (approximately 30%), compared with almost 70% observed in other cell types (20,22,36). Cells exhibiting a high degree of PLC-γ tyrosine phosphorylation also express high levels of the EGF receptor, hence the lower degree of PLC-γ tyrosine phosphorylation in hepatocytes may reflect a lower EGF receptor density. An alternative explanation for both the transient kinetics of PLC-γ tyrosine phosphorylation and its low stoichometry is that the EGF receptor coupling pathway to PLC-γ is different in hepatocytes from the other cells that have been investigated. This was suggested by the original finding of Johnson *et al* (12) and Johnson & Garrison (13) that in hepatocytes the effects of EGF on inositol phosphate production and Ca^{2+} mobilization were selectively inhibited by pertussis toxin, thereby indicating a role for a Gi protein in the signal transduction pathway from the activated EGF receptor to phospholipase C. A similar pertussis toxin sensitive, G-protein-linked pathway for EGF-induced phosphoinositide hydrolysis has also been demonstrated recently in renal epithelial cells (32).

Studies presented here with single hepatocytes loaded with fura-2 have confirmed that pertussis toxin treatment abolishes the EGF-induced Ca^{2+} transient without affecting the Ca^{2+} responses elicited by angiotensin II or phenylephrine. The EGF receptor was not down regulated by the pertussis toxin treatment since EGF caused receptor autophosphorylation. However, pertussis toxin completely blocked the tyrosine phosphorylation of PLC-γ, suggesting that a causal relationship exists between tyrosine phosphorylation of PLC-γ and the mobilization of Ca^{2+} elicited by the generation of Ins 1,4,5-P_3. Nevertheless, the relationship is not a simple one since phosphorylation of purified PLC-γ *in vitro* on tyrosine (16) or serine (17) residues does not alter the catalytic activity of the enzyme. These findings have led to the suggestion that tyrosine or serine phosphorylations alter the activity of PLC-γ in a manner dependent on other modulatory protein factors that were not present in the *in vitro* system (15,17). PLC-γ is a cytosolic enzyme, hence a key factor involved in the signal transduction response must involve a mechanism for membrane-association of the enzyme, thereby bringing it into proximity with its substrate PIP_2 in the plasma membrane (33). In hepatocytes, the modulatory factor appears to be a Gi-protein. Thus, in nonstimulated hepatocytes, a Gi-protein is apparently complexed with the EGF receptor, as evidenced by the fact that a 41kDa protein in the immunoprecipitate produced by anti-EGF receptor antibodies was ADP-ribosylated by [^{32}P]NAD^+ and activated pertussis toxin *in vitro*. Furthermore, this complex was dissociated upon EGF addition to the hepatocytes. EGF also enhanced the formation of a complex between a Gi-protein α subunit and PLC-γ.

It is evident, therefore, that when EGF binds to its receptor in hepatocytes, a number of processes are initiated. Firstly, PLC-γ binds to the activated EGF receptor via *Src* homology in SH_2 domains in order for it to become tyrosine phosphorylated (1) and secondly, a Giα subunit dissociates from the receptor and forms a complex with PLC-γ. These events are associated with an activation of the enzyme, which is evidenced by a rapid increase of cytosolic free Ca^{2+} with a delay time of about 20s at maximum concentrations of EGF. The relationship between tyrosine phosphorylation of PLC-γ and the onset of PIP_2 hydrolysis is presently not clear, but the simplest explanation is that tyrosine phosphorylation of the enzyme promotes its association with Gi-protein α-subunits that remain anchored in the plasma membrane by their attached myristoyl groups (14). Thus tyrosine phosphorylation of PLC-γ may be required for its translocation to the membrane rather than for an enhancement of its catalytic activity. This hypothesis is similar in principle to that proposed by Kim *et al* (16), who suggested that one of the proteins that are coimmunoprecipititated with PLC-γ by antibodies to PLC-γ may modulate the translocation of PLC-γ to the membrane. Recent studies, however, suggest that an activator protein is present in the PLC-γ immunoprecipitate from A-431 cells (26). Whether the Gi protein α subunit is responsible for activation of PLC-γ in hepatocytes or whether it merely serves as a mechanism for attachment of the enzyme to the plasma mambrane remains an open question. Further studies are also required to determine whether the α subunit dissociates from the βγ subunit complex of the heterotrimeric Gi-protein after EGF addition.

The present results demonstrate that the mechanism of EGF receptor coupling in hepatocytes is different from that employed by a number of other cell types. The most striking difference is the obligatory requirement for a Gi-protein in the coupling mechanism leading to activation of phospholipase C. Characteristic features of receptor-G-protein-phospholipase C coupling are the formation of a receptor-G-protein complex and a rapid production of inositol phosphates upon agonist binding (3,7). Johnson & Garrison (13) have shown that the EGF-mediated stimulation of Ins 1,4,5-P_3, although of smaller magnitude, was as rapid as that induced by angiotensin II stimulation of hepatocytes. This is in marked contrast to the much slower accumulation of inositol phosphates observed with EGF than angiotensin II in WB cells (10). The identification of an EGF receptor-Gi protein complex in hepatocytes in the present study is a novel finding. Interaction of the Gi-protein with the EGF receptor may be facilitated by the presence of an amphiphilic α-helical structure in the cytoplasmic domain of the EGF and other growth factor receptors that is also present and responsible for interaction with G-proteins in the wasp toxin mastoparan and in the amino terminal region of the third intracytoplasmic loop of the β_2-adrenergic receptor (8). It is clear, however, that in other cell types such as A-431 cells (27), 3T3 fibroblasts (34) and WB cells (10), the EGF receptor is not coupled to a Gi-like protein since inositol phosphate production is insensitive to inhibition by pertussis toxin. The possible involvement of a G-protein that is insensitive to bacterial toxins for growth factor-induced activation of phospholipase C is more difficult to demonstrate and relies on the ability of guanine nucleotides to stimulate an agonist-induced activation of phosphoinositide hydrolysis in broken cell suspensions. Thus, loss of PLC-γ from the cells prior to its tyrosine phosphorylation upon cell permeation may account for the failure of stable guanine nucleotide analogs to stimulate PDGF-induced phosphoinositide hydrolysis in fibroblasts (5) or EGF-induced activation of phospholipase C in WB cells (10) and NIH 3T3 cells (34). Alternatively, EGF may induce activation of PLC-γ by distinctly different mechanisms in different cell types.

The present study highlights the importance of a G-protein in the overall mechanism of EGF receptor coupling in hepatocytes and raises the possibility that different cell types may utilize other G-proteins (including small molecular weight non heterotrimeric G-proteins) to induce translocation of tyrosine phosphorylated PLC-γ to the membrane and enhance its catalytic activity upon activation of the EGF receptor. Tyrosine phosphorylation of PLC-γ thus becomes necessary but not sufficient for increased production of inositol phosphates after treatment of cells with EGF. The relative ease by which Gi-proteins can be detected and assayed makes hepatocytes an excellent system for further studies on the mechanism of EGF-induced activation of PLC-γ and the role of its phosphorylation in protein-protein interactions.

ACKNOWLEDGEMENTS
This work was supported by DK-15120 from the National Institutes of Health.

REFERENCES

1. Anderson, D. J., Koch, C. A., Grey, L., Ellis, C., Moran, M. & Pawson, T. (1990) *Science* **250**: 979-982.
2. Berridge, M. J. (1987) *Biochim. Biophys. Acta* **907**: 33-45.
3. Boyer, J. L., Hepler, J. R. & Harden, T. K. (1989) *TIPS* **10**: 360-364.
4. Carlson, K. E., Brass, L. F. & Manning, D. R. (1989) *J. Biol. Chem.* **264**: 13298-13305.
5. Cattaneo, M. G. & Vicentini, L. M. (1989) *Biochem. J.* **262**: 665-668.
6. Church, J. G. & Buick, R. N. (1988) *J. Biol. Chem.* **263**: 4242-4246.
7. Cockcroft, S. (1987) *Trends Biochem. Sci* **12**: 75-78.
8. Feinstein, D. L. & Larhammar, D. (1990) *FEBS Lett.* **272**: 7-11.
9. Grynkiewicz, G., Poenie, M. & Tsien, R. Y. (1985) *J. Biol. Chem.* **260**: 3440-3450.
10. Hepler, J. R., Jeffs, R. A., Huckle, W. R., Outlaw, H. E., Rhee, S. G., Earp, H. S. & Harden, T. K. (1990) *Biochem. J.* **270**: 337-344.
11. Huang, N., Wang, D. & Heppel, L. A. (1989) *Proc. Natl. Acad. Sci. USA* **86**: 7904-7908.
12. Johnson, R. M., Connelly, P., Sisk, R., Pobiner, B. F., Hewlett, E. L. & Garrison, J. C. (1986) *Proc. Natl. Acad. Sci. USA* **83**: 2032-2036.
13. Johnson, R. M. & Garrison, J. C. (1987) *J. Biol. Chem.* **262**: 17285-17293.
14. Jones, T. L., Simonds, W. F., Merendino, J. J. , Jr., Brann, M. R. & Spiegel, A. M. (1990) *Proc. Natl. Acad. Sci. USA* **87**: 568-572.
15. Kim, J. W., Sim, S. S., Kim, U. H., Nishibe, S., Wahl, M. I., Carpenter, G. & Rhee, S. G. (1990) *J. Biol. Chem.* **265**: 3940-3943.
16. Kim, U. H., Kim, H. S. & Rhee, S. G. (1990) *FEBS Lett.* **270**: 33-36.
17. Kim, U. H., Kim, J. W. & Rhee, S. G. (1989) *J. Biol. Chem.* **264**: 20167-20170.
18. Letterio, J. L., Coughlin, S. R. & Williams, L. T. (1986) *Science* **234**: 1117-1119.
19. Margolis, B. L., Bellot, F., Honegger, A. M., Ullrich, A., Schlessinger, J. & Zilberstein, A. (1990) *Mol. Cell. Biol.* **10**: 435-441.
20. Margolis, B. L., Rhee, S. G., Felder, S., Mervic, M., Lyall, R. M., Levitski, A., Ullrich, A., Zilberstein, A. & Schlessinger, J. (1989) *Cell* **57**: 1101-1107.
21. Meijer, A. J., Gimpel, J. A., Deleeuw, G. A., Tager, J. M. & Williamson, J. R. (1975) *J. Biol. Chem.* **250**: 7728-7738.
22. Meisenhelder, J., Suh, P. G., Rhee, S. G. & Hunter, T. (1989) *Cell* **57**: 1109-1122.
23. Mitsui, K. & Iwashita, S. (1990) *FEBS Lett.* **268**: 157-160.
24. Monck, J. R., Reynolds, E. E., Thomas, A. P. & Williamson, J. R. (1988) *J. Biol. Chem.* **263**: 4569-4575.
25. Nishibe, S., Wahl, M. I., Rhee, S. G. & Carpenter, G. (1989) *J. Biol. Chem.* **264**: 10335-10338.
26. Nishibe, S., Wahl, M. I., Wedegaertner, P. B., Kim, J. J., Rhee, S. G. & Carpenter, G. (1990) *Proc. Natl. Acad. Sci. USA* **87**: 424-428.
27. Pike, L. J. & Eakes, A. T. (1987) *J. Biol. Chem.* **262**: 1644-1651.
28. Pobiner, B. F., Hewlett, E. L. & Garrison, J. C. (1985) *J. Biol. Chem.* **260**: 16200-16209.
29. Rhee, S. G., Suh, P. G., Ryu, S. H. & Lee, S. Y. (1989) *Science* **244**: 546-550.
30. Suh, P. G., Ryu, S. H., Choi, W. C., Lee, K. Y. & Rhee, S. G. (1988) *J. Biol. Chem.* **263**: 14497-14504.

31. Taylor, C. W., Blakeley, D. M., Corps, A. N., Berridge, M. J. & Brown, K. D. (1988) *Biochem. J.* **249**: 917-920.
32. Teitelbaum, I., Strasheim, A. & Berl, T. (1990) *J. Clin. Invest.* **85**: 1044-1050.
33. Todderud, G., Wahl, M. I., Rhee, S. G. & Carpenter, G. (1990) *Science* **249**: 296-298.
34. Verheijden, G. F., Verlaan, I., Schlessinger, J. & Moolenaar, W. H. (1991) *Cell Regulation* **1**: 615-620.
35. Wahl, M. I., Nishibe, S. & Carpenter, G. (1989) *Cancer Cells* **1**: 101-107.
36. Wahl, M. I., Nishibe, S., Kim, J. W., Kim, H., Rhee, S. G. & Carpenter, G. (1990) *J. Biol. Chem.* **265**: 3944-3948.
37. Wahl, M. I., Nishibe, S., Suh, P. G., Rhee, S. G. & Carpenter, G. (1989) *Proc. Natl. Acad. Sci. USA* **86**: 1568-1572.
38. Whitman, M. & Cantley, L. C. (1988) *Biochim. Biophys. Acta* **948**: 327-344.

INDEX

NATO ASI Series H

Vol. 1: **Biology and Molecular Biology of Plant-Pathogen Interactions.** Edited by J.A. Bailey. 415 pages. 1986.

Vol. 2: **Glial-Neuronal Communication in Development and Regeneration.** Edited by H.H. Althaus and W. Seifert. 865 pages. 1987.

Vol. 3: **Nicotinic Acetylcholine Receptor: Structure and Function.** Edited by A. Maelicke. 489 pages. 1986.

Vol. 4: **Recognition in Microbe-Plant Symbiotic and Pathogenic Interactions.** Edited by B. Lugtenberg. 449 pages. 1986.

Vol. 5: **Mesenchymal-Epithelial Interactions in Neural Development.** Edited by J.R. Wolff, J. Sievers, and M. Berry. 428 pages. 1987.

Vol. 6: **Molecular Mechanisms of Desensitization to Signal Molecules.** Edited by T.M. Konijn, P.J.M. Van Haastert, H. Van der Starre, H. Van der Wel, and M.D. Houslay. 336 pages. 1987.

Vol. 7: **Gangliosides and Modulation of Neuronal Functions.** Edited by H. Rahmann. 647 pages. 1987.

Vol. 8: **Molecular and Cellular Aspects of Erythropoietin and Erythropoiesis.** Edited by I.N. Rich. 460 pages. 1987.

Vol. 9: **Modification of Cell to Cell Signals During Normal and Pathological Aging.** Edited by S. Govoni and F. Battaini. 297 pages. 1987.

Vol. 10: **Plant Hormone Receptors.** Edited by D. Klämbt. 319 pages. 1987.

Vol. 11: **Host-Parasite Cellular and Molecular Interactions in Protozoal Infections.** Edited by K.-P. Chang and D. Snary. 425 pages. 1987.

Vol. 12: **The Cell Surface in Signal Transduction.** Edited by E. Wagner, H. Greppin, and B. Millet. 243 pages. 1987.

Vol. 13: **Toxicology of Pesticides: Experimental, Clinical and Regulatory Perspectives.** Edited by L.G. Costa, C.L. Galli, and S.D. Murphy. 320 pages. 1987.

Vol. 14: **Genetics of Translation. New Approaches.** Edited by M.F. Tuite, M. Picard, and M. Bolotin-Fukuhara. 524 pages. 1988.

Vol. 15: **Photosensitisation. Molecular, Cellular and Medical Aspects.** Edited by G. Moreno, R.H. Pottier, and T.G. Truscott. 521 pages. 1988.

Vol. 16: **Membrane Biogenesis.** Edited by J.A.F. Op den Kamp. 477 pages. 1988.

Vol. 17: **Cell to Cell Signals in Plant, Animal and Microbial Symbiosis.** Edited by S. Scannerini, D. Smith, P. Bonfante-Fasolo, and V. Gianinazzi-Pearson. 414 pages. 1988.

Vol. 18: **Plant Cell Biotechnology.** Edited by M.S.S. Pais, F. Mavituna, and J.M. Novais. 500 pages. 1988.

Vol. 19: **Modulation of Synaptic Transmission and Plasticity in Nervous Systems.** Edited by G. Hertting and H.-C. Spatz. 457 pages. 1988.

Vol. 20: **Amino Acid Availability and Brain Function in Health and Disease.** Edited by G. Huether. 487 pages. 1988.

NATO ASI Series H

Vol. 21: **Cellular and Molecular Basis of Synaptic Transmission.**
Edited by H. Zimmermann. 547 pages. 1988.

Vol. 22: **Neural Development and Regeneration. Cellular and Molecular Aspects.**
Edited by A. Gorio, J.R. Perez-Polo, J. de Vellis, and B. Haber. 711 pages. 1988.

Vol. 23: **The Semiotics of Cellular Communication in the Immune System.**
Edited by E.E. Sercarz, F. Celada, N.A. Mitchison, and T. Tada. 326 pages. 1988.

Vol. 24: **Bacteria, Complement and the Phagocytic Cell.**
Edited by F.C. Cabello und C. Pruzzo. 372 pages. 1988.

Vol. 25: **Nicotinic Acetylcholine Receptors in the Nervous System.**
Edited by F. Clementi, C. Gotti, and E. Sher. 424 pages. 1988.

Vol. 26: **Cell to Cell Signals in Mammalian Development.**
Edited by S.W. de Laat, J.G. Bluemink, and C.L. Mummery. 322 pages. 1989.

Vol. 27: **Phytotoxins and Plant Pathogenesis.**
Edited by A. Graniti, R.D. Durbin, and A. Ballio. 508 pages. 1989.

Vol. 28: **Vascular Wilt Diseases of Plants. Basic Studies and Control.**
Edited by E.C. Tjamos and C.H. Beckman. 590 pages. 1989.

Vol. 29: **Receptors, Membrane Transport and Signal Transduction.**
Edited by A.E. Evangelopoulos, J.P. Changeux, L. Packer, T.G. Sotiroudis, and K.W.A. Wirtz. 387 pages. 1989.

Vol. 30: **Effects of Mineral Dusts on Cells.**
Edited by B.T. Mossman and R.O. Bégin. 470 pages. 1989.

Vol. 31: **Neurobiology of the Inner Retina.**
Edited by R. Weiler and N.N. Osborne. 529 pages. 1989.

Vol. 32: **Molecular Biology of Neuroreceptors and Ion Channels.**
Edited by A. Maelicke. 675 pages. 1989.

Vol. 33: **Regulatory Mechanisms of Neuron to Vessel Communication in Brain.**
Edited by F. Battaini, S. Govoni, M.S. Magnoni, and M. Trabucchi. 416 pages. 1989.

Vol. 34: **Vectors as Tools for the Study of Normal and Abnormal Growth and Differentiation.**
Edited by H. Lother, R. Dernick, and W. Ostertag. 477 pages. 1989.

Vol. 35: **Cell Separation in Plants: Physiology, Biochemistry and Molecular Biology.**
Edited by D.J. Osborne and M.B. Jackson. 449 pages. 1989.

Vol. 36: **Signal Molecules in Plants and Plant-Microbe Interactions.**
Edited by B.J.J. Lugtenberg. 425 pages. 1989.

Vol. 37: **Tin-Based Antitumour Drugs.** Edited by M. Gielen. 226 pages. 1990.

Vol. 38: **The Molecular Biology of Autoimmune Disease.**
Edited by A.G. Demaine, J-P. Banga, and A.M. McGregor. 404 pages. 1990.

Vol. 39: **Chemosensory Information Processing.** Edited by D. Schild. 403 pages. 1990.

Vol. 40: **Dynamics and Biogenesis of Membranes.**
Edited by J.A.F. Op den Kamp. 367 pages. 1990.

Vol. 41: **Recognition and Response in Plant-Virus Interactions.**
Edited by R.S.S. Fraser. 467 pages. 1990.

NATO ASI Series H